COMBINATORICS
An Invitation

COMBINATORICS
An Invitation

H. Joseph Straight

State University of New York at Fredonia

Brooks/Cole Publishing Company
Pacific Grove, California

Brooks/Cole Publishing Company
A Division of Wadsworth, Inc.

Printed in the United States of America

10 9 8 7 6 5 4 3 2 1

Library of Congress Cataloging in Publication Data
Straight, H. Joseph, [date–   ]
    Combinatorics : an invitation / H. Joseph Straight
        p.   cm.
    Includes bibliographical references and index.
    ISBN 0-534-19926-7
    1. Combinatorial analysis.   I. Title.
QA164.S82   1993
511'.6–dc20
                                                                            92-36178
                                                                            CIP

Sponsoring Editor: *Jeremy Hayhurst*
Marketing Representative: *Dawn Burnam*
Editorial Associate: *Nancy Champlin*
Production Coordinator: *Marlene Thom*
Manuscript Editor: *Betty Berenson*
Interior Design: *Roy R. Neuhaus*
Cover Design: *Susan Haberkorn*
Art Coordinator: *Lisa Torri*
Interior Illustration: *Lori Heckelman*
Typesetting: *Archetype Publishing, Inc.*
Printing and Binding: *Arcata Graphics/Fairfield*

In memory of my father,
Harry J. Straight (1918–1992)

Preface

*T*he title for this book was suggested by one of my favorite musical compositions, Carl Maria von Weber's *Aufforderung zum Tanze* (*Invitation to the Dance*). Even though, generally, I have two left feet when it comes to dancing and have never waltzed, this music makes me want to try! Likewise, this book invites students to sample an exciting and challenging (even fun!) area of mathematics. I hope that many of them will be encouraged to take further courses or do additional reading in the area.

This book is intended to be used primarily as a text for a course in combinatorics, with the expected audience being upper-level undergraduates. It is not meant to be encyclopedic, but the organization and selection of topics provide a great deal of flexibility when designing a course. In the one-semester, three-credit course at SUNY Fredonia, for example, we cover Chapter 1, Sections 2.1 through 2.4 from Chapter 2, and additional topics from Sections 2.5, 2.6, and Chapter 4 as time permits. We do not cover any topics from Chapter 3 in our combinatorics course because we have a separate course in graph theory. Chapter 3 could be and has been used as the basis for such a course, supplemented by additional material from other sources. In the graph theory course (usually taken by a small number of senior undergraduates), I have found it successful to have the students prepare and present lectures on topics of their choice during the last few weeks of the semester; some read and report on research articles. The same approach could be used in a graduate-level combinatorics course.

Having extensively class-tested this book, I am confident that it can, for the most part, be read by students. Every effort has been made to ensure that the writing is clear and precise; that examples are numerous and well placed; and that there are plenty of carefully worded exercises ranging in difficulty from the routine to the challenging. To me the exercises are the most important part of any textbook—students learn mathematics best by doing it! In addition to the exercises at the end of each section, there are problems at the end of Chapters 1, 2, 3, and 4. These chapter problems not only provide helpful review but also deepen exposure to key concepts. Often a chapter problem will integrate concepts from several previous chapters, and occasionally a problem invites students to explore some aspect of a topic not covered in detail in the text. Exercises and problems that are especially challenging are prefixed with a star ($\star$).

Answers to most exercises are provided in the appendix. It is always difficult to decide how many answers to give and how detailed solutions should be. Providing detailed solutions to nearly all exercises may make students happy (over the short term), but instructors find it more difficult then to assign problems to be collected and graded. I have decided to compromise by giving only a numeric or "short" answer whenever possible. For instance, rather than giving the answer to a combinatorics exercise as $C(6,2)(C(4,2) + C(4,1))$, the answer is given as 150. This allows students to check their answers to most exercises without giving away the solution strategies to be used. Additional, more detailed solutions and/or hints can be obtained by contacting me or Brooks/Cole; ask for *Additional Answers and Hints to Selected Exercises and Problems*. In particular, I hope to have this available as a (La)T$_{E}$X file that can be requested and sent by electronic mail.

The book begins with Chapter 0, "Preliminaries." Section 0.1 introduces the set notation to be used in the book and reviews basic definitions and results concerning the operations of power set, intersection, union, difference, and product. Section 0.2 covers one-to-one, onto, and inverse functions, setting the stage for the bijective combinatorial proofs encountered in Chapters 1 and 2. Several specific functions, such as the floor and ceiling functions, are introduced, and the (extended) euclidean algorithm is reviewed. Also, big-oh and theta notations are defined. Relations and their properties are studied in Section 0.3, and graphs are introduced as a means of representing relations. This section also includes a discussion of equivalence relations and partial-order relations. Section 0.4 reviews the principle of mathematical induction (both the weak and strong forms) and gives several examples of induction proofs; some of these involve the notion of a recursively defined function. Finally, Section 0.5 discusses the construction of finite fields, which are needed in Chapter 4 on combinatorial designs.

Depending on course prerequisites, students should be familiar with much of the material in this preliminary chapter. Therefore, I suggest that the course begin straightaway with Chapter 1, and that material from Chapter 0 be reviewed as needed. For example, some instructors may find early on that their students lack experience and facility with induction proofs and so will need to review Section 0.4. In any case, it has been my experience that students find such a chapter helpful and appreciate the convenience of being able to easily refer to background definitions and results.

Chapter 1, "Basic Combinatorics," begins with an introductory section focusing on Euler's famous problem of the 36 officers. I have found that this problem and the story of its solution, as well as the visual nature of Latin squares, grab the students' interest and attention. The chapter then proceeds to the basic concepts of combinatorial enumeration: addition and multiplication principles in Section 1.2, permutations and combinations in Section 1.3, the pigeonhole principle in Section 1.4, ordered partitions and distributions in Section 1.5 (in how many ways can a given number of balls be distributed to a given number of different boxes?), and the binomial and multinomial theorems in Section 1.6.

Chapter 2 is entitled "Intermediate Combinatorics." My experience has been that students find these topics to be an order of magnitude more difficult than those in Chapter 1, and the pace of the course must necessarily slow to give them time to work through the exercises and thereby gain understanding of the material. Section 2.1 covers the principle of inclusion–exclusion; recurrence equations and generating functions are covered in Sections 2.2 and 2.3, respectively; and Section 2.4 looks at unordered partitions and distributions (for example, in how many ways can a given number of balls be distributed to a given number of identical boxes?). In the last two (optional) sections, I introduce the important area of combinatorial algorithms—algorithms for listing, ranking, and unranking (the inverse of ranking) the objects in some combinatorial family, and for selecting one of the objects at random. Specifically, these algorithms are presented for permutations in Section 2.5 and for combinations in Section 2.6.

As is true also for Chapter 3, the algorithms in Sections 2.5 and 2.6 are first described informally or mathematically and are illustrated with one or several examples. Then the discussion proceeds to a more formal presentation, in which the family of combinatorial objects is considered as an abstract data type and the algorithms are described using an Ada-like pseudocode (whose syntax should be clear to anyone having experience with a modern, imperative programming language such as Pascal or any of its descendants). This more formal presentation of the algorithms is provided for those with a computer science orientation and can easily be skipped if so desired.

Chapter 3 gives an overview of graph theory with an emphasis on algorithms and applications, taking advantage of the strong connection between algorithms and constructive proofs. Section 3.1 introduces graphs, and Section 3.2 discusses eulerian graphs, informally describing an algorithm for finding an eulerian circuit in an eulerian graph. Section 3.3 introduces directed graphs and proceeds to a more formal description of the eulerian algorithm, now given in the context of directed graphs. Section 3.4 covers strong components, reachability (Warshall's algorithm), and distance (Dijkstra's algorithm). Acyclic digraphs and trees are met in Section 3.5, along with topological sort, depth-first search, and Prim's algorithm. Section 3.6, "Matchings," places the standard problem of finding a system of distinct representatives for a list of sets in the more general setting of finding a maximum matching in a graph. Section 3.7 looks at hamiltonian graphs, while Section 3.8 surveys vertex coloring and planar graphs, including the famous four-color theorem. For those wishing to include some graph theory in a combinatorics course, I suggest covering the first five sections. As for the later sections, 3.6 and 3.8 depend only on 3.1, and 3.7 can be covered after 3.1 and 3.2.

Chapter 4, "Combinatorial Designs," is intended to give the student a brief encounter with a fascinating and aesthetically pleasing subject. The sections are ordered to proceed from simpler to more complex designs: finite geometries in 4.1; Latin squares in 4.2; graph factorizations in 4.3 (in my opinion, these belong more naturally here than in Chapter 3); and block designs in 4.4.

Finally, a word of thanks to the following individuals for their helpful comments and reviews: Edward J. Carney, University of Rhode Island; John Haverhals, Bradley University; Fritz Kronk, State University of New York, Binghamton; Craig Rasmussen, Naval Postgraduate School; Richard Rockwell, Pacific Union College; and Edward Scheinerman, Johns Hopkins University. Last but not least, I would like to express my appreciation especially to my colleague Y. H. Harris Kwong, who used a preliminary draft of this book and made many valuable suggestions.

I hope you enjoy using this text, and I welcome your comments and suggestions! Please let me hear from you—my email address is straight@mary.cs.fredonia.edu.

H. Joseph Straight

Contents

Chapter 3   Graph Theory   289

Chapter 4   Combinatorial Designs   417

COMBINATORICS
An Invitation

Chapter 0

Preliminaries

0.1 SETS

This preliminary chapter reviews some basic concepts used throughout this text. As mentioned in the preface, this chapter could be covered at the beginning of a course on combinatorics. However, depending on course prerequisites, you may already be familiar with much of this material. Thus, I suggest that you begin straight away with Chapter 1 and review parts of Chapter 0 as required. For example, the material in Section 0.5 on finite fields is used in Chapter 4 on combinatorial designs and so you may wish to read it prior to beginning that chapter.

This first section introduces basic terminology and notation for sets. A **set** can be thought of as a collection of objects, referred to as the **elements** or **members** of the set. Normally, we treat the terms *set* and *member* as undefined or primitive notions. All of us have an intuitive understanding what these terms mean.

Certain sets of numbers are used frequently in this book, so special notations are used for them. The numbers $0, 1, 2, 3, \ldots$ are called the **natural numbers**, and the notation $\mathbb{N}$ is used to denote the set of natural numbers. The natural numbers, together with the negative whole numbers $-1, -2, -3, \ldots$ form the set of **integers**; this set is denoted by $\mathbb{Z}$. A **rational number** is any number expressible in the form m/n, where m and n are integers and $n \neq 0$. For example, $2/3$, $-5/11$, $17 = 17/1$, and $0.222\ldots = 2/9$ are rational numbers. The set of rational numbers is denoted by $\mathbb{Q}$. Numbers such as $\sqrt{2}, \sqrt[3]{5}$, and π are not rational; in general, such numbers are referred to as **irrational numbers**. The rational numbers and the irrational numbers together make up the set of **real numbers**, which is denoted by $\mathbb{R}$. A **complex number** is an ordered pair (a, b) of real numbers, usually written as $a + bi$, where i is a special complex number to be defined shortly. The set of complex numbers is denoted by $\mathbb{C}$.

The operations of addition $+$ and multiplication $\cdot$ are defined on $\mathbb{C}$ as follows: For complex numbers $a + bi$ and $c + di$,

$$(a + bi) + (c + di) = (a + c) + (b + d)i$$
$$(a + bi) \cdot (c + di) = (ac - bd) + (ad + bc)i$$

1

Note that

$$(a + 0i) + (c + 0i) = (a + c) + 0i$$
$$(a + 0i) \cdot (c + 0i) = (ac) + 0i$$

Hence, the set of complex numbers of the form $a + 0i$ behaves just like the set $\mathbb{R}$ under addition and multiplication. For this reason it is usual to identify the complex number $a + 0i$ with the real number a; that is,

$$a + 0i = a$$

Next notice that the complex number $i = 0 + 1i$ has the interesting property that

$$(0 + 1i) \cdot (0 + 1i) = -1 + 0i = -1$$

namely, i is a "square root" of -1. Historically, this special number arose as a solution to the quadratic equation

$$x^2 + 1 = 0$$

which has no solutions in the set of real numbers. With this understanding, we have

$$(a, b) = (a, 0) + (0, b) = (a, 0) + ((b, 0) \cdot (0, 1)) = a + bi$$

so that the complex number (a, b) may alternately be written as $a + bi$; this is perhaps the most familiar form of a complex number.

To repeat, then, we adopt the following notational conventions:

$\mathbb{N}$ = the set of natural numbers

$\mathbb{Z}$ = the set of integers

$\mathbb{Q}$ = the set of rational numbers

$\mathbb{R}$ = the set of real numbers

$\mathbb{C}$ = the set of complex numbers

The basic relation between an element x and a set A is that of *membership*. If x is an element of A, then we write $x \in A$; if x is not an element of A, we write $x \notin A$. For instance,

$$-3 \in \mathbb{Z}, \quad -3 \notin \mathbb{N} \quad \sqrt{2} \in \mathbb{R}, \quad \sqrt{2} \notin \mathbb{Q} \quad i \in \mathbb{C}, \quad i \notin \mathbb{R}$$

Except for certain special sets such as $\mathbb{N}$, sets are denoted in this book by uppercase italic letters such as A, B, C, and elements by lowercase italic letters such as a, b, c.

If a set A consists of a "small" number of elements, then we can exhibit A by explicitly listing its elements between braces. For example, if A is the set of odd integers between -3 and 11, inclusive, then we write

$$A = \{-3, -1, 1, 3, 5, 7, 9, 11\}$$

However, some sets contain too many elements to be listed in this way. In many such cases, three-dot notation (ellipses) is used to mean "and so on" or "and so on up to," depending on the context. For instance, the set $\mathbb{N}$ can be exhibited as

$$\mathbb{N} = \{0, 1, 2, 3, \ldots\}$$

the set $\mathbb{Z}$ as

$$\mathbb{Z} = \{\ldots, -3, -2, -1, 0, 1, 2, 3, \ldots\}$$

and the set B of integers between 17 and 93 as

$$B = \{17, 18, 19, \ldots, 93\}$$

Often a set A is described as consisting of those elements x in some set B that satisfy a specified property. As an example, let E be the set of even integers. Then we may write

$$E = \{\ldots, -4, -2, 0, 2, 4, \ldots\}$$

or

$$E = \{x \in \mathbb{Z} \mid x \text{ is even}\}$$

Here the symbol $\mid$ is translated as "such that."

Note that every natural number is an integer, every integer is a rational number, and every rational number is a real number. In general, for two sets A and B, it is possible that each element of A is also an element of B.

DEFINITION 0.1.1 A set A is called a **subset** of a set B, denoted $A \subseteq B$, provided that every element of A is also an element of B. If A is a subset of B and there is an element b such that $b \in B$ and $b \notin A$, then A is called a **proper subset** of B and we write $A \subset B$. ∎

Special subsets of $\mathbb{R}$ called **intervals** are used occasionally in this book. For example, let a and b be real numbers with $a < b$. Then

$$(a, b) = \{x \in \mathbb{R} \mid a < x < b\}$$
$$[a, b] = \{x \in \mathbb{R} \mid a \leq x \leq b\}$$
$$[a, b) = \{x \in \mathbb{R} \mid a \leq x < b\}$$
$$(a, b] = \{x \in \mathbb{R} \mid a < x \leq b\}$$

The interval (a, b) is called an **open interval**, $[a, b]$ is called a **closed interval**, and $[a, b)$ and $(a, b]$ are called **half-open intervals**. In each case, a and b are called the **endpoints** of the interval. Some other intervals are

$$(a, \infty) = \{x \in \mathbb{R} \mid x > a\}$$
$$[a, \infty) = \{x \in \mathbb{R} \mid x \geq a\}$$
$$(-\infty, b) = \{x \in \mathbb{R} \mid x < b\}$$
$$(-\infty, b] = \{x \in \mathbb{R} \mid x \leq b\}$$
$$(-\infty, \infty) = \mathbb{R}$$

Given a subset S of $\mathbb{R}$, we sometimes wish to conveniently denote the set of positive members of S, the set of negative members of S, or the set of nonzero members of S; thus, we let

$$S^+ = \{x \in S \mid x > 0\}$$
$$S^- = \{x \in S \mid x < 0\}$$
$$S^* = \{x \in S \mid x \neq 0\}$$

For example, with this notation we can denote the set of positive integers by $\mathbb{Z}^+$, the set of negative integers by $\mathbb{Z}^-$, and the set of nonzero rational numbers by $\mathbb{Q}^*$.

In any particular discussion of sets, we are usually concerned with subsets of some specified set U. The set U is called the **universal set**. For instance, in a calculus lecture it may be that $U = \mathbb{R}$, whereas for a lecture in combinatorics it might be that $U = \mathbb{N}$.

There is a special set that is often encountered in any mathematics course; this is the set with no elements. It is aptly named the **empty set** and is denoted by $\{\ \}$ or by $\emptyset$. Unlike U, the empty set is unique, although there are numerous examples of it. For example, consider the set of female presidents of the United States prior to 1993, or the set of integers m such that m^2 is negative.

A set A is called **finite** provided it consists of n elements for some natural number n; in this case n is called the **cardinality** of A and we write $|A| = n$. In particular, the empty set is finite, as is the set

$$\{0, 1, 2, \ldots, n - 1\}$$

of natural numbers less than a given positive integer n. Sets that are not finite are called **infinite**; in particular, the sets $\mathbb{N}$, $\mathbb{Z}$, $\mathbb{Q}$, $\mathbb{R}$, and $\mathbb{C}$ are infinite.

Perhaps the most fundamental relation that can exist between two sets is that of **equality**.

DEFINITION 0.1.2   Two sets A and B are called **equal**, written $A = B$, provided they consist of the same elements. If A and B are not equal, we write $A \neq B$. ∎

Set equality may be restated as follows: Sets A and B are equal provided $A \subseteq B$ and $B \subseteq A$. Often we are given two sets A and B and asked to prove they are equal. It is common practice to divide such a proof into two parts: first, we show that $A \subseteq B$; second, we show that $B \subseteq A$. Alternately, we may apply Definition 0.1.2 directly and show, for an arbitrary element $x \in U$, that

$$x \in A \text{ if and only if } x \in B$$

We shall consider several examples of such proofs momentarily.

Given a set A, we are often interested in the collection of all subsets of A.

DEFINITION 0.1.3   The set consisting of all subsets of a given set A is called the **power set** of A and is denoted $\mathcal{P}(A)$. ∎

■ *Example 0.1.1*   Describe $\mathcal{P}(A)$ if the set A has:

(a) no elements   (b) one element   (c) two elements

Solution

(a) If the set A has no elements, then $A = \{\ \}$ and the only subset of A is itself. So $\mathcal{P}(\{\ \}) = \{\{\ \}\}$. (Note that $\{\{\ \}\}$ is not the empty set; it is a set with exactly one element, that element being $\{\ \}$.)

(b) If A has exactly one element, then the subsets of A are $\{\ \}$ and A, and so $\mathcal{P}(A) = \{\{\ \}, A\}$.

(c) Suppose $A = \{a, b\}$. Then, besides $\{\ \}$ and A, the power set of A includes two nonempty proper subsets of A, namely, $\{a\}$ and $\{b\}$. Thus, $\mathcal{P}(A) = \{\{\ \}, \{a\}, \{b\}, A\}$.   ■

Two important properties of $\mathcal{P}(A)$ to be remembered are that $\{\ \} \in \mathcal{P}(A)$ and $A \in \mathcal{P}(A)$.

We next turn to several common operations on sets; these are intersection, union, difference, and product.

DEFINITION 0.1.4   Let A and B be subsets of some universal set U. The set operations of intersection, union, and difference are defined as follows:

1. The **intersection** of A and B is the set $A \cap B$ defined by

$$A \cap B = \{x \mid x \in A \text{ and } x \in B\}$$

2. The **union** of A and B is the set $A \cup B$ defined by

$$A \cup B = \{x \mid x \in A \text{ or } x \in B\}$$

3. The **difference set**, $A - B$, is defined by

$$A - B = \{x \mid x \in A \text{ and } x \notin B\}$$ ■

The set $A - B$ is also called the **relative complement** of B in A. It is important to realize that, in general, $A - B \neq B - A$. The relative complement of A in the universal set U is called the **complement** of A and is denoted by $\overline{A}$. That is, $\overline{A} = U - A$.

■ *Example 0.1.2*   Given $U = \mathbb{N}$, $A = \{1, 3, 5, 7, \ldots\}$, $P = \{p \in \mathbb{N} \mid p \text{ is prime}\}$, and $T = \{0, 3, 6, 9, 12, \ldots\}$, find:

(a) $\overline{A}$ (b) $\overline{P}$ (c) $A \cap P$ (d) $A - P$

(e) $P - A$ (f) $\overline{A} \cap T$ (g) $A \cup \overline{T}$

Solution

Note that A is the set of odd natural numbers, $P = \{2, 3, 5, 7, 11, 13, 17, 19, \ldots\}$, and T is the set of natural numbers that are multiples of 3. Thus:

(a) $\bar{A} = \{0, 2, 4, 6, \ldots\}$ is the set of even natural numbers.

(b) $\bar{P} = \{n \in \mathbb{N} \mid n \text{ is not prime }\} = \{0, 1, 4, 6, 8, 9, 10, 12, 14, 15, 18, \ldots\}$.

(c) The set $A \cap P$ is the set of odd primes: $A \cap P = \{3, 5, 7, 11, 13, 17, 19, \ldots\} = P - \{2\}$.

(d) The set of odd natural numbers that are not prime is $A - P = \{1, 9, 15, 21, 25, 27, \ldots\}$.

(e) $P - A$ is the set of primes that are not odd, namely, $P - A = \{2\}$.

(f) $\bar{A} \cap T$ consists of those elements of T that are even, so $\bar{A} \cap T = \{0, 6, 12, 18, 24, \ldots\}$.

(g) $A \cup \bar{T}$ includes any natural number that is either odd or is not a multiple of 3, so $A \cup \bar{T} = \{1, 2, 3, 4, 5, 7, 8, 9, 10, 11, 13, \ldots\}$. Note that

$$A \cup \bar{T} = \overline{\{6, 12, 18, 24, \ldots\}} = \overline{\bar{A} \cap T} \qquad \blacksquare$$

■ **Example 0.1.3** Let $U = \{0, 1, 2, 3, 4, 5, 6\}$, $A = \{0, 2, 4, 6\}$, $B = \{2, 3, 5\}$, and $C = \{0, 1, 5, 6\}$. Find and compare each pair of sets:

(a) $A \cap (B \cup C)$ and $(A \cap B) \cup (A \cap C)$

(b) $A \cup (B \cap C)$ and $(A \cup B) \cap (A \cup C)$

(c) $\overline{A \cup B}$ and $\bar{A} \cap \bar{B}$

(d) $\overline{A \cap B}$ and $\bar{A} \cup \bar{B}$

Solution

(a) We find that

$$A \cap (B \cup C) = \{0, 2, 4, 6\} \cap \{0, 1, 2, 3, 5, 6\} = \{0, 2, 6\}$$

and

$$(A \cap B) \cup (A \cap C) = \{2\} \cup \{0, 6\} = \{0, 2, 6\}$$

Note that $A \cap (B \cup C) = (A \cap B) \cup (A \cap C)$.

(b) Now verify that

$$A \cup (B \cap C) = (A \cup B) \cap (A \cup C) = \{0, 2, 4, 5, 6\}$$

(c) Here we have

$$A \cup B = \overline{\{0, 2, 3, 4, 5, 6\}} = \{1\}$$

and

$$\bar{A} \cap \bar{B} = \{1, 3, 5\} \cap \{0, 1, 4, 6\} = \{1\}$$

so again the two sets are equal.

(d) Similarly, note that

$$\overline{A \cap B} = \overline{A} \cup \overline{B} = \{0, 1, 3, 4, 5, 6\}$$ ■

Example 0.1.3 illustrates some general properties of the set operations; these are among those presented in the first two theorems.

THEOREM 0.1.1   The following properties hold for any subsets A, B, and C of a universal set U:

1. Commutative properties
    (a) $A \cup B = B \cup A$ (b) $A \cap B = B \cap A$
2. Associative properties
    (a) $(A \cup B) \cup C = A \cup (B \cup C)$ (b) $(A \cap B) \cap C = A \cap (B \cap C)$
3. Distributive properties
    (a) $A \cup (B \cap C) = (A \cup B) \cap (A \cup C)$
    (b) $A \cap (B \cup C) = (A \cap B) \cup (A \cap C)$
4. Idempotent laws
    (a) $A \cup A = A$ (b) $A \cap A = A$
5. De Morgan's laws
    (a) $\overline{A \cup B} = \overline{A} \cap \overline{B}$ (b) $\overline{A \cap B} = \overline{A} \cup \overline{B}$
6. Law of the excluded middle
    (a) $A \cup \overline{A} = U$ (b) $A \cap \overline{A} = \{\ \}$
7. $A \subseteq B$ if and only if $\overline{B} \subseteq \overline{A}$
8. (a) $A \cup U = U$ (b) $A \cup \{\ \} = A$
    (c) $A \cap U = A$ (d) $A \cap \{\ \} = \{\ \}$

Proof   We prove here parts 2(b), 3(a), and 5(b). You are asked to prove the remaining parts in Exercise 2.
  Proof of 2(b): Let x be an arbitrary element of U. We show that

$$x \in (A \cap B) \cap C \leftrightarrow x \in A \cap (B \cap C)$$

($\leftrightarrow$ stands for "if and only if"). Now,

$$
\begin{aligned}
x \in (A \cap B) \cap C &\leftrightarrow x \in A \cap B \text{ and } x \in C \\
&\leftrightarrow (x \in A \text{ and } x \in B) \text{ and } x \in C \\
&\leftrightarrow x \in A \text{ and } (x \in B \text{ and } x \in C) \\
&\leftrightarrow x \in A \text{ and } x \in (B \cap C) \\
&\leftrightarrow x \in A \cap (B \cap C)
\end{aligned}
$$

Therefore, $(A \cap B) \cap C = A \cap (B \cap C)$.

Proof of 3(a): For $x \in U$,

$$x \in A \cup (B \cap C) \leftrightarrow x \in A \text{ or } x \in B \cap C$$
$$\leftrightarrow x \in A \text{ or } (x \in B \text{ and } x \in C)$$
$$\leftrightarrow (x \in A \text{ or } x \in B) \text{ and } (x \in A \text{ or } x \in C)$$
$$\leftrightarrow (x \in A \cup B) \text{ and } (x \in A \cup C)$$
$$\leftrightarrow x \in (A \cup B) \cap (A \cup C)$$

Therefore, $A \cup (B \cap C) = (A \cup B) \cap (A \cup C)$.

Proof of 5(b): Here we have, for any x,

$$x \in \overline{A \cap B} \leftrightarrow x \notin A \cap B$$
$$\leftrightarrow x \notin A \text{ or } x \notin B$$
$$\leftrightarrow x \in \overline{A} \text{ or } x \in \overline{B}$$
$$\leftrightarrow x \in \overline{A} \cup \overline{B}$$

Hence, $\overline{A \cap B} = \overline{A} \cup \overline{B}$.   □

Another way to express De Morgan's law $\overline{B \cap C} = \overline{B} \cup \overline{C}$ is

$$U - (B \cap C) = (U - B) \cup (U - C)$$

Given subsets A, B, and C of U, the above expression can be generalized by considering those elements of A that are not in $B \cap C$, namely, $A - (B \cap C)$. A similar expression arises from considering $A - (B \cup C)$. The resulting properties are called the *Generalized De Morgan Laws*.

THEOREM 0.1.2 (Generalized De Morgan Laws) The following two properties hold for any subsets A, B, and C of a universal set U.

1. $A - (B \cup C) = (A - B) \cap (A - C)$
2. $A - (B \cap C) = (A - B) \cup (A - C)$

Proof We prove here part 1. You are asked to prove part 2 in Exercise 4. For any $x \in U$,

$$x \in A - (B \cup C) \leftrightarrow x \in A \text{ and } x \notin B \cup C$$
$$\leftrightarrow x \in A \text{ and } (x \notin B \text{ and } x \notin C)$$
$$\leftrightarrow (x \in A \text{ and } x \notin B) \text{ and } (x \in A \text{ and } x \notin C)$$
$$\leftrightarrow x \in A - B \text{ and } x \in A - C$$
$$\leftrightarrow x \in (A - B) \cap (A - C)$$

So we have $A - (B \cup C) = (A - B) \cap (A - C)$.   □

In the preceding statements of definitions and theorems we have exercised care by using qualifying phrases such as, "Let A, B, and C be subsets of a universal set U." Henceforth, unless the situation demands it, we will omit explicit reference to U and implicitly assume that all sets in a given discussion are subsets of some universal set. Thus, for example, in place of the above phrase, we will write, "Let A, B, and C be any sets."

The order in which the elements of a set are listed is of no importance in the definition of a set; for example, it is clear that $\{1, 2\} = \{2, 1\}$. Thus, it really makes no sense to speak of the "first element" of $\{1, 2\}$. In many cases, however, it turns out to be important to distinguish the order of appearance of two elements. This leads to the notion of an **ordered pair** of two elements a and b, written (a, b), where a is the **first element** and b is the **second element**. One then states that $(a, b) = (c, d)$ if and only if $a = c$ and $b = d$.

DEFINITION 0.1.5   Given two sets A and B, the **(cartesian) product** of A with B is the set $A \times B$ defined by

$$A \times B = \{(a, b) \mid a \in A \text{ and } b \in B\}$$ ■

■ ***Example 0.1.4***   Given $A = \{0, 1, 2\}$ and $B = \{1, 3\}$, find $A \times B$ and $B \times A$.

Solution

$$A \times B = \{(0, 1), (0, 3), (1, 1), (1, 3), (2, 1), (2, 3)\}$$
$$B \times A = \{(1, 0), (1, 1), (1, 2), (3, 0), (3, 1), (3, 2)\}$$

Note that $A \times B \neq B \times A$, although both sets have six elements. ■

We see in Example 0.1.4 that the product operation does not obey a commutative law. The next theorem presents several distributive laws that are satisfied when the product operation is combined with the operations of union, intersection, and difference.

THEOREM 0.1.3   For any sets A, B, and C, the following properties are satisfied.

1. $A \times (B \cup C) = (A \times B) \cup (A \times C)$
2. $A \times (B \cap C) = (A \times B) \cap (A \times C)$
3. $A \times (B - C) = (A \times B) - (A \times C)$

Proof   We prove property 1. You are asked to prove properties 2 and 3 in Exercise 10.

For any x and y,

$$(x, y) \in A \times (B \cup C) \leftrightarrow x \in A \text{ and } y \in B \cup C$$
$$\leftrightarrow x \in A \text{ and } (y \in B \text{ or } y \in C)$$
$$\leftrightarrow (x \in A \text{ and } y \in B) \text{ or } (x \in A \text{ and } y \in C)$$
$$\leftrightarrow (x, y) \in A \times B \text{ or } (x, y) \in A \times C$$
$$\leftrightarrow (x, y) \in (A \times B) \cup (A \times C)$$

Therefore, $A \times (B \cup C) = (A \times B) \cup (A \times C)$.   □

The idea of an ordered pair can be extended to more than two elements. Given k elements $x_1, x_2, \ldots, x_k$, $k \geq 2$, we define the **ordered k-tuple** $(x_1, x_2, \ldots, x_k)$, in which x_1 is the first element (or **first coordinate**), x_2 is the second element, and so on, and x_k is the kth element. It is then possible to generalize the definition of the product operation.

DEFINITION 0.1.6   Given any $k \geq 2$ sets $A_1, A_2, \ldots, A_k$, their (k-fold) **product** is the set $A_1 \times A_2 \times \cdots \times A_k$ defined by

$$A_1 \times A_2 \times \cdots \times A_k = \{(a_1, a_2, \ldots, a_k) \mid a_i \in A_i \text{ for each } i, 1 \leq i \leq k\}$$ ■

We remark that $(a_1, a_2, \ldots, a_k) = (b_1, b_2, \ldots, b_k)$ if and only if $a_i = b_i$ for each i, $1 \leq i \leq k$. Also, when there is no confusion, we agree to simplify notation by writing the k-tuple $(x_1, x_2, \ldots, x_k)$ as a string: $x_1 x_2 \cdots x_k$.

■ *Example 0.1.5*   Let $B = \{0, 1\}$. Then $B \times B \times B =$

$$\{(0,0,0), (0,0,1), (0,1,0), (0,1,1), (1,0,0), (1,0,1), (1,1,0), (1,1,1)\}$$

or, using our simplified notation, $B \times B \times B =$

$$\{000, 001, 010, 011, 100, 101, 110, 111\}$$ ■

In Example 0.1.5, if we think of B as the set of binary digits, then $B \times B \times B$ can be thought of as the set of binary strings (or words) of length three. In general, an **alphabet** is a nonempty set of characters, for example, the ASCII character set or the set $\{0, 1\}$ of binary digits. Given an alphabet A, a **string** over A is any finite sequence $(x_1, x_2, \ldots, x_n)$ of characters from A. We usually denote the string $(x_1, x_2, \ldots, x_n)$ by "$x_1 x_2 \ldots x_n$" or, more simply, by $x_1 x_2 \ldots x_n$ when there is no confusion. The **length** of a string is the number of characters it contains, so that the string $x_1 x_2 \ldots x_n$ has length n. There is a unique string of length 0, the **empty string**, and it is denoted by ϵ.

Just as Definition 0.1.6 generalizes the product operation for an arbitrary finite number of sets, the same thing can be done for the intersection and union operations. For example, if A_1, A_2, and A_3 are any three sets, then by the associative laws we have:

$$(A_1 \cup A_2) \cup A_3 = A_1 \cup (A_2 \cup A_3)$$
$$(A_1 \cap A_2) \cap A_3 = A_1 \cap (A_2 \cap A_3)$$

Hence, there is no confusion if we simply write $A_1 \cup A_2 \cup A_3$ or $A_1 \cap A_2 \cap A_3$. We then have

$$A_1 \cup A_2 \cup A_3 = \{x \mid x \in A_1 \text{ or } x \in A_2 \text{ or } x \in A_3\}$$
$$= \{x \mid x \in A_i \text{ for some } i, \ 1 \le i \le 3\}$$

and

$$A_1 \cap A_2 \cap A_3 = \{x \mid x \in A_1 \text{ and } x \in A_2 \text{ and } x \in A_3\}$$
$$= \{x \mid x \in A_i \text{ for each } i, \ 1 \le i \le 3\}$$

DEFINITION 0.1.7   Given any n sets $A_1, A_2, \ldots, A_n$, we define

1.   their **union** to be the set

$$A_1 \cup A_2 \cup \cdots \cup A_n = \{x \mid x \in A_i \text{ for some } i, \ 1 \le i \le n\}$$

2.   their **intersection** to be the set

$$A_1 \cap A_2 \cap \cdots \cap A_n = \{x \mid x \in A_i \text{ for each } i, \ 1 \le i \le n\}$$

These sets are expressed as

$$\bigcup_{i=1}^{n} A_i \qquad \text{and} \qquad \bigcap_{i=1}^{n} A_i$$

respectively.   ∎

■ **Example 0.1.6**   For $i \in \{1, 2, \ldots, 10\}$, define

$$A_i = [-i, 10 - i]$$

Then $A_1 = [-1, 9]$, $A_2 = [-2, 8], \ldots, A_{10} = [-10, 0]$. Hence,

$$\bigcup_{i=1}^{10} A_i = [-10, 9] \qquad \text{and} \qquad \bigcap_{i=1}^{10} A_i = [-1, 0]$$ ∎

■   ***Example 0.1.7***   For $k \in \{1, 2, \ldots, 100\}$, define $B_k = \{r \in \mathbb{Q} \mid -1 \leq k \cdot r \leq 1\}$.
Then

$$B_1 = \{r \in \mathbb{Q} \mid -1 \leq r \leq 1\}$$
$$B_2 = \{r \in \mathbb{Q} \mid -1 \leq 2r \leq 1\} = \{r \in \mathbb{Q} \mid -1/2 \leq r \leq 1/2\}$$
$$\vdots$$
$$B_{100} = \{r \in \mathbb{Q} \mid -1 \leq 100r \leq 1\} = \{r \in \mathbb{Q} \mid -1/100 \leq r \leq 1/100\}$$

Note that

$$B_{100} \subset B_{99} \subset \cdots \subset B_2 \subset B_1$$

So it follows that

$$\bigcup_{k=1}^{100} B_k = B_1 \qquad \text{and} \qquad \bigcap_{k=1}^{100} B_k = B_{100}$$ ■

In Definition 0.1.7, there is a set A_i for each $i \in \{1, 2, \ldots, n\}$, where n is a positive integer. In this case we call the set $\{1, 2, \ldots, n\}$ the **index set** for the collection $\{A_1, A_2, \ldots, A_n\}$. In general, any nonempty set can serve as an index set, although in this text we are nearly always concerned with finite index sets such as $\{1, 2, \ldots, n\}$ and $\{0, 1, \ldots, n-1\}$.

DEFINITION 0.1.8   Let I be a nonempty set, and suppose that for each element $i \in I$ there is associated a set A_i. We then call I the **index set** for the collection of sets

$$\mathcal{A} = \{A_i \mid i \in I\}$$

1.   The **union** of the collection $\mathcal{A}$ is defined to be the set

$$\bigcup_{i \in I} A_i = \{x \mid x \in A_i \text{ for some } i \in I\}$$

2.   The **intersection** of the collection $\mathcal{A}$ is defined to be the set

$$\bigcap_{i \in I} A_i = \{x \mid x \in A_i \text{ for each } i \in I\}$$ ■

Note: In the case that $I = \{n \in \mathbb{Z} \mid n \geq k\}$, where k is a fixed integer, we write, as an alternate notation

$$\bigcup_{n \in I} A_n = \bigcup_{n=k}^{\infty} A_n \qquad \text{and} \qquad \bigcap_{n \in I} A_n = \bigcap_{n=k}^{\infty} A_n$$

THEOREM 0.1.4 (Extended De Morgan Laws) Let $\mathcal{A} = \{A_i \mid i \in I\}$ be a collection of sets indexed by the nonempty set I. Then

1. $\overline{\bigcup_{i \in I} A_i} = \bigcap_{i \in I} \overline{A_i}$

2. $\overline{\bigcap_{i \in I} A_i} = \bigcup_{i \in I} \overline{A_i}$

Proof We prove part 1. You are asked to prove part 2 in Exercise 16. For each x,

$$x \in \overline{\bigcup_{i \in I} A_i} \leftrightarrow x \notin \bigcup_{i \in I} A_i$$

$$\leftrightarrow x \notin A_i \text{ for all } i \in I$$

$$\leftrightarrow x \in \overline{A_i} \text{ for all } i \in I$$

$$\leftrightarrow x \in \bigcap_{i \in I} \overline{A_i}$$

This completes the proof. □

DEFINITION 0.1.9 Two sets A and B are called **disjoint** provided $A \cap B = \{\ \}$. More generally, if $\mathcal{A} = \{A_i \mid i \in I\}$ is a collection of sets indexed by the index set I and $A_i \cap A_j = \{\ \}$ for all $i, j \in I$, $i \neq j$, then we call the collection $\mathcal{A}$ a **pairwise disjoint collection** of sets. ∎

■ **Example 0.1.8** Let $A = \{1, 2, 3\}$, $B = \{2, 3, 4\}$, $C = \{0, 1, 4\}$, $D = \{4, 5\}$, and $E = \{6, 7, 8, 9\}$. Then A and B are not disjoint and so $\{A, B, C\}$ is not a pairwise disjoint collection, even though

$$A \cap B \cap C = \{\ \}$$

On the other hand, $\{A, D, E\}$ is a pairwise disjoint collection. ∎

Note that the sets A and B in Example 0.1.8 are not disjoint; however, $A - B = \{1\}$ and $A \cap B = \{2, 3\}$ are disjoint and $A = (A - B) \cup (A \cap B)$. This relation holds in general and turns out to be quite useful. It is stated in Theorem 0.1.5, with the proof left to Exercise 8, parts (d) and (e).

THEOREM 0.1.5 For any sets A and B, the sets $A - B$ and $A \cap B$ are disjoint and

$$A = (A - B) \cup (A \cap B)$$ □

EXERCISE SET 0.1

1. Determine which of the following assertions are true and which are false.
   (a) $\{\ \} \in \{\ \}$
   (b) $1 \in \{1\}$
   (c) $\{1,2\} = \{2,1\}$
   (d) $\{\ \} = \{\{\ \}\}$
   (e) $\{\ \} \subset \{\{\ \}\}$
   (f) $1 \subseteq \{1\}$
   (g) $\{1\} \subset \{1,2\}$
   (h) $\{\ \} \in \{\{\ \}\}$

2. Prove the remaining parts of Theorem 0.1.1:
   (a) parts 1(a) and (b)
   (b) part 2(a)
   (c) part 3(b)
   (d) part 4
   (e) part 5(a)
   (f) parts 6(a) and (b)
   (g) part 7
   (h) parts 8(a), (b), (c), and (d)

3. Find the power set of each of the following sets:
   (a) $\{\ \}$
   (b) $\{1\}$
   (c) $\{1,2\}$
   (d) $\{1,2,3\}$
   (e) $\{\{\ \},\{1\}\}$
   (f) $\mathcal{P}(\mathcal{P}(\{\ \}))$

4. Prove Theorem 0.1.2, part 2.

5. Use the results of Exercise 3 to guess the number of subsets of a finite set having n elements.

6. Give examples of sets A, B, and C such that:
   (a) $A \in B$ and $B \in C$ and $A \notin C$
   (b) $A \in B$ and $B \in C$ and $A \in C$
   (c) $A \in B$ and $A \subset B$

7. Given $U = \mathbb{Z}$, $A = \{\ldots,-4,-2,0,2,4,\ldots\}$, $B = \{\ldots,-6,-3,0,3,6,\ldots\}$, and $C = \{\ldots,-8,-4,0,4,8,\ldots\}$, find the following sets:
   (a) $A \cap B$
   (b) $B - A$
   (c) $A - C$
   (d) $A \cap \overline{C}$
   (e) $C - A$
   (f) $B \cup C$
   (g) $(A \cup B) \cap C$
   (h) $(A \cup B) - C$

8. Let A and B be any two sets.
   (a) Prove that $A \cap B \subseteq A \cup B$.
   (b) Under what condition does $A \cap B = A \cup B$?
   (c) Prove that $A - B = A \cap \overline{B}$.
   (d) Show that $A - B$ and $A \cap B$ are disjoint.
   (e) Show that $A = (A - B) \cup (A \cap B)$.

9. Each of the following statements concerns arbitrary sets A and B. Complete the statement to make it true.
   (a) $A \subseteq B \leftrightarrow A \cap B =$ _____
   (b) $A \subseteq B \leftrightarrow A \cup B =$ _____
   (c) $A \subseteq B \leftrightarrow A - B =$ _____
   (d) $A \subset B \leftrightarrow (A - B =$ _____ and $B - A \neq$ _____)
   (e) $A \subset B \leftrightarrow (A \cap B =$ _____ and $A \cap B \neq$ _____)
   (f) $A - B = B - A \leftrightarrow$ _____

10. Prove the following parts of Theorem 0.1.3:
    (a) part 2
    (b) part 3

11. Let $A = \{-1, 1\}$, $B = \{0, 1\}$, and $C = \{-1, 0, 1\}$. Find the following sets:

(a) $A \times B$ (b) $B \times C$ (c) $A \times B \times C$

(d) $(A \times B) \times C$ (e) $A \times (B \times C)$ (f) $B \times B \times B \times B$

(g) $\mathcal{P}(B \times B)$ (h) $\mathcal{P}(B) \times \mathcal{P}(B)$

12. Let A and B be subsets of a universal set U. Prove or disprove the following assertions:

(a) $(U \times U) - (A \times B) = (U - A) \times (U - B)$ (b) $A \times A = A$

(c) $A \times U = U$ (d) $A \times \{\,\} = \{\,\}$

13. Let A, B, and C be arbitrary sets. Prove that if $B \subseteq C$, then $A \times B \subseteq A \times C$.

14. Let A and B be arbitrary nonempty sets.

(a) Under what condition does $A \times B = B \times A$?

(b) Under what condition is $(A \times B) \cap (B \times A)$ empty?

15. For any sets A, B, and C, prove: If $A \subseteq B$ and $B \subseteq C$, then $A \subseteq C$.

16. Prove Theorem 0.1.4, part 2.

17. For arbitrary nonempty sets A, B, C, and D, prove that

$$(A \subseteq C \text{ and } B \subseteq D) \leftrightarrow (A \times B \subseteq C \times D)$$

18. Let A, B, C, and D be sets such that $A \subseteq C$ and $B \subseteq D$. Prove that

(a) $A \cap B \subseteq C \cap D$ (b) $A \cup B \subseteq C \cup D$ (c) $A - D \subseteq C - B$

19. For sets A, B, and C, consider the following conditions: (i) $A \cap B = A \cap C$, (ii) $A \cup B = A \cup C$, and (iii) $B - A = C - A$. For each pair of these conditions,

(a) (i) and (ii)   (b) (i) and (iii)   (c) (ii) and (iii)

either prove that the two conditions together imply that $B = C$, or give an example in which the two conditions hold but $B \neq C$.

20. For sets A, B, and C, prove or disprove that

(a) $(A - B) - C = A - (B - C)$ (b) $(A - B) - C = (A - C) - B$

(c) $(A - B) - C = (A - C) - (B - C)$

21. For sets A and B, prove or disprove that

(a) $\mathcal{P}(A \cap B) = \mathcal{P}(A) \cap \mathcal{P}(B)$ (b) $\mathcal{P}(A \cup B) = \mathcal{P}(A) \cup \mathcal{P}(B)$

(c) $\mathcal{P}(A - B) = \mathcal{P}(A) - \mathcal{P}(B)$

22. For sets A and B, the **symmetric difference** of A and B is the set $A * B$ defined by

$$A * B = (A - B) \cup (B - A)$$

Prove:

(a) $A * B = B * A$ ⋆(b) $(A * B) * C = A * (B * C)$

(c) $A * B = (A \cup B) - (A \cap B)$ (d) $A \cap (B * C) = (A \cap B) * (A \cap C)$

(e) $A \cap B = \{\,\} \leftrightarrow A * B = A \cup B$ (f) $A * A = \{\,\}$

23. For sets A, B, and C, prove each of the following:

(a) If $C \subseteq A$ and $C \subseteq B$, then $C \subseteq A \cap B$.

(b) If $A \subseteq C$ and $B \subseteq C$, then $A \cup B \subseteq C$.

24. Prove: If $A_1 \subseteq A_2 \subseteq \cdots \subseteq A_k$, then

$$\bigcap_{n=1}^{k} A_n = A_1 \qquad \text{and} \qquad \bigcup_{n=1}^{k} A_n = A_k$$

25. For $n \in \{1, 2, \ldots, 50\}$, define $C_n = (-1/n, 2n)$. Find

$$\bigcup_{n=1}^{50} C_n \qquad \text{and} \qquad \bigcap_{n=1}^{50} C_n$$

26. For each collection $\{A_n \mid n \in \mathbb{N}\}$ below, find

$$\bigcup_{n=0}^{\infty} A_n \qquad \text{and} \qquad \bigcap_{n=0}^{\infty} A_n$$

(a) $A_n = \{k \in \mathbb{Z} \mid -n \le k \le 2n\}$
(b) $A_n = (-1/(n+1), (n+2)/(n+1))$ if n is even; $A_n = (-n/(n+1), 1/(n+1))$ if n is odd
(c) $A_0 = A_1 = \mathbb{N}$ and, for $n \ge 2$, $A_n = \{m \in \mathbb{N} \mid m \ge n \text{ and } m/n \notin \mathbb{N}\}$
(d) $A_n = \{kn \mid k \in \mathbb{Z}\}$

27. Let $I = (0, 1)$; for $i \in I$, define $A_i = (-i, i)$. Find

$$\bigcup_{i \in I} A_i \qquad \text{and} \qquad \bigcap_{i \in I} A_i$$

Hint: If $0 < i < j < 1$, then $\{0\} \subset A_i \subset A_j \subset (-1, 1)$.

28. Let $I = \mathbb{R}$. For $r \in \mathbb{R}$, define $A_r = \{(x, y) \mid y = rx\}$. Geometrically, A_r is a line in the coordinate plane that passes through the origin and has slope r. Find

$$\bigcup_{r \in \mathbb{R}} A_r \qquad \text{and} \qquad \bigcap_{r \in \mathbb{R}} A_r$$

29. Let S denote the set of ASCII character strings, and for $s \in S$, let C_s be the set of ASCII character strings that contain the string s as a substring. For instance, if $s = \text{MA}$, then the strings MATRIX and KALAMAZOO both belong to C_s. Find

$$\bigcup_{s \in S} C_s \qquad \text{and} \qquad \bigcap_{s \in S} C_s$$

30. For $n \in \{1, 2, \ldots, 100\}$, define $A_n = \{-n, -n+1, \ldots, n\}$. Find

$$\bigcup_{n=1}^{100} A_n \qquad \text{and} \qquad \bigcap_{n=1}^{100} A_n$$

31. For $n \in \mathbb{N}$, define $A_n = \{n, n+1, \ldots, 2n+1\}$. Find

$$\bigcup_{n=0}^{\infty} A_n \qquad \text{and} \qquad \bigcap_{n=0}^{\infty} A_n$$

32. For $n \in \mathbb{N}$, define $A_n = \{-n, -n+1, \ldots, n^2\}$. Find

$$\bigcup_{n=0}^{\infty} A_n \quad \text{and} \quad \bigcap_{n=0}^{\infty} A_n$$

33. For $n \in \mathbb{Z}^+$, define $A_n = \{m/n \mid m \in \mathbb{Z}\}$; that is, A_n is the set of fractions with denominator n. Find

$$\bigcup_{n=1}^{\infty} A_n \quad \text{and} \quad \bigcap_{n=1}^{\infty} A_n$$

34. For $n \in \mathbb{Z}^+$, define $A_n = \left(\dfrac{n+1}{2n}, \dfrac{3n-1}{n} \right)$. Find

$$\bigcup_{n=1}^{\infty} A_n \quad \text{and} \quad \bigcap_{n=1}^{\infty} A_n$$

35. Let $I = (0, \infty)$. For $i \in I$, define $A_i = \{(x, y) \mid y = ix^2\}$; that is, A_i is the set of points on the parabola $y = ix^2$. Find

$$\bigcup_{i \in I} A_i \quad \text{and} \quad \bigcap_{i \in I} A_i$$

36. Let A, B, and C be sets and consider the sets $A \times B \times C$ and $(A \times B) \times C$. Explain why, according the definition of product, these two sets are not equal. (Although the sets $(A \times B) \times C$, $A \times (B \times C)$, and $A \times B \times C$ are not formally equal, they are usually treated as being essentially the same.)

0.2 FUNCTIONS

In this section we review what is undoubtedly the most fundamental type of relation used in mathematics.

DEFINITION 0.2.1   Given sets X and Y, a **function from X to Y** is a subset f of $X \times Y$ with the property that, for every $x \in X$, there is a unique $y \in Y$ such that $(x, y) \in f$. ∎

We denote the fact that f is a function from X to Y by writing

$$f : X \to Y$$

The set X is called the **domain** of the function f and is written dom f. The set Y is called the **codomain** of f. Also, the subset of Y, written im f and defined by

$$\text{im } f = \{y \in Y \mid \text{ there exists } x \in X \text{ such that } (x, y) \in f\}$$

is called the **image** of f. If $X = Y$, then we call f a **function on X**.

■ *Example 0.2.1*   Determine which of the following are functions and find the image of each function.

(a) The subset f_0 of $\{1,2,3\} \times \{1,2,3\}$ given by $f_0 = \{(1,2),(2,3)\}$.

(b) The subset f_1 of $\{1,2,3\} \times \{1,2,3\}$ given by $f_1 = \{(1,2),(2,1),(3,2)\}$.

(c) The subset f_2 of $\{1,2,3\} \times \{1,2,3\}$ given by $f_2 = \{(1,1),(1,3),(2,3),(3,1)\}$.

(d) The subset f_3 of $\mathbb{Z} \times \mathbb{Z}$ defined by $f_3 = \{(m,n) \mid n \text{ is a multiple of } m\}$.

(e) The subset f_4 of $\mathbb{Z} \times \mathbb{Z}$ defined by $f_4 = \{(m,n) \mid n = 2m + 1\}$.

Solution

(a) f_0 is not a function because there is no ordered pair in f_0 of the form $(3,y)$. This violates the condition of the definition that, for each $x \in \{1,2,3\}$, there must exist a $y \in \{1,2,3\}$ such that $(x,y) \in f_0$.

(b) f_1 is a function on $\{1,2,3\}$; im $f_1 = \{1,2\}$.

(c) f_2 is not a function on $\{1,2,3\}$ because both $(1,1) \in f_2$ and $(1,3) \in f_2$. This violates the condition of the definition that, for each $x \in \{1,2,3\}$, there should be a unique $y \in \{1,2,3\}$ for which $(x,y) \in f_2$.

(d) The subset f_3 is not a function on $\mathbb{Z}$ because, for instance, $(2,2) \in f_3$ and $(2,-2) \in f_3$.

(e) The subset f_4 is a function on $\mathbb{Z}$; im f_4 is precisely the set of odd integers. ■

Suppose that f is a function from X to Y. Again, the condition of Definition 0.2.1 is that:

Given $x \in X$, there is a unique $y \in Y$ such that $(x,y) \in f$.

We call y the **image of x under f** and write $y = f(x)$; the notation $f(x)$ is read "f of x." Moreover, x is said to be a **preimage of y under f**. So

$$\text{im } f = \{f(x) \mid x \in X\}$$

It is also common to refer to $y = f(x)$ as the **value of f at x** and to say that f **maps** x to y. (A function is sometimes called a **mapping**.)

It is usual to define a function by writing: "Define $f : X \to Y$ by $y = f(x)$." Here, the "defining equation" $y = f(x)$ is viewed as a rule specifying how to compute the image of a given $x \in X$.

■ *Example 0.2.2*   Define $f : \mathbb{Q}^+ \to \mathbb{Q}^+$ by $f(x) = 1/x$. This f is called the **reciprocal function** (on $\mathbb{Q}^+$), because each positive rational number is mapped to its reciprocal. It is clear that f is a function since each positive rational number has a unique reciprocal. For example, $f(2) = 1/2$ and $f(2/3) = 1/(2/3) = 3/2$. Also, each positive rational number is the reciprocal of its reciprocal, so that im $f = \mathbb{Q}^+$. ■

■ *Example 0.2.3*   Define $g : \mathbb{R} \to \mathbb{Z}$ by $g(x) = \lfloor x \rfloor$, where $\lfloor x \rfloor$ denotes the greatest integer that is less than or equal to x. This g is called the **greatest integer function** or **floor function**. It is clear that g is a function since the process of "rounding down" a given real number x to the largest integer m such that $m \leq x$ determines a unique m. For example, $g(2) = 2$, $g(\pi) = 3$, $g(-2) = -2$, and $g(-\pi) = -4$. Note that for each integer m, $g(m) = m$, so that im $g = \mathbb{Z}$. ∎

■ ***Example 0.2.4*** Define $l : \mathbb{R} \to \mathbb{Z}$ by $l(x) = \lceil x \rceil$, where $\lceil x \rceil$ denotes the least integer that is greater than or equal to x. This l is called the **least integer function** or **ceiling function**. It is clear that l is a function since the process of "rounding up" a given real number x to the smallest integer m such that $x \leq m$ determines a unique m. For example, $l(2) = 2$, $l(\pi) = 4$, $l(-2) = -2$, and $l(-\pi) = -3$. Note that for each integer m, $l(m) = m$, so that im $l = \mathbb{Z}$. ∎

In the next several examples we need to recall the following idea. Given integers m and n with $n > 0$, there exist integers q and r, uniquely determined by m and n, such that

$$m = nq + r \qquad \text{and} \qquad 0 \leq r < n$$

This result is known as the **division algorithm**, and the numbers m, n, q, and r are called the **dividend**, **divisor**, **quotient**, and **remainder**, respectively. In this context we define the operations **div** and **mod** by

$$m \text{ div } n = q \qquad \text{and} \qquad m \text{ mod } n = r$$

It should be emphasized that $m \bmod n \in \{0, 1, \ldots, n-1\}$. In general, the set $\{0, 1, \ldots, n-1\}$ is called the **set of integers modulo** n and is denoted by $\mathbb{Z}_n$.

■ ***Example 0.2.5*** Define $g : \mathbb{Z}_7 \to \mathbb{Z}_7$ by $g(x) = 4x \bmod 7$. Then $g(0) = 0$, $g(1) = 4$,

$$g(2) = (4 \cdot 2) \bmod 7 = 8 \bmod 7 = 1$$
$$g(3) = (4 \cdot 3) \bmod 7 = 12 \bmod 7 = 5$$

and, similarly, $g(4) = 2$, $g(5) = 6$, and $g(6) = 3$. Note that im $g = \mathbb{Z}_7$. ∎

■ ***Example 0.2.6*** Define $h : \mathbb{Z}_6 \to \mathbb{Z}_3$ by $h(x) = x \bmod 3$. Then $h(0) = 0$, $h(1) = 1$, $h(2) = 2$, $h(3) = 3 \bmod 3 = 0$, $h(4) = 4 \bmod 3 = 1$, and $h(5) = 5 \bmod 3 = 2$. So im $h = \mathbb{Z}_3$. ∎

Given positive integers m and n, we say that n **divides** m provided $m \bmod n = 0$. The fact that n divides m is denoted by writing $n \mid m$. If m_1 and m_2 are positive integers and both $n \mid m_1$ and $n \mid m_2$, then n is called a **common divisor** of m_1 and m_2. We then define the **greatest common divisor** of m_1 and m_2 to be the largest common divisor of m_1 and m_2; it is denoted by $\gcd(m_1, m_2)$. Note that this gcd is a function from $\mathbb{Z}^+ \times \mathbb{Z}^+$ to $\mathbb{Z}^+$.

There is a famous algorithm for finding $\gcd(m_1, m_2)$, called the **euclidean algorithm**. It is based on the following observation. Assume without loss of generality that $m_1 \leq m_2$ and let $r = m_2 \bmod m_1$. If $r = 0$, then $m_1 \mid m_2$ and so $\gcd(m_1, m_2) = m_1$. On the other hand, if $r > 0$, then

$$\gcd(m_1, m_2) = \gcd(r, m_1)$$

We now have the problem of finding $\gcd(r, m_1)$, but since $r < m_1 < m_2$, this problem is "easier" than the one we started with! In fact, the euclidean algorithm provides a very efficient method for computing greatest common divisors; see, for example, Exercise 27 in Exercise Set 0.4.

■   ***Example 0.2.7***   We employ the euclidean algorithm to compute $\gcd(147, 546)$. First note that

$$546 \bmod 147 = 105$$

so that $\gcd(147, 546) = \gcd(105, 147)$. Next,

$$147 \bmod 105 = 42$$

so $\gcd(105, 147) = \gcd(42, 105)$. Next,

$$105 \bmod 42 = 21$$

so $\gcd(42, 105) = \gcd(21, 42)$. Finally, $42 \bmod 21 = 0$, and so $\gcd(21, 42) = 21$. It thus follows that $\gcd(147, 546) = 21$.   ■

Given (positive) integers a and b, a **linear combination** of a and b (over $\mathbb{Z}$) is any expression of the form

$$as + bt$$

where s and t are integers. An important characterization of $\gcd(a, b)$ is that it is the smallest positive linear combination of a and b. Thus, given integers a and b with $1 \leq a \leq b$, we often wish to determine $d = \gcd(a, b)$ and find integers s and t such that $d = as + bt$. A variation of the euclidean algorithm can be used to do this; the resulting algorithm is called the **extended euclidean algorithm**.

■   ***Example 0.2.8***   We illustrate the idea of the extended euclidean algorithm by computing $d = \gcd(147, 546)$ and finding integers s and t such that

$$d = 147s + 546t$$

Recall that d is the last nonzero remainder obtained in the application of the euclidean algorithm. If we take care to express, at each stage of the process, the current remainder as a linear combination of 147 and 546, then d is so expressed when the process terminates.

First of all,

$$546 \text{ div } 147 = 3 \qquad \text{and} \qquad 546 \text{ mod } 147 = 105$$

so that

$$105 = 546 - (147)(3) = 147(-3) + 546(1)$$

Next,

$$147 \text{ div } 105 = 1 \qquad \text{and} \qquad 147 \text{ mod } 105 = 42$$

so

$$42 = 147 - (105)(1) = 147 - [147(-3) + 546(1)] = 147(4) + 546(-1)$$

Next,

$$105 \text{ div } 42 = 2 \qquad \text{and} \qquad 105 \text{ mod } 42 = 21$$

so

$$21 = 105 - (42)(2)$$
$$= [147(-3) + 546(1)] - [147(4) + 546(-1)](2)$$
$$= 147(-11) + 546(3)$$

Finally, $42 \text{ mod } 21 = 0$, so we have that $\gcd(147, 546) = 21$. Also, we see from the above that $s = -11$ and $t = 3$. ∎

The information contained in the preceding example is conveniently organized in the following table:

r	s	t	q
546	0	1	
147	1	0	3
105	-3	1	1
42	4	-1	2
21	-11	3	

The remainders are placed in the column labeled r; the quotients in the column labeled q. Each remainder is expressed as $147s + 546t$, and the values of s and t are placed in the appropriate columns. In general, a table such as this would look as follows:

r	s	t	q
b	0	1	
a	1	0	q_1
$\vdots$			
r_{i-1}	s_{i-1}	t_{i-1}	q_i
r_i	s_i	t_i	q_{i+1}
r_{i+1}	s_{i+1}	t_{i+1}	
$\vdots$			

where $q_{i+1} = r_{i-1}$ div r_i and $r_{i+1} = r_{i-1}$ mod r_i. The question is, how are the values of s_{i+1} and t_{i+1} obtained? Well,

$$r_{i+1} = r_{i-1} - r_i q_{i+1}$$
$$= (a s_{i-1} + b t_{i-1}) - (a s_i + b t_i) q_{i+1}$$
$$= a(s_{i-1} - s_i q_{i+1}) + b(t_{i-1} - t_i q_{i+1})$$

It follows that

$$s_{i+1} = s_{i-1} - s_i q_{i+1} \qquad \text{and} \qquad t_{i+1} = t_{i-1} - t_i q_{i+1}$$

For a more complete discussion of the euclidean algorithm, see Chapter 3 of *Foundations of Discrete Mathematics*, 2nd ed., by Polimeni and Straight.

Now let $f : X \to Y$ be a function. Again, for each $x \in X$ there is a unique $y \in Y$ for which $(x, y) \in f$. However, it need not be the case that for each $y \in Y$ there is a unique $x \in X$ such that $f(x) = y$. In fact, for some y there may be no x such that $f(x) = y$, or there may be two distinct elements $x_1, x_2 \in X$ with $f(x_1) = y = f(x_2)$. For example, consider the function $f_4 : \mathbb{Z} \to \mathbb{Z}$ (from Example 0.2.1) defined by $f_4(m) = 2m + 1$. It is clear that $f_4(m)$ is odd for all $m \in \mathbb{Z}$, so, for instance, there is no m for which $f_4(m) = 0$. On the other hand, consider the function $h : \mathbb{Z}_6 \to \mathbb{Z}_3$ of Example 0.2.4 defined by $h(x) = x$ mod 3. Here each $y \in \mathbb{Z}_3$ has two preimages; for instance, $h(0) = 0 = h(3)$.

Some functions $f : X \to Y$ satisfy the property that for each $y \in Y$ there is at most one $x \in X$ such that $f(x) = y$; that is, each $y \in Y$ has at most one preimage under f. This condition may be rephrased as follows:

For $x_1, x_2 \in X$, if $f(x_1) = f(x_2)$, then $x_1 = x_2$

For example, the function $f_4 : \mathbb{Z} \to \mathbb{Z}$ defined above has this property. To show this, suppose that $f(m_1) = f(m_2)$ for some $m_1, m_2 \in \mathbb{Z}$. Then $2m_1 + 1 = 2m_2 + 1$, from which it easily follows that $m_1 = m_2$.

DEFINITION 0.2.2   A function $f : X \to Y$ is called **one-to-one** provided that, for all $x_1, x_2 \in X$, the following condition holds:

$$f(x_1) = f(x_2) \text{ implies } x_1 = x_2 \qquad \blacksquare$$

It is sometimes convenient to use the condition in Definition 0.2.2 in the following form:

For all $x_1, x_2 \in X$, if $x_1 \neq x_2$, then $f(x_1) \neq f(x_2)$

This is obtained by taking the contrapositive of the implication in the definition.

When is the function $f : X \rightarrow Y$ not one-to-one? Taking the negation of the condition in Definition 0.2.2 gives

f is not one-to-one if and only if, for some $x_1, x_2 \in X$, $x_1 \neq x_2$ and $f(x_1) = f(x_2)$

■ *Example 0.2.9*   Determine which of the following functions are one-to-one.

(a) $f_1 : \{1, 2, 3\} \rightarrow \{1, 2, 3\}$; $f_1(1) = 2, f_1(2) = 2, f_1(3) = 1$

(b) $f_2 : \{1, 2, 3\} \rightarrow \{1, 2, 3\}$; $f_2(1) = 3, f_2(2) = 2, f_2(3) = 1$

(c) $f_3 : \mathbb{Z} \rightarrow \mathbb{Z}$; $f_3(m) = m - 1$

(d) $f_4 : \mathbb{Z} \rightarrow \mathbb{Z}$; $f_4(m) = 2m + 1$

(e) $h : \mathbb{Z} \rightarrow \mathbb{Z}^+$; $h(m) = |m| + 1$

(f) $p : \mathbb{Q} - \{1\} \rightarrow \mathbb{Q}$; $p(r) = r/(1 - r)$

Solution

(a) This function f_1 is not one-to-one since $f_1(1) = f_1(2) = 2$.

(b) This function f_2 is one-to-one since no two elements of $\{1, 2, 3\}$ have the same image under f_2.

(c) This function f_3 maps each integer to its predecessor, and it is easily seen to be one-to-one.

(d) This is the function f_4 of Example 0.2.1 and it was shown to be one-to-one prior to Definition 0.2.2.

(e) If $h(m_1) = h(m_2)$, then $|m_1| + 1 = |m_2| + 1$, which implies that $|m_1| = |m_2|$. But this does not imply that $m_1 = m_2$, which leads one to suspect that h is not one-to-one. Indeed, if we let $m_1 = -1$ and $m_2 = 1$, then we see that $h(-1) = h(1) = 2$, which shows that h is not one-to-one.

(f) Let $r_1 \neq 1$ and $r_2 \neq 1$ be rational numbers. Then we have

$$p(r_1) = p(r_2) \rightarrow \frac{r_1}{1 - r_1} = \frac{r_2}{1 - r_2}$$

$$\rightarrow r_1(1 - r_2) = r_2(1 - r_1)$$

$$\rightarrow r_1 - r_1 r_2 = r_2 - r_2 r_1$$

$$\rightarrow r_1 = r_2$$

(Here $\rightarrow$ stands for "implies.") This shows that the function p is one-to-one.   ■

Given a function $f : X \rightarrow Y$, what can be said about im f? Knowing nothing else, all that can be said is that im $f \subseteq Y$. One extreme possibility is provided by the example $g : X \to Y$ defined by $g(x) = y_0$, where y_0 is a fixed element of Y. In this case im $f = \{y_0\}$, and g is called a **constant function** (the value of g is constant at y_0). The other extreme case is that of a function $f : X \to Y$ for which im $f = Y$.

DEFINITION 0.2.3   A function $f : X \to Y$ is called **onto** provided im $f = Y$.

■

Observe that a function $f : X \to Y$ is onto provided, for each $y \in Y$, there exists an $x \in X$ such that $f(x) = y$; that is, each $y \in Y$ has at least one preimage under f. This condition provides a common method for proving that a given function $f : X \to Y$ is onto. We choose an arbitrary element $y \in Y$, set $f(x) = y$, and then attempt to solve this equation for x in terms of y. If a solution x exists and is in X, then f is onto. On the other hand, if for some $y \in Y$ there is no solution $x \in X$ to the equation $f(x) = y$, then f is not onto.

■   ***Example 0.2.10***   Determine which of the following functions are onto.

(a) $f_1 : \{1, 2, 3\} \to \{1, 2, 3\}$; $f_1(1) = 2, f_1(2) = 2, f_1(3) = 1$

(b) $f_2 : \{1, 2, 3\} \to \{1, 2, 3\}$; $f_2(1) = 3, f_2(2) = 2, f_2(3) = 1$

(c) $f_3 : \mathbb{Z} \to \mathbb{Z}$; $f_3(m) = m - 1$

(d) $f_4 : \mathbb{Z} \to \mathbb{Z}$; $f_4(m) = 2m + 1$

(e) $h : \mathbb{Z} \to \mathbb{Z}^+$; $h(m) = |m| + 1$

(f) $p : \mathbb{Q} - \{1\} \to \mathbb{Q}$; $p(r) = r/(1 - r)$

(g) $g_4 : \mathbb{Q} \to \mathbb{Q}$; $g_4(r) = 2r + 1$

Solution

(a)   This function f_1 is not onto since im $f_1 = \{1, 2\} \neq \{1, 2, 3\}$.

(b)   The function f_2 is onto since im $f_2 = \{1, 2, 3\}$, the codomain of f_2.

(c)   The function f_3 is onto since, for any $n \in \mathbb{Z}$,

$$f(n + 1) = (n + 1) - 1 = n$$

(d)   As determined in Example 0.2.1, im f_4 is the set of odd integers:

$$\text{im } f_4 = \{2m + 1 \mid m \in \mathbb{Z}\}$$

Since im $f_4 \neq \mathbb{Z}$, f_4 is not onto.

(e)   For $n \in \mathbb{Z}^+$,

$$h(m) = n \leftrightarrow |m| + 1 = n$$
$$\leftrightarrow |m| = n - 1$$
$$\leftrightarrow m = n - 1 \quad \text{or} \quad m = 1 - n$$

Thus, each $n \geq 2$ is the image of two integers, namely, $n - 1$ and $1 - n$. Also, 1 is the image of 0. This shows that h is onto.

(f) For $s \in \mathbb{Q}$,

$$p(r) = s \leftrightarrow \frac{r}{1-r} = s$$

$$\leftrightarrow r = s(1-r)$$

$$\leftrightarrow r + rs = s$$

$$\leftrightarrow r(1+s) = s$$

$$\leftrightarrow r = \frac{s}{1+s}$$

Thus, if $s \neq -1$, then $r = s/(1+s) \in \mathbb{Q} - \{1\}$ and $p(r) = s$. However, there does not exist $r \in \mathbb{Q} - \{1\}$ such that $p(r) = -1$. Therefore, im $p = \mathbb{Q} - \{-1\}$, and the function p "just misses" being onto.

(g) Note that this function has the same rule as the function f_4 of part (d), but the domain and codomain have been changed from $\mathbb{Z}$ to $\mathbb{Q}$. Let's see what happens. Let $s \in \mathbb{Q}$; we wish to find $r \in \mathbb{Q}$ such that $g_4(r) = s$. Now,

$$g_4(r) = s \leftrightarrow 2r + 1 = s \leftrightarrow r = \frac{s-1}{2}$$

This shows that g_4 is onto; for each rational number s, the image of the rational number $r = (s-1)/2$ is s. This example illustrates the important point that whether a given function is onto depends not only on the defining equation of the function but on the domain and codomain as well. ∎

We have seen examples of functions that are one-to-one and not onto, and the reverse possibility, functions that are onto but not one-to-one. Under what conditions does the existence of one condition imply the other? One very important case is supplied by the following theorem.

THEOREM 0.2.1   Let X and Y be finite sets and let f be a function from X to Y.

1. If f is one-to-one, then $|X| \leq |Y|$.
2. If f is onto, then $|X| \geq |Y|$.
3. If f is one-to-one and onto, then $|X| = |Y|$.
4. If $|X| = |Y|$, then f is one-to-one if and only if f is onto.

Proof   We prove part 4. You are asked to prove parts 1 and 2 in Exercise 8; note that part 3 follows immediately from parts 1 and 2. For part 4, let $|X| = |Y| = n$; suppose $X = \{x_1, x_2, \ldots, x_n\}$.

We first assume that f is one-to-one and show that f is onto. The image of f is the set

$$\text{im } f = \{f(x) \mid x \in X\} = \{f(x_1), f(x_2), \ldots, f(x_n)\}$$

Suppose $f(x_i) = f(x_j)$ for some i and j. Since f is one-to-one, $f(x_i) = f(x_j)$ implies that $x_i = x_j$, and hence that $i = j$. This shows that $f(x_1)$, $f(x_2)$, $\ldots$, $f(x_n)$ are distinct elements of Y. Thus, we have both im $f \subseteq Y$ and $|\text{im } f| = |Y|$. We may conclude that im $f = Y$, thus proving that f is onto.

Next we assume that f is onto and show that f is one-to-one. Since f is onto, im $f = Y$. Thus, $\{f(x_1), f(x_2), \ldots, f(x_n)\} = Y$ and $|Y| = n$, so it must be that $f(x_1)$, $f(x_2), \ldots, f(x_n)$ are distinct. This shows that f is one-to-one. $\qquad\qquad\square$

Theorem 0.2.1, part 3, is often applied to prove that two finite sets X and Y have the same cardinality; to show this we may construct a one-to-one and onto function from X to Y. This may seem like a rather roundabout way to do things, but it is often quite enlightening. Such a proof is called a **bijective proof**, and examples of such proofs are given in Chapter 1.

The following example illustrates the application of Theorem 0.2.1, part 4.

■ *Example 0.2.11*   Consider the function $f : \mathbb{Z}_{30} \to \mathbb{Z}_{30}$ defined by $f(x) = 7x$ mod 30. Let $x_1, x_2 \in \mathbb{Z}_{30}$. The following steps show that f is one-to-one:

$$f(x_1) = f(x_2) \to 7x_1 \text{ mod } 30 = 7x_2 \text{ mod } 30$$

$$\to 7x_1 - 30q_1 = 7x_2 - 30q_2$$

(for some $q_1, q_2 \in \mathbb{Z}$)

$$\to 7(x_1 - x_2) = 30(q_1 - q_2)$$

$$\to 7(x_1 - x_2) \text{ is a multiple of } 30$$

$$\to x_1 - x_2 \text{ is a multiple of } 30$$

(since $\gcd(7, 30) = 1$)

$$\to x_1 - x_2 = 0$$

$$\to x_1 = x_2$$

Thus, f is one-to-one. We now see that f is onto with no additional work—we simply apply Theorem 0.2.1, part 4! ■

You should be aware that to apply Theorem 0.2.1, part 4, to a function f, the domain and codomain of f must be finite sets with the same cardinality. For example, define f and g on $\mathbb{Z}$ by $f(m) = 2m$ and $g(m) = \lfloor m/2 \rfloor$. We can then check that f is one-to-one but not onto, whereas g is onto but not one-to-one.

Now let $X = \{x_1, x_2, \ldots, x_n\}$ and suppose that $f : X \to X$ is a one-to-one function. Then it follows that the n-tuple $(f(x_1), f(x_2), \ldots, f(x_n))$ is simply an ordered arrangement of the elements of X. Indeed, in this sense any one-to-one and onto function on a set can regarded as selecting the elements of the set in some order, or "permuting" the elements of the set.

DEFINITION 0.2.4 Let X and Y be nonempty sets. A function $f : X \to Y$ that is one-to-one and onto is called a **bijection** from X to Y. If $X = Y$, then f is called a **permutation** of X. ∎

■ ***Example 0.2.12*** Consider again the functions in Examples 0.2.9 and 0.2.10. The function f_2 is a permutation of $\{1, 2, 3\}$. The function $f_3 : \mathbb{Z} \to \mathbb{Z}$ defined by $f_3(m) = m - 1$ is a permutation of $\mathbb{Z}$. In Example 0.2.6, part (g), we saw that the function $g_4 : \mathbb{Q} \to \mathbb{Q}$ defined by $g_4(r) = 2r + 1$ is onto. It is not hard to show that this function is also one-to-one. Therefore, g_4 is a permutation of $\mathbb{Q}$. Moreover, if we modify the function p by changing its codomain to $\mathbb{Q} - \{-1\}$; that is, we define $p : \mathbb{Q} - \{1\} \to \mathbb{Q} - \{-1\}$ by $p(r) = r/(1 - r)$, then p is a bijection from $\mathbb{Q} - \{1\}$ to $\mathbb{Q} - \{-1\}$. ∎

Suppose now that f is a function from X to Y; recall that $\operatorname{im} f = \{f(x) \mid x \in X\}$. It seems natural to write $\operatorname{im} f = f(X)$. More generally, for $A \subseteq X$, we define the **image of A under f** to be the set

$$f(A) = \{f(x) \mid x \in A\}$$

Similarly, for $B \subseteq Y$, we may wish to refer to the set of preimages of elements of B. Formally, we define the **preimage of B under f** to be the set

$$f^{-1}(B) = \{x \in X \mid f(x) \in B\}$$

■ ***Example 0.2.13*** Let g be the permutation of $\mathbb{Q}$ defined by $g(r) = 2r + 1$. Find the following sets, where E is the set of even integers and D is the set of multiples of 3:

(a) $g(\mathbb{Z})$ (b) $g(E)$ (c) $g^{-1}(\mathbb{N})$ (d) $g^{-1}(D)$

Solution

(a) For $m \in \mathbb{Z}$, $g(m) = 2m + 1$. Thus, $g(\mathbb{Z}) = \{2m + 1 \mid m \in \mathbb{Z}\}$; namely, $g(\mathbb{Z})$ is the set of odd integers.

(b) We have

$$k \in g(E) \leftrightarrow k = g(2m) \quad \text{for some } m \in \mathbb{Z}$$
$$\leftrightarrow k = 2(2m) + 1$$
$$\leftrightarrow k = 4m + 1$$

Hence, $g(E) = \{4m + 1 \mid m \in \mathbb{Z}\} = \{\ldots, -7, -3, 1, 5, 9, \ldots\}$.

(c) Here we have

$$r \in g^{-1}(\mathbb{N}) \leftrightarrow 2r + 1 = n \quad \text{for some } n \in \mathbb{N}$$
$$\leftrightarrow r = (n - 1)/2$$

Hence,

$$g^{-1}(\mathbb{N}) = \{(n-1)/2 \mid n \in \mathbb{N}\} = \left\{\frac{-1}{2}, 0, \frac{1}{2}, 1, \frac{3}{2}, 2, \dots\right\}$$

(d) Similarly,

$$r \in g^{-1}(D) \leftrightarrow 2r + 1 = 3m \qquad \text{for some } m \in \mathbb{Z}$$
$$\leftrightarrow r = \frac{3m-1}{2}$$

So

$$g^{-1}(D) = \left\{\frac{3m-1}{2} \mid m \in \mathbb{Z}\right\} = \left\{\dots, \frac{-7}{2}, -2, \frac{-1}{2}, 1, \frac{5}{2}, \dots\right\} \qquad \blacksquare$$

THEOREM 0.2.2 Given $f : X \to Y$, let A_1 and A_2 be subsets of X and let B_1 and B_2 be subsets of Y. Then the following relationships hold:

1. (a) $f(A_1 \cup A_2) = f(A_1) \cup f(A_2)$
   (b) $f^{-1}(B_1 \cup B_2) = f^{-1}(B_1) \cup f^{-1}(B_2)$
2. (a) $f(A_1 \cap A_2) \subseteq f(A_1) \cap f(A_2)$
   (b) $f^{-1}(B_1 \cap B_2) = f^{-1}(B_1) \cap f^{-1}(B_2)$
3. (a) $f(A_1) - f(A_2) \subseteq f(A_1 - A_2)$
   (b) $f^{-1}(B_1) - f^{-1}(B_2) = f^{-1}(B_1 - B_2)$
4. (a) If $A_1 \subseteq A_2$, then $f(A_1) \subseteq f(A_2)$.
   (b) If $B_1 \subseteq B_2$, then $f^{-1}(B_1) \subseteq f^{-1}(B_2)$.

Proof We will prove 1(a) and 3(b). You are asked to prove the remaining parts in Exercise 2.

To show that $f(A_1 \cup A_2) = f(A_1) \cup f(A_2)$, we show that each side is a subset of the other. If $y \in f(A_1 \cup A_2)$, then there is some $x \in A_1 \cup A_2$ such that $y = f(x)$. This element x is such that $x \in A_1$ or $x \in A_2$. If $x \in A_1$, then $y = f(x) \in f(A_1)$. Similarly, if $x \in A_2$, then $y \in f(A_2)$. Hence, $y \in f(A_1)$ or $y \in f(A_2)$, so that $y \in f(A_1) \cup f(A_2)$. This shows that $f(A_1 \cup A_2) \subseteq f(A_1) \cup f(A_2)$. To show the reverse inclusion, suppose $y \in f(A_1) \cup f(A_2)$, so that $y \in f(A_1)$ or $y \in f(A_2)$. This means that $y = f(x)$, where $x \in A_1$ or $x \in A_2$. Thus, $x \in A_1 \cup A_2$, which shows that $y \in f(A_1 \cup A_2)$. Therefore, $f(A_1) \cup f(A_2) \subseteq f(A_1 \cup A_2)$.

The proof that $f^{-1}(B_1) - f^{-1}(B_2) = f^{-1}(B_1 - B_2)$ is easily done using a string of biconditionals:

$$x \in f^{-1}(B_1) - f^{-1}(B_2) \leftrightarrow x \in f^{-1}(B_1) \text{ and } x \notin f^{-1}(B_2)$$
$$\leftrightarrow f(x) \in B_1 \text{ and } f(x) \notin B_2$$
$$\leftrightarrow f(x) \in B_1 - B_2$$
$$\leftrightarrow x \in f^{-1}(B_1 - B_2)$$

Therefore, $f^{-1}(B_1) - f^{-1}(B_2) = f^{-1}(B_1 - B_2)$. $\qquad \square$

Given $f : X \to Y$, note the following:

f is one-to-one if and only if $|f^{-1}(\{y\})| \le 1$ for every $y \in Y$

f is onto if and only if $|f^{-1}(\{y\})| \ge 1$ for every $y \in Y$

Additional properties involving the notions of image and preimage are addressed in the exercises.

Suppose that $f : X \to Y$ is both one-to-one and onto. Since f is onto, given any $y \in Y$, there is an element $x \in X$ such that $f(x) = y$. Moreover, since f is one-to-one, this element x is uniquely determined. Thus, for each $y \in Y$, there is exactly one $x \in X$ such that $y = f(x)$. We can then define a new function $g : Y \to X$ as follows: For $y \in Y$,

$$g(y) = x \text{ if and only if } f(x) = y$$

In other words, $g(y)$ is that unique element $x \in X$ for which $f(x) = y$.

DEFINITION 0.2.5 Let $f : X \to Y$ be one-to-one and onto. The function $g : Y \to X$ defined by

$$g(y) = x \text{ if and only if } f(x) = y$$

is called the **inverse function** of f and is denoted by f^{-1}. ■

THEOREM 0.2.3 If $f : X \to Y$ is a bijection, then the inverse function $f^{-1} : Y \to X$ is also a bijection.

Proof We first show that f^{-1} is one-to-one. Suppose $f^{-1}(y_1) = x = f^{-1}(y_2)$ for some $y_1, y_2 \in Y$ and $x \in X$. Then, by definition, $y_1 = f(x)$ and $y_2 = f(x)$. Since f is a function, it follows that $y_1 = y_2$. Thus, f^{-1} is one-to-one.

Next we show that f^{-1} is onto. Let $x \in X$, and suppose $f(x) = y$. Then, again by definition, $x = f^{-1}(y)$, so f^{-1} is onto. □

COROLLARY 0.2.4 Let X be a nonempty set. If f is a permutation of X, then f^{-1} is also a permutation of X. □

If $f : X \to Y$ is a bijection, how do we find its inverse function? For example, given $f : \mathbb{Q} \to \mathbb{Q}$ defined by $f(x) = 2x + 1$, let's attempt to find f^{-1}. Given $y \in \mathbb{Q}$, we want to find $x \in \mathbb{Q}$ such that $f^{-1}(y) = x$. This means that $y = f(x)$, so that $y = 2x + 1$. Solving for x we obtain $x = (y - 1)/2$, so that $f^{-1} : \mathbb{Q} \to \mathbb{Q}$ is defined by

$$f^{-1}(y) = \frac{y - 1}{2}$$

In general, given $y = f(x)$, we solve for x in terms of y to obtain

$$x = f^{-1}(y)$$

■   *Example 0.2.14*   We saw previously that the function $p : \mathbb{Q} - \{1\} \rightarrow \mathbb{Q} - \{-1\}$, defined by $p(x) = x/(1 - x)$, is a bijection. Now find p^{-1}.

Solution
If $y \in \mathbb{Q} - \{-1\}$ and $p^{-1}(y) = x$, where $x \in \mathbb{Q} - \{1\}$, then $p(x) = y$. Thus, $x/(1-x) = y$, so $x = (1 - x) y$ or $x + xy = y$. Now we solve for x to obtain $x = y/(1 + y)$. Therefore, $p^{-1} : \mathbb{Q} - \{-1\} \rightarrow \mathbb{Q} - \{1\}$ is given by

$$p^{-1}(y) = \frac{y}{1 + y}$$ ■

■   *Example 0.2.15*   We saw previously that the function $f : \mathbb{Z}_{30} \rightarrow \mathbb{Z}_{30}$, defined by $f(x) = 7x \bmod 30$, is a permutation of $\mathbb{Z}_{30}$. Now find f^{-1}.

Solution
We have

$$f^{-1}(y) = x \leftrightarrow f(x) = y$$
$$\leftrightarrow 7x \bmod 30 = y$$
$$\leftrightarrow 7x - 30q = y$$
$$\leftrightarrow 7x - y = 30q$$

for some $q \in \mathbb{Z}$. To find $f^{-1}(y)$, we must rewrite the equation $7x - y = 30q$ in the form $x - sy = 30q_1$ for some integers q_1 and s. It then follows that $x = sy \bmod 30$. Now, since $\gcd(7, 30) = 1$, there exist integers s and t such that $7s + 30t = 1$. In fact, the euclidean algorithm can be used to find that $7(13) + 30(-3) = 1$. Thus, continuing from above, we have

$$7x - y = 30q \leftrightarrow (7 \cdot 13)x - 13y = 30q \cdot 13$$
$$\leftrightarrow (1 + 3 \cdot 30)x - 13y = 30q \cdot 13$$
$$\leftrightarrow x - 13y = 30(13q - 3x)$$
$$\leftrightarrow x = 13y \bmod 30$$

Therefore, $f^{-1} : \mathbb{Z}_{30} \rightarrow \mathbb{Z}_{30}$ is defined by $f^{-1}(y) = 13y \bmod 30$.   ■

There are various ways in which two functions may be combined to produce a third function. One of the most common and important of these is called **composition**. Suppose $f : X \rightarrow Y$ and $g : Y \rightarrow Z$ are given. Then, for any $x \in X$, there is a unique $y \in Y$ such that $y = f(x)$. For this element y, there is a unique $z \in Z$ such that $z = g(y) = g(f(x))$. Hence, for each $x \in X$, there is associated a unique element $z \in Z$, namely, $z = g(f(x))$. This association allows us to define a new function $h : X \rightarrow Z$ by $h(x) = g(f(x))$.

DEFINITION 0.2.6   Given $f : X \to Y$ and $g : Y \to Z$, the **composition of f with g** (or **composite function**) is the function $g \circ f : X \to Z$ defined by

$$(g \circ f)(x) = g(f(x)) \qquad \blacksquare$$

■ **Example 0.2.16**   For E the set of even integers, let $f : \mathbb{Z} \to E$ be defined by $f(m) = 2m$, and let $g : E \to \mathbb{Z}^+$ be defined by $g(n) = |n|/2 + 1$. Then $g \circ f : \mathbb{Z} \to \mathbb{Z}^+$ is given by

$$(g \circ f)(m) = g(f(m)) = g(2m) = \frac{|2m|}{2} + 1 = |m| + 1 \qquad \blacksquare$$

■ **Example 0.2.17**   Let $f : \mathbb{Q} - \{0\} \to \mathbb{Q} - \{1\}$ be defined by $f(x) = (x+1)/x$ and $g : \mathbb{Q} - \{1\} \to \mathbb{Q} - \{2\}$ be defined by $g(x) = 3x - 1$. Then $g \circ f : \mathbb{Q} - \{0\} \to \mathbb{Q} - \{2\}$ is given by

$$\begin{aligned}
(g \circ f)(x) = g(f(x)) &= g\left(\frac{x+1}{x}\right) \\
&= 3\frac{x+1}{x} - 1 \\
&= \frac{3(x+1)}{x} - \frac{x}{x} \\
&= \frac{2x+3}{x} \qquad \blacksquare
\end{aligned}$$

■ **Example 0.2.18**   Let $f : \mathbb{Z} \to \mathbb{Z}$ be defined by $f(m) = m + 3$ and $g : \mathbb{Z} \to \mathbb{Z}$ be defined by $g(m) = -m$. Then $f \circ g : \mathbb{Z} \to \mathbb{Z}$ is given by

$$(f \circ g)(m) = f(g(m)) = f(-m) = -m + 3$$

and $g \circ f : \mathbb{Z} \to \mathbb{Z}$ is given by

$$(g \circ f)(m) = g(f(m)) = g(m + 3) = -(m + 3) = -m - 3$$

Note that $f \circ g \neq g \circ f$; also note that f, g, $f \circ g$, and $g \circ f$ are all permutations of $\mathbb{Z}$. ■

Let f and g be functions from X to Y. When is it the case that $f = g$? Since f and g are functions, each is a subset of $X \times Y$, and we already know when two sets are equal. We thus say that $f = g$ provided the condition

$$f(x) = g(x)$$

holds for every $x \in X$.

There are several interesting results that involve composition of functions and the properties onto and one-to-one. For example, suppose $f : X \rightarrow Y$ is a bijection; then $f^{-1} : Y \rightarrow X$ exists. Given $x \in X$ with $f(x) = y$, then $f^{-1}(y) = x$ and, hence, $f^{-1}(f(x)) = x$. So the composite function $f^{-1} \circ f : X \rightarrow X$ satisfies the property $(f^{-1} \circ f)(x) = x$ for all $x \in X$. In a similar fashion, we can determine that $f \circ f^{-1} : Y \rightarrow Y$ satisfies the condition $(f \circ f^{-1})(y) = y$ for every $y \in Y$. Note that both $f \circ f^{-1}$ and $f^{-1} \circ f$ are functions of the type $h : A \rightarrow A$, where $h(a) = a$.

DEFINITION 0.2.7 For any nonempty set A, the function $i_A : A \rightarrow A$ defined by

$$i_A(a) = a$$

is called the **identity function** on A. ∎

In view of the preceding discussion, if $f : X \rightarrow Y$ is a bijection, then $f^{-1} \circ f = i_X$ and $f \circ f^{-1} = i_Y$. A very basic and easily verified property of identity functions is contained in the following theorem, whose proof is left to Exercise 14.

THEOREM 0.2.5 Let X and Y be nonempty sets. For any function $f : X \rightarrow Y$,

$$i_Y \circ f = f \quad \text{and} \quad f \circ i_X = f \qquad \square$$

THEOREM 0.2.6 Given $f : X \rightarrow Y$ and $g : Y \rightarrow Z$, the following hold:

1. If f and g are both one-to-one, then $g \circ f$ is one-to-one.
2. If f and g are both onto, then $g \circ f$ is onto.

Proof We will prove part 2. You are asked to prove part 1 in Exercise 16.
  Assume f and g are both onto; to prove that $g \circ f$ is onto, we begin with an arbitrary element $z \in Z$. Since g is onto, there is an element $y \in Y$ such that $g(y) = z$. Since $y \in Y$ and f is onto, there is element $x \in X$ such that $f(x) = y$. Thus, $(g \circ f)(x) = g(f(x)) = g(y) = z$, and it follows that $g \circ f$ is onto. $\square$

COROLLARY 0.2.7 If $f : X \rightarrow Y$ and $g : Y \rightarrow Z$ are both bijections, then the function $g \circ f : X \rightarrow Z$ is a bijection. In particuliar, if $X = Y = Z$ so that f and g are both permutations of X, then $g \circ f$ is a permutation of X. $\square$

For each of the statements in Theorem 0.2.6, the converse is false (see Exercises 18 and 20). However, a partial converse does hold.

THEOREM 0.2.8   Given $f : X \to Y$ and $g : Y \to Z$, the following hold:

1. If $g \circ f$ is one-to-one, then f is one-to-one.
2. If $g \circ f$ is onto, then g is onto.

Proof   We will prove part 1. You are asked to prove part 2 in Exercise 22.
Suppose that $f(x_1) = f(x_2)$ for some $x_1, x_2 \in X$. Then, since $f(x_1) \in Y$, we have $g(f(x_1)) = g(f(x_2))$, namely, $(g \circ f)(x_1) = (g \circ f)(x_2)$. Since $g \circ f$ is given to be one-to-one, it may be concluded that $x_1 = x_2$. Therefore, f is one-to-one.   □

Given functions $f : A \to B$, $g : B \to C$, and $h : C \to D$, notice that $h \circ g$ is a function from B to D and $g \circ f$ is a function from A to C. Thus, $(h \circ g) \circ f$ and $h \circ (g \circ f)$ are both functions from A to D. In fact, they are equal functions.

THEOREM 0.2.9   Given $f : A \to B$, $g : B \to C$, and $h : C \to D$, we have

$$(h \circ g) \circ f = h \circ (g \circ f)$$

Proof   Both $(h \circ g) \circ f$ and $h \circ (g \circ f)$ have domain A and codomain D. To show equality, we must show that the two functions have the same value at each $x \in A$. Now,

$$
\begin{aligned}
[(h \circ g) \circ f](x) &= (h \circ g)[f(x)] \\
&= h(g(f(x))) \\
&= h[(g \circ f)(x)] \\
&= [h \circ (g \circ f)](x)
\end{aligned}
$$

Therefore, $(h \circ g) \circ f = h \circ (g \circ f)$.   □

For the last topic of this section, we look at the so-called big-oh and theta notations, which are used to measure the time complexity of algorithms. The running time of an algorithm is usually measured as a function of the size n of the problem. For example, suppose we have an algorithm that sorts a list of n integers in $f(n) = n^2 + bn + c$ steps, where b and c are positive constants and $b \le c$. Then, for $n \ge c$, we have

$$f(n) = n^2 + bn + c \le n^2 + n^2 + n^2 = 3n^2$$

In this case we would say that $f(n)$ is $O(n^2)$; in fact, we could make the stronger statement that $f(n)$ is $\Theta(n^2)$. Essentially, this notation emphasizes that, for n sufficiently large, $f(n)$ is bounded by some constant times n^2.

DEFINITION 0.2.8   Given functions $f : \mathbb{Z}^+ \to \mathbb{R}^+$ and $g : \mathbb{Z}^+ \to \mathbb{R}^+$, we write $f(n)$ is $O(g(n))$ provided there is some positive constant C and some positive integer n_0 such that

$$f(n) \leq Cg(n)$$

for all $n \geq n_0$. We read $f(n)$ is $O(g(n))$ as "$f(n)$ is big-O of $g(n)$." Similarly, we write $f(n)$ is $\Theta(g(n))$ provided there exist positive constants C_1 and C_2 and some positive integer n_0 such that

$$C_1 g(n) \leq f(n) \leq C_2 g(n)$$

for all $n \geq n_0$. We read $f(n)$ is $\Theta(g(n))$ as "$f(n)$ is theta of $g(n)$."   ■

In other words, $f(n)$ is $O(g(n))$ provided, for large enough values of n, $f(n)$ is bounded above by some constant times $g(n)$. Also note that $f(n)$ is $\Theta(g(n))$ if and only if both $f(n)$ is $O(g(n))$ and $g(n)$ is $O(f(n))$; see Exercise 52.

In general, if $f(n)$ is $O(g(n))$, then we like $g(n)$ to have a simple form and be as small as possible. In estimating running times of algorithms, some of the most commonly used functions are the following: $1, n, n^2, n^3, \log_2 n, n \log_2 n$, and 2^n. To compare the relative sizes of these functions, note that, for n sufficiently large,

$$1 < \log_2 n < n < n \log_2 n < n^2 < n^3 < 2^n$$

Thus, each function in the above list is big-oh to the function succeeding it. Of course, other functions, such as $n^2 / \log_2 n$, $\sqrt{n}$, and $\sqrt{n} \log_2 n$, also arise naturally in estimating running times.

■ ***Example 0.2.19***   Each of the following parts gives two functions $f(n)$ and $g(n)$ from $\mathbb{Z}^+$ to $\mathbb{R}^+$. Choose the statement $f(n)$ is $O(g(n))$ or $f(n)$ is $\Theta(g(n))$ that most accurately describes the relationship between f and g.

(a) $f(n) = (n^2 + 5)(n + 2)$, $g(n) = n^3$
(b) $f(n) = \sqrt{n^2 + 3}$, $g(n) = n$
(c) $f(n) = 4n^{1.9} + 15n + 9$, $g(n) = n^2$
(d) $f(n) = n^2 + 2^n$, $g(n) = 2^n$
(e) $f(n) = 5n \log_2 n + 4n + 3 \log_2 n + 2$, $g(n) = n \log_2 n$

Solution

(a) For $n \geq 4$,

$$(n^2 + 5)(n + 2) \leq (n^2 + 5)(n + n)$$
$$= (n^2 + 5)(2n)$$
$$= 2n^3 + 10n$$
$$\leq 2n^3 + n^3$$
$$= 3n^3$$

Thus, for $n \geq 4$,

$$n^3 \leq f(n) \leq 3n^3$$

so that $f(n)$ is $\Theta(n^3)$.

(b) For $n \geq 1$,

$$n = \sqrt{n^2} \leq f(n) = \sqrt{n^2 + 3} \leq \sqrt{n^2 + 3n^2} = \sqrt{4n^2} = 2n$$

so that $f(n)$ is $\Theta(n)$.

(c) For $n \geq 15$,

$$f(n) = 4n^{1.9} + 15n + 9 \leq 4n^2 + n^2 + n^2 = 6n^2$$

so $f(n)$ is $O(n^2)$. Moreover, for any positive constant C_1,

$$C_1 n^2 > f(n)$$

for n sufficiently large. Therefore, it is not the case that $f(n)$ is $\Theta(n^2)$.

(d) It can be proved (by induction on n) that $n^2 \leq 2^n$ for $n \geq 4$. Hence, for $n \geq 4$,

$$2^n \leq f(n) = n^2 + 2^n \leq 2^n + 2^n = 2 \cdot 2^n$$

showing that $f(n)$ is $\Theta(2^n)$.

(e) In this part, $f(n)$ is $\Theta(n \log_2 n)$. ■

EXERCISE SET 0.2

1. Let $S = \{1, 2, 3, 4\}$. Determine which of the following subsets of $S \times S$ are functions.

(a) $f_1 = \{(1, 2), (3, 4), (4, 1)\}$
(b) $f_2 = \{(1, 3), (2, 3), (3, 3), (4, 3)\}$
(c) $f_3 = \{(1, 1), (2, 2), (3, 3), (3, 4), (4, 4)\}$
(d) $f_4 = \{(1, 3), (2, 4), (3, 1), (4, 2)\}$

2. Prove the remaining parts of Theorem 0.2.2:
   (a) part 1(b)   (b) part 2(a)   (c) part 2(b)
   (d) part 3(a)   (e) part 4(a)   (f) part 4(b)

3. Determine which of the following functions are one-to-one.

   (a) $f : \mathbb{Z} \rightarrow \mathbb{Z}^+$; $f(m) = m^2 + 1$
   (b) $g : \mathbb{Q} \rightarrow \mathbb{Q}$; $g(x) = x^3$
   (c) $h : \mathbb{Q} \rightarrow \mathbb{Q}$; $h(x) = x^3 - x$
   (d) $p : \mathbb{Q} \rightarrow \mathbb{R}$; $p(x) = 2^x$
   (e) the "cardinality function" n from $\mathcal{P}(U)$, where U is a nonempty finite set, to $\{0, 1, \ldots, |U|\}$; $n(X) = |X|$
   (f) the "complement function" c on $\mathcal{P}(U)$; $c(X) = U - X$

4. List all the one-to-one functions from $\{1, 2\}$ to $\{1, 2, 3, 4\}$.

5. Determine which of the functions in Exercise 3 are onto.

6. List the permutations of the following sets.

   (a) $\{1\}$
   (b) $\{1, 2\}$
   (c) $\{1, 2, 3\}$

7. Each of the following parts gives sets X and Y and a function from X to Y. Determine whether the function is one-to-one.

   (a) $X = \{1, 2, 3, 4\}$, $Y = \{1, 2, 3\}$; $f_1(1) = 2, f_1(2) = 3, f_1(3) = f_1(4) = 1$
   (b) $X = \{1, 2, 3\}$, $Y = \{1, 2, 3, 4\}$; $f_2(1) = 3, f_2(2) = 2, f_2(3) = 1$
   (c) $X = Y = \{1, 2, 3, 4\}$; $f_3(1) = f_3(3) = 2, f_3(2) = f_3(4) = 1$
   (d) $X = Y = \{1, 2, 3, 4\}$; $f_4(1) = 3, f_4(2) = 4, f_4(3) = 1, f_4(4) = 2$
   (e) $X = Y = \mathbb{Z}$; $f_5(m) = -m$
   (f) $X = Y = \mathbb{Z}$; if $m \geq 0, f_6(m) = 2m$; if $m < 0, f_6(m) = 3m$
   (g) $X = Y = \mathbb{Z}^+$; if n is odd, $f_7(n) = (n + 1)/2$; if n is even, $f_7(n) = n/2$
   (h) $X = Y = \mathbb{Z}^+$; if n is odd, $f_8(n) = n + 1$; if n is even, $f_8(n) = n - 1$

8. Prove parts 1 and 2 of Theorem 0.2.1.

9. Determine which of the functions in Exercise 7 are onto.

10. Give an example of a function on $\mathbb{N}$ that is:
   (a) neither one-to-one nor onto   (b) one-to-one but not onto
   (c) onto but not one-to-one   (d) both one-to-one and onto

11. Determine which of the following functions are one-to-one.

   (a) $f_1 : \mathbb{Z}_{10} \rightarrow \mathbb{Z}_{10}$; $f_1(n) = 3n \bmod 10$
   (b) $f_2 : \mathbb{Z}_{10} \rightarrow \mathbb{Z}_{10}$; $f_2(n) = 5n \bmod 10$
   (c) $f_3 : \mathbb{Z}_{36} \rightarrow \mathbb{Z}_{36}$; $f_3(n) = 3n \bmod 36$
   (d) $f_4 : \mathbb{Z}_{36} \rightarrow \mathbb{Z}_{36}$; $f_4(n) = 5n \bmod 36$
   (e) $f_5 : \mathbb{Z}_{10} \rightarrow \mathbb{Z}_{10}$; $f_5(n) = (n + 3) \bmod 10$
   (f) $f_6 : \mathbb{Z}_{10} \rightarrow \mathbb{Z}_{10}$; $f_6(n) = (n + 5) \bmod 10$
   (g) $f_7 : \mathbb{Z}_{12} \rightarrow \mathbb{Z}_8$; $f_7(n) = 2n \bmod 8$

(h) $f_8 : \mathbb{Z}_8 \rightarrow \mathbb{Z}_{12}$; $f_8(n) = 3n$ mod 12

(i) $f_9 : \mathbb{Z}_6 \rightarrow \mathbb{Z}_{12}$; $f_9(n) = 2n$ mod 12

(j) $f_0 : \mathbb{Z}_{12} \rightarrow \mathbb{Z}_{36}$; $f_0(n) = 6n$ mod 36

12. Determine which of the following functions on $\mathbb{Z}_{12}$ are permutations. In each case, find the image of $A = \{1, 5, 7, 11\}$ and the preimage of $B = \{4, 8\}$.

(a) $f_1(n) = 2n$ mod 12   (b) $f_2(n) = 4n$ mod 12

(c) $f_3(n) = 5n$ mod 12

13. Determine which of the functions in Exercise 11 are onto.

14. Prove Theorem 0.2.5.

15. Each of the following parts refers to the corresponding function defined in Exercise 7. For the given $A \subseteq X$ and $B \subseteq Y$, find the image of A and the preimage of B.

(a) $A = \{1, 2\} = B$

(b) $A = \{1, 3\}, B = \{2, 4\}$

(c) $A = \{1, 3\}, B = \{1\}$

(d) $A = \{2\} = B$

(e) $A = \{2m \mid m \in \mathbb{Z}\}, B = \mathbb{Z}^+$

(f) $A = \mathbb{Z}^+, B = \{2m \mid m \in \mathbb{Z}\}$

(g) $A = B = \{2n \mid n \in \mathbb{Z}^+\}$

(h) $A = B = \{2n \mid n \in \mathbb{Z}^+\}$

16. Prove Theorem 0.2.6, part 1.

17. Find the inverse of each of the following functions.

(a) $f_1 : \mathbb{Q} \rightarrow \mathbb{Q}$; $f_1(x) = 4x + 2$

(b) $f_2 : \mathbb{Q} - \{1\} \rightarrow \mathbb{Q} - \{1\}$; $f_2(x) = \dfrac{x}{x - 1}$

(c) $f_3 : \mathbb{Z}_{12} \rightarrow \mathbb{Z}_{12}$; $f_3(x) = 5x$ mod 12

(d) $f_4 : \mathbb{Z}_{39} \rightarrow \mathbb{Z}_{39}$; $f_4(x) = (5x + 2)$ mod 39

(e) $f_5 : \mathbb{Z} \rightarrow \mathbb{Z}$; $f_5(m) = m + 1$

(f) $f_6 : \mathbb{Z} \rightarrow \mathbb{N}$; $f_6(m) = \begin{cases} 2m - 1, & m > 0 \\ -2m, & m \leq 0 \end{cases}$

(g) $f_7 : \{1, 2, 3, 4\} \rightarrow \{1, 2, 3, 4\}$; $f_7(1) = 4, f_7(2) = 1, f_7(3) = 2, f_7(4) = 3$

(h) $f_8 : \{1, 2, 3, 4\} \rightarrow \{1, 2, 3, 4\}$; $f_8(1) = 3, f_8(2) = 4, f_8(3) = 1, f_8(4) = 2$

18. Give an example of sets X, Y, and Z, and of functions $f : X \rightarrow Y$ and $g : Y \rightarrow Z$, such that $g \circ f$ and f are both one-to-one, but g is not.

19. Find $g \circ f$.

(a) $f : \mathbb{Z} \rightarrow \mathbb{Z}^+$; $f(m) = |m| + 1$, $g : \mathbb{Z}^+ \rightarrow \mathbb{Q}^+$; $g(n) = 1/n$

(b) $f : \mathbb{R} \rightarrow (0, 1)$; $f(x) = 1/(x^2 + 1)$, $g : (0, 1) \rightarrow (0, 1)$; $g(x) = 1 - x$

(c) $f : \mathbb{Q} - \{2\} \rightarrow \mathbb{Q} - \{0\}$; $f(x) = 1/(x - 2)$, $g : \mathbb{Q} - \{0\} \rightarrow \mathbb{Q} - \{0\}$; $g(x) = 1/x$

(d) $f : \mathbb{R} \to (1, \infty)$; $f(x) = x^2 + 1$, $g : [1, \infty) \to [0, \infty)$; $g(x) = \sqrt{x-1}$

(e) $f : \mathbb{Q} - \{10/3\} \to \mathbb{Q} - \{3\}$; $f(r) = 3r - 7$, $g : \mathbb{Q} - \{3\} \to \mathbb{Q} - \{2\}$; $g(r) = \dfrac{2r}{r-3}$

(f) $f : \mathbb{Z} \to \mathbb{Z}_5$; $f(m) = m \bmod 5$, $g : \mathbb{Z}_5 \to \mathbb{Z}_5$; $g(m) = (m+1) \bmod 5$

(g) $f : \mathbb{Z}_8 \to \mathbb{Z}_{12}$; $f(m) = 3m \bmod 12$, $g : \mathbb{Z}_{12} \to \mathbb{Z}_6$; $g(m) = 2m \bmod 6$

(h) $f : \{1,2,3,4\} \to \{1,2,3,4\}$; $f(1) = 4$, $f(2) = 1$, $f(3) = 2$, $f(4) = 3$, $g : \{1,2,3,4\} \to \{1,2,3,4\}$; $g(1) = 3$, $g(2) = 4$, $g(3) = 1$, $g(4) = 2$

20. Give an example of sets X, Y, and Z, and of functions $f : X \to Y$ and $g : Y \to Z$, such that $g \circ f$ and g are both onto, but f is not.

21. Given the permutations f and g, find f^{-1}, g^{-1}, $f \circ g$, $(f \circ g)^{-1}$, and $g^{-1} \circ f^{-1}$.

(a) $f : \mathbb{Z} \to \mathbb{Z}$; $f(m) = m + 1$, $g : \mathbb{Z} \to \mathbb{Z}$; $g(m) = 2 - m$

(b) $f : \mathbb{Z}_7 \to \mathbb{Z}_7$; $f(m) = (m + 3) \bmod 7$, $g : \mathbb{Z}_7 \to \mathbb{Z}_7$; $g(m) = 2m \bmod 7$

(c) $f : \{1,2,3,4\} \to \{1,2,3,4\}$; $f(1) = 4$, $f(2) = 1$, $f(3) = 2$, $f(4) = 3$, $g : \{1,2,3,4\} \to \{1,2,3,4\}$; $g(1) = 3$, $g(2) = 4$, $g(3) = 1$, $g(4) = 2$

(d) $f : \{1,2,3,4\} \to \{1,2,3,4\}$; $f(1) = 2$, $f(2) = 4$, $f(3) = 3$, $f(4) = 1$, $g : \{1,2,3,4\} \to \{1,2,3,4\}$; $g(1) = 1$, $g(2) = 3$, $g(3) = 4$, $g(4) = 2$

(e) $f : \mathbb{Q} \to \mathbb{Q}$; $f(x) = 4x$, $g : \mathbb{Q} \to \mathbb{Q}$; $g(x) = (x - 3)/2$

(f) $f : \mathbb{Q} - \{1\} \to \mathbb{Q} - \{1\}$; $f(x) = 2x - 1$, $g : \mathbb{Q} - \{1\} \to \mathbb{Q} - \{1\}$; $g(x) = x/(x - 1)$

22. Prove Theorem 0.2.8, part 2.

23. For the permutations f and g given in Exercise 21, find $g \circ f$ and $f^{-1} \circ g^{-1}$.

24. Let $f : X \to Y$ and $g : Y \to Z$ both be bijections. Prove that

$$(g \circ f)^{-1} = f^{-1} \circ g^{-1}$$

25. Define the functions f and g on your family tree by $f(x) =$ the father of x and $g(x) =$ the eldest child of the father of x. Describe the following functions.
     (a) $f \circ f$ (b) $f \circ g$ (c) $g \circ f$ (d) $g \circ g$

26. Let $f : X \to Y$ be a bijection. Prove that $(f^{-1})^{-1} = f$.

27. Give an example of a function $f : [0, 1] \to [0, 1]$ that is:
     (a) both one-to-one and onto   (b) one-to-one but not onto
     (c) onto but not one-to-one   (d) neither one-to-one nor onto

28. Let $f : X \to Y$ and $g : Y \to Z$ be bijections. Prove that if $g \circ f = i_X$ (or $f \circ g = i_Y$), then $g = f^{-1}$.

29. Show that there are infinitely many pairs of distinct functions f and g on $\mathbb{Q}$ such that none of f, g, and $f \circ g$ is the identity function on $\mathbb{Q}$ and that $f \circ g = g \circ f$.

30. Let $f : X \to Y$, where X and Y are subsets of $\mathbb{R}$. The function f is said to be **increasing** provided the condition

$$x_1 < x_2 \to f(x_1) < f(x_2)$$

holds for all $x_1, x_2 \in X$. Similarly, the function f is said to be **decreasing** provided the condition

$$x_1 < x_2 \to f(x_1) > f(x_2)$$

holds for all $x_1, x_2 \in X$. If f is either increasing or decreasing, then we say that f is **monotonic**.

(a) Prove that a monotonic function is one-to-one.
(b) Apply the result of part (a) to show that $f : \mathbb{R} \to \mathbb{R}$ defined by $f(x) = x^3 + x - 2$ is one-to-one.

31. Let X and Y be nonempty sets. Prove: A function $f : X \to Y$ is one-to-one if and only if the condition

$$f(A_1 \cap A_2) = f(A_1) \cap f(A_2)$$

holds for all subsets A_1 and A_2 of X.

32. Let X and Y be nonempty sets and let $f : X \to Y$. Prove the following results.

(a) f is onto if and only if $f(f^{-1}(B)) = B$ for every $B \subseteq Y$.
(b) f is one-to-one if and only if $f^{-1}(f(A)) = A$ for every $A \subseteq X$.

33. Let f, g, and h be functions on $\mathbb{Z}$ defined as follows: $f(m) = m + 1$, $g(m) = 2m$, and

$$h(m) = \begin{cases} 0 & \text{if } m \text{ is even} \\ 1 & \text{if } m \text{ is odd} \end{cases}$$

Determine the following composite functions.
(a) $f \circ g$ (b) $g \circ f$ (c) $f \circ h$ (d) $h \circ f$
(e) $g \circ h$ (f) $h \circ g$ (g) $g \circ g$ (h) $h \circ f \circ g$

34. Let U be a nonempty universal set. For $A \subseteq U$, define the function $\chi_A : U \to \{0, 1\}$ by

$$\chi_A(x) = \begin{cases} 0 & \text{if } x \notin A \\ 1 & \text{if } x \in A \end{cases}$$

The function χ_A is called the **characteristic function** of A. For $A, B \in \mathcal{P}(U)$, let $C = A \cap B$, $D = A \cup B$, and $E = A - B$. Prove the following hold for all $x \in U$.

(a) $\chi_C(x) = \chi_A(x) \cdot \chi_B(x)$
(b) $\chi_D(x) = \chi_A(x) + \chi_B(x) - [\chi_A(x) \cdot \chi_B(x)]$
(c) $\chi_U(x) = 1$
(d) $\chi_{\{\}}(x) = 0$

(e) $\chi_{\bar{B}}(x) = 1 - \chi_B(x)$

(f) $\chi_E(x) = \chi_A(x)[1 - \chi_B(x)]$

35. Let X be a nonempty set, let i denote the identity function on X, and let f be a function on X. Define $f^0 = i$, $f^1 = f$, $f^2 = f \circ f$, $f^3 = f \circ f \circ f$, and so on. (Recursively, $f^0 = i$ and $f^n = f \circ f^{n-1}$, for $n \geq 1$.) Take the case $X = \mathbb{Z}$; give an example of a function $f : \mathbb{Z} \to \mathbb{Z}$ such that:

(a) $f \neq i$ but $f^2 = i$

(b) $f^2 \neq i$ but $f^3 = i$

(c) Generalize parts (a) and (b); for each $n > 1$, give an example of a function $f : \mathbb{Z} \to \mathbb{Z}$ such that $f \neq i, \ldots, f^{n-1} \neq i$ but $f^n = i$.

36. Let $\mathcal{F}$ denote the set of functions on $\mathbb{R}$, let $\mathcal{C} = \{f \in \mathcal{F} \mid f$ is continuous$\}$, and let $\mathcal{D} = \{f \in \mathcal{F} \mid f$ is differentiable$\}$. Consider the function $\Delta : \mathcal{D} \to \mathcal{F}$ that maps each function $f \in \mathcal{D}$ to its derivative f' in $\mathcal{F}$, that is, $\Delta(f) = f'$.

(a) Is the function Δ one-to-one?

(b) Let $f \in \mathcal{D}$. What is $\Delta^{-1}(\{f'\})$?

$\star$(c) Show that $\mathcal{C} \subset \text{im } \Delta$.

37. Define $f : \mathbb{Z}_{119} \to \mathbb{Z}_{119}$ by $f(x) = 15x \bmod 119$. Show that f is a permutation of $\mathbb{Z}_{119}$ and find f^{-1}.

38. Let m and n be positive integers, and let $A = \{0, 1, \ldots, m - 1\}$, $B = \{0, 1, \ldots, n-1\}$, and $C = \{0, 1, \ldots, mn - 1\}$. Construct a bijection $f : A \times B \to C$.

39. Define $f : \mathbb{Q} - \{1/4\} \to \mathbb{Q} - \{0\}$ by $f(x) = 1 - 4x$ and $g : \mathbb{Q} - \{0\} \to \mathbb{Q} - \{3/2\}$ by $g(x) = (3x - 1)/(2x)$. Determine each of the functions $g \circ f$, $(g \circ f)^{-1}$, f^{-1}, g^{-1}, and $f^{-1} \circ g^{-1}$.

40. Find a bijection:

(a) from $(-1, 1)$ to $\mathbb{R}$ 　　　(b) from $\mathbb{Z}$ to $\mathbb{N}$

41. Let a and b be real numbers with $a < b$. Find a bijection from $(0, 1)$ to (a, b).

$\star$**42.** Find a function $f : \mathbb{Q}^+ \to \mathbb{Z}^+$ such that f is one-to-one.

43. Find a bijection from $[0, 1)$ to $[0, \infty)$.

$\star$**44.** For any set A, show that there does not exist a bijection from A to $\mathcal{P}(A)$.

45. Given real numbers a, b, c, and d such that $a < b$ and $c < d$, find a bijection from (a, b) to (c, d).

46. Let A_1, B_1, A_2, and B_2 be nonempty sets such that $A_1 \cap B_1 = \{ \} = A_2 \cap B_2$. Given bijections $f : A_1 \to A_2$ and $g : B_1 \to B_2$, construct a bijection $h : A_1 \cup B_1 \to A_2 \cup B_2$.

47. Let A_1, B_1, A_2, and B_2 be nonempty sets. Given bijections $f : A_1 \to A_2$ and $g : B_1 \to B_2$, construct a bijection $h : A_1 \times B_1 \to A_2 \times B_2$.

48. Prove: If $f(n)$ is $O(g(n))$ and $g(n)$ is $O(h(n))$, then $f(n)$ is $O(h(n))$.

49. Each part gives two functions $f(n)$ and $g(n)$ from $\mathbb{Z}^+$ to $\mathbb{R}^+$. Choose the statement $f(n) = O(g(n))$ or $f(n) = \Theta(g(n))$ that most accurately describes the relationship between f and g.

(a) $f(n) = \dfrac{\log_2 n}{n}$, $\quad g(n) = 1$

(b) $f(n) = \dfrac{n^2 + 3}{n}$, $\quad g(n) = n$

(c) $f(n) = \log_2(\log_2 n))$, $\quad g(n) = \log_2 n$

(d) $f(n) = \begin{cases} n, & \text{if } n \text{ is prime} \\ 1, & \text{otherwise} \end{cases}$, $\quad g(n) = n$

$\star$(e) $f(n) = n^{99}$, $\quad g(n) = 2^{\sqrt{n}}$

(f) $f(n) = \dfrac{n^4 + 2n^3 - 6n + 3}{3n^2 + 4}$, $\quad g(n) = n^2$

(g) $f(n) = 3n + n \log_2 n$, $\quad g(n) = n^2$

(h) $f(n) = 1^2 + 2^2 + \cdots + n^2$, $\quad g(n) = n^3$

$\star$(i) $f(n) = 2^{\sqrt{n}}$, $\quad g(n) = (1.01)^n$

$\star$(j) $f(n) = \log_2(2^n + n^2)$, $\quad g(n) = n$

50. Prove: If $f_1(n)$ is $O(g_1(n))$ and $f_2(n)$ is $O(g_2(n))$, then $f_1(n) + f_2(n)$ is $O(g_1(n) + g_2(n))$.

51. Prove: If $f_1(n)$ is $O(g(n))$ and $f_2(n)$ is $O(h(n))$, then $f_1(n)f_2(n)$ is $O(g(n)h(n))$.

52. Show that $f(n)$ is $\Theta(g(n))$ if and only if both $f(n)$ is $O(g(n))$ and $g(n)$ is $O(f(n))$.

53. Prove: If $f_1(n)$ is $\Theta(g(n))$ and $f_2(n)$ is $O(g(n))$, then $f_1(n) + f_2(n)$ is $\Theta(g(n))$.

$\star$**54.** Let $f(n) = 2^n$ and $g(n) = n!$ (n-factorial). Show that $f(n)$ is $O(g(n))$.

55. Apply the extended euclidean algorithm to find $d = \gcd(a, b)$ and integers s and t such that $d = as + bt$.
(a) $a = 119$, $b = 154$ $\qquad$ (b) $a = 357$, $b = 629$
(c) $a = 405$, $b = 1380$ $\qquad$ (d) $a = 812$, $b = 2800$

56. Prove: If

$$\lim_{n \to \infty} \frac{f(n)}{g(n)} = c$$

where c is a positive constant, then $f(n)$ is $\Theta(g(n))$. Use this result to redo Example 0.2.18 and Exercise 49. (If $c = 1$, then we say that $f(n)$ is **asymptotic** to $g(n)$ and write $f(n) \approx g(n)$. If $c = 0$, then we say that $f(n)$ is $o(g(n))$; this is read, "$f(n)$ is little-oh of $g(n)$." Note that $f(n)$ is $o(g(n))$ implies that $f(n)$ is $O(g(n))$ and $f(n)$ is not $\Theta(g(n))$.)

0.3 RELATIONS AND GRAPHS

In this section we review the notion of a relation between two sets and a related idea, that of a graph.

A rather nice example of a relation is provided by the idea of divisibility for positive integers. Given two positive integers a and b, we can ask whether a divides b (b is a multiple of a); if so, we say that "a is related to b." It turns out that "divides" is relation from $\mathbb{Z}^+$ to $\mathbb{Z}^+$. Note that it is possible to have a related to b but b not related to a; this happens, for example, when $a = 2$ and $b = 6$. Because of this, it makes good sense to single out the ordered pair (a, b) when a is related to b. Thus, the relation determines a unique set of ordered pairs, namely,

$R = \{(a, b) \mid a, b \in \mathbb{Z}^+ \text{ and } a \text{ divides } b\}$

DEFINITION 0.3.1   A **relation** from a set X to a set Y is a subset of $X \times Y$. If R is a relation from X to Y and $(a, b) \in R$, then we say that a **is related to** b and write $a \, R \, b$.   ■

Since the preceding definition involves two sets X and Y, we sometimes refer to such a relation as a **binary relation**.

■ *Example 0.3.1*   Let

$X = \{$Brinkerhoff, Chan, McKenna, Slonneger, Will, Yellen$\}$

be the set of computer science instructors at a small college, and let

$Y = \{$CS105, CS260, CS261, CS360, CS460$\}$

be the set of computer science courses offered next semester at that college. Then $X \times Y$ gives all possible pairings of instructors and courses. Let the relation R from X to Y be given by

$R = \{$(Brinkerhoff, CS105), (Brinkerhoff, CS360), (Chan, CS260),
(Chan, CS360), (Chan, CS460), (McKenna, CS105), (McKenna, CS260),
(Slonneger, CS260), (Slonneger, CS261), (Will, CS105), (Yellen, CS261),
(Yellen, CS460)$\}$

Then R might tell us, for example, which instructors are assigned to teach which courses. ∎

■ *Example 0.3.2*   Let P be the set of primes and define the relation R from P to $\mathbb{Z}^+$ by

$p R n \leftrightarrow p$ divides n

Find:

(a) all primes p such that $p R$ 126
(b) all n such that $3 R n$

Solution

(a) Since $126 = 2 \cdot 3 \cdot 3 \cdot 7$, we have $2 R$ 126, $3 R$ 126, and $7 R$ 126.
(b) Note that $3 R n$ if and only if n is a multiple of 3. Thus, the set of positive integers to which 3 is related under R is $\{3, 6, 9, \ldots\}$. ∎

Suppose that X and Y are finite sets—say, $X = \{x_1, x_2, \ldots, x_m\}$ and $Y = \{y_1, y_2, \ldots, y_n\}$—and R is a relation from X to Y. One way to represent R is to form a matrix with m rows and n columns in which the ith row corresponds to x_i, $1 \le i \le m$, and the jth column corresponds to y_j, $1 \le j \le n$. Specifically, if $x_i R y_j$, then a 1 is placed in the position of the matrix corresponding to the ith row and jth column; if it is not the case that $x_i R y_j$, then we place a 0 in this position. We call this matrix the **matrix of R**. Denoting the element in the ith row and jth column of this matrix by r_{ij}, note that

$$r_{ij} = \begin{cases} 1 & \text{if } (x_i, y_j) \in R \\ 0 & \text{if } (x_i, y_j) \notin R \end{cases}$$

For example, the matrix of the relation R of Example 0.3.1 is

$$\begin{bmatrix} 1 & 0 & 0 & 1 & 0 \\ 0 & 1 & 0 & 1 & 0 \\ 1 & 1 & 0 & 0 & 0 \\ 0 & 1 & 1 & 0 & 0 \\ 1 & 0 & 0 & 0 & 0 \\ 0 & 0 & 1 & 0 & 1 \end{bmatrix}$$

where the rows correspond to Brinkerhoff, Chan, McKenna, Slonneger, Will, and Yellen, respectively, and the columns to CS105, CS260, CS261, CS360, and CS460, respectively.

In general, recall that a rectangular array of numbers of the form

$$
C = \begin{bmatrix}
c_{11} & c_{12} & \cdots & c_{1n} \\
c_{21} & c_{22} & \cdots & c_{2n} \\
 & & \vdots & \\
c_{m1} & c_{m2} & \cdots & c_{mn}
\end{bmatrix}
$$

is called an *m* **by** *n* **matrix**. For $1 \le i \le m$ and $1 \le j \le n$, we call c_{ij} the **(*i*,*j*)-entry** of the matrix. We often use the customary shorthand notation $[c_{ij}]$ to denote C.

■ *Example 0.3.3* In a certain tennis tournament involving five players, each contestant plays each of the others exactly once. If we denote the contestants by v_1, v_2, v_3, v_4, and v_5, then the results of the matches in the tournament can be represented as a relation A from the set $V = \{v_1, v_2, v_3, v_4, v_5\}$ to itself, where

$$v_i \, A \, v_j \leftrightarrow v_i \text{ beats } v_j$$

Suppose that the resulting relation is

$$A = \{(v_1, v_2), (v_1, v_3), (v_1, v_5), (v_2, v_3), (v_2, v_4), (v_3, v_4), (v_3, v_5), (v_4, v_1), (v_5, v_2), (v_5, v_4)\}$$

The matrix of A is then

$$
\begin{bmatrix}
0 & 1 & 1 & 0 & 1 \\
0 & 0 & 1 & 1 & 0 \\
0 & 0 & 0 & 1 & 1 \\
1 & 0 & 0 & 0 & 0 \\
0 & 1 & 0 & 1 & 0
\end{bmatrix}
$$

■

■ *Example 0.3.4* Define a relation R from the set $V = \{2, 3, 7, 9, 14, 24, 27\}$ to itself by

$$u \, R \, v \leftrightarrow \gcd(u, v) > 1$$

(where $\gcd(u, v)$ denotes the greatest common divisor of u and v). In other words, u and v are related if and only if they are not relatively prime. Determine the relation R and its matrix.

Solution
The relation R is given by

$$
\begin{aligned}
R = \{&(2, 2), (2, 14), (2, 24), (3, 3), (3, 9), (3, 24), (3, 27), (7, 7), (7, 14), (9, 3), \\
&(9, 9), (9, 24), (9, 27), (14, 2), (14, 7), (14, 14), (14, 24), (24, 2), (24, 3), \\
&(24, 9), (24, 14), (24, 24), (24, 27), (27, 3), (27, 9), (27, 24), (27, 27)\}
\end{aligned}
$$

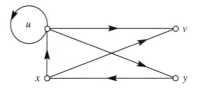

Figure 0.3.1   A directed graph.

The matrix of R is then

$$\begin{bmatrix} 1 & 0 & 0 & 0 & 1 & 1 & 0 \\ 0 & 1 & 0 & 1 & 0 & 1 & 1 \\ 0 & 0 & 1 & 0 & 1 & 0 & 0 \\ 0 & 1 & 0 & 1 & 0 & 1 & 1 \\ 1 & 0 & 1 & 0 & 1 & 1 & 0 \\ 1 & 1 & 0 & 1 & 1 & 1 & 1 \\ 0 & 1 & 0 & 1 & 0 & 1 & 1 \end{bmatrix}$$

■

Notice that the entries of the matrix C in Example 0.3.4 satisfy the condition

$$c_{ij} = c_{ji}$$

for $1 \le i, j \le 7$. In general, an n by n matrix $C = [c_{ij}]$ is called a **symmetric matrix** provided $c_{ij} = c_{ji}$ for $1 \le i, j \le n$.

One final remark regarding the matrix of a relation. If R is a relation from $X = \{x_1, x_2, \ldots, x_m\}$ to $Y = \{y_1, y_2, \ldots, y_n\}$ and we are given the matrix C of R (with x_i corresponding to row i and y_j corresponding to column j), but are not given R explicitly, then we can determine the pairs of R from C. Thus, assuming some ordering of the elements in X and in Y (that is, we know which elements correspond to the rows and columns of the matrix C), we can think of this matrix as a "matrix representation" of R.

The relations given in Examples 0.3.3 and 0.3.4 are examples of relations from a set to itself. In general, if R is a relation from a set V to itself, then we call R a **relation on V**.

Suppose now that A is a relation on a finite set V. Another representation of A, a geometric one, is obtained using the following scheme. Each element of V corresponds to a point in the plane; these points are called **vertices**. Further, if $v_1 \, A \, v_2$ for some $v_1, v_2 \in A$, then a directed simple curve can be drawn from the point corresponding to v_1 to the point corresponding to v_2; such a directed simple curve is called a **directed edge** or **arc**. In the case $v_1 = v_2$, a special type of arc called a **loop** is drawn from the point corresponding to v_1 to itself. The resulting structure is referred to as a **directed graph**. For example, Figure 0.3.1 shows a directed graph with vertices u, v, x, and y and arcs (u, u), (u, v), (u, y), (x, u), (x, v), and (y, x); the arc (u, u) is a loop.

Mathematically, we think of a directed graph as being composed of a finite set V together with a relation A on V; formally then, we have the following definition.

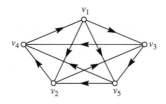

Figure 0.3.2   The directed graph of the relation of Example 0.3.3.

DEFINITION 0.3.2   A **directed graph** (or **digraph**) G consists of a nonempty finite set V together with a relation A on V. We call V the **vertex set** of G and A the **arc set** of G, and write $G = (V, A)$ to denote the fact that G is a digraph with vertex set V and arc set A.   ∎

Given a digraph $G = (V, A)$, the elements of V are called **vertices** (the plural of **vertex**) and the elements of A are called **arcs** (or **directed edges**). An alternate term for vertex is **node**. An arc of the form (v, v), where v is a vertex, is called a **loop**. Given $(u, v) \in A$, we say that u is **adjacent to** v, that v is **adjacent from** u, and that u and v are **incident** with the arc (u, v).

As illustrated by Figure 0.3.1, a digraph has a geometric representation or "drawing." However, we shall not make a distinction between the mathematical representation of a directed graph and its geometric representation.

■   *Example 0.3.5*   Figure 0.3.1 shows the digraph

$$G_1 = (\{u, v, x, y\}, \{(u, u), (u, v), (u, y), (x, u), (x, v), (y, x)\})$$

Let $G_2 = (\{v_1, v_2, v_3, v_4, v_5\}, A)$, where

$$A = \{(v_1, v_2), (v_1, v_3), (v_1, v_5), (v_2, v_3), (v_2, v_4), (v_3, v_4), (v_3, v_5), (v_4, v_1), (v_5, v_2), (v_5, v_4)\}$$

is the relation of Example 0.3.3. The digraph G_2 is drawn in Figure 0.3.2. Note, for instance, that v_1 is adjacent to v_2 and that v_1 is adjacent from v_4.   ∎

■   *Example 0.3.6*   Let U be a set and define the relation A on $V = \mathcal{P}(U)$ as follows:

$$(X, Y) \in A \leftrightarrow X \subseteq Y$$

(In words, X is related to Y if and only if X is a subset of Y.) In particular, if U is finite, then $G = (V, A)$ is a digraph. Figure 0.3.3 shows a drawing of G where $U = \{1, 2\}$.   ∎

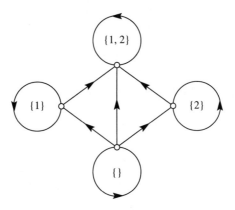

Figure 0.3.3   The digraph G of Example 0.3.6 where $U = \{1, 2\}$.

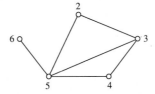

Figure 0.3.4   The digraph G of Example 0.3.7.

For distinct vertices u and v of a digraph G, if both u is not adjacent to v and v is not adjacent to u, then u and v are said to be **nonadjacent** vertices. If both u is adjacent to v and v is adjacent to u, then the arcs (u, v) and (v, u) are called **symmetric arcs**. In this case we can simplify the drawing of G by replacing the symmetric arcs (u, v) and (v, u) by an undirected simple curve joining u and v; this is called an **edge**. The use of edges is illustrated by the next example.

■   *Example 0.3.7*   Let $V = \{2, 3, 4, 5, 6\}$ and define the relation A on V by

$$(u, v) \in A \leftrightarrow \gcd(u, v) = 1$$

Then $A = \{(2, 3), (2, 5), (3, 2), (3, 4), (3, 5), (4, 3), (4, 5), (5, 2), (5, 3), (5, 4), (5, 6), (6, 5)\}$. Figure 0.3.4 shows the digraph $G = (V, A)$. Notice that the arcs of G occur in symmetric pairs; thus, the drawing of G has edges joining those distinct vertices u and v for which u and v are relatively prime.   ■

If R is a relation from a set X to a set Y, then R is a subset of $X \times Y$. Thus, R determines a subset of X, namely, the set of all first coordinates of the ordered pairs in R. Similarly, a subset of Y is determined by the set of all second coordinates of the ordered pairs in R. These subsets generalize the idea of the domain and image of a function, as defined in the last section.

DEFINITION 0.3.3 Let R be a relation from the set X to the set Y. The **domain** of R is the set dom R defined by

$$\text{dom } R = \{x \in X \mid (x, y) \in R \text{ for some } y \in Y\}$$

The **image** of R is the set im R defined by

$$\text{im } R = \{y \in Y \mid (x, y) \in R \text{ for some } x \in X\} \qquad \blacksquare$$

Recall that a relation on a set V is defined as a relation from V to itself. We now discuss several key properties that such a relation may possess.

DEFINITION 0.3.4 If A is a relation on a set V, then:

1. A is **reflexive** provided $(v, v) \in A$ for all $v \in V$.
2. A is **irreflexive** provided $(v, v) \notin A$ for all $v \in V$.
3. A is **symmetric** provided $(u, v) \in A$ implies $(v, u) \in A$ for all $u, v \in V$.
4. A is **antisymmetric** provided $(u, v) \in A$ and $(v, u) \in A$ implies $u = v$ for all $u, v \in V$.
5. A is **transitive** provided $(u, v) \in A$ and $(v, w) \in A$ implies $(u, w) \in A$ for all $u, v, w \in V$. $\qquad \blacksquare$

Making use of the partial contrapositive, note that a relation A on V is antisymmetric if and only if the condition

$$(u, v) \in A \text{ and } u \neq v \text{ implies } (v, u) \notin A$$

holds for all $u, v \in V$. In other words, A is antisymmetric provided that, for any two distinct elements u and v of V, at most one of the ordered pairs (u, v) and (v, u) is an element of A.

■ *Example 0.3.8* Let V be a nonempty set. Then the empty set $\{\ \}$ is a relation on V, called the **empty relation**. It is clearly irreflexive. Is it symmetric? In other words, is it true that, for all $u, v \in V$, $(u, v) \in \{\ \}$ implies $(v, u) \in \{\ \}$? Since $(u, v) \in \{\ \}$ is always false, this implication is true. So the relation is symmetric. Similarly, it is antisymmetric and transitive.

Next, consider the relation

$$I_V = \{(v, v) \mid v \in V\}$$

which is called the **identity relation** on V. It is readily seen to be reflexive, symmetric, antisymmetric, and transitive, but not irreflexive. It should be noted, in fact, that a relation A on V is both symmetric and antisymmetric if and only if A is a subset I_V. To see the necessity of this, suppose $(u, v) \in A$. Then, since A is symmetric, $(v, u) \in A$. Moreover, since A is antisymmetric, $(u, v) \in A$ and $(v, u) \in A$ imply that $u = v$. So $(u, v) = (u, u) \in I_V$. Thus, $A \subseteq I_V$.

Finally, consider the relation $A = V \times V$. We call this the **complete relation** on V. Notice that A is reflexive, symmetric, and transitive, but not irreflexive. It is antisymmetric if and only if $|V| = 1$. ∎

As a general comment, note that no relation on a nonempty set can be both reflexive and irreflexive.

■ **Example 0.3.9** For each of the following relations on $\{1, 2, 3, 4\}$, determine which of the properties of Definition 0.3.4 it satisfies.

(a) $A_1 = \{(1, 1), (1, 3), (1, 4), (2, 2), (3, 1), (3, 3), (3, 4), (4, 1), (4, 3), (4, 4)\}$

(b) $A_2 = \{(1, 1), (1, 2), (1, 3), (1, 4), (2, 2), (2, 3), (2, 4), (3, 3), (3, 4), (4, 4)\}$

(c) $A_3 = \{(1, 1), (1, 2), (2, 1), (2, 2), (2, 3), (3, 2), (3, 3), (3, 4)\}$

(d) $A_4 = \{(1, 3), (3, 1), (2, 4), (4, 2)\}$

Solution

(a) It is easy to check that A_1 is reflexive and symmetric; with some effort, it can also be checked that A_1 is transitive.

(b) Note that

$$A_2 = \{(x, y) \mid x \leq y\}$$

The relation A_2 is reflexive, antisymmetric, and transitive.

(c) The relation A_3 satisfies none of the properties. It is not reflexive since $(4, 4) \notin A_3$; it is not irreflexive since, for instance, $(1, 1) \in A_3$; it is not symmetric since $(3, 4) \in A_3$ but $(4, 3) \notin A_3$; it is not antisymmetric since, for instance, both $(1, 2) \in A_3$ and $(2, 1) \in A_3$; it is not transitive since, for instance, $(1, 2) \in A_3$ and $(2, 3) \in A_3$ but $(1, 3) \notin A_3$.

(d) The relation A_4 is irreflexive and symmetric. It is not transitive, because $(1, 3) \in A_4$ and $(3, 1) \in A_4$, but $(3, 3) \notin A_4$. ∎

■ **Example 0.3.10** Define the relation A on $\mathbb{Z}$ by

$$(u, v) \in A \leftrightarrow 3 \mid (u - v)$$

(where $\mid$ denotes "divides"). Show that A is reflexive, symmetric, and transitive.

Solution
Since $3 \mid (v - v)$ for any integer v, the relation A is reflexive. Given $u, v \in \mathbb{Z}$, if $3 \mid (u - v)$, then $3 \mid (v - u)$. It follows that $(u, v) \in A$ implies $(v, u) \in A$, so that A is symmetric. Also, by a well-known property of multiples, if $3 \mid (u - v)$ and $3 \mid (v - w)$ for some $u, v, w \in \mathbb{Z}$, then $3 \mid [(u - v) + (v - w)]$, namely, $3 \mid (u - w)$. Hence, $(u, v) \in A$ and $(v, w) \in A$ implies $(u, w) \in A$. This shows that A is transitive.

An alternate solution can be based on the fact that

$$3 \mid (u - v) \leftrightarrow u \bmod 3 = v \bmod 3$$

that is, u and v yield the same remainder when divided by 3. This is left for you to check.

We note that A is not antisymmetric since, for instance, $(6,9) \in A$ and $(9,6) \in A$, but $6 \neq 9$. ∎

■ **Example 0.3.11** Let U be a nonempty set and, as in Example 0.3.6, define the relation A on the power set of U by the following rule:

$$(X,Y) \in A \leftrightarrow X \subseteq Y$$

Show that A is reflexive, antisymmetric, and transitive, but not symmetric.

Solution
Let $X, Y, Z \in \mathcal{P}(U)$. Since $X \subseteq X$, we have $(X,X) \in A$ and the relation A is reflexive. Next, if $(X,Y) \in A$ and $(Y,X) \in A$, then $X \subseteq Y$ and $Y \subseteq X$, so $X = Y$. Hence, A is antisymmetric. Also, if $(X,Y) \in A$ and $(Y,Z) \in A$, then $X \subseteq Z$ by Exercise 15 in Exercise Set 0.1. It follows that A is transitive. The relation A is not symmetric; if $X, Y \in \mathcal{P}(U)$ and X is a proper subset of Y, then $(X,Y) \in A$ but $(Y,X) \notin A$. ∎

In the next example we examine how the reflexive, symmetric, and antisymmetric properties manifest themselves in the matrix and directed graph of a relation.

■ **Example 0.3.12** Consider first the relation A of Example 0.3.11 in the particular case $U = \{1,2\}$, where $V = \mathcal{P}(U) = \{\{ \}, \{1\}, \{2\}, \{1,2\}\}$. As shown in Example 0.3.11, the relation A is reflexive and antisymmetric. The directed graph $G = (V,A)$ is shown in Figure 0.3.3. The reflexive property of A is reflected in G by the presence of a loop at each vertex. The antisymmetric property of A is reflected in G by the lack of any symmetric pair of arcs; that is, there is at most one arc joining any pair of distinct vertices. The matrix of A is

$$C = \begin{bmatrix} 1 & 1 & 1 & 1 \\ 0 & 1 & 1 & 1 \\ 0 & 0 & 1 & 1 \\ 0 & 0 & 0 & 1 \end{bmatrix}$$

The reflexive property is shown by C by the fact that every entry on the main diagonal is 1. As a consequence of the antisymmetric property of A, we have, for $i \neq j$, that either the (i,j)-entry or the (j,i)-entry of C is 0.

Next consider the relation A on $V = \{2,3,4,5,6\}$ given in Example 0.3.7:

$$(u,v) \in A \leftrightarrow \gcd(u,v) = 1$$

In this case the directed graph $G = (V, A)$ is shown in Figure 0.3.4 and the matrix of A is readily seen to be

$$C = \begin{bmatrix} 0 & 1 & 0 & 1 & 0 \\ 1 & 0 & 1 & 1 & 0 \\ 0 & 1 & 0 & 1 & 0 \\ 1 & 1 & 1 & 0 & 1 \\ 0 & 0 & 0 & 1 & 0 \end{bmatrix}$$

Note that A is irreflexive and symmetric. That A is irreflexive is reflected by the fact that G has no loops and by the fact that each entry on the main diagonal of C is 0. That A is symmetric is reflected by the fact that C is a symmetric matrix. The symmetric property is shown in G by the following property: for distinct vertices u and v, either both of the arcs (u, v) and (v, u) are present (and represented by an edge joining u and v) or neither of them is present. ∎

As indicated by the observations in Example 0.3.12, the following general remarks can be made regarding properties satisfied by a relation A on a finite set V, its matrix representation C, and its digraph $G = (V, A)$:

A is reflexive $\leftrightarrow$ each entry on the main diagonal of C is 1

A is irreflexive $\leftrightarrow$ each entry on the main diagonal of C is 0

A is symmetric $\leftrightarrow$ C is a symmetric matrix

A is antisymmetric $\leftrightarrow$ for $i \neq j$, either the (i, j)-entry
or the (j, i)-entry of C is 0

A is reflexive $\leftrightarrow$ there is a loop at each vertex of G

A is irreflexive $\leftrightarrow$ there are no loops in G

A is symmetric $\leftrightarrow$ any two distinct vertices of G either are
joined by a symmetric pair of arcs
(or an edge) or are nonadjacent

A is antisymmetric $\leftrightarrow$ any two distinct vertices of G are joined
by at most one arc

Let V be a finite set, let A be a relation on V, and suppose that A is irreflexive and symmetric. Consider the digraph $G = (V, A)$. Since A is irreflexive, G has no loops. Since A is symmetric, for distinct vertices $u, v \in V$, if $(u, v) \in A$, then $(v, u) \in A$, and we have already agreed to replace the symmetric pair of arcs (u, v) and (v, u) by a single undirected edge, denoted uv (or vu). The resulting structure is called a (**simple**) **graph** and is also said to be the graph of the relation A. Mathematically, we have the following formal definition.

DEFINITION 0.3.5 A (**simple**) **graph** G consists of a finite nonempty set V and a set E of two-element subsets of V. The set V is called the **vertex set** of

G, the set E is called the **edge set** of G, and we write $G = (V, E)$ to denote the graph G with vertex set V and edge set E.   ■

As a general remark, it can be said that much of the terminology for graphs and digraphs corresponds in a natural way. For instance, given a graph $G = (V, E)$, the elements of V are called **vertices** (the plural of **vertex**), whereas the elements of E are called edges. An alternate term for vertex is **node**. If $e = \{u, v\}$ is an edge, we agree to denote it simply by $e = uv$ (or $e = vu$). In this case, u and v are referred to as **adjacent vertices**, u and v are said to be **incident** with e, and e is **incident** with u and v.

Just as with digraphs, a graph can be "drawn" in a plane, and we make no distinction between a graph $G = (V, E)$ and its drawing—sometimes a graph is presented by giving its vertex and edge sets and sometimes by giving a drawing of the graph. For example, we can now say that Figure 0.3.4 shows a graph. This graph has vertex set $V = \{2, 3, 4, 5, 6\}$ and edge set

$$E = \{\{2, 3\}, \{2, 5\}, \{3, 4\}, \{3, 5\}, \{4, 5\}, \{5, 6\}\}$$

A relation A on a set V may have several of the properties of being reflexive, irreflexive, symmetric, antisymmetric, or transitive. Certain combinations of these properties lead to important kinds of relations for which a general theory has been developed.

DEFINITION 0.3.6   A relation A on a set V is called an **equivalence relation** provided it is reflexive, symmetric, and transitive.   ■

■ *Example 0.3.13*   Let $V = \{1, 2, 3\}$. Determine which of the following relations on V are equivalence relations:

(a) $A_1 = \{(1, 2), (1, 3), (2, 1), (2, 3), (3, 1), (3, 2)\}$
(b) $A_2 = \{(1, 1), (1, 2), (2, 1), (2, 2), (2, 3), (3, 2), (3, 3)\}$
(c) $A_3 = \{(1, 1), (1, 2), (1, 3), (2, 2), (2, 3), (3, 3)\}$
(d) $A_4 = \{(1, 1), (1, 2), (2, 1), (2, 2), (3, 3)\}$

Solution

(a) It is clear that A_1 is symmetric and not reflexive. It is also not transitive. To see this, let $u = 1$, $v = 2$, and $w = 1$ in the definition. Then we have $(u, v) \in A_1$ and $(v, w) \in A_1$, but $(u, w) \notin A_1$.
(b) It is evident that A_2 is both reflexive and symmetric. However, A_2 is not transitive because $(1, 2) \in A_2$ and $(2, 3) \in A_2$, but $(1, 3) \notin A_2$.
(c) The relation A_3 is reflexive and transitive but not symmetric.
(d) The relation A_4 is reflexive, symmetric, and transitive. Hence, of these four relations, only A_4 is an equivalence relation on V.   ■

Consider again the relation A of Example 0.3.10, defined for integers u and v by

$$(u, v) \in A \leftrightarrow 3 \mid (u - v)$$

We saw in that example that A is reflexive, symmetric, and transitive; hence, A is an equivalence relation. If $(u, v) \in A$, then there is some integer k such that $3k = u - v$. Suppose that $u \bmod 3 = r$, $r \in \{0, 1, 2\}$. Then there is some integer q such that $u = 3q + r$. Since $u = v + 3k$, we have

$$v + 3k = 3q + r$$

so

$$v = 3q - 3k + r = 3(q - k) + r$$

This shows that $v \bmod 3 = r$. Hence, if $(u, v) \in A$, then $u \bmod 3 = v \bmod 3$. In other words, u and v yield the same remainder upon division by 3. In this case we say that "u is congruent to v modulo 3," and we employ the notation

$$u \equiv v \pmod 3$$

This relation is called "congruence modulo 3." There is nothing special about the use of 3 in this example; we can just as well use any positive integer n.

DEFINITION 0.3.7 Let n be a positive integer. Given integers u and v, we say that **u is congruent to v modulo n**, denoted $u \equiv v \pmod n$, provided $u - v$ is a multiple of n. This relation on $\mathbb{Z}$ is referred to as **congruence modulo n**. ■

As illustrated for the case $n = 3$, we can alternately define u to be congruent to v modulo n provided

$$u \bmod n = v \bmod n$$

This form makes it easy to derive the following result.

THEOREM 0.3.1 For any $n \in \mathbb{Z}^+$, the relation congruence modulo n is an equivalence relation on $\mathbb{Z}$. □

You are asked to prove Theorem 0.3.1 in Exercise 16. The relation of congruence modulo n gives rise to some interesting results in number theory and provides a number of very useful and fundamental examples in the area of abstract algebra.

Consider once again the relation congruence modulo 3 on $\mathbb{Z}$. The very use of the word *relation* suggests asking for the set of all "relatives" of a given fixed integer v; that is, we ask for the set of all those integers u such that $u \equiv v \pmod 3$. For an integer u, we note that $u \bmod 3$ is exactly one of 0, 1, or 2, so it seems reasonable to consider the set of relatives of each of 0, 1, and 2. Thus, we consider the following sets:

$$[0] = \{u \in \mathbb{Z} \mid u \equiv 0 \ (\mathrm{mod}\,3)\}$$
$$[1] = \{u \in \mathbb{Z} \mid u \equiv 1 \ (\mathrm{mod}\,3)\}$$
$$[2] = \{u \in \mathbb{Z} \mid u \equiv 2 \ (\mathrm{mod}\,3)\}$$

What are the elements of [0]? This is easily determined in the following string:

$$u \in [0] \leftrightarrow u \equiv 0 \ (\mathrm{mod}\ 3) \leftrightarrow 3 \mid u \leftrightarrow u = 3q \qquad \text{for some } q \in \mathbb{Z}$$

Thus, we see that $[0] = \{3q \mid q \in \mathbb{Z}\}$. In a similar fashion, we obtain

$$[1] = \{3q + 1 \mid q \in \mathbb{Z}\} \qquad \text{and} \qquad [2] = \{3q + 2 \mid q \in \mathbb{Z}\}$$

More explicitly,

$$[0] = \{\ldots, -6, -3, 0, 3, 6, \ldots\}$$
$$[1] = \{\ldots, -5, -2, 1, 4, 7, \ldots\}$$
$$[2] = \{\ldots, -4, -1, 2, 5, 8, \ldots\}$$

Here are some relevant observations to be made about the sets [0], [1], and [2]:

1.   [0], [1], and [2] are each nonempty.
2.   [0], [1], and [2] are pairwise disjoint.
3. $[0] \cup [1] \cup [2] = \mathbb{Z}$

Facts 2 and 3 follow from the division algorithm, because each $u \in \mathbb{Z}$ is uniquely expressible in the form $u = 3q + r$, with $r = 0$, 1, or 2.

An analogous discussion can be applied to any equivalence relation.

DEFINITION 0.3.8   Let V be a nonempty set and let A be an equivalence relation on V. For each $v \in V$, the **equivalence class** of v is the set

$$[v] = \{u \in V \mid (u, v) \in A\} \qquad\qquad \blacksquare$$

■   *Example 0.3.14*   Find the equivalence classes for each of the following equivalence relations:

(a)   the relation A_4 of Example 0.3.13

(b)   the relation A_5 defined on $V = \{1, 2, 3, 4, 5\}$ by

$$A_5 = \{(1, 1), (1, 2), (1, 3), (2, 1), (2, 2), (2, 3), (3, 1), (3, 2), (3, 3),$$
$$(4, 4), (4, 5), (5, 4), (5, 5)\}$$

(c)   the relation A_6 defined on $\mathbb{Z}^* = \mathbb{Z} - \{0\}$ by

$$A_6 = \{(m, n) \mid mn > 0\}$$

Solution

(a) For this equivalence relation, we observe that

$$[1] = [2] = \{1, 2\} \qquad \text{and} \qquad [3] = \{3\}$$

(b) We can verify in a routine manner that A_5 is an equivalence relation on V, with $[1] = [2] = [3] = \{1, 2, 3\}$ and $[4] = [5] = \{4, 5\}$.

(c) The relation A_6 is an equivalence relation with

$$[1] = [2] = \cdots = \mathbb{Z}^+ \qquad \text{and} \qquad [-1] = [-2] = \cdots = \mathbb{Z}^- \qquad ■$$

In each part of Example 0.3.14, we are given an equivalence relation A on a set V and we obtain (from the equivalence classes) a collection of pairwise disjoint nonempty subsets of V whose union is V. Such a collection has a special name.

DEFINITION 0.3.9 Let V be a nonempty set. A set $\mathcal{P}$ of subsets of V is called a **partition** of V provided the following conditions hold:

1. Each subset X in $\mathcal{P}$ is nonempty.
2. Any two distinct subsets X and Y in $\mathcal{P}$ are disjoint.
3. The union of the subsets in $\mathcal{P}$ is V. ■

■ ***Example 0.3.15*** The relation congruence modulo 3 on $\mathbb{Z}$ yields the partition $\mathcal{P} = \{[0], [1], [2]\}$ of $\mathbb{Z}$, where $[r] = \{3q + r \mid q \in \mathbb{Z}\}$.

The relation A_5 of Example 0.3.14(b) yields the partition $\mathcal{P} = \{\{1, 2, 3\}, \{4, 5\}\}$ of $\{1, 2, 3, 4, 5\}$.

The relation A_6 of Example 0.3.14(c) yields the partition of $\mathbb{Z}^* = \mathbb{Z} - \{0\}$ into the set $\mathbb{Z}^+$ of positive integers and the set $\mathbb{Z}^-$ of negative integers. ■

In looking at the definitions of equivalence relation and partition, it is difficult to see any strong connection between the two concepts. However, as we have seen in the preceding examples, there is indeed a very strong relationship.

THEOREM 0.3.2 (Fundamental Theorem on Equivalence Relations) Let V be a nonempty set. If A is an equivalence relation on V, then the set

$$\mathcal{P} = \{[v] \mid v \in V\}$$

of equivalence classes is a partition of V. Conversely, if $\mathcal{P}$ is a partition of V, then the relation A defined on V by

$$(u, v) \in A \leftrightarrow u \in X \text{ and } v \in X \text{ for some } X \in \mathcal{P}$$

is an equivalence relation on V.

Proof We will prove that if A is an equivalence relation on V, then the set $\mathcal{P}$ of equivalence classes is a partition of V. You are asked to prove the other part in Exercise 18.

First, it is clear that $[v]$ is nonempty for each $v \in V$; in particular, $v \in [v]$ since A is reflexive. Next, to show that distinct equivalence classes are disjoint, we prove the contrapositive: if $[u]$ and $[v]$ are not disjoint, then $[u] = [v]$. So suppose $[u] \cap [v] \neq \{ \ \}$, say, $w \in [u] \cap [v]$. We show that $[u] \subseteq [v]$ and $[v] \subseteq [u]$; in fact, since the proofs of these inclusions are analogous, we just show that $[u] \subseteq [v]$. Now we let $x \in [u]$. Then $(x, u) \in A$, $(w, u) \in A$, and $(w, v) \in A$. Since A is an equivalence relation, it follows that $(u, w) \in A$. Now we have $(x, u) \in A$, $(u, w) \in A$, and $(w, v) \in A$, and the transitive property can be applied (twice) to obtain $(x, v) \in A$. Therefore, $x \in [v]$, showing that $[u] \subseteq [v]$. So $\mathcal{P}$ is a pairwise disjoint collection. Lastly, to show that the union of all equivalence classes is V, it suffices to show that each $v \in V$ belongs to some equivalence class. But this is obvious because $v \in [v]$. Therefore, $\mathcal{P}$ is a partition of V. $\qquad\square$

To repeat, if A is an equivalence relation on a set V, then the set $\mathcal{P} = \{[v] \mid v \in V\}$ of equivalence classes is a partition of V. Conversely, given a partition $\mathcal{C}$ of V, the partition $\mathcal{C}$ determines an equivalence relation A on V, where $(u, v) \in A$ if and only if u and v belong to the same set in $\mathcal{C}$. We call A the **equivalence relation on V induced by $\mathcal{C}$**. In fact, if A is an equivalence relation on a set V and $\mathcal{P} = \{[v] \mid v \in V\}$, then the equivalence relation on V induced by $\mathcal{P}$ is precisely A. Thus, in a sense, the notions of equivalence relation and partition are the same.

■ ***Example 0.3.16*** For each $b \in \mathbb{R}$, let L_b denote the line in the xy-plane with equation $y = x + b$. Then the set $\mathcal{L} = \{L_b \mid b \in \mathbb{R}\}$ is a partition of the plane $\mathbb{R}^2 = \mathbb{R} \times \mathbb{R}$. Determine the equivalence relation on $\mathbb{R}^2$ induced by $\mathcal{L}$.

Solution
Let $\sim$ denote the equivalence relation induced on $\mathbb{R}^2$ by $\mathcal{L}$ and let (x_1, y_1) and (x_2, y_2) be elements of $\mathbb{R}^2$. Then, according to the discussion preceding the example, $(x_1, y_1) \sim (x_2, y_2)$ if and only if (x_1, y_1) and (x_2, y_2) are points on the same line L_b for some $b \in \mathbb{R}$. This happens if and only if $y_1 - x_1 = b = y_2 - x_2$. Hence the relation $\sim$ on $\mathbb{R}^2$ is given by

$$(x_1, y_1) \sim (x_2, y_2) \leftrightarrow y_1 - x_1 = y_2 - x_2$$

Incidentally, it can be noted that if we start with the equivalence relation $\sim$ on $\mathbb{R}^2$ as found, then the set of equivalence classes of $\sim$ is precisely $\mathcal{L}$. ■

■ ***Example 0.3.17*** Each of the following parts gives a partition of the set $V = \{1, 2, 3, 4, 5, 6\}$. Find the equivalence relation on V induced by the partition.

(a) $\mathcal{P}_1 = \{\{1, 2\}, \{3, 4\}, \{5, 6\}\}$

(b) $\mathcal{P}_2 = \{\{1\}, \{2\}, \{3, 4, 5, 6\}\}$

(c) $\mathcal{P}_3 = \{\{1, 2, 3\}, \{4, 5, 6\}\}$

Solution

(a) The equivalence relation induced by $\mathcal{P}_1$ is $A_1 =$

$$\{(1,1),(1,2),(2,1),(2,2),(3,3),(3,4),(4,3),(4,4),(5,5),(5,6),(6,5),(6,6)\}$$

(b) The equivalence relation induced by $\mathcal{P}_2$ is $A_2 =$

$$\{(1,1),(2,2)\} \cup (\{3,4,5,6\} \times \{3,4,5,6\})$$

(c) Similarly, the equivalence relation induced by $\mathcal{P}_3$ is $A_3 =$

$$(\{1,2,3\} \times \{1,2,3\}) \cup (\{4,5,6\} \times \{4,5,6\}) \qquad ∎$$

In the special case of congruence modulo n on $\mathbb{Z}$, the implied partition is $\{[r] \mid r \in \mathbb{Z}\}$, where $[r] = \{u \in \mathbb{Z} \mid u \equiv r \pmod{n}\}$. What are the distinct equivalence classes of congruence modulo n? For any $u \in \mathbb{Z}$, recall that there is a unique integer r, with $0 \le r \le n - 1$, such that $u \bmod n = r$. Hence, $u \equiv r \pmod{n}$, and we have $u \in [r]$. Thus, every integer belongs to exactly one of the equivalence classes $[0], [1], \ldots, [n-1]$. It follows that these classes are pairwise disjoint and that their union is $\mathbb{Z}$. Thus, $\{[0], [1], \ldots, [n-1]\}$ is a partition of $\mathbb{Z}$.

Consider the subset relation $\subseteq$ on the power set $\mathcal{P}(U)$ of a (nonempty) set U. As shown in Example 0.3.11, this relation is reflexive, antisymmetric, and transitive. Such relations play an important role in mathematics and are defined next.

> **DEFINITION 0.3.10** A relation A on a nonempty set V that is reflexive, antisymmetric, and transitive is called a **partial-order relation**. We also refer to A as a **partial ordering** of V and call (V, A) a **partially ordered set** (or **poset**). ∎

So the subset relation is a partial-order relation on $\mathcal{P}(U)$. Another important example is the following.

■ *Example 0.3.18* Consider the relation divides on $\mathbb{Z}^+$. Show that divides is a partial ordering of $\mathbb{Z}^+$. Is divides a partial ordering of $\mathbb{Z} - \{0\}$?

Solution
We must verify that divides on $\mathbb{Z}^+$ is reflexive, antisymmetric, and transitive. The reflexive property is obvious. For antisymmetry, we let $u, v \in \mathbb{Z}^+$ and suppose $u \mid v$ and $u \ne v$. Then there exists an integer $q > 1$ such that $v = uq$. This means that $v > u > 0$, and so it is not possible for v to divide u. This shows antisymmetry. For the transitive property, we let $u, v, w \in \mathbb{Z}^+$ and suppose $u \mid v$ and $v \mid w$. Then there exist $q_1, q_2 \in \mathbb{Z}^+$ such that $v = uq_1$ and $w = vq_2$. Thus,

$$w = vq_2 = (uq_1)q_2 = u(q_1q_2)$$

so that $u \mid w$. This verifies transitivity.

The relation divides is not a partial ordering of $\mathbb{Z} - \{0\}$ because it is not antisymmetric; for example, $-2 \mid 2$ and $2 \mid -2$, but $-2 \ne 2$. ∎

Perhaps the partial ordering most familiar to you is the standard ordering "less than or equal to," denoted by $\leq$, on the set of real numbers. It is the prototype of a partial-order relation and, for this reason, it is common to use the symbol $\preceq$ to denote an abstract partial-order relation.

Some authors use the term **ordering** of a set V to mean a relation $\preceq$ on V that is transitive. If this relation is also antisymmetric, then given $u \preceq v$ and $u \neq v$, we write $u \prec v$ and say that u **is less than** v (or u **precedes** v). The term **partial** is used to describe an ordering in which it is not necessary that every two elements are related.

■ *Example 0.3.19* Let V be a set on which a relation $\prec$ is defined that is irreflexive and transitive. For $u, v \in V$, if $u \preceq v$ is defined to mean that

$$u \prec v \qquad \text{or} \qquad u = v$$

show that $\preceq$ is a partial-order relation on V.

Solution
The statement $v \prec v$ or $v = v$ is clearly true for all $v \in V$, so that $\preceq$ is reflexive. Next, let $u, v \in V$ and suppose that $u \preceq v$ and $v \preceq u$ both hold. Then both $u \prec v$ or $u = v$ and $v \prec u$ or $v = u$. If $u \prec v$ and $v \prec u$, then the transitivity of $\prec$ implies $u \prec u$, contradicting the irreflexive property of $\prec$. Thus, we must have $u = v$, showing that $\preceq$ is antisymmetric. The transitive property of $\preceq$ follows directly from the transitivity of both $\prec$ and $=$. ■

We have already discussed the use of a directed graph as a means of representing a relation A on a finite set V. When $\preceq$ is a partial ordering of V, this method of represention can be simplified. First, since a partial ordering is understood to be reflexive, it is customary to omit the loops at the vertices. Second, we adopt the convention of omitting any arc that is implied by transitivity; in other words, there is an arc from u to w if and only if $u \prec w$ and there does not exist $v \in V$ with $u \prec v$ and $v \prec w$. The resulting representation is a digraph and is called the **Hasse diagram** of the poset $(V, \preceq)$. It should be noted that some people prefer to construct a Hasse diagram so that, if $u \prec v$, the vertex v is placed above the vertex u—then all arcs are replaced by undirected edges, with the understanding that the orientation of all edges is from bottom to top.

■ *Example 0.3.20* The Hasse diagram of the poset $(\mathcal{P}(\{1, 2, 3\}), \subseteq)$ is shown in Figure 0.3.5(a).
Let $V = \{1, 2, 3, 4, 6, 8, 12, 24\}$ denote the set of positive divisors of 24 and consider the relation divides on V. Using an argument analogous to that given in Example 0.3.18, we can readily check that $(V, |)$ is a poset. Its Hasse diagram is given in Figure 0.3.5(b). ■

As mentioned before, not every two elements in a poset are necessarily comparable. In $(\mathbb{Z}^+, |)$, for example, neither of the integers 3 nor 8 divides the other; hence,

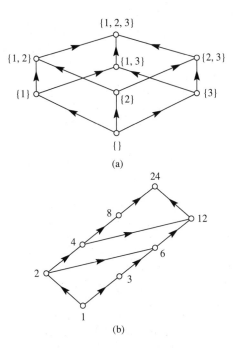

Figure 0.3.5 Two Hasse diagrams.

3 and 8 are not comparable in this poset. However, in $(\mathbb{Z}^+, \leq)$ (here $\leq$ has its usual interpretation, "less than or equal to"), if $u, v \in \mathbb{Z}^+$ and $u \neq v$, then we have that either $u < v$ or $v < u$. This brings us to the next definition.

> **DEFINITION 0.3.11** A partial-order relation $\preceq$ on a set V is called a **total ordering** provided that, for any two distinct elements $u, v \in V$, either $u \prec v$ or $v \prec u$. ∎

A set V, together with a total ordering $\preceq$, is called a **totally ordered set**.

∎ ***Example 0.3.21*** Let V be the set of positive divisors of 24. The poset $(V, |)$, as shown in Figure 0.3.5(b), is not totally ordered; for example, 4 and 6 are not comparable. Now let X be the set of positive divisors of 81. The poset $(X, |)$, as shown in Figure 0.3.6(a), is a totally ordered set.

As shown in Figure 0.3.5(a), the poset $(\mathcal{P}(\{1, 2, 3\}), \subseteq)$ is not a totally ordered set. However, let $\mathcal{P}' = \{\{\ \}, \{1\}, \{1, 2\}, \{1, 2, 3\}\}$. Then $(\mathcal{P}', \subseteq)$ is a totally ordered set; its Hasse diagram is shown in Figure 0.3.6(b). ∎

For the final topic of this section, consider the digraph $T = (V, A)$, shown in Figure 0.3.7, where $V = \{0, 1, \ldots, 9\}$ and $A = \{(0, 1), (0, 2), (0, 3), (1, 4), (1, 5), (3, 6), (5, 7),$ $(6, 8), (6, 9)\}$. This digraph is an example of a **rooted tree**. We may consider T to be the Hasse diagram of a partial-order relation $\preceq$ on V. Then $\preceq$ is not a total ordering;

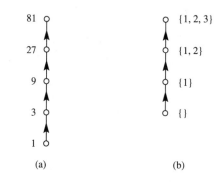

Figure 0.3.6   Totally ordered sets.

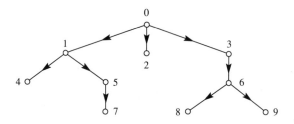

Figure 0.3.7   A rooted tree.

however, it does have more structure than an ordinary partial ordering. For instance, the vertex 0 has the property that $0 \preceq v$ for all $v \in V$ and is the only vertex with this property. For this reason, we call 0 the **root** of T. Moreover, for any $v \in V, v \neq 0$, there is a unique $u \in V$ such that $(u, v) \in A$. We call u the **parent** of v; for example, 0 is the parent of 1, 1 is the parent of 5, and 5 is the parent of 7.

> **DEFINITION 0.3.12**   Let $T = (V, A)$ be a digraph such that T is the Hasse diagram of a partial-order relation $\preceq$ on V. Then T is a **rooted tree** provided the following properties are satisfied:
>
> 1.   There is a unique $r \in V$ with the property that $r \preceq v$ for all $v \in V$; the vertex r is called the **root** of T.
> 2.   For every vertex $v, v \neq r$, there is a unique $u \in V$ with $(u, v) \in A$; the vertex u is called the **parent** of the vertex v.   ∎

There is quite a bit of standard terminology associated with rooted trees. For example, consider T a rooted tree and u a vertex. Vertices whose parent is u are called **children** of u; if u has no children, then u is called a **leaf** of T. In the tree of Figure 0.3.7, for example, 4 and 5 are children of 1, and vertices 2, 4, 7, 8, and 9 are leaves. Other terminology is left to Chapter 3.

EXERCISE SET 0.3

1. For each of the given relations on $\{1,2,3,6\}$, determine its matrix and directed graph.

   (a) $\{(x,y) \mid x \text{ divides } y\}$ (b) $\{(x,y) \mid x < y\}$
   (c) $\{(x,y) \mid x = y\}$ (d) $\{(x,y) \mid x \neq y\}$
   (e) $\{(x,y) \mid x^2 \leq y\}$ (f) $\{(x,y) \mid x+y \text{ is odd}\}$

2. Determine the matrix and directed graph of the relation A defined on $V = \{-3, -2, -1, 0, 1, 2, 3\}$ by

   $$x \, A \, y \leftrightarrow 3 \text{ divides } x - y$$

3. Find the domain and image of each relation in Exercise 1.

4. Determine the matrix and directed graph of the relation A defined on $V = \{1, 2, 4, 5, 10, 20\}$ by

   $$x \, A \, y \leftrightarrow (x < y \text{ and } x \text{ divides } y)$$

5. For each of the following relations on $\{1,2,3,4,5\}$, determine which of the properties (i) reflexive, (ii) irreflexive, (iii) symmetric, (iv) antisymmetric, and (v) transitive are satisfied.

   (a) $A_1 = \{(1,1),(2,2),(2,3),(3,2),(3,3),(3,4),(4,3),(4,4),(5,5)\}$
   (b) $A_2 = \{(1,2),(1,4),(1,5),(2,4),(2,5),(3,4),(3,5),(4,5)\}$
   (c) $A_3 = \{(1,3),(1,5),(2,4),(3,1),(3,5),(4,2),(5,1),(5,3)\}$
   (d) $A_4 = A_3 \cup \{(1,1),(2,2),(3,3),(4,4),(5,5)\}$
   (e) $A_5 = \{(1,1),(1,2),(2,1),(2,2),(4,3),(5,4)\}$

6. For each of the relations in Exercise 5, draw its digraph (or graph).

7. Follow the directions of Exercise 5 for the following relations on $\{-2, -1, 0, 1, 2\}$.

   (a) $A_1 = \{(m,n) \mid m+n \leq 6\}$ (b) $A_2 = \{(m,n) \mid n = m+1\}$
   (c) $A_3 = \{(m,n) \mid m \leq n^2\}$ (d) $A_4 = \{(m,n) \mid |m| = |n|\}$

8. For each of the relations in Exercise 7, draw its digraph (or graph).

9. For each of the following relations on $\mathbb{Z}^+ - \{1\}$, determine which of the properties (i) reflexive, (ii) irreflexive, (iii) symmetric, (iv) antisymmetric, and (v) transitive are satisfied.

   (a) $A_1 = \{(m,n) \mid m \text{ and } n \text{ are relatively prime}\}$
   (b) $A_2 = \{(m,n) \mid m \text{ and } n \text{ are not relatively prime}\}$
   (c) $A_3 = \{(m,n) \mid m \text{ divides } n\}$
   (d) $A_4 = \{(m,n) \mid m+n \text{ is even}\}$
   (e) $A_5 = \{(m,n) \mid m+n \text{ is odd}\}$
   (f) $A_6 = \{(m,n) \mid mn \text{ is odd}\}$

10. Let A be a relation on a finite set $V = \{v_1, v_2, \ldots, v_n\}$. Characterize the transitive property in terms of the matrix and directed graph of A.

11. For each of the following relations on $\mathbb{Z} - \{0\}$, determine which of the properties (i) reflexive, (ii) irreflexive, (iii) symmetric, (iv) antisymmetric, and (v) transitive are satisfied.

(a) $A_1 = \{(m, n) \mid m \leq n\}$
(b) $A_2 = \{(m, n) \mid m - n \text{ is even}\}$
(c) $A_3 = \{(m, n) \mid m \text{ divides } n\}$
(d) $A_4 = \{(m, n) \mid m - n \text{ is odd}\}$
(e) $A_5 = \{(m, n) \mid 3 \text{ divides } m - n\}$
(f) $A_6 = \{(m, n) \mid mn > 0\}$

12. Let U be a nonempty set and define the relation A on the power set of U by

$$(X, Y) \in A \leftrightarrow X \cap Y = \{\ \}$$

Clearly, A is symmetric.

(a) Explain why A is not reflexive.
(b) Explain why A is not irreflexive.
(c) Show that A is not transitive.
(d) Let $U = \{1, 2, 3\}$; draw the graph of A.

13. Each part of Figure 0.3.8 shows the directed graph of a relation on $\{1, 2, 3, 4\}$; for each of these relations, determine which of the properties (i) reflexive, (ii) irreflexive, (iii) symmetric, (iv) antisymmetric, and (v) transitive it satisfies.

14. Let A_1 and A_2 be relations on the nonempty set V. Then each of $A_1 \cap A_2$, $A_1 \cup A_2$, and $A_1 - A_2$ is also a relation on V. (Why?) Indicate whether each of the following statements is true or false. (Optional: If the statement is true, prove it; if it is false, give a counterexample.)

(a) If both A_1 and A_2 are reflexive, then $A_1 \cap A_2$ is reflexive.
(b) If both A_1 and A_2 are reflexive, then $A_1 \cup A_2$ is reflexive.
(c) If both A_1 and A_2 are reflexive, then $A_1 - A_2$ is irreflexive.
(d) If both A_1 and A_2 are irreflexive, then $A_1 \cap A_2$ is irreflexive.
(e) If both A_1 and A_2 are irreflexive, then $A_1 \cup A_2$ is irreflexive.
(f) If both A_1 and A_2 are irreflexive, then $A_1 - A_2$ is irreflexive.
(g) If both A_1 and A_2 are symmetric, then $A_1 \cap A_2$ is symmetric.
(h) If both A_1 and A_2 are symmetric, then $A_1 \cup A_2$ is symmetric.
(i) If both A_1 and A_2 are symmetric, then $A_1 - A_2$ is symmetric.
(j) If both A_1 and A_2 are antisymmetric, then $A_1 \cap A_2$ is antisymmetric.
(k) If both A_1 and A_2 are antisymmetric, then $A_1 \cup A_2$ is antisymmetric.
(l) If both A_1 and A_2 are antisymmetric, then $A_1 - A_2$ is antisymmetric.
(m) If both A_1 and A_2 are transitive, then $A_1 \cap A_2$ is transitive.
(n) If both A_1 and A_2 are transitive, then $A_1 \cup A_2$ is transitive.
(o) If both A_1 and A_2 are transitive, then $A_1 - A_2$ is transitive.

15. Determine which of the relations defined in Exercise 5 are equivalence relations. Find the equivalence classes of those that are.

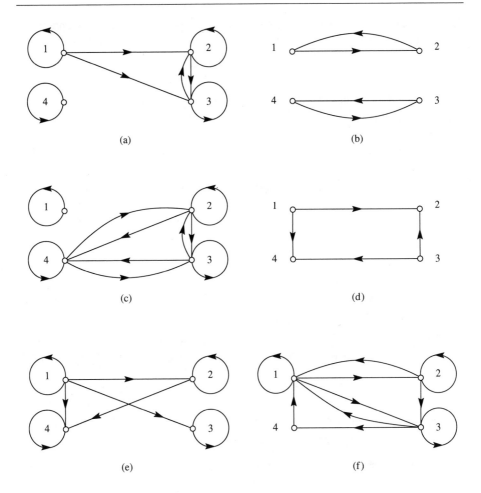

(a)   (b)   (c)   (d)   (e)   (f)

Figure 0.3.8

16. Prove Theorem 0.3.1.

17. Determine which of the relations defined in Exercise 7 are equivalence relations. Find the equivalence classes of those that are.

18. Complete the proof of Theorem 0.3.2.

19. Determine which of the relations defined in Exercise 9 are equivalence relations. Find the equivalence classes of those that are.

20. Define the relation $\sim$ on $\mathbb{Z}$ by

$$u \sim v \leftrightarrow |u - 2| = |v - 2|$$

Verify that $\sim$ is an equivalence relation and find the equivalence classes.

21. Determine which of the relations defined in Exercise 11 are equivalence relations. Find the equivalence classes of those that are.

22. Determine which of the numbers $-22, -12, -6, -3, -2, 3, 6, 8, 39$, and 44 are congruent modulo:

    (a) 2   (b) 3   (c) 5   (d) 9

23. Determine which of the relations defined in Exercise 13 are equivalence relations. Find the equivalence classes of those that are.

24. Let A_1 and A_2 be relations on the nonempty set V. Then each of $A_1 \cap A_2$, $A_1 \cup A_2$, and $A_1 - A_2$ is also a relation on V. (See Exercise 14.) Indicate whether each of the following statements is true or false. (Optional: If the statement is true, prove it; if it is false, give a counterexample.)

    (a) If A_1 and A_2 are equivalence relations, then so is $A_1 \cap A_2$.
    (b) If A_1 and A_2 are equivalence relations, then so is $A_1 \cup A_2$.
    (c) If A_1 and A_2 are equivalence relations, then so is $A_1 - A_2$.

25. Let U be a nonempty set and let W be a fixed subset of U. Define the relation $\sim$ on $\mathcal{P}(U)$ by

    $$X \sim Y \leftrightarrow X \cap W = Y \cap W$$

    (a) Show that $\sim$ is an equivalence relation.
    (b) In particular, if $U = \{1, 2, 3, 4, 5\}$, $W = \{1, 2, 5\}$, and $X = \{2, 4, 5\}$, find $[X]$.

26. Each of the following parts gives a partition of the set $V = \{1, 2, 3, 4, 5, 6\}$. Find the equivalence relation on V induced by the partition.

    (a) $\mathcal{P}_1 = \{\{1\}, \{2\}, \{3\}, \{4\}, \{5\}, \{6\}\}$ (b) $\mathcal{P}_2 = \{\{1\}, \{2, 3\}, \{4, 5, 6\}\}$
    (c) $\mathcal{P}_3 = \{\{1, 3, 5\}, \{2, 4, 6\}\}$ (d) $\mathcal{P}_4 = \{\{1, 2, 3, 4, 5, 6\}\}$

27. Determine the number of equivalence relations on each of the following sets:

    (a) $\{1\}$ (b) $\{1, 2\}$ (c) $\{1, 2, 3\}$ (d) $\{1, 2, 3, 4\}$

28. In Exercise 24 it is observed that if A_1 and A_2 are both equivalence relations on a nonempty set V, then $A_1 \cap A_2$ is an equivalence relation on V. How are the equivalence classes for $A_1 \cap A_2$ determined from the equivalence classes for A_1 and those for A_2?

29. Each of the following parts gives a relation on $V = \{a, b, c, d, e, f\}$. Determine whether the relation is a partial ordering. If it is, then draw its Hasse diagram.

    (a) $A_1 = \{(f,f), (f,e), (f,d), (f,c), (f,b), (f,a), (e,e), (e,c), (e,a), (d,d), (d,c),$
    $(d,b), (d,a), (c,c), (c,a), (b,b), (b,a), (a,a)\}$
    (b) $A_2 = \{(f,f), (f,d), (f,c), (f,b), (f,a), (e,e), (e,d), (e,c), (e,b), (e,a), (d,d),$
    $(d,c), (d,b), (d,a), (c,c), (c,b), (c,a), (b,b), (b,a), (a,a)\}$
    (c) $A_3 = A_1 \cup \{(e,b), (d,c)\}$
    (d) $A_4 = \{(f,f), (f,e), (f,d), (f,c), (f,b), (f,a), (e,e), (e,d), (e,c), (e,b), (e,a),$
    $(d,d), (d,b), (d,a), (c,c), (c,b), (c,a), (b,b), (b,a), (a,a)\}$
    (e) $A_5 = \{(f,d), (f,b), (e,c), (e,a), (d,b), (c,a)\}$
    (f) $A_6 = \{(f,f), (e,e), (d,d), (c,c), (b,b), (a,a)\}$
    (g) $A_7 = A_5 \cup A_6$
    (h) $A_8 = A_7 \cup \{(d,c)\}$

(i) $A_9 = A_8 \cup \{(f,c)\}$

(j) $A_{10} = A_9 \cup \{(a,c)\}$

30. If $(X, \preceq)$ is a poset and $Y \subseteq X$, show that $(Y, \preceq)$ is a poset. $(Y, \preceq)$ is called a **subposet** of $(X, \preceq)$.

31. Determine which of the relations defined in Exercise 13 are partial-order relations. For each such relation, draw its Hasse diagram.

32. Each of the following parts gives a subset V of $\mathbb{Z}^+$. Draw the Hasse diagram of the (sub)poset $(V, |)$.

(a) $V = \{n \mid n \text{ divides } 30\}$

(b) $V = \{n \mid n \text{ divides } 45\}$

(c) $V = \{n \mid n \text{ divides } 36\}$

(d) $V = \{n \mid n \text{ divides } 36 \text{ or } n \text{ divides } 45\}$

33. Define the relation $\prec$ on $\mathbb{Z} - \{0\}$ by

$$m \prec n \leftrightarrow 2m \mid n$$

(a) Show that $\prec$ is irreflexive and transitive.

(b) Use the result of Example 0.3.19 to define a partial-order relation $\preceq$ on $\mathbb{Z} - \{0\}$.

34. Let n_0 be a fixed positive integer, and let $V_{n_0} = \{n \in \mathbb{N} \mid n \text{ divides } n_0\}$. Show that $(V_{n_0}, |)$ is a poset.

35. Consider the poset $(\mathbb{Z} - \{0\}, \preceq)$, as defined in Exercise 33. Let $V_{24} = \{n \in \mathbb{Z}^+ \mid n \text{ divides } 24\}$. Draw the Hasse diagram of $(V_{24}, \preceq)$.

36. Prove or disprove: If $(X, \preceq)$ is a totally ordered set and Y is a subset of X, then $(Y, \preceq)$ is a totally ordered set.

37. Let $V = \{a, b, c, d, e, f, g\}$ and define the relation A on V by

$$A = I_V \cup \{(g,f), (g,e), (g,d), (g,c), (g,b), (g,a), (f,c), (f,a), (e,b), (e,a),$$
$$(d,c), (d,b), (d,a), (c,a), (b,a)\}$$

Verify that A is a partial-order relation, and draw the Hasse diagram of the poset (V, A).

38. Consider the poset $(V_{n_0}, |)$, as defined in Exercise 34. For what values of n_0 is $(V_{n_0}, |)$ a totally ordered set?

39. Let A be a relation on a nonempty set V. Complete each of the following statements.

(a) A is not reflexive provided _____ .

(b) A is not irreflexive provided _____ .

(c) A is not symmetric provided _____ .

(d) A is not antisymmetric provided _____ .

(e) A is not transitive provided _____ .

40. Let A be a symmetric and transitive relation on a nonempty set V. Prove: If dom $A = V$, then A is reflexive.

41. Let A_1 and A_2 be relations on a nonempty set V. Indicate whether each of the following statements is true or false. (Optional: If the statement is true, prove it; if it is false, give a counterexample.)

    (a) If $A_1 \cap A_2$ is reflexive, then both A_1 and A_2 are reflexive.
    (b) If A_1 is reflexive, then $A_1 \cup A_2$ is reflexive.
    (c) If A_1 is reflexive and A_2 is irreflexive, then $A_1 - A_2$ is reflexive.
    (d) If A_1 is irreflexive, then $A_1 \cap A_2$ is irreflexive.
    (e) If $A_1 \cup A_2$ is irreflexive, then both A_1 and A_2 are irreflexive.
    (f) If A_1 is irreflexive, then $A_1 - A_2$ is irreflexive.
    (g) If $A_1 \cap A_2$ is symmetric, then both A_1 and A_2 are symmetric.
    (h) If $A_1 \cup A_2$ is symmetric, then both A_1 and A_2 are symmetric.
    (i) If A_1 is symmetric, then $A_1 - A_2$ is symmetric.
    (j) If A_1 is antisymmetric, then $A_1 \cap A_2$ is antisymmetric.
    (k) If $A_1 \cup A_2$ is antisymmetric, then both A_1 and A_2 are antisymmetric.
    (l) If A_1 is antisymmetric, then $A_1 - A_2$ is antisymmetric.
    (m) If $A_1 \cap A_2$ is transitive, then both A_1 and A_2 are transitive.
    (n) If $A_1 \cup A_2$ is transitive, then both A_1 and A_2 are transitive.

42. Let $\sim$ be a reflexive relation on a nonempty set V. Prove that $\sim$ is an equivalence relation if and only if the following condition holds for any elements $u, v, w \in V$ (u, v, and w are not necessarily distinct):

    $$(u \sim v \text{ and } u \sim w) \rightarrow v \sim w$$

43. For each of the following relations $\sim$ on $\mathbb{R} \times \mathbb{R}$, determine whether it is an equivalence relation. If it is, describe (geometrically) the equivalence class $[(a, b)]$.

    (a) $(x_1, y_1) \sim (x_2, y_2) \leftrightarrow x_1 + y_2 = x_2 + y_1$
    (b) $(x_1, y_1) \sim (x_2, y_2) \leftrightarrow (x_1 - 1)^2 + y_1^2 = (x_2 - 1)^2 + y_2^2$
    (c) $(x_1, y_1) \sim (x_2, y_2) \leftrightarrow (x_1 - x_2)(y_1 - y_2) = 0$
    (d) $(x_1, y_1) \sim (x_2, y_2) \leftrightarrow |x_1| + |y_1| = |x_2| + |y_2|$
    (e) $(x_1, y_1) \sim (x_2, y_2) \leftrightarrow x_1 y_1 = x_2 y_2$

44. For each of the following relations $\sim$ on $\mathbb{R}$, verify that $\sim$ is an equivalence relation and describe the equivalence classes of $\sim$.

    (a) $x \sim y \leftrightarrow \lfloor x \rfloor = \lfloor y \rfloor$, where $\lfloor x \rfloor$ denotes the largest integer m such that $m \leq x$.
    (b) $x \sim y \leftrightarrow \lfloor x + 0.5 \rfloor = \lfloor y + 0.5 \rfloor$
    (c) $x \sim y \leftrightarrow |x| = |y|$
    (d) $x \sim y \leftrightarrow x - y \in \mathbb{Z}$
    (e) $x \sim y \leftrightarrow x - y \in \mathbb{Q}$

45. Define the relation $\sim$ on $\mathbb{Z} \times \mathbb{Z}^+$ by

    $$(m_1, n_1) \sim (m_2, n_2) \leftrightarrow m_1 n_2 = m_2 n_1$$

    (a) Show that $\sim$ is an equivalence relation.

(b) Describe the equivalence class $[(m, n)]$ in a nice way. (Hint: Think of (m, n) as the fraction m/n.)

46. Each of the following parts gives a set X, a relation $\preceq$ on X, and a subset Y of X. Verify that $\preceq$ is a partial ordering of X and draw the Hasse diagram of the subposet $(Y, \preceq)$.

(a) $X = \mathbb{Z}$, $x \preceq y \leftrightarrow (x = y$ or $|x| < |y|)$, $Y = \{-3, -2, -1, 0, 1, 2, 3\}$

(b) $X = \mathbb{Z}$, $x \preceq y \leftrightarrow (x = y$ or $x \bmod 4 < y \bmod 4)$, $Y = \{-4, -3, -2, -1, 0, 1, 2, 3, 4\}$

(c) $X = \mathbb{Z} \times \mathbb{Z}$, $(a, b) \preceq (c, d) \leftrightarrow [(a, b) = (c, d)$ or $a^2 + b^2 < c^2 + d^2]$, $Y = \{0, 1, 2\} \times \{0, 1, 2\}$

47. Let $V = \{(a, b) \mid a, b \in \mathbb{Z}^+$ and a and b are relatively prime$\}$. Define a relation $\preceq$ on V by

$$(a, b) \preceq (c, d) \leftrightarrow ad \leq bc$$

Show that $\preceq$ defines a total ordering of V.

48. Let $\mathbb{Q}^+$ denote the set of positive rational numbers. Define the relation $\preceq$ on $\mathbb{Q}^+$ by

$$r \preceq s \leftrightarrow \frac{s}{r} \in \mathbb{Z}^+$$

(a) Show that $\preceq$ is a partial ordering of $\mathbb{Q}^+$.

(b) Let $V = \{\frac{1}{6}, \frac{1}{3}, \frac{1}{2}, 1, 2, 3, 6\}$. Draw the Hasse diagram of the subposet $(V, \preceq)$.

49. Define the relation $\sim$ on $\mathbb{Z}$ by

$$a \sim b \leftrightarrow 3 \mid (a + 2b)$$

(a) Show that $\sim$ is an equivalence relation.

(b) Determine the equivalence classes of $\sim$.

50. Define the relation $\prec$ on $\mathbb{Z}^+ - \{1\}$ by

$$m \prec n \leftrightarrow m^2 \mid n$$

(a) Show that $\prec$ is irreflexive and transitive.

(b) Apply the result of Example 0.3.19 to define a partial ordering $\preceq$ on $\mathbb{Z}^+ - \{1\}$. Extend this partial ordering to all of $\mathbb{Z}^+$ by defining $1 \preceq n$ for all $n \in \mathbb{Z}^+$.

(c) Let $V = \{1, 2, 3, 4, 6, 9, 16, 36, 81, 1296\}$. Draw the Hasse diagram of the subposet $(V, \preceq)$.

51. Each of the following parts gives a set V and a relation A on V. Determine which of the properties (i) reflexive, (ii) irreflexive, (iii) symmetric, (iv) antisymmetric, and (v) transitive A satisfies.

(a) $V = \mathcal{P}(\{1, 2, 3, 4\})$, $(X, Y) \in A \leftrightarrow X \subseteq Y \cup \{1\}$

(b) $V = \{0, 1, 2, 3, 4, 5, 6\}$, $(x, y) \in A \leftrightarrow x - y = 1$ or $x - y = 6$

(c) $V = \mathbb{Z}$, $(x, y) \in A \leftrightarrow |x - y| > 2$

(d) $V = \mathbb{Z}$, $(x, y) \in A \leftrightarrow xy \geq 0$

(e) $V = (0, 1)$, $(x, y) \in A \leftrightarrow xy \in \mathbb{Q}$

52. Give an example of a relation on $\mathbb{Z}$ that is:

(a) irreflexive and symmetric
(b) irreflexive and transitive
(c) reflexive and symmetric but not transitive
(d) reflexive and transitive but not symmetric

53. Let A be a relation on V. Define the relation A^r on V by

$$(u, v) \in A^r \leftrightarrow [(u, v) \in A \text{ or } u = v]$$

that is, $A^r = A \cup I_V$.

(a) Show that A^r is reflexive.
(b) Under what condition does $A^r = A$?
(c) Prove: If A' is a reflexive relation on V and $A \subseteq A'$, then $A^r \subseteq A'$.

The relation A^r is called the **reflexive closure** of A. Find the reflexive closure of each of the following relations:

(d) $A_1 = \{(1, 1), (1, 2), (2, 1), (2, 2), (3, 4), (3, 5), (4, 3), (4, 5), (5, 3), (5, 4)\}$ on $\{1, 2, 3, 4, 5\}$
(e) $<$ on $\mathbb{Z}$
(f) $\sim$ on $\mathbb{Z}$ defined by $m \sim n \leftrightarrow mn$ is odd

54. Let A be a relation on V. Define the relation A^s on V by

$$(u, v) \in A^s \leftrightarrow [(u, v) \in A \text{ or } (v, u) \in A]$$

(a) Show that A^s is symmetric.
(b) Note that $A \subseteq A^s$. Under what condition does $A = A^s$?
(c) Prove: If A' is a symmetric relation on V and $A \subseteq A'$, then $A^s \subseteq A'$.

The relation A^s is called the **symmetric closure** of A. Find the symmetric closure of each of the following relations:

(d) $A_1 = \{(1, 1), (1, 2), (1, 3), (1, 4), (2, 2), (3, 3), (4, 4)\}$ on $\{1, 2, 3, 4\}$
(e) $<$ on $\mathbb{Z}$

55. Let A be a relation on V. The **transitive closure** of A is the relation A^t on V that is defined by the following properties: (i) A^t is transitive; (ii) $A \subseteq A^t$; (iii) if A' is a transitive relation on V and $A \subseteq A'$, then $A^t \subseteq A'$. Informally, A^t is the smallest transitive relation on V that contains A.

(a) Find the transitive closure of the relation $A_1 = \{(1, 2), (2, 1), (2, 3), (3, 4)\}$ on $\{1, 2, 3, 4\}$.
(b) Find the transitive closure of the relation A_2 defined on $\mathbb{Z}$ by $(m, n) \in A_2 \leftrightarrow n = m + 1$.
(c) Define the relation A^2 on V by

$$(u, w) \in A^2 \leftrightarrow [(u, w) \in A \text{ or } [(u, v) \in A \text{ and } (v, w) \in A \text{ for some } v \in V]]$$

In view of Exercises 53 and 54, it is tempting to define A^t by $A^t = A^2$. However, show that A^2 is not transitive in general.

56. Let A_1 and A_2 be relations on a nonempty set V; suppose that A_1 is an equivalence relation and A_2 is a partial ordering.

(a) For example, suppose that $V = \mathbb{Z}^+, A_1$ is the relation of congruence modulo 2, and A_2 is the relation of divides. Find $A_1 \cap A_2$ in this case.

(b) Show, in general, that $A_1 \cap A_2$ is a partial ordering of V.

57. Let A be a relation on V. Prove or disprove each of the following statements:

(a) If A is symmetric and transitive, then so is A^r. (See Exercise 53.)

(b) If A is reflexive and transitive, then so is A^s. (See Exercise 54.)

(c) If A is reflexive and symmetric, then so is A^t. (See Exercise 55.)

(d) The relation $((A^r)^s)^t$ is an equivalence relation on V.

(e) If A is reflexive and antisymmetric, then A^t is a partial ordering of V.

58. Let A be a relation on X and let B be a relation on Y. Define the relations A_1 and A_2 on $X \times Y$ as follows:

$$((x_1, y_1), (x_2, y_2)) \in A_1 \leftrightarrow ((x_1, x_2) \in A \text{ and } (y_1, y_2) \in B)$$
$$((x_1, y_1), (x_2, y_2)) \in A_2 \leftrightarrow [(x_1 \neq x_2 \text{ and } (x_1, x_2) \in A) \text{ or}$$
$$(x_1 = x_2 \text{ and } (y_1, y_2) \in B)]$$

Indicate whether each of the following statements is true or false. (Optional: If the statement is true, prove it; if it is false, give a counterexample.)

(a) If both A and B are reflexive, then A_1 is reflexive.

(b) If B is reflexive, then A_2 is reflexive.

(c) If both A and B are irreflexive, then A_1 is irreflexive.

(d) If B is irreflexive, then A_2 is irreflexive.

(e) If both A and B are symmetric, then A_1 is symmetric.

(f) If both A and B are symmetric, then A_2 is symmetric.

(g) If both A and B are antisymmetric, then A_1 is antisymmetric.

(h) If both A and B are antisymmetric, then A_2 is antisymmetric.

(i) If both A and B are transitive, then A_1 is transitive.

(j) If both A and B are transitive, then A_2 is transitive.

59. Let $V = \{a, b, c, d, e, f, g, h, i, j\}$ and define the relation A on V by

$$A = I_V \cup \{(j, i), (j, h), (j, g), (j, f), (j, e), (j, d), (j, c), (j, b), (j, a), (i, f), (i, d), (i, c),$$
$$(i, a), (h, e), (h, d), (h, b), (h, a), (g, c), (g, b), (g, a), (f, c), (f, a), (e, b),$$
$$(e, a), (d, a), (c, a), (b, a)\}$$

Verify that A is a partial-order relation and draw the Hasse diagram of the poset (V, A).

60. Define the relation $\sim$ on the set $\mathbb{R}^+$ of positive real numbers by

$$x \sim y \leftrightarrow x/y \in \mathbb{Q}^+$$

Show that $\sim$ is an equivalence relation and discuss the equivalence classes of $\sim$.

61. Let X and Y be nonempty sets, and consider a function $f : X \to Y$ such that f is onto. Define the relation $\sim$ on X by

$$x_1 \sim x_2 \leftrightarrow f(x_1) = f(x_2)$$

(a) Show that $\sim$ is an equivalence relation on X.

Let $\mathcal{P}$ be the partition of X induced by $\sim$. Define $q : X \to \mathcal{P}$ by $q(x) = [x]$, and define $\bar{f} : \mathcal{P} \to Y$ by $\bar{f}([x]) = f(x)$.

(b) Show that the function $\bar{f}$ is well defined; that is, if $[x_1] = [x_2]$, then $\bar{f}([x_1]) = \bar{f}([x_2])$.

(c) Show that $\bar{f}$ is one-to-one and onto.

(d) Show that $f = \bar{f} \circ q$.

Thus, every onto mapping (f) may be "factored" as the composition of a "quotient mapping" (q) and a one-to-one, onto mapping ($\bar{f}$). This result is the set-theoretic analogue of the fundamental morphism theorem of abstract algebra.

62. Show that the relation is theta of is an equivalence relation on the set of functions from $\mathbb{Z}^+$ to $\mathbb{R}^+$.

0.4 INDUCTION

One of the most basic principles used in discrete mathematics, especially in combinatorics, is the *Principle of Well-ordering*:

> **PRINCIPLE OF WELL-ORDERING**   Every nonempty set of nonnegative integers has a smallest element.

It is not possible to prove this statement using the familiar properties satisfied by the integers under addition and multiplication. However, after a little thought, the principle should seem truly self-evident. Hence, it is usually accepted as an axiom. We will see shortly that the principle of mathematical induction is a logical consequence of the principle of well-ordering; in fact, it turns out that these two principles are logically equivalent.

In general, a subset T of $\mathbb{R}$ is said to be **well-ordered** provided any nonempty subset S of T has a smallest element. Thus, the principle of well-ordering states that $\mathbb{N}$ is well-ordered. Note that any subset of a well-ordered set is itself well-ordered (see Exercise 2). Therefore, we can also state that $\mathbb{Z}^+$ is well-ordered.

■ ***Example 0.4.1***   Find the smallest element in each of the following subsets of $\mathbb{Z}^+$:

(a) $S_1 = \{n \in \mathbb{Z}^+ \mid n \text{ is prime}\}$

(b) $S_2 = \{n \in \mathbb{Z}^+ \mid n \text{ is a multiple of } 7\}$

(c) $S_3 = \{n \in \mathbb{Z}^+ \mid n = 110 - 17m \text{ for some } m \in \mathbb{Z}\}$

(d) $S_4 = \{n \in \mathbb{N} \mid n = 12s + 18t \text{ for some } s, t \in \mathbb{Z}\}$

Solution

(a) The smallest prime number is 2.

(b) The smallest positive multiple of 7 is 7.

(c) Here we must find the smallest positive number n of the form $110 - 17m$, where m is an integer. The number $110 = 110 - (17)(0)$ is of this form, and as m increases, n decreases. As m takes on the values $1, 2, \ldots$, the values of n form the sequence

$$93, 76, 56, \ldots, 8, -9, \ldots$$

Hence, the smallest element of S_3 is 8. The number 8 just happens to be the remainder when 110 is divided by 17. This is more than just a coincidence; it is a consequence of the division algorithm.

(d) We are looking for the smallest positive number n of the form $12s + 18t$, where s and t are integers. Note that $12s + 18t = 6(2s + 3t)$, so that an element of S_4 must be a multiple of 6. Also, $6 = (12)(-1) + (18)(1)$, so that 6 is an element of S_4. This shows that 6 is the smallest element of S_4. The number 6 happens to be the greatest common divisor of 12 and 18. It is a well-known result that the greatest common divisor of two integers a and b (not both 0) is the smallest positive integer d such that

$$d = as + bt$$

for some integers s and t.   ∎

■ *Example 0.4.2*   The following are two examples of subsets of $\mathbb{R}$ that are not well-ordered. First consider the set $\mathbb{Z}$ of integers; $\mathbb{Z}$ is not well-ordered because $\mathbb{Z}$ itself does not have a smallest element. In particular, given any integer m, the integer $m - 1$ is less than m. Second, consider the set $T = \{x \in \mathbb{Q} \mid 0 \leq x \leq 1\}$ of rational numbers between 0 and 1, inclusive. The set T has a smallest element, namely 0, but this does not mean that T is well-ordered. In fact, the set $S = T - \{0, 1\}$ is a subset of T that does not have a smallest element. To see this, note that if x is a rational number such that $0 < x < 1$, then $0 < x/2 < x$.   ∎

In mathematics there are many cases in which we wish to prove that a statement $P(n)$ is true for every positive integer (or natural number) n. Many examples of such problems will be encountered as you progress through this book. Let us now look at the *Principle of Mathematical Induction*, which is perhaps the most common technique applied to handle such problems.

Let $P(n)$ be a statement about the positive integer n. Assume we can prove that the following two properties hold:

1. $P(1)$ is true.

2. For any $k \in \mathbb{Z}^+$, if $P(k)$ is true, then $P(k + 1)$ is true.

How do these two properties help to prove that $P(n)$ is true for every $n \in \mathbb{Z}^+$?

Property 1 tells us that $P(1)$ is true. Hence, property 2, applied with $k = 1$, states that $P(2)$ is true. Then, since $P(2)$ is true, we may apply condition 2 again, this time with $k = 2$, to yield that $P(3)$ is true. Since $P(3)$ is true, $P(4)$ is true, and so on. It seems reasonable to conclude that $P(n)$ is true for every $n \in \mathbb{Z}^+$. This is exactly what the principle of mathematical induction allows us to do.

THEOREM 0.4.1 (Principle of Mathematical Induction) Let S be a subset of $\mathbb{Z}^+$ satisfying the following two properties:

1. $1 \in S$
2. For any $k \in \mathbb{Z}^+$, if $k \in S$, then $k + 1 \in S$.

Then $S = \mathbb{Z}^+$.

Proof Let $S \subseteq \mathbb{Z}^+$ satisfy properties 1 and 2; to show that $S = \mathbb{Z}^+$, we proceed by contradiction and employ the principle of well-ordering. So suppose $S \neq \mathbb{Z}^+$; then $\mathbb{Z}^+ - S$ is a nonempty set of positive integers. Hence, the principle of well-ordering states that $\mathbb{Z}^+ - S$ has a smallest element, say, n. Since $1 \in S$, it must be that $n \geq 2$. Also, since n is the smallest element of $\mathbb{Z}^+ - S$, we have $n - 1 \in S$. But now apply property 2 with $k = n - 1$: since $k \in S$, we have $k + 1 = n \in S$, a contradiction. It follows that $S = \mathbb{Z}^+$. $\qquad\square$

In Exercise 6, you are asked to show that the principle of well-ordering is implied by the principle of mathematical induction, thereby showing that these two principles are, indeed, logically equivalent.

To apply the principle of mathematical induction to prove that $P(n)$ is true for all $n \in \mathbb{Z}^+$, we let $S = \{n \in \mathbb{Z}^+ \mid P(n) \text{ is true}\}$. It is common to divide the proof into two steps. The first step is to verify that $1 \in S$, namely, that $P(1)$ is true. This step is commonly called the **anchor step**; once it has been shown that $1 \in S$, we often say that the induction is "anchored." The second step is to let k be an arbitrary positive integer, assume $k \in S$, and proceed to show that $k + 1 \in S$. This step is called the **inductive step**, and the assumption that $k \in S$ is called the **induction hypothesis**. It is crucial that we carefully and clearly identify the induction hypothesis, because the main idea is to use it to prove that $k + 1 \in S$.

We next consider several examples that illustrate the power and usefulness of mathematical induction as a proof technique.

■ *Example 0.4.3* Use mathematical induction to prove that the following formula $P(n)$ holds for every $n \in \mathbb{Z}^+$:

$$\sum_{i=1}^{n} \frac{1}{(2i - 1)(2i + 1)} = \frac{n}{2n + 1}$$

Here the sigma notation is used to indicate a sum. In general, given a function f whose domain is the set of positive integers, the notation

$$\sum_{i=1}^{n} f(i)$$

denotes the sum $f(1) + f(2) + \cdots + f(n)$. Thus,

$$\sum_{i=1}^{n} \frac{1}{(2i-1)(2i+1)} = \frac{1}{1 \cdot 3} + \frac{1}{3 \cdot 5} + \cdots + \frac{1}{(2n-1)(2n+1)}$$

Solution

We proceed by induction on n; let $S = \{n \in \mathbb{N} \mid P(n) \text{ is true}\}$.

Since

$$\frac{1}{1 \cdot 3} = \frac{1}{3} = \frac{1}{2(1)+1}$$

we have $1 \in S$.

Let k be an arbitrary positive integer and assume $k \in S$. Then the induction hypothesis is that

$$\sum_{i=1}^{k} \frac{1}{(2i-1)(2i+1)} = \frac{k}{2k+1}$$

We must show that $k + 1 \in S$, namely, that

$$\sum_{i=1}^{k+1} \frac{1}{(2i-1)(2i+1)} = \frac{k+1}{2k+3}$$

Adding $1/[(2k+1)(2k+3)]$ to both sides of the identity given by the induction hypothesis yields the following:

$$\frac{1}{1 \cdot 3} + \frac{1}{3 \cdot 5} + \cdots + \frac{1}{(2k-1)(2k+1)} + \frac{1}{(2k+1)(2k+3)}$$

$$= \frac{k}{2k+1} + \frac{1}{(2k+1)(2k+3)}$$

$$= \frac{k(2k+3)}{(2k+1)(2k+3)} + \frac{1}{(2k+1)(2k+3)}$$

$$= \frac{k(2k+3)+1}{(2k+1)(2k+3)}$$

$$= \frac{2k^2 + 3k + 1}{(2k + 1)(2k + 3)}$$

$$= \frac{(2k + 1)(k + 1)}{(2k + 1)(2k + 3)}$$

$$= \frac{k + 1}{2k + 3}$$

Thus, $k + 1 \in S$, and we may conclude that $S = \mathbb{Z}^+$ by the principle of mathematical induction. ∎

■ *Example 0.4.4* Let $n \in \mathbb{Z}^+$ and let $A_1, A_2, \ldots, A_n$ be any n sets. Use mathematical induction to prove the statement $P(n)$ holds for every $n \in \mathbb{Z}^+$, where $P(n)$ is De Morgan's law:

$$\overline{\bigcup_{i=1}^{n} A_i} = \bigcap_{i=1}^{n} \overline{A_i}$$

Solution

We proceed by induction on n. Let S be the set of positive integers n for which $P(n)$ is true.

For $n = 1$, note that $P(1)$ is simply the statement that $\overline{A_1} = \overline{A_1}$. So $P(1)$ is clearly true, and it follows that $1 \in S$. In fact, we also know that $P(2)$ is true by Theorem 0.1.1, part 5(a).

Now let k be an arbitrary positive integer, $k \geq 2$, and assume $k \in S$. Then the induction hypothesis is that

$$\overline{\bigcup_{i=1}^{k} A_i} = \bigcap_{i=1}^{k} \overline{A_i}$$

holds for any k sets $A_1, A_2, \ldots, A_k$. We must show that $k + 1 \in S$, namely, that

$$\overline{\bigcup_{i=1}^{k+1} A_i} = \bigcap_{i=1}^{k+1} \overline{A_i}$$

holds for any $k + 1$ sets $A_1, A_2, \ldots, A_{k+1}$. So,

$$\overline{\bigcup_{i=1}^{k+1} A_i} = \overline{\bigcup_{i=1}^{k} A_i \cup A_{k+1}}$$

(by the associative property for ∪)

$$= \overline{\bigcup_{i=1}^{k} A_i} \cap \overline{A_{k+1}}$$

(by Theorem 0.1.1, part 5(a))

$$= \bigcap_{i=1}^{k} \overline{A_i} \cap \overline{A_{k+1}}$$

(by using the induction hypothesis)

$$= \bigcap_{i=1}^{k+1} \overline{A_i}$$

(by the associative property for ∩). Thus, $k+1 \in S$, and we may conclude that $S = \mathbb{Z}^+$ by the principle of mathematical induction. ∎

For the record, we mention that the implication

$$k \in S \rightarrow k+1 \in S$$

in the second part of an induction proof may be established by using an indirect proof. For example, if a proof by contradiction is used, then we assume $k \in S$ and $k+1 \notin S$ and attempt to derive a contradiction. This technique is illustrated in the next example.

■ ***Example 0.4.5*** Use mathematical induction to prove that the following inequality $P(n)$ holds for every $n \in \mathbb{Z}^+$:

$$\sum_{i=1}^{n} \frac{1}{\sqrt{i}} \leq 2\sqrt{n} - 1$$

Note that

$$\sum_{i=1}^{n} \frac{1}{\sqrt{i}} = \frac{1}{\sqrt{1}} + \frac{1}{\sqrt{2}} + \cdots + \frac{1}{\sqrt{n}}$$

Solution
We proceed by induction on n. Let $S = \{n \in \mathbb{N} \mid P(n) \text{ is true}\}$.

Since

$$\frac{1}{\sqrt{1}} = 1 \le 2(1) - 1$$

we have $1 \in S$.

Now let k be an arbitrary positive integer and assume $k \in S$. Then the induction hypothesis is that

$$\frac{1}{\sqrt{1}} + \frac{1}{\sqrt{2}} + \cdots + \frac{1}{\sqrt{k}} \le 2\sqrt{k} - 1$$

We must show that $k + 1 \in S$; we proceed by contradiction, supposing that $k + 1 \notin S$, namely, that

$$\frac{1}{\sqrt{1}} + \frac{1}{\sqrt{2}} + \cdots + \frac{1}{\sqrt{k}} + \frac{1}{\sqrt{k+1}} > 2\sqrt{k+1} - 1$$

By using the induction hypothesis, it follows that

$$\frac{1}{\sqrt{k+1}} > 2\sqrt{k+1} - 2\sqrt{k}$$

Then multiplying by $\sqrt{k+1}$ gives

$$1 > 2(k+1) - 2\sqrt{k(k+1)}$$

Isolating the radical and then squaring both sides yields

$$4k^2 + 4k > 4k^2 + 4k + 1$$

but this implies that $0 > 1$, a contradiction. Thus, $k + 1 \in S$, and we may conclude that $S = \mathbb{Z}^+$ by the principle of mathematical induction. ∎

In some cases we are asked to prove that a statement $P(n)$ is true for all integers $n \ge n_0$, where n_0 is some given fixed integer. For instance, we may be asked to prove $n^2 \le 2^n$ for every integer $n \ge 4$. (Note that this inequality fails for $n = 3$.) In such cases the following corollary to Theorem 0.4.1 provides the necessary proof technique.

COROLLARY 0.4.2   Let $n_0 \in \mathbb{Z}$ and let $M = \{n \in \mathbb{Z} \mid n \ge n_0\}$. Let S be a subset of M satisfying the following two properties:

1. $n_0 \in S$
2.   For any $k \in M$, if $k \in S$, then $k + 1 \in S$.

Then $S = M$. □

You are asked to prove Corollary 0.4.2 in Exercise 8.

■ ***Example 0.4.6***   Use Corollary 0.4.2 to prove that

$$n^2 \le 2^n$$

holds for every integer $n \ge 4$.

Solution
We proceed by induction on n. Let $M = \{n \in \mathbb{Z} \mid n \ge 4\}$ and let $S = \{n \in M \mid n^2 \le 2^n\}$.

Since $4^2 = 16 = 2^4$, it follows that $4 \in S$.

Now let k be an arbitrary integer, $k \ge 4$, and assume $k \in S$. Hence, the induction hypothesis is that

$$k^2 \le 2^k$$

We must show $k + 1 \in S$, namely, that

$$(k + 1)^2 \le 2^{k+1}$$

We proceed as follows:

$$(k + 1)^2 = k^2 + 2k + 1 \le k^2 + k^2 = 2k^2 \le 2(2^k) = 2^{k+1}$$

This shows that $k + 1 \in S$. It follows by Corollary 0.4.2 that $S = M$. (Note that the inductive step uses the inequality $2n + 1 \le n^2$, which holds for $n \ge 3$; this can also be proven by induction.)   ■

Consider the function $f : \mathbb{N} \to \mathbb{Z}^+$ defined "recursively" as follows:

1. $f(0) = f(1) = 1$
2. $f(n) = f(n - 2) + f(n - 1)$ for $n \ge 2$

In what sense does the above definition determine a function? We see that the values $f(0) = 1$ and $f(1) = 1$ are given explicitly. Moreover, for $n \ge 2$, the relation in part 2 of the definition can be used to compute $f(n)$. For example, suppose we want to know $f(4)$. We proceed as follows:

$$f(2) = f(0) + f(1) = 1 + 1 = 2$$
$$f(3) = f(1) + f(2) = 1 + 2 = 3$$
$$f(4) = f(2) + f(3) = 2 + 3 = 5$$

so that $f(4) = 5$. As noted, the function f is an example of a function that is recursively defined. In fact, you may recognize the sequence of numbers $(1, 1, 2, 3, 5, \ldots, f(n), \ldots)$ as the famous **Fibonacci sequence**.

> *DEFINITION 0.4.1* A function $f : \mathbb{N} \to \mathbb{C}$ is said to be **recursively** (or **inductively**) **defined** provided:
>
> 1. For some $n_0 \in \mathbb{N}$, the values $f(0), f(1), \ldots, f(n_0)$ are explicitly given.
> 2. For $n > n_0$, $f(n)$ is defined in terms of $f(0), f(1), \ldots, f(n-1)$.
>
> We call $f(0), f(1), \ldots, f(n_0)$ the **initial values** of f and refer to the identity that defines $f(n)$ in terms of $f(0), f(1), \ldots, f(n-1)$ as the **recurrence equation** (or **recurrence relation**) for f. ∎

■ *Example 0.4.7* For $n \in \mathbb{Z}^+$, ***n*-factorial** is denoted $n!$ and is defined as the product of the integers between 1 and n, inclusive:

$$n! = \prod_{k=1}^{n} k$$

So $1! = 1$, $2! = 2$, $3! = 6$, $4! = 24$, and so on. It is also natural to define 0-factorial by $0! = 1$. Thus, we may define the function $p : \mathbb{N} \to \mathbb{Z}^+$ by $p(n) = n!$. Recursively, the function p is defined as follows:

1. $p(0) = 1$
2. $p(n) = np(n-1)$ for $n \geq 1$ ■

In some cases, the method of induction we've employed in previous examples cannot be applied so nicely. To see this, consider again the "Fibonacci function" $f : \mathbb{N} \to \mathbb{Z}^+$ defined by:

1. $f(0) = f(1) = 1$
2. $f(n) = f(n-2) + f(n-1)$ for $n \geq 2$

Suppose we wish to prove that $f(n) > (4/3)^n$ for every $n \geq 2$. The statement of the problem suggests using induction; let's see what happens.

As usual, we let S be the set of integers $n \geq 2$ for which the relation $f(n) > (4/3)^n$ holds. It is easy to check that $2 \in S$. We let k be an arbitrary integer, $k \geq 2$, and assume $k \in S$; that is, we assume $f(k) > (4/3)^k$. It must be shown that $k+1 \in S$, namely, that $f(k+1) > (4/3)^{k+1}$. What we do know is that $f(k+1) = f(k-1) + f(k)$ for $k \geq 2$. However, it appears we have a slight problem, as we have no direct information concerning $f(k-1)$.

Problems such as this can be handled by using an alternate form of mathematical induction, called the **strong form of induction**. Like the regular principle of induction, the strong form of induction is equivalent to the principle of well-ordering; the proof of the next theorem is left to Exercise 12.

THEOREM 0.4.3 (Strong Form of Induction)   Let $n_0 \in \mathbb{Z}$ and let $M = \{n \in \mathbb{Z} \mid n \geq n_0\}$. Let S be a subset of M satisfying the following two properties:

1. $n_0 \in S$
2. For any $k \geq n_0$, if $n \in S$ for all integers n such that $n_0 \leq n \leq k$, then $k + 1 \in S$.

Then $S = M$.   □

■ ***Example 0.4.8***   Consider again the Fibonacci function $f : \mathbb{N} \to \mathbb{Z}^+$ defined recursively by:

1. $f(0) = f(1) = 1$
2. $f(n) = f(n-2) + f(n-1)$ for $n \geq 2$

Use induction (the strong form) to prove that

$$f(n) > (4/3)^n$$

for all $n \geq 2$.

Solution
We proceed by induction on n. Let $M = \{2, 3, 4, \dots\}$ and let $S = \{n \in M \mid f(n) > (4/3)^n\}$.

   Note that $f(2) = 2 > (4/3)^2$ and $f(3) = 3 > 64/27 = (4/3)^3$. So $2 \in S$ and $3 \in S$. (The reason for anchoring the induction at both 2 and 3 is that we need to have $k \geq 3$ in the inductive step. It is typical that a proof using the strong form of induction must anchor the induction for several values.)

   For an arbitrary integer $k \geq 3$, assume $\{2, 3, \dots, k\} \subseteq S$; that is, the induction hypothesis is that the relation

$$f(n) > (4/3)^n$$

holds for every integer n between 2 and k, inclusive. We must show that $k + 1 \in S$, namely, that $f(k+1) > (4/3)^{k+1}$. We proceed as follows:

$$\begin{aligned} f(k+1) &= f(k-1) + f(k) \\ &> (4/3)^{k-1} + (4/3)^k \end{aligned}$$

(by the induction hypothesis)

$$= \left(\frac{4}{3}\right)^{k-1} \left(1 + \frac{4}{3}\right)$$

$$= \left(\frac{7}{3}\right) \left(\frac{4}{3}\right)^{k-1}$$

$$> \left(\frac{16}{9}\right)\left(\frac{4}{3}\right)^{k-1}$$

$$= \left(\frac{4}{3}\right)^{k+1}$$

This shows that $k + 1 \in S$ and completes the proof.   ∎

■   ***Example 0.4.9***   In Section 0.2 we stated the **division algorithm**—given integers m and n with $n > 0$, there exist integers q and r (uniquely determined by m and n) such that

$$m = nq + r \qquad \text{and} \qquad 0 \le r < n$$

Make use of the strong form of induction to prove (the existence part of) this result.

Solution

We first prove the result for $m \ge 0$ by induction on m. Let n be a fixed but arbitrary positive integer, and let

$$S = \{m \in \mathbb{N} \mid \text{there exist } q, r \in \mathbb{Z} \text{ such that } m = nq + r \text{ and } 0 \le r < n\}$$

We anchor the induction for $0 \le m < n$; in this case note that

$$m = n \cdot 0 + m$$

so that $q = 0$ and $r = m$. Hence, $\{0, 1, \dots, n - 1\} \subseteq S$.

Now let k be an arbitrary integer, $k \ge n - 1$, and assume that $m \in S$ for every m between 0 and k, inclusive; that is, for any such m there exist integers q' and r' such that

$$m = nq' + r' \qquad \text{and} \qquad 0 \le r' < n$$

We must show that $k + 1 \in S$. Note that $0 \le k + 1 - n \le k$; hence, by the induction hypothesis, there exist $q', r' \in \mathbb{Z}$ such that

$$k + 1 - n = nq' + r' \qquad \text{and} \qquad 0 \le r' < n$$

Thus,

$$k + 1 = nq' + n + r' = n(q' + 1) + r'$$

so we may let $q = q' + 1$ and $r = r'$. This shows that $k + 1 \in S$. Therefore, $S = \mathbb{N}$.

To complete the proof, we must handle the case $m < 0$; again let $n > 0$ be fixed but arbitrary. If $m < 0$, then $-m > 0$, so by the first part of the proof there exist integers q' and r' such that

$$-m = nq' + r' \quad \text{and} \quad 0 \le r' < n$$

If $r' = 0$, then

$$m = n(-q')$$

and we have $q = -q'$ and $r = 0$. On the other hand, if $r' > 0$, then

$$m = n(-q') - r' = n(-q') - n + n - r' = n(-q' - 1) + (n - r')$$

and $0 < n - r' < n$. So we let $q = q' - 1$ and $r = n - r'$ in this case. This completes the proof. ∎

EXERCISE SET 0.4

1. Determine whether the following subsets of $\mathbb{R}$ are well-ordered.
   (a) T_1 is a finite subset of $\mathbb{R}$. (In particular, what if $T = \{ \ \}$?)
   (b) $T_2 = \mathbb{Q}^+$
   (c) T_3 is the set of even integers.
   (d) T_4 is a subset of $\mathbb{Z}$, and T_4 itself has a smallest element n_0.

2. Let T_1 and T_2 be subsets of $\mathbb{R}$ such that $T_1 \subseteq T_2$. Prove: If T_2 is well-ordered, then T_1 is well-ordered.

3. Find the smallest element of each of the following subsets of $\mathbb{N}$.
   (a) $A = \{n \in \mathbb{N} \mid n = m^2 - 10m + 28 \text{ for some integer } m\}$
   (b) $B = \{n \in \mathbb{N} \mid n = 5q + 2 \text{ for some integer } q\}$
   (c) $C = \{n \in \mathbb{N} \mid n = -150 - 19m \text{ for some integer } m\}$
   (d) $D = \{n \in \mathbb{Z}^+ \mid n = 5s + 8t \text{ for some integers } s \text{ and } t\}$

4. Prove that the division algorithm also works for any nonzero divisor: Given integers m and n with $n \ne 0$, there exist integers q and r (uniquely determined by m and n) such that $m = nq + r$ and $0 \le r < |n|$. In fact, for $n < 0$, show that

   $$m \text{ div } n = -(m \text{ div } (-n)) \quad \text{and} \quad m \text{ mod } n = m \text{ mod } (-n)$$

5. Use mathematical induction to prove that

   $$1 + 3 + \cdots + (2n - 1) = n^2$$

   holds for every $n \in \mathbb{Z}^+$.

6. Show that the principle of well-ordering is implied by the principle of mathematical induction.

7. Use mathematical induction to prove that the following formulas hold for all $n \in \mathbb{Z}^+$.

   (a) $1 + 2 + \cdots + n = n(n + 1)/2$
   (b) $1^2 + 2^2 + \cdots + n^2 = n(n + 1)(2n + 1)/6$
   (c) $1^3 + 2^3 + \cdots + n^3 = n^2(n + 1)^2/4$

8. Prove Corollary 0.4.2.

9. Use mathematical induction to prove that $n^3 + 5n$ is a multiple of 6 for every $n \in \mathbb{N}$.

10. For $n \in \mathbb{N}$, let $P_1(n)$ be the statement that $n^2 + n + 11$ is prime and let $P_2(n)$ be the statement that 3 divides $(3n + 2)$.

    (a) Note that $P_1(0), P_1(1), \ldots, P_1(9)$ are all true. Is $P_1(n)$ true for every $n \in \mathbb{N}$?
    (b) Note that the implication $P_2(k) \rightarrow P_2(k + 1)$ holds for every $k \in \mathbb{N}$. Can we conclude that $P_2(n)$ is true for every $n \in \mathbb{N}$? Why or why not?

11. Use mathematical induction to prove that the following formulas hold for any $n \in \mathbb{N}$.

    (a) $1 + \dfrac{1}{3} + \cdots + \dfrac{1}{3^n} = \dfrac{3}{2}\left(1 - \dfrac{1}{3^{n+1}}\right)$

    (b) $a + ar + \cdots + ar^n = \dfrac{a(r^{n+1} - 1)}{r - 1}$

    where a and r are real numbers and $r \neq 1$. (Note that, since a is a factor of both sides of the formula, it suffices to prove it for the case $a = 1$.)

12. Prove Theorem 0.4.3.

13. Use mathematical induction to prove that the following relations hold for all $n \in \mathbb{Z}^+$.

    (a) $1 + \dfrac{1}{4} + \cdots + \dfrac{1}{n^2} \leq 2 - \dfrac{1}{n}$

    (b) $2 + \left(1 + \dfrac{1}{\sqrt{2}} + \cdots + \dfrac{1}{\sqrt{n}}\right) > 2\sqrt{n + 1}$

14. Use mathematical induction to prove that

    $$2\left(1 + \dfrac{1}{8} + \cdots + \dfrac{1}{n^3}\right) < 3 - \dfrac{1}{n^2}$$

    holds for all $n \geq 2$.

15. Use mathematical induction to prove that $n! > 2^n$ for all $n \geq 4$.

16. For the Fibonacci function (see Example 0.4.8), prove that $f(n) \leq 2^n$ for all $n \in \mathbb{Z}^+$.

17. Use the strong form of induction to prove the existence part of the fundamental theorem of arithmetic: Every positive integer $n \geq 2$ can be factored as a product of primes.

18. Use mathematical induction to prove that the number of primes is infinite. (Hint: Let $P(n)$ be the statement that there exist at least n distinct primes; if $p_1, p_2, \ldots, p_k$ are k distinct primes, consider a prime factor of the integer $(p_1 p_2 \cdots p_k) + 1$.)

19. Use the strong form of induction to prove that any integer $n \geq 24$ can be expressed as $n = 5x + 7y$, where x and y are nonnegative integers.

20. Define the function $t : \mathbb{N} \to \mathbb{N}$ recursively by

$$t(0) = 0, t(1) = 1, t(2) = 2$$
$$t(n) = t(n-3) + t(n-2) + t(n-1) \qquad \text{for all } n \geq 3$$

Use the strong form of induction to prove that $t(n) < 2^n$ for every $n \in \mathbb{N}$.

21. Let m and n be integers with $1 \leq n \leq m$ and let $d = \gcd(n, m)$. Prove that there exist integers s and t such that

$$d = ns + mt$$

(Hint: Use the strong form of induction and the euclidean algorithm; if $r = m \bmod n$, then $r < n$ and $d = \gcd(r, n)$.)

22. Let m and n be integers with $1 \leq n \leq m$, let $d = \gcd(n, m)$, and let $e \in \mathbb{Z}$. Prove: There exist integers s and t such that $e = ns + mt$ if and only if e is a multiple of d.

23. A professional football team may score a field goal for 3 points or a touchdown (with conversion) for 7 points. (A safety, for 2 points, is also possible, but occurs rarely.) Use the strong form of induction to prove that, theoretically, it is possible for a football team to score some number of field goals and some number of touchdowns totaling n points for any integer $n \geq 12$.

24. Let a, b, and c be positive integers with $2 \leq a < b$ and $\gcd(a, b) = 1$. Consider the equation

$$ax + by = c \qquad\qquad\qquad (0.4.1)$$

(a) Discuss how to obtain a particular solution of Equation (0.4.1) in integers x_1 and y_1.

(b) Let (x_1, y_1) be a particular solution of (0.4.1). Show that (x, y) is a solution, $x, y \in \mathbb{Z}$, if and only if $x = x_1 + bt$ and $y = y_1 - at$ for some integer t.

(c) Show that (0.4.1) has a solution in nonnegative integers x and y if and only if, for any particular solution (x_1, y_1), the interval $[-x_1/b, y_1/a]$ contains an integer.

(d) Show that the equation $ax + by = ab - a - b$ does not have a solution in nonnegative integers.

(e) Show that (0.4.1) has a solution in nonnegative integers if $c > ab - a - b$. (Hint: Write $c = ab - a - b + n, n \in \mathbb{Z}^+$, and use the strong form of induction on n. Anchor the induction by showing that (0.4.1) has a solution in nonnegative integers for $n \in \{1, 2, \ldots, a\}$. For $n \in \{1, 2, \ldots, a - 1\}$, let x_0 and y_0 be integers such that $n = ax_0 + by_0$. Note that y_0 may be chosen so that $1 \leq y_0 < a$. Then $c = a(x_0 + b - 1) + b(y_0 - 1)$. Show that $x_0 + b - 1 \geq 0$.)

25. Use mathematical induction to prove that the following statements hold for all $n \in \mathbb{Z}^+$:

(a) $1 \cdot 1! + 2 \cdot 2! + 3 \cdot 3! + \cdots + n \cdot n! = (n+1) \cdot n! - 1$

(b) $(2^n)^2 - 1$ is a multiple of 3.

(c) $1^2 - 2^2 + \cdots + (-1)^{n+1} n^2 = \dfrac{(-1)^{n+1} n(n+1)}{2}$

(d) $\dfrac{1}{1 \cdot 2} + \dfrac{1}{2 \cdot 3} + \cdots + \dfrac{1}{n(n+1)} = \dfrac{n}{n+1}$

(e) $1 \cdot 2 + 2 \cdot 3 + \cdots + n(n+1) = \dfrac{n(n+1)(n+2)}{3}$

26. Let a and d be real numbers. Use induction to prove the following formula for the sum of a finite arithmetic series:

$$a + (a+d) + (a+2d) + \cdots + (a+nd) = \frac{(n+1)(2a+nd)}{2}$$

27. Let a and b be integers with $1 \le a \le b$, and let $r = b \bmod a$. Recall that the euclidean algorithm is based on the fact that $\gcd(a,b) = \gcd(r,a)$.

(a) Show that $2r < b$.

Let $P(b)$ be the statement that, if $1 < a < b$, then the number of divisions required by the euclidean algorithm to compute $\gcd(a,b)$ is less than $2 \log_2 b$.

(b) Use the result of (a) and the strong form of induction to prove that $P(b)$ holds for all integers $b \ge 3$.

(c) Use the result of (b) to prove that, if $1 < a < b$, then the number of divisions required by the euclidean algorithm to compute $\gcd(a,b)$ is less than $2 \log_2 a + 1$.

28. Let n be a positive integer. The purpose of this exercise is to develop efficient algorithms to do arithmetic modulo n; i.e., given integers a and b with $0 \le a < n$, $0 \le b < n$, and a nonnegative integer m, we wish to compute $(a+b) \bmod n$, $ab \bmod n$, and $a^m \bmod n$.

(a) Develop an algorithm to efficiently compute $(a+b) \bmod n$. (Note that, if $a + b \ge n$, then $(a+b) \bmod n = a + b - n$.)

(b) Develop an algorithm to efficiently compute $ab \bmod n$. Use the technique known as repeated doubling: Compute the terms $a \bmod n$, $2a \bmod n$, $4a \bmod n$, and so on, by doubling; then use the binary representation of b to determine which terms are needed for the final result. For example, $(a \cdot 101) \bmod n$ is computed as

$$(((((a + (4a \bmod n)) \bmod n) + (32a \bmod n)) \bmod n) + (64a \bmod n)) \bmod n$$

(c) Develop an algorithm to efficiently compute $a^m \bmod n$. Use the technique known as repeated squaring: Compute the factors $a \bmod n$, $a^2 \bmod n$, $a^4 \bmod n$, and so on, by squaring; then use the binary representation of m to determine which factors are needed for the final result. For example, $a^{101} \bmod n$ is computed as

$$((((((a \cdot (a^4 \bmod n)) \bmod n) \cdot (a^{32} \bmod n)) \bmod n) \cdot (a^{64} \bmod n)) \bmod n$$

29. Define the function $L : \mathbb{N} \to \mathbb{Z}^+$ recursively by

$$L(0) = 1, \quad L(1) = 3$$
$$L(n) = L(n-2) + L(n-1), \qquad n \geq 2$$

Prove that $L(n) = f(n-1) + f(n)$ for $n \geq 1$, where f is the Fibonacci function. The numbers $L(0), L(1), L(2), \ldots$ are called the **Lucas numbers**.

0.5 FINITE FIELDS

In Section 0.1 we introduced the set $\mathbb{C}$ of complex numbers. These are numbers of the form $x + yi$ where x and y are real numbers. Operations of addition and multiplication are defined on $\mathbb{C}$ as follows: For $z_1 = x_1 + y_1 i$ and $z_2 = x_2 + y_2 i$,

$$z_1 + z_2 = (x_1 + y_1 i) + (x_2 + y_2 i) = (x_1 + x_2) + (y_1 + y_2)i$$
$$z_1 \cdot z_2 = z_1 z_2 = (x_1 + y_1 i)(x_2 + y_2 i) = (x_1 x_2 - y_1 y_2) + (x_1 y_2 + x_2 y_1)i$$

THEOREM 0.5.1   The operations of addition and multiplication on $\mathbb{C}$ satisfy the following properties:

1.   The **associative laws**: For any $z_1, z_2, z_3 \in \mathbb{C}$,
     (a) $(z_1 + z_2) + z_3 = z_1 + (z_2 + z_3)$
     (b) $(z_1 \cdot z_2) \cdot z_3 = z_1 \cdot (z_2 \cdot z_3)$
2.   The **commutative laws**: For any $z_1, z_2 \in \mathbb{C}$,
     (a) $z_1 + z_2 = z_2 + z_1$
     (b) $z_1 \cdot z_2 = z_2 \cdot z_1$
3.   The **distributive laws**: For any $z_1, z_2, z_3 \in \mathbb{C}$,
     (a) $z_1 \cdot (z_2 + z_3) = (z_1 \cdot z_2) + (z_1 \cdot z_3)$
     (b) $(z_1 + z_2) \cdot z_3 = (z_1 \cdot z_3) + (z_2 \cdot z_3)$
4.   The number $0 = 0 + 0i$ is the **additive identity**; that is, $0 + z = z + 0 = z$ for any $z \in \mathbb{C}$.
5.   The number $1 = 1 + 0i$ is the **multiplicative identity**; that is, $1 \cdot z = z \cdot 1 = z$ for any $z \in \mathbb{C}$.
6.   Every number has an **(additive) inverse**; the inverse of the number $z = x + yi$ is $-z = -x - yi$.

7. Every number except 0 has a **reciprocal** (or **mutiplicative inverse**); the reciprocal of $z = x + yi$ is

$$z^{-1} = \frac{x}{r^2} - \frac{y}{r^2}i$$

where $r^2 = x^2 + y^2$. $\qquad\qquad\square$

Because it has the properties listed in Theorem 0.5.1, we call the algebraic structure $(\mathbb{C}, +, \cdot)$ a **field**. In particular, $(\mathbb{C}, +, \cdot)$ is called the **field of complex numbers**.

Let x be a real number and notice that $x = x + 0i$ is also a complex number. Thus, $\mathbb{R} \subseteq \mathbb{C}$. Furthermore, for $x, y \in \mathbb{R}$, we have $x + y \in \mathbb{R}$ and $xy \in \mathbb{R}$. Thus, we can consider the algebraic structure $(\mathbb{R}, +, \cdot)$. Immediately we see that the associative, commutative, and distributive laws hold; in fact, these are inherited from $(\mathbb{C}, +, \cdot)$. Since 0 and 1 are real, 0 and 1 are the additive and multiplicative identities of $(\mathbb{R}, +, \cdot)$, respectively. Also, $x \in \mathbb{R}$ has inverse $-x \in \mathbb{R}$ and the reciprocal of x is $1/x \in \mathbb{R}$, provided $x \neq 0$. Therefore, $(\mathbb{R}, +, \cdot)$ is a field, called the **field of real numbers**. Since $\mathbb{R} \subseteq \mathbb{C}$, we say that $(\mathbb{R}, +, \cdot)$ is a subfield of $(\mathbb{C}, +, \cdot)$, and that $(\mathbb{C}, +, \cdot)$ is an extension field of $(\mathbb{R}, +, \cdot)$. In general, if $(F, +, \cdot)$ and $(E, +, \cdot)$ are fields and $F \subseteq E$, then F is a **subfield** of E and E is an **extension field** of F. In particular, any field is a subfield and an extension field of itself.

Just as $(\mathbb{R}, +, \cdot)$ is a field, it can be shown that $(\mathbb{Q}, +, \cdot)$ is a field, called the **field of rational numbers**. It also makes sense to consider the algebraic structure $(\mathbb{Z}, +, \cdot)$; however, this is *not* a field. Note that it satifies all the properties listed in Theorem 0.5.1 except property 7; for example, the reciprocal of 2 is $1/2$, but $1/2$ is not an integer.

The fields discussed above are infinite fields because the sets $\mathbb{Q}$, $\mathbb{R}$, and $\mathbb{C}$ are infinite. In combinatorics, particularly in working with certain combinatorial designs, it is useful to have a catalog of finite fields. The main purpose of this section is to describe how such fields are constructed.

Recall the set $\mathbb{Z}_n = \{0, 1, 2, \ldots, n-1\}$ of **integers modulo n**, where n is a positive integer. We define operations of addition, denoted $\oplus$, and multiplication, denoted $\odot$, on this set as follows: For $x, y \in \mathbb{Z}_n$:

$x \oplus y = (x + y) \bmod n$

$x \odot y = (xy) \bmod n$

It is clear that, since $x + y$ and xy are integers, $(x+y) \bmod n \in \mathbb{Z}_n$ and $(xy) \bmod n \in \mathbb{Z}_n$. So $\oplus$ and $\odot$ are binary operations on $\mathbb{Z}_n$. Note that

$x \oplus y = (x + y) \bmod n = (y + x) \bmod n = y \oplus x$

and

$x \odot y = (xy) \bmod n = (yx) \bmod n = y \odot x$

so that both $\oplus$ and $\odot$ are commutative. To prove that the associative and distributive laws hold, we need the following lemma.

LEMMA 0.5.2   Let $n \in \mathbb{Z}^+$ and let $m_1, m_2 \in \mathbb{Z}$. Then:

1. $(m_1 + m_2) \bmod n = [(m_1 \bmod n) + (m_2 \bmod n)] \bmod n$
2. $(m_1 m_2) \bmod n = [(m_1 \bmod n)(m_2 \bmod n)] \bmod n$

Proof   Suppose $m_1 \text{ div } n = q_1$, $m_1 \bmod n = r_1$, $m_2 \text{ div } n = q_2$, and $m_2 \bmod n = r_2$. Then $m_1 = nq_1 + r_1$ and $m_2 = nq_2 + r_2$, so

$$m_1 + m_2 = (nq_1 + r_1) + (nq_2 + r_2) = n(q_1 + q_2) + (r_1 + r_2)$$

and

$$m_1 m_2 = (nq_1 + r_1)(nq_2 + r_2) = n(nq_1 q_2 + q_1 r_2 + r_1 q_2) + r_1 r_2$$

Thus,

$$
\begin{aligned}
(m_1 + m_2) \bmod n &= [n(q_1 + q_2) + (r_1 + r_2)] \bmod n \\
&= (r_1 + r_2) \bmod n \\
&= [(m_1 \bmod n) + (m_2 \bmod n)] \bmod n
\end{aligned}
$$

which proves part 1. And,

$$
\begin{aligned}
(m_1 m_2) \bmod n &= [n(nq_1 q_2 + q_1 r_2 + r_1 q_2) + r_1 r_2] \bmod n \\
&= (r_1 r_2) \bmod n \\
&= [(m_1 \bmod n)(m_2 \bmod n)] \bmod n
\end{aligned}
$$

which proves part 2. $\square$

Now, to prove that $\oplus$ and $\odot$ are associative, we let $x, y, z \in \mathbb{Z}_n$; then

$$
\begin{aligned}
(x \oplus y) \oplus z &= [(x + y) \bmod n] \oplus z \\
&= ([(x + y) \bmod n] + z) \bmod n \\
&= [(x + y) + z] \bmod n
\end{aligned}
$$

(by part 1 of Lemma 0.5.2)

$$
\begin{aligned}
&= [x + (y + z)] \bmod n \\
&= (x + [(y + z) \bmod n]) \bmod n \\
&= x \oplus [(y + z) \bmod n] \\
&= x \oplus (y \oplus z)
\end{aligned}
$$

and

$$
\begin{aligned}
(x \odot y) \odot z &= [(xy) \bmod n] \odot z \\
&= ([(xy) \bmod n]z) \bmod n \\
&= [(xy)z] \bmod n
\end{aligned}
$$

(by part 2 of Lemma 0.5.2)

$$= [x(yz)] \bmod n$$
$$= (x[(yz) \bmod n]) \bmod n$$
$$= x \odot [(yz) \bmod n]$$
$$= x \odot (y \odot z)$$

Next, for $x, y, z \in \mathbb{Z}_n$:

$$x \odot (y \oplus z) = x \odot [(y + z) \bmod n]$$
$$= (x[(y + z) \bmod n]) \bmod n$$
$$= [x(y + z)] \bmod n$$
$$= (xy + xz) \bmod n$$
$$= ([(xy) \bmod n] + [(xz) \bmod n]) \bmod n$$
$$= ([x \odot y] + [x \odot z]) \bmod n$$
$$= (x \odot y) \oplus (x \odot z)$$

which proves one of the distributive laws, and the other one follows from this and commutativity:

$$(x \oplus y) \odot z = z \odot (x \oplus y) = (z \odot x) \oplus (z \odot y) = (x \odot z) \oplus (y \odot z)$$

Next we observe, for any $x \in \mathbb{Z}_n$, the following three properties:

$$x \oplus 0 = (x + 0) \bmod n = x \bmod n = x$$
$$x \odot 1 = (x \cdot 1) \bmod n = x \bmod n = x$$
$$x \oplus (-x \bmod n) = (x + -x) \bmod n = 0 \bmod n = 0$$

It follows that 0 is the additive identity, 1 is the multiplicative identity, and the inverse of x is $-x \bmod n$. For $x \in \mathbb{Z}_n^* = \mathbb{Z}_n - \{0\}$, note that

$$-x \bmod n = n - x$$

At this point we have shown that $(\mathbb{Z}_n, \oplus, \odot)$ has all the properties of a field, with the possible exception of the existence of reciprocals for elements of $\mathbb{Z}_n^*$. This brings us to the following important result:

THEOREM 0.5.3   For $n \in \mathbb{Z}^+, n \geq 2$, $(\mathbb{Z}_n, \oplus, \odot)$ is a field if and only if n is prime.

Proof   Let $x \in \mathbb{Z}_n^*$; for sufficiency, we show that x has a reciprocal if x and n are relatively prime.
  If $\gcd(x, n) = 1$, then there exist integers s and t such that

$$1 = xs + nt$$

Hence, $xs = 1 - nt$, so that $(xs) \bmod n = 1$. This implies that

$$x \odot (s \bmod n) = 1$$

Therefore, the reciprocal of x is $s \bmod n$.

For necessity, suppose n is composite; say, $n = xy$ where $x, y \in \mathbb{Z}$ and $0 < x < y < n$. We proceed by contradiction and suppose $(\mathbb{Z}_n, \oplus, \odot)$ is a field. Then, in particular, x has a reciprocal, say, x'. Now

$$
\begin{aligned}
xy = n &\rightarrow x \odot y = 0 \\
&\rightarrow x' \odot (x \odot y) = x' \odot 0 = 0 \\
&\rightarrow (x' \odot x) \odot y = 0 \\
&\rightarrow 1 \odot y = 0 \\
&\rightarrow y = 0
\end{aligned}
$$

which is a contradiction. □

Thus, we have an example of a finite field with p elements when p is prime, namely, $(\mathbb{Z}_p, \oplus, \odot)$. It becomes cumbersome at this point to continue to use special symbols such as $\oplus$ and $\odot$ to denote the operations of this and other fields; we henceforth agree to denote the addition and multiplication for any field F by the familiar $+$ and $\cdot$. For $x \in \mathbb{Z}_p$, the inverse of x is denoted simply by $-x$, the reciprocal of x is denoted by x^{-1}, and, for $y \in F$, we write $x + (-y)$ as $x - y$.

■ *Example 0.5.1* For $x \in \mathbb{Z}_p$, note again the following facts:

$$x + 0 = 0 + x = x$$
$$x \cdot 0 = 0 \cdot x = 0, \quad \text{and} \quad x \cdot 1 = 1 \cdot x = x$$

This tells us everything about the addition and multiplication in the field $(\mathbb{Z}_2, +, \cdot)$, because $\mathbb{Z}_2 = \{0, 1\}$.

For the field $(\mathbb{Z}_3, +, \cdot)$, we have the following additional results:

$$1 + 1 = 2, \quad 1 + 2 = 2 + 1 = 0, \quad 2 + 2 = 1$$
$$2 \cdot 2 = 1$$

For p prime, $p \geq 5$, and $x, y \in \mathbb{Z}_p$, it is convenient to present the values $x + y$ and $x \cdot y$ by giving addition and multiplication tables; these are like the familiar tables we all used in elementary school. Because of the facts noted above, we may exclude 0 from the addition table and exclude both 0 and 1 from the multiplication table. For instance, the addition and multiplication tables for $\mathbb{Z}_5$ are shown in Figure 0.5.1. ■

In $\mathbb{Z}_p$, we have

$$-0 = 0 \quad \text{and} \quad 1^{-1} = 1$$

+	1	2	3	4		·	2	3	4
1	2	3	4	0		2	4	1	3
2	3	4	0	1		3	1	4	2
3	4	0	1	2		4	3	2	1
4	0	1	2	3					

Figure 0.5.1   Addition and multiplication tables for the field $(\mathbb{Z}_5, +, \cdot)$.

Moreover,

$$-(-x) = x$$

and, for $x \neq 0$,

$$(x^{-1})^{-1} = x$$

Thus, in $\mathbb{Z}_3$, once we have determined that $-1 = 2$, then it follows immediately that $-2 = 1$. Also, $2^{-1} = 2$ in $\mathbb{Z}_3$. In $\mathbb{Z}_5$, the operation tables show us that $-1 = 4$, $-2 = 3$, $2^{-1} = 3$, and $4^{-1} = 4$. In stating that $2^{-1} = 3$, we are also implicitly stating that $3^{-1} = 2$.

If a finite field F has n elements, then we say that F has **order** n. For what values of n does there exist a finite field of order n? This question is answered by the next theorem.

THEOREM 0.5.4   There exists a finite field of order n if and only if $n = p^k$ for some prime p and positive integer k.   □

When $k = 1$, we already have an example of a field of order p, namely, $(\mathbb{Z}_p, +, \cdot)$. The rest of this section is devoted to showing how to construct a field of order $n = p^k$ when p is prime and k is a positive integer, $k \geq 2$. The proof that there does not exist a field of order n when n has distinct prime factors is developed in the exercises.

If $(F, +, \cdot)$ is a field and $x \in F$, then integer **multiples** and **powers** of x are defined in the usual way. Multiples of x are defined recursively, for $m \in \mathbb{Z}^+$, as follows:

1. $0x = 0$
2. $mx = x + (m - 1)x$
3. $(-m)x = -(mx)$

Thus, for example, $2x = x + x$, $3x = x + 2x = x + x + x$, and $(-2)x = -(2x) = -(x + x)$.
Powers of x are defined recursively by:

1. $x^0 = 1$
2. $x^m = x \cdot x^{m-1}$
3. $x^{-m} = (x^m)^{-1}$

where $m \in \mathbb{Z}^+$ is called an **exponent**. Thus, for example, $x^2 = x \cdot x$, $x^3 = x \cdot x^2 = x \cdot x \cdot x$, and $x^{-2} = (x^2)^{-1}$. Multiples and exponents obey the usual properties; for instance, given $x \in F$ and $m_1, m_2 \in \mathbb{Z}$, we have

$$(m_1 x) + (m_2 x) = (m_1 + m_2)x \qquad \text{and} \qquad (x^{m_1})(x^{m_2}) = x^{m_1 + m_2}$$

DEFINITION 0.5.1   Given a field $(F, +, \cdot)$, a **polynomial over F** is a function g determined by choosing $n \in \mathbb{N}$, then choosing $n + 1$ elements $c_0, c_1, \ldots, c_n \in F$, $c_n \neq 0$, and then letting

$$g(x) = c_0 + c_1 x + \cdots + c_n x^n$$

The nonnegative integer n is called the **degree** of g, and the elements $c_0, c_1, \ldots, c_n$ are called **coefficients** of g. Moreover, there is a special polynomial z, called the **zero polynomial**, defined by

$$z(x) = 0$$

The set of polynomials over F is written $F[x]$.   ∎

Note that we can consider the polynomial g over F to be a function on E for any extension field E of F. Polynomials over F are added and multiplied in the usual fashion, remembering that the coefficients are operated on as elements of F. For example, let $s, t \in \mathbb{Z}_5[x]$ be defined by

$$s(x) = 1 + 2x + 4x^2 \qquad \text{and} \qquad t(x) = 2 + 3x$$

Then

$$(s + t)(x) = s(x) + t(x) = (1 + 2) + (2 + 3)x + (0 + 4)x^2 = 3 + 0x + 4x^2 = 3 + 4x^2$$

and

$$
\begin{aligned}
(s \cdot t)(x) &= s(x) \cdot t(x) \\
&= 1 \cdot 2 + [(1 \cdot 3) + (2 \cdot 2)]x + [(1 \cdot 0) + (2 \cdot 3) + (4 \cdot 2)]x^2 + (4 \cdot 3)x^3 \\
&= 2 + (3 + 4)x + (0 + 1 + 3)x^2 + 2x^3 \\
&= 2 + 2x + 4x^2 + 2x^3
\end{aligned}
$$

In general, let $s, t \in F[x]$, with

$$s(x) = a_0 + a_1 x + \cdots + a_m x^m \qquad \text{and} \qquad t(x) = b_0 + b_1 x + \cdots + b_n x^n$$

Then

$$(s + t)(x) = s(x) + t(x) = (a_0 + b_0) + (a_1 + b_1)x + \cdots + (a_k + b_k)x^k$$

where $k = \max(m, n)$, and

$$(s \cdot t)(x) = s(x) \cdot t(x) = c_0 + c_1 x + \cdots + c_{m+n} x^{m+n}$$

where

$$c_i = \sum_{j=0}^{i} a_j b_{i-j}$$

Note that, if $a_m \neq 0$ and $b_n \neq 0$, then $c_{m+n} = a_m b_n \neq 0$, so that the degree of $s \cdot t$ is $m + n$.

For $c \in F$, the polynomial

$$g(x) = c$$

is called a **constant polynomial**. If $c = 0$, then g is the zero polynomial. Note that constant polynomials can be added and multiplied just like elements of F; in this sense we may consider F to be a subset of $F[x]$.

The proof of the next result is developed in Exercise 4.

THEOREM 0.5.5 (Division Algorithm for Polynomials)   Given polynomials $s(x)$ and $t(x)$ in $F[x]$ with $s(x)$ not the zero polynomial, there exist polynomials $q(x), r(x) \in F[x]$, uniquely determined by s and t, such that

$$t(x) = s(x)q(x) + r(x)$$

and either $r(x)$ is the zero polynomial or the degree of $r(x)$ is less than the degree of $s(x)$.   □

In the context of Theorem 0.5.5, we define the operations **div** and **mod** by:

$t(x)$ div $s(x) = q(x)$

$t(x)$ mod $s(x) = r(x)$

Analogous to the situation for integers, we call $t(x)$, $s(x)$, $q(x)$, and $r(x)$ the **dividend**, **divisor**, **quotient**, and **remainder**, respectively.

A nonconstant polynomial $t(x) \in \mathbb{Z}_p[x]$ is called **prime** (or **irreducible**) over $\mathbb{Z}_p$ provided it cannot be factored as $s(x) \cdot q(x)$ for nonconstant polynomials $s(x)$, $q(x) \in \mathbb{Z}_p[x]$. For example, we saw previously that, in $\mathbb{Z}_5[x]$,

$$t_1(x) = 2 + 2x + 4x^2 + 2x^3 = (1 + 2x + 4x^2) \cdot (2 + 3x)$$

so t_1 is not prime over $\mathbb{Z}_5$. On the other hand, any polynomial of degree 1 is prime.

Given a polynomial $g(x)$ over F and an extension field E of F, an element $c \in E$ is called a **root** of g provided $g(c) = 0$.

THEOREM 0.5.6 (Factor Theorem)   Given $g(x) \in F[x]$ and $c \in F$, we have that c is a root of g if and only if

$$g(x) = (x - c) \cdot q(x)$$

for some polynomial $q(x) \in F[x]$.

Proof   For necessity, assume c is a root of $g(x)$ and let $s(x) = x - c$. By Theorem 0.5.5, there exist polynomials $q(x), r(x) \in F[x]$ such that

$$g(x) = s(x)q(x) + r(x)$$

and the degree of r is less than the degree of s. Since the degree of s is 1, this means that r is a constant polynomial. Then

$$0 = g(c) = s(c)q(c) + r(c) = (0 \cdot q(c)) + r(c) = r(c)$$

which implies that r is the zero polynomial.

For sufficiency, if $g(x) = (x - c) \cdot q(x)$, then

$$g(c) = (c - c) \cdot q(c) = 0 \cdot q(c) = 0$$

showing that c is a root of g.   □

For polynomials of degree 2 or 3, we have the following useful result.

THEOREM 0.5.7   Let $g(x) \in \mathbb{Z}_p[x]$ be a polynomial of degree 2 or 3. Then $g(x)$ is prime over $\mathbb{Z}_p$ if and only if $g(x)$ has no roots in $\mathbb{Z}_p$.

Proof   For necessity, we assume that $g(x)$ has a root $c \in \mathbb{Z}_p$. Then, by the factor theorem, $g(x) = (x - c) \cdot q(x)$ with $q(x) \in \mathbb{Z}_p[x]$, so clearly $g(x)$ is not prime. Conversely, assume $g(x)$ is not prime, say, $g(x) = s(x) \cdot t(x)$ for nonconstant $s(x), t(x) \in \mathbb{Z}_p[x]$. Then, since g has degree 2 or 3 and the degree of g is the sum of the degrees of s and t, it must be that s or t has degree 1. Say, $s(x) = bx - c$. Then $b^{-1}c$ is a root of $g(x)$ since $s(b^{-1}c) = 0$.   □

We are now ready to demonstrate the construction of a finite field F_n of order $n = p^k$ for p prime and k an integer, $k \geq 2$; this F_n will turn out to be an extension of $\mathbb{Z}_p$. The general method makes use of a prime polynomial $g(x)$ of degree k over $\mathbb{Z}_p$.

■   ***Example 0.5.2***   Before looking at the case of finite fields, it may be illustrative to see how the field $(\mathbb{C}, +, \cdot)$ can be constructed as an extension of $(\mathbb{R}, +, \cdot)$. We begin with a prime polynomial $g(x)$ over $\mathbb{R}$; perhaps the simplest example is $g(x) = x^2 + 1$. The polynomial g has no roots in $\mathbb{R}$. Let us suppose for the moment that g does have a root in some extension field of $\mathbb{R}$; historically, this "imaginary" root has been denoted by i. (At this point, the number i is just an abstract concept; however, we can use it to explicitly construct the extension field F of $\mathbb{R}$ whose existence is being postulated.) Let

$$F = \{x + yi \mid x, y \in \mathbb{R}\}$$

(F is the set of polynomials over $\mathbb{R}$ (with variable i) of degree less than 2.) How are the elements of F to be added and multiplied? We add elements of F as polynomials, that is,

$$(x_1 + y_1 i) + (x_2 + y_2 i) = (x_1 + x_2) + (y_1 + y_2)i$$

We also multiply the elements of F as polynomials; however, this multiplication must be done "modulo g." That is,

$$(x_1 + y_1 i)(x_2 + y_2 i) = [x_1 x_2 + (x_1 y_2 + x_2 y_1)i + y_1 y_2 i^2] \bmod g(i)$$

It turns out that this can be done fairly easily using the fact that $g(i) = 0$, namely, that $i^2 = -1$. Thus,

$$(x_1 + y_1 i)(x_2 + y_2 i) = x_1 x_2 + (x_1 y_2 + x_2 y_1)i + y_1 y_2 i^2 = (x_1 x_2 - y_1 y_2) + (x_1 y_2 + x_2 y_1)i$$

It can then be checked that $(F, +, \cdot)$ is a field. In fact, observe that F is the field of complex numbers! ∎

■ *Example 0.5.3*   To construct a field F_4 of order $4 = 2^2$, we require a prime polynomial of degree 2 over $\mathbb{Z}_2$. The distinct polynomials of degree 2 over $\mathbb{Z}_2$ are:

$$x^2 \qquad 1 + x^2 \qquad x + x^2 \qquad \text{and} \qquad 1 + x + x^2$$

Of these, only $g(x) = 1 + x + x^2$ is prime. To verify this, we employ Theorem 0.5.7. Note that 0 is a root of both x^2 and $x + x^2$ and 1 is a root of $1 + x^2$; however, $g(0) = 1 = g(1)$, so that g has no roots in $\mathbb{Z}_2$. Let r be a root of $g(x)$. (As in Example 0.5.2, we assume that r comes from some extension field of $\mathbb{Z}_2$; in fact, we will see shortly that $r \in F_4$.) Let

$$F_4 = \{0, 1, r, 1 + r\}$$

be the set of polynomials over $\mathbb{Z}_2$, with variable r, of degree less than 2. We must define an addition $+$ and a multiplication $\cdot$ on F_4 so that $(F_4, +, \cdot)$ is a field. The trick is to add and multiply the elements of F_4 as polynomials, with the added proviso that multiplication be done modulo g. It turns out that this can be done fairly easily using the following two facts:

1. $2y = 0$ for any $y \in F_4$
2. $r^2 = 1 + r$

The first fact is shown as follows:

$$2y = y + y = y \cdot (1 + 1) = y \cdot 0 = 0$$

+	1	r	$1+r$		$\cdot$	r	$1+r$
1	0	$1+r$	r		r	$1+r$	1
r	$1+r$	0	1		$1+r$	1	r
$1+r$	r	1	0				

Figure 0.5.2   Addition and multiplication tables for a field of order 4.

This fact can also be stated as $-y = y$. The second fact follows from the first and $g(r) = 0$. Figure 0.5.2 shows the addition and multiplication tables for $(F_4, +, \cdot)$. Note, for example, that

$$r + (1 + r) = 1 + 2r = 1 + 0 = 1$$

and

$$r \cdot (1 + r) = r + r^2 = r + (1 + r) = 1$$

It is left for you to check that $(F_4, +, \cdot)$ is a field.   ∎

Note that the field F_4 of order $4 = 2^2$ constructed in Example 0.5.2 is an extension field of $\mathbb{Z}_2$. In general, we wish to construct F_n of order $n = p^k$ as an extension field of $\mathbb{Z}_p$. We begin with a prime polynomial $g(x)$ over $\mathbb{Z}_p$ of degree k. It is interesting that such a polynomial is known to exist although there is no general algorithm for its construction. (See, for instance, page 429 of *Abstract Algebra*, by David M. Burton.) However, extensive tables of such polynomials have been computed, particularly for the case $p = 2$ (the case arising most often in applications). Once we have g, we let r be a root of g and F_n be the set of all polynomials in r over $\mathbb{Z}_p$ with degree less than k; that is,

$$F_n = \{c_0 + c_1 r + \cdots + c_{k-1} r^{k-1} \mid c_0, c_1, \ldots, c_{k-1} \in \mathbb{Z}_p\}$$

The elements of F_n are added and mulitplied as polynomials, modulo $g(r)$; this is done by adding and multiplying in the usual fashion and then using the following two facts to simplify the result:

1. $py = 0$ for any $y \in F_n$
2. $g(r) = 0$

Fact 1 is a consequence of the fact that F_n is an extension field of $\mathbb{Z}_p$:

$py = y + y + \cdots + y$ (with p terms)

$\quad = y \cdot (1 + 1 + \cdots + 1)$ (by the distributive law)

$\quad = y \cdot (p1)$

$\quad = y \cdot 0$ (since $p1 = 0$ in $\mathbb{Z}_p$)

$\quad = 0$

By looking at the algebraic structure $(F_n, +, \cdot)$, it is not difficult to show that $+$ is associative and commutative, that 0 is the additive identity, and that the inverse of $f(r)$ is $-f(r)$. It is also not hard to show that $\cdot$ is associative and commutative, that 1 is the multiplicative identity, and that the distributive laws hold. To show that $f(r) \neq 0$ has a reciprocal, we need an additional notion and result—since $g(x)$ is prime, it follows that $f(x)$ and $g(x)$ are "relatively prime," and hence there exist $s(x)$ and $t(x)$ such that

$$1 = f(x)s(x) + g(x)t(x)$$

Then, looking at this identity modulo $g(x)$, we have

$$1 = f(r) \cdot (s(r) \bmod g(r))$$

so that the reciprocal of $f(r)$ is $s(r) \bmod g(r)$. In fact, there is a "euclidean algorithm" for polynomials that can be used to find $s(x)$ and $t(x)$. The complete details can be found in a course in abstract algebra.

■ ***Example 0.5.4***   Construct a field $(F_8, +, \cdot)$ of order 8.

Solution
Since $8 = 2^3$, F_8 is to be an extension field of $\mathbb{Z}_2$. We need a prime polynomial

$$g(x) = c_0 + c_1 x + c_2 x^2 + c_3 x^3$$

over $\mathbb{Z}_2$ of degree 3. Thus, $c_3 = 1$ and, to avoid having 0 as a root, $c_0 = 1$. Then, to avoid having 1 as a root, exactly one of c_1 and c_2 must be 1. So there are two choices for $g(x)$,

$$1 + x + x^3 \qquad \text{or} \qquad 1 + x^2 + x^3$$

Let us take $g(x) = 1 + x + x^3$. Now let r be a root of g so that $g(r) = 0$. Then

$$F_8 = \{0, 1, r, 1 + r, r^2, 1 + r^2, r + r^2, 1 + r + r^2\}$$

The two facts used to do addition and multiplication are

$$2y = 0 \qquad \text{and} \qquad r^3 = 1 + r$$

where y is any element of F_8. Thus, for example,

$$(1 + r) + (r + r^2) = 1 + 2r + r^2 = 1 + r^2$$

and

$$(1 + r) \cdot (r + r^2) = r + r^2 + r^2 + r^3$$
$$= r + 2r^2 + r^3$$
$$= r + 0 + (1 + r)$$
$$= 1 + 2r$$
$$= 1$$

The complete addition and multiplication tables are left for you to provide in Exercise 1. ∎

EXERCISE SET 0.5

1. Finish Example 0.5.4 by giving the addition and multiplication tables for $(F_8, +, \cdot)$.

2. Given a finite field $(F, +, \cdot)$, the **characteristic** of F is the smallest positive integer c such that $c1 = 0$.

   (a) Show that the characteristic of $\mathbb{Z}_p$ is p.
   (b) Let $x, y \in F$. Prove: If $x \neq 0$ and $y \neq 0$, then $xy \neq 0$.
   (c) Show that the characterisitc of any finite field F must be prime.

   (If the characteristic of F is the prime p, then the theorems of Lagrange and Cauchy from the theory of finite groups can be used to prove that the order of F must be p^k for some positive integer k.)

3. Redo Example 0.5.4 using the prime polynomial $g(x) = 1 + x^2 + x^3$ to construct a field F_8' of order 8.

4. Prove Theorem 0.5.5.

   (a) To prove existence, let m and n be the degrees of $t(x)$ and $s(x)$, respectively, and proceed by induction on m. If $m < n$, then let $q(x) = 0$ and $r(x) = t(x)$. For $m \geq n$, suppose $s(x)$ has leading term $a_n x^n$ and $t(x)$ has leading term $b_m x^m$. Let $q_1(x) = b_m a_n^{-1} x^{m-n}$ and $t_1(x) = t(x) - s(x)q_1(x)$. Then the degree of $t_1(x)$ is less than m, so apply the induction hypothesis to t_1.
   (b) To prove uniqueness, suppose $t(x) = s(x)q_1(x) + r_1(x) = s(x)q_2(x) + r_2(x)$. Then

   $$r_1(x) - r_2(x) = s(x)[q_2(x) - q_1(x)]$$

   and either both r_1 and r_2 are the zero polynomial (in the case s is a constant polynomial) or both r_1 and r_2 have degree less than s. Use this to show that $r_1 - r_2$ is the zero polynomial, so that $r_1 = r_2$. It then follows that $q_1 = q_2$.

5. Construct a field $(F_9, +, \cdot)$ of order 9.

6. With regard to Theorem 0.5.7, let $g(x)$ be a polynomial over $\mathbb{Z}_p$.

   (a) Prove: If g has a root in $\mathbb{Z}_p$, then g is not prime over $\mathbb{Z}_p$.
   (b) Disprove: If g is not prime over $\mathbb{Z}_p$, then g has a root in $\mathbb{Z}_p$. (Note that a counterexample must have degree at least 4.)

7. Outline the construction of a field of order: (a) 16, (b) 25, (c) 27. List the elements and work out several representative sums and products.

8. Let $(F, +, \cdot)$ and $(F', +, \cdot)$ be finite fields of the same order. An **isomorphism** between F and F' is a bijection $\phi : F \to F'$ that satisfies the following two properties for any $x, y \in F$:

$$\phi(x + y) = \phi(x) + \phi(y)$$
$$\phi(x \cdot y) = \phi(x) \cdot \phi(y)$$

In the case that such an isomorphism exists, we say that F and F' are **isomorphic**.

(a) Show that the fields F_8 and F'_8, constructed in Example 0.5.3 and Exercise 3, respectively, are isomorphic.

(b) In constructing a field of order 9 (see Exercise 5), there are several choices for the prime polynomial of degree 2 over $\mathbb{Z}_3$; among them are $g_1(x) = x^2 + 1$ and $g_2(x) = x^2 + x + 2$. Show that the corresponding fields F_9 and F'_9 are isomorphic.

(It can be shown, in fact, that any two finite fields of the same order are isomorphic; in this sense, there is a unique field of order p^k for a given prime p and positive integer k.)

Chapter 1

Basic Combinatorics

1.1 INTRODUCTION TO COMBINATORICS

Although its historical roots are found in mathematical recreations and games, the field of combinatorics has recently found a variety of applications both within and outside of mathematics. Its importance in computer science, operations research, and probability is easy to document, and now there are conferences on combinatorial algebra and important results in analysis whose proofs are distinctly combinatorial in nature. Moreover, combinatorics has become these days one of the more active areas of mathematical research, with more than a dozen international journals devoted to publishing new results. For these reasons, a course in combinatorics is strongly recommended for undergraduates majoring in the mathematical sciences.

Basically, combinatorics is concerned with arranging the elements of sets into definite patterns. Typically, the sets under consideration are "discrete," including finite sets such as the set $\{0, 1, \ldots, n-1\}$ of all natural numbers less than the positive integer n, and infinite discrete sets such as $\mathbb{N}$ and $\mathbb{Z}$. Thus, combinatorics is placed under the broad heading of discrete mathematics.

In attempting to arrange the elements of a set into a pattern that satisfies certain specified conditions, a number of issues generally arise:

1. *Existence.* Does such an arrangement exist?

2. *Enumeration or classification.* How many valid arrangements are there? Can they be classified in some way?

3. *Algorithms.* If such arrangements exist, is there a definite method for constructing one, or all, of them?

4. *Generalization.* Does the problem under consideration suggest other, related problems?

We begin by considering a famous combinatorial problem known as the "problem of the 36 officers," first considered in the eighteenth century by the great Swiss mathematician Leonard Euler. The problem is as follows: given 36 military officers,

0	1	2	3	4	5
1	2	3	4	5	0
2	3	4	5	0	1
3	4	5	0	1	2
4	5	0	1	2	3
5	0	1	2	3	4

Figure 1.1.1   A possible arrangement for the ranks of the 36 officers.

0	1	2	3	4	5
2	3	4	5	0	1
4	5	0	1	2	3
1	2	3	4	5	0
3	4	5	0	1	2
5	0	1	2	3	4

Figure 1.1.2   A possible arrangement for the regiments of the 36 officers.

representing each possible combination of six ranks with six regiments, can they be arranged in a square marching formation of six rows and six columns so that each rank and each regiment is represented in every row and in every column?

To clarify the problem a bit, note that the rank and regiment of an officer can be represented as an ordered pair (r, s), where r denotes the rank and s the regiment of the officer. For simplicity, let us denote the ranks and regiments by the numbers $0, 1, 2, 3, 4, 5$; thus, for instance, the ordered pair $(2, 5)$ denotes the officer with rank 2 and regiment 5. There are 36 officers, one for each of the 36 ordered pairs of the form (r, s), with $r, s \in \{0, 1, 2, 3, 4, 5\}$.

First, consider just the ranks of the officers in any proposed solution to our problem. There are six officers of each rank, so a solution to our problem must have the ranks $0, 1, 2, 3, 4, 5$ arranged in a square array of six rows and six columns so that each number occurs exactly once in any row and in any column. One possible arrangement for the officer ranks is the array shown in Figure 1.1.1. For convenience, we number the rows and columns of this array with the numbers 0, 1, 2, 3, 4, 5 and let r_{ij} denote the rank of the officer in row i and column j. Thus, for example, $r_{03} = 3$, $r_{25} = 1$, and $r_{44} = 2$.

Next, let us focus our attention on the regiments. Again, a proposed solution to our problem must have the six regiments $0, 1, 2, 3, 4, 5$ arranged in an array of six rows and six columns so that each regiment occurs exactly once in any row and in any column. A possible arrangement for the officer regiments is the array shown in Figure 1.1.2. Let s_{ij} denote the regiment of the officer in row i and column j of this array, $0 \leq i, j \leq 5$.

We can now consider the array of 36 ordered pairs (r_{ij}, s_{ij}), where $0 \leq i, j \leq 5$, shown in Figure 1.1.3. The problem of the 36 officers is solved in the affirmative if the 36 ordered pairs in this array are distinct, giving us each of the 36 possible rank-regiment pairs. Unfortunately, this is not the case, as can be observed. Notice

(0,0)	(1,1)	(2,2)	(3,3)	(4,4)	(5,5)
(1,2)	(2,3)	(3,4)	(4,5)	(5,0)	(0,1)
(2,4)	(3,5)	(4,0)	(5,1)	(0,2)	(1,3)
(3,1)	(4,2)	(5,3)	(0,4)	(1,5)	(2,0)
(4,3)	(5,4)	(0,5)	(1,0)	(2,1)	(3,2)
(5,5)	(0,0)	(1,1)	(2,2)	(3,3)	(4,4)

Figure 1.1.3 The rank and regiment arrays combined.

that the ordered pairs in the first row are repeated in the last row, and that the ordered pairs $(0,3)$, $(1,4)$, $(2,5)$, $(3,0)$, $(4,1)$, and $(5,2)$ do not appear.

Consideration of the problem of the 36 officers led Euler to generalize the problem by defining "Latin squares." For a positive integer n, a **Latin square of order n** is an arrangement of the integers $0, 1, \ldots, n-1$ into a square array of n rows and n columns such that each number occurs exactly once in any row and in any column. Given a Latin square R of order n, let r_{ij} denote the entry in row i and column j, where the rows and columns are numbered $0, 1, \ldots, n-1$. Let S be another Latin square of order n with typical entry s_{ij}. We say that R and S are **orthogonal Latin squares** provided the n^2 ordered pairs (r_{ij}, s_{ij}), $0 \leq i, j \leq n-1$, are distinct, thus giving all n^2 elements in the product set $\{0, 1, \ldots, n-1\} \times \{0, 1, \ldots, n-1\}$. In this case the n by n array (R, S) of ordered pairs (r_{ij}, s_{ij}) is called a **Græco-Latin square of order n**. (The names Latin square and Græco-Latin square come from Euler's practice of using Latin letters a, b, c, ... to denote the elements of one square and the corresponding Greek letters $\alpha, \beta, \gamma, \ldots$ to denote the elements of an orthogonal square.) Then the problem of the 36 officers can be restated as follows:

Do there exist two orthogonal Latin squares of order 6?

or

Does there exist a Græco-Latin square of order 6?

Euler gave methods for constructing a pair of orthogonal Latin squares of order $n \geq 2$ whenever $n \bmod 4 \in \{0, 1, 3\}$. For instance, if there were 16 officers in Euler's problem instead of 36, with one officer of each possible combination of four ranks with four regiments, then we would be interested in finding two orthogonal Latin squares of order 4. An example of these is shown in Figure 1.1.4, along with the corresponding Græco-Latin square of order 4.

It can be checked that there do not exist two orthogonal Latin squares of order 2 (see Exercise 1). Euler searched for, but could not find, a pair of orthogonal Latin squares of order 6. Notice that $2 \bmod 4 = 2 = 6 \bmod 4$. This led Euler to conjecture that there do not exist two orthogonal Latin squares of order n when $n \bmod 4 = 2$. Around 1900, using an exhaustive search, G. Tarry [*Le problème des 36 officiers*, *Comptes Rendus de L'Association Francaise pour L'Avancement de Science* 1 (1900): 122–123; 2 (1901): 170–203] showed that Euler's conjecture is true for $n = 6$. This answers the problem of the 36 officers in the negative. The cases $n = 10, 14, 18, 22, \ldots$ remained unsettled until around 1960, when R. C. Bose,

0	1	2	3	0	1	2	3
1	0	3	2	2	3	0	1
2	3	0	1	3	2	1	0
3	2	1	0	1	0	3	2

(a) Two orthogonal Latin squares of order 4.

(0,0)	(1,1)	(2,2)	(3,3)
(1,2)	(0,3)	(3,0)	(2,1)
(2,3)	(3,2)	(0,1)	(1,0)
(3,1)	(2,0)	(1,3)	(0,2)

(b) The corresponding Græco-Latin square of order 4.

Figure 1.1.4

(0,0)	(4,7)	(1,8)	(7,6)	(2,9)	(9,3)	(8,5)	(3,4)	(6,1)	(5,2)
(8,6)	(1,1)	(5,7)	(2,8)	(7,0)	(3,9)	(9,4)	(4,5)	(0,2)	(6,3)
(9,5)	(8,0)	(2,2)	(6,7)	(3,8)	(7,1)	(4,9)	(5,6)	(1,3)	(0,4)
(5,9)	(9,6)	(8,1)	(3,3)	(0,7)	(4,8)	(7,2)	(6,0)	(2,4)	(1,5)
(7,3)	(6,9)	(9,0)	(8,2)	(4,4)	(1,7)	(5,8)	(0,1)	(3,5)	(2,6)
(6,8)	(7,4)	(0,9)	(9,1)	(8,3)	(5,5)	(2,7)	(1,2)	(4,6)	(3,0)
(3,7)	(0,8)	(7,5)	(1,9)	(9,2)	(8,4)	(6,6)	(2,3)	(5,0)	(4,1)
(1,4)	(2,5)	(3,6)	(4,0)	(5,1)	(6,2)	(0,3)	(7,7)	(8,8)	(9,9)
(2,1)	(3,2)	(4,3)	(5,4)	(6,5)	(0,6)	(1,0)	(8,9)	(9,7)	(7,8)
(4,2)	(5,3)	(6,4)	(0,5)	(1,6)	(2,0)	(3,1)	(9,8)	(7,9)	(8,7)

Figure 1.1.5   A Græco-Latin square of order 10.

S. S. Shrikhande, and E. T. Parker [Further results on the construction of mutually orthogonal Latin squares and the falsity of Euler's conjecture, *Canadian Journal of Mathematics* 12 (1960): 189–203] showed how to construct a pair of orthogonal Latin squares of order n whenever $n \geq 10$ and $n \bmod 4 = 2$. This amazing discovery showed that Euler's conjecture is false for $n > 6$, so that $n = 2$ and $n = 6$ are the only cases when a pair of orthogonal Latin squares fails to exist.

The story of the Euler conjecture and its solution is a beautiful example of how new mathematics is discovered and communicated. (If you are interested in learning more about this problem, see, as a start, the November 1959 issue of *Scientific American*. The cover of this issue shows a color-coded version of the Græco-Latin square of order 10 depicted in Figure 1.1.5. Martin Gardner's "Mathematical Games" column in that issue also discusses the solution of Euler's conjecture.)

Why is there so much interest in what appears to be a mathematical recreation? It turns out that Latin squares and other types of combinatorial designs are important in the design of certain kinds of statistical studies. There is an example of this application in Exercise 2, and a further discussion of combinatorial designs in Chapter 4.

EXERCISE SET 1.1

1. Find all Latin squares of order 2 and verify that no two of them are orthogonal.

2. A statistical study is to be conducted to test the effect of three different kinds of fertilizer on the yield of three different varieties of corn. A large rectangular field to be used for the study is divided into nine plots arranged into three rows and three columns—one plot for each of the nine combinations of fertilizer type with corn variety. However, growing conditions may not be uniform across the field; for example, one side is bordered by a busy highway, while another side is near a river and may receive additional moisture in the form of dew. To minimize the effect of such factors on the study, it is desired that each kind of fertilizer be applied, and each variety of corn be planted, in exactly one plot in any row and in any column of the field. Explain how to use orthogonal Latin squares to determine how the varieties of corn should be planted and the kinds of fertilizer should be applied.

3. Let n be an odd positive integer. For $0 \le i, j \le n - 1$, define n by n arrays R and S by

   $$r_{ij} = (i + j) \bmod n \quad \text{and} \quad s_{ij} = (2i + j) \bmod n$$

   (a) Find R and S in the case $n = 5$ and verify that they are orthogonal Latin squares of order 5.
   (b) Verify, in general, that R and S are Latin squares of order n.
   (c) Verify, in general, that R and S are orthogonal.

4. Let A be a Latin square of order 4 and let B be a Latin square of order m. We can define a Latin square C of order $n = 4m$ as follows. For $0 \le i, j \le 4m - 1$, let

   $$c_{ij} = a_{uv} + 4b_{xy}$$

   where $u = i$ div m, $v = j$ div m, $x = i \bmod m$, and $y = j \bmod m$. The Latin square C is called the **composition** of A with B; we write $C = A \circ B$.

   (a) Consider the case $m = 3$. Let A_1 be a Latin square of order 4 and let B_1 be a Latin square of order 3. Construct the Latin square $C_1 = A_1 \circ B_1$ of order 12.
   (b) Let A_2 be a Latin square of order 4 orthogonal to A_1 and let B_2 be a Latin square of order 3 orthogonal to B_1. Construct $C_2 = A_2 \circ B_2$ and verify that C_1 and C_2 are orthogonal Latin squares of order 12.

   (It might be worthwhile to write a computer program to generate the Latin squares C_1 and C_2 and to check that they are orthogonal.)

5. The following two problems were considered before Euler's work on Latin squares.

   (a) Arrange the 16 court cards (aces, kings, queens, jacks) of a standard deck into a square array of four rows and four columns so that each row and column contains exactly one card of every rank and of every suit.

(b) Here's a tougher problem: Arrange these 16 cards so that each row, each column, and each of the two diagonals contains exactly one card of each rank and suit.

⋆**6.** The idea of composition introduced in Exercise 4 can be generalized. Let A be a Latin square of order k and let B be a Latin square of order m. The **composition** of A with B is the Latin square $C = A \circ B$ of order $n = km$ defined as follows. For $0 \le i, j \le km - 1$, let

$$c_{ij} = a_{uv} + kb_{xy}$$

where $u = i$ div m, $v = j$ div m, $x = i$ mod m, and $y = j$ mod m.

(a) Show that C is, indeed, a Latin square.
(b) Let A_1 and A_2 be orthogonal Latin squares of order k and let B_1 and B_2 be orthogonal Latin squares of order m. Let $C_1 = A_1 \circ B_1$ and $C_2 = A_2 \circ B_2$. Show that C_1 and C_2 are orthogonal.
(c) It is known that there exist two orthogonal Latin squares of order 2^d for each $d \ge 2$. Use this fact and the result of Exercise 3 to prove Euler's result that there exist two orthogonal Latin squares of order n if n mod $4 = 0$.

1.2 ADDITION AND MULTIPLICATION PRINCIPLES

As mentioned in the last section, if S is the set of solutions to some combinatorial problem and S is finite, then we may be interested in enumerating the elements of S or in finding the cardinality of S. It often helps to express S as the union of two nonempty disjoint sets, that is, to express S as $S_1 \cup S_2$, where $S_1 \ne \{ \ \}$, $S_2 \ne \{ \ \}$, and $S_1 \cap S_2 = \{ \ \}$. Or, it may be that S can be expressed as the product of two sets, that is, $S = S_1 \times S_2$. Thus, given that S_1 and S_2 are disjoint finite sets and that $|S_1|$ and $|S_2|$ are known, how do we find $|S_1 \cup S_2|$? Similarly, given that S_1 and S_2 are finite sets, how is $|S_1 \times S_2|$ determined from $|S_1|$ and $|S_2|$? These questions are answered by the addition principle and the multiplication principle, which are introduced in this section.

We begin with the addition principle for two sets. Let S_1 have cardinality m and let S_2 have cardinality n, where m and n are positive integers and $S_1 \cap S_2 = \{ \ \}$; say, $S_1 = \{x_1, x_2, \dots, x_m\}$ and $S_2 = \{y_1, y_2, \dots, y_n\}$. Then $S_1 \cup S_2 = \{x_1, x_2, \dots, x_m, y_1, y_2, \dots, y_n\}$, and it is easy to see that $S_1 \cup S_2$ has cardinality $m + n$. This observation yields the following theorem:

THEOREM 1.2.1   If S_1 and S_2 are nonempty, disjoint finite sets, then

$$|S_1 \cup S_2| = |S_1| + |S_2| \qquad\qquad \square$$

Theorem 1.2.1 is obvious if $|S_2| = 1$. This case can be used to anchor a rigorous proof by induction on $|S_2|$ (see Exercise 12).

In Theorem 1.2.1, let $S = S_1 \cup S_2$. Then S_1 and S_2 are two nonempty disjoint subsets of S whose union is S. More generally, we may have n nonempty, pairwise-disjoint subsets $S_1, S_2, \ldots, S_n$ of a set S such that $S = S_1 \cup S_2 \cup \cdots \cup S_n$. We wish to generalize Theorem 1.2.1 to this case.

DEFINITION 1.2.1 Let S be a nonempty set and let n be a positive integer. Given that $S_1, S_2, \ldots, S_n$ are n nonempty, pairwise-disjoint subsets of S such that $S = S_1 \cup S_2 \cup \cdots \cup S_n$, then the collection $\{S_1, S_2, \ldots, S_n\}$ is called a **partition** of S, more particularly, $\{S_1, S_2, \ldots, S_n\}$ is called a **partition of S into n parts**. The number n of subsets in a partition is called the **degree** of the partition. ■

■ **Example 1.2.1** The collection $\{\{1, 2, 3\}, \{4, 5\}, \{6, 7\}\}$ is a partition of degree 3 of $\{1, 2, 3, 4, 5, 6, 7\}$. The collection $\{\{1\}, \{2, 3, 5, 7\}, \{4, 6\}\}$ is a different partition of degree 3 of $\{1, 2, 3, 4, 5, 6, 7\}$. Note that, because a partition of a set S is a set (of subsets of S), we know when two partitions of S are, in fact, the same. For instance, $\{\{0, 1, 2, 9\}, \{3, 4, 8\}, \{5, 6\}, \{7\}\}$ and $\{\{7\}, \{5, 6\}, \{3, 4, 8\}, \{0, 1, 2, 9\}\}$ are equal partitions of degree 4 of the set $\{0, 1, 2, 3, 4, 5, 6, 7, 8, 9\}$. Also note that, given a nonempty set S, there is only one partition of S with degree 1, namely, $\{S\}$. ■

THEOREM 1.2.2 (Addition Principle) Let S be a nonempty finite set, let $n \geq 2$ be an integer, and let $\{S_1, S_2, \ldots, S_n\}$ be a partition of S. Then

$$|S| = |S_1| + |S_2| + \cdots + |S_n|$$

Proof We proceed by induction on n. Let T be the set of those integers $n \geq 2$ for which the statement of the theorem is true. We have $2 \in T$ by Theorem 1.2.1.

Now let k be an arbitrary integer, $k \geq 2$, and assume that $k \in T$. More precisely, the induction hypothesis is that if $\{S_1', S_2', \ldots, S_k'\}$ is any partition of degree k of any nonempty finite set S', then

$$|S'| = |S_1'| + |S_2'| + \cdots + |S_k'|$$

To show that $k+1 \in T$, we let S be a nonempty finite set and let $\{S_1, S_2, \ldots, S_k, S_{k+1}\}$ be a partition of S. Then $\{S_1 \cup \cdots \cup S_k, S_{k+1}\}$ is a partition of S of degree 2. Hence, by using Theorem 1.2.1 and the induction hypothesis (with $S' = S_1 \cup S_2 \cup \cdots \cup S_k$), we have

$$|S| = |S_1 \cup S_2 \cup \cdots \cup S_k| + |S_{k+1}|$$
$$= |S_1| + |S_2| + \cdots + |S_k| + |S_{k+1}|$$

which shows that $k + 1 \in T$. Therefore, $T = \{2, 3, 4, \ldots\}$. □

■ **Example 1.2.2** The set S of 52 cards in a standard deck may be partitioned as $\{S_1, S_2, S_3, S_4\}$, where S_1 is the set of spades, S_2 is the set of hearts, S_3 is the set of diamonds, and S_4 is the set of clubs. Note that $|S_1| = |S_2| = |S_3| = |S_4| = 13$, and $4 \cdot 13 = 52$. Another way to partition S is $\{R_1, R_2, R_3\}$, where R_1 is the set of aces, R_2 is the set of face cards (jacks, queens, kings), and $R_3 = S - (R_1 \cup R_2)$, that is, R_3 is the set of cards that are neither aces nor face cards (twos, threes, ... , nines, tens). Note that $|R_1| = 4$, $|R_2| = 12$, $|R_3| = 36$, and $4 + 12 + 36 = 52 = |S|$. ∎

■ **Example 1.2.3** The addition principle allows us to use a "divide and conquer" approach to solving certain counting problems. For example, consider baseball's World Series, in which the American League champion and the National League champion play a series of games, the winner being the first team to win four games. Let us denote the outcome of the series as a k-tuple, $4 \leq k \leq 7$; for instance, (N, N, A, N, N) (or NNANN) indicates a series that the National League team wins in five games, with the American League team winning the third game, and (A, A, N, N, A, N, A) (or AANNANA) indicates a series that the American League team wins in seven games, with the National League team winning the third, fourth, and sixth games. How many outcomes are there for such a series?

Let S be the set of outcomes. First, we partition S as $\{S_1, S_2\}$, where S_1 is the set of outcomes won by the American League team and S_2 is the set of outcomes won by the National League team. By appealing to symmetry, we can certainly say that $|S_1| = |S_2|$, so that $|S| = 2|S_1|$. Now the problem will be solved if we can find $|S_1|$. Let's partition S_1 as $\{A_4, A_5, A_6, A_7\}$, where A_i is the set of outcomes won by the American League team in i games, $4 \leq i \leq 7$. Then $A_4 = \{AAAA\}$ and $A_5 = \{NAAAA, ANAAA, AANAA, AAANA\}$, so that $|A_4| = 1$ and $|A_5| = 4$. With a bit of work, you can show that $|A_6| = 10$ and $|A_7| = 20$. (A method for computing $|A_i|$ is developed in the next section.) Thus, we have

$$|S| = 2|S_1| = 2(|A_4| + |A_5| + |A_6| + |A_7|) = 2(1 + 4 + 10 + 20) = 70$$ ∎

An important instance of the addition principle is the following: Let S be a finite set such that $|S|$ is known and let T be a nonempty subset of S such that $|T|$ is to be determined. Often it may be easier to determine $|S - T|$ and then use the addition principle to compute $|T|$. This is done by observing that $\{T, S - T\}$ is a partition of S, so that

$$|T| = |S| - |S - T|$$

■ **Example 1.2.4** Refer to Example 1.2.3 and find the number of outcomes for the World Series in which the American League team wins at least two games.

Solution
Let S be the set of outcomes for the World Series and let T be the set of outcomes in which the American League team wins at least two games. Example 1.2.3 shows that $|S| = 70$; we are asked to find $|T|$. Note that $S - T$ is the set of outcomes such that the American League team wins at most one game; that is, $S - T$ consists of the one outcome in which the National League team sweeps the series in four games, plus the

four outcomes won by the National League team in five games. Thus, $|S - T| = 5$, and so

$$|T| = |S| - |S - T| = 70 - 5 = 65$$ ∎

Next we turn to the multiplication principle, beginning with the case of two finite sets S_1 and S_2. As an example, let $S_1 = \{0, 1\}$ and $S_2 = \{0, 1, 2\}$. Then

$$S_1 \times S_2 = \{(0,0), (0,1), (0,2), (1,0), (1,1), (1,2)\}$$

Note that

$$|S_1 \times S_2| = 6 = 2 \cdot 3 = |S_1| \cdot |S_2|$$

This example suggests the following result:

THEOREM 1.2.3   For finite sets S_1 and S_2,

$$|S_1 \times S_2| = |S_1| \cdot |S_2|$$

Proof   If S_1 or S_2 is empty, then the result of the theorem is immediate. So we may assume that $|S_1| = m > 0$ and $|S_2| = n > 0$, say, $S_1 = \{x_1, x_2, \ldots, x_m\}$ and $S_2 = \{y_1, y_2, \ldots, y_n\}$. We proceed by induction on n, letting T be the set of those positive integers n for which the statement of the theorem is true.

If $n = 1$, then $S_2 = \{y_1\}$ and $S_1 \times S_2 = \{(x_1, y_1), (x_2, y_1), \ldots, (x_m, y_1)\}$. Thus, $|S_1 \times S_2| = m = |S_1| \cdot |S_2|$, so $1 \in T$.

Now we let k be an arbitrary positive integer and assume that $k \in T$. More precisely, the induction hypothesis is that if S_1' is any finite set and S_2' is any set of cardinality k, then $|S_1' \times S_2'| = |S_1'| \cdot |S_2'|$.

To show that $k + 1 \in T$, we let S_1 be any finite set and let S_2 be a set of cardinality $k + 1$, say, $S_2 = \{y_1, y_2, \ldots, y_k, y_{k+1}\}$. We partition S_2 as $\{S_2', \{y_{k+1}\}\}$, where $S_2' = S_2 - \{y_{k+1}\} = \{y_1, y_2, \ldots, y_k\}$. Note that $|S_2'| = k$ and that

$$S_1 \times S_2 = S_1 \times (S_2' \cup \{y_{k+1}\}) = (S_1 \times S_2') \cup (S_1 \times \{y_{k+1}\})$$

(using the result of Theorem 0.1.3, part 1). In fact, $\{S_1 \times S_2', S_1 \times \{y_{k+1}\}\}$ is a partition of $S_1 \times S_2$. Thus, by using the addition principle and the induction hypothesis, we have

$$
\begin{aligned}
|S_1 \times S_2| &= |(S_1 \times S_2') \cup (S_1 \times \{y_{k+1}\})| \\
&= |S_1 \times S_2'| + |S_1 \times \{y_{k+1}\}| \\
&= |S_1| \cdot |S_2'| + |S_1| \cdot |\{y_{k+1}\}| \\
&= |S_1| \cdot (|S_2'| + |\{y_{k+1}\}|) \\
&= |S_1| \cdot |S_2' \cup \{y_{k+1}\}| \\
&= |S_1| \cdot |S_2|
\end{aligned}
$$

So $k + 1 \in T$. Therefore, $T = \mathbb{N}$.   □

Theorem 1.2.3 may be generalized to the product of $n \geq 2$ finite sets, as in Theorem 1.2.4. We can prove this result by induction on n in a manner similar to the proof of Theorem 1.2.2; the proof also uses the fact that, for $n \geq 3$,

$$|S_1 \times S_2 \times \cdots \times S_n| = |(S_1 \times S_2 \times \cdots \times S_{n-1}) \times S_n|$$

The complete proof is left for you to develop in Exercise 14.

THEOREM 1.2.4 (Multiplication Principle)   For $n \geq 2$ finite sets $S_1 S_2, \ldots, S_n$, if $S = S_1 \times S_2 \times \cdots \times S_n$, then

$$|S| = |S_1| \cdot |S_2| \cdot \cdots \cdot |S_n| \qquad \qquad \square$$

COROLLARY 1.2.5   For a finite set T, if $S = T \times T \times \cdots \times T$ is the n-fold product of T with itself, then

$$|S| = |T|^n \qquad \qquad \square$$

The statement of the multiplication principle in the preceding results no doubt seems quite abstract. An alternate formulation applies to a variety of combinatorial problems in which a sequence of elements is chosen.

Multiplication Principle. Suppose that a sequence $(x_1, x_2, \ldots, x_n)$ of $n \geq 2$ elements is to be chosen. If x_1 may be chosen in k_1 ways and, for each choice of x_1, the element x_2 may be chosen in k_2 ways, and so on, until finally, for each choice of $(x_1, x_2, \ldots, x_{n-1})$, the element x_n may be chosen in k_n ways, then the sequence $(x_1, x_2, \ldots, x_n)$ may be chosen in $k_1 k_2 \cdots k_n$ ways.

■ *Example 1.2.5*   To play a certain state lottery, a person must choose a sequence of four digits, for instance, 3738 or 0246.

(a) How many ways are there to play?

(b) How many ways are there to play if no digit is repeated?

(c) How many ways are there to play if the digit 0 is used at least once?

(d) How many ways are there to play if the last digit cannot be 0 nor can two 0s be consecutive?

Solution

Let $D = \{0, 1, \ldots, 9\}$ be the set of digits. In playing the lottery, a person chooses a sequence (x_1, x_2, x_3, x_4) with each $x_i \in D$.

(a) The number of ways to play the lottery is

$$|D \times D \times D \times D| = |D|^4 = 10^4 = 10,000$$

by Corollary 1.2.5.

(b) Let us use the alternate formulation of the multiplication principle to find the number of ways to play the lottery if no digit is repeated. There are 10 choices for x_1. Once x_1 has been chosen, then x_2 must be selected from the set $D - \{x_1\}$. So there are 9 choices for x_2. Then x_3 must be chosen from the set $D - \{x_1, x_2\}$, so there are 8 choices for x_3. Finally, x_4 must be chosen from $D - \{x_1, x_2, x_3\}$, so there are 7 choices for x_4. Thus, altogether, there are $10 \cdot 9 \cdot 8 \cdot 7 = 5040$ ways to play.

(c) Let $S = D \times D \times D \times D$ be the set of ways to play the lottery and let T be the set of ways to play using the digit 0 at least once. Here we use the technique of finding $|S - T|$. Note that $S - T$ is the set of ways to play the lottery without using the digit 0, so that

$$S - T = (D - \{0\}) \times (D - \{0\}) \times (D - \{0\}) \times (D - \{0\})$$

Thus,

$$|S - T| = |D - \{0\}|^4 = 9^4$$

so that

$$|T| = |S| - |S - T| = 10^4 - 9^4 = 3439$$

(d) Here the restriction is to disallow sequences such as 9870 that end with a 0, and also to disallow sequences such as 5002 that have two consecutive 0s. We let X be the set of ways to play the lottery under this restriction. Suppose we try to find $|X|$ using the multiplication principle. There are 10 ways to choose x_1. However, having chosen x_1, the number of choices for x_2 depends on whether or not $x_1 = 0$; if $x_1 \neq 0$, then there are 10 choices for x_2, but if $x_1 = 0$, then there are only 9 choices for x_2 since x_1 and x_2 cannot both be 0. When a situation of this kind arises, the solution can often be found by partitioning X and employing the addition principle. Here we partition X as $\{X_1, X_2, X_3\}$, where

$$X_1 = \{(x_1, x_2, x_3, x_4) \in X \mid x_1 = 0\}$$
$$X_2 = \{(x_1, x_2, x_3, x_4) \in X \mid x_1 \neq 0 \text{ and } x_2 = 0\}$$
$$X_3 = \{(x_1, x_2, x_3, x_4) \in X \mid x_1 \neq 0 \text{ and } x_2 \neq 0\}$$

By using the multiplication principle, you can check that

$$|X_1| = 9^2 \cdot 10 \qquad |X_2| = 9^3 \qquad \text{and} \qquad |X_3| = 9^3 \cdot 10$$

(Let's check $|X_1|$, for instance. Since $x_1 = 0$, we have $x_2 \in D - \{0\}$, $x_3 \in D$, and $x_4 \in D - \{0\}$. Thus, there are 9 choices for x_2, 10 choices for x_3, and 9 choices for x_4. So $|X_1| = 9^2 \cdot 10$.) Now, by the addition principle,

$$|X| = |X_1| + |X_2| + |X_3| = 9^2(10 + 9 + 90) = 8829$$ ∎

■ *Example 1.2.6* Find the number of positive divisors of $n = 21,168 = 2^4 \cdot 3^3 \cdot 7^2$.

Solution

The key observation here is that k is a positive divisor of n if and only if

$$k = 2^r \cdot 3^s \cdot 7^t$$

with integers r, s, and t satisfying $0 \leq r \leq 4$, $0 \leq s \leq 3$, and $0 \leq t \leq 2$. Thus, we choose a positive divisor of n by choosing values for r, s, and t. For example, if we choose $r = 1$, $s = 0$, and $t = 2$, we obtain $k = 98$, whereas choosing $r = 4$, $s = 3$, and $t = 0$ results in $k = 432$. Hence, the number of positive divisors of $21,168$ is the number of sequences (r, s, t) with $r, s, t \in \mathbb{Z}$ and $0 \leq r \leq 4$, $0 \leq s \leq 3$, and $0 \leq t \leq 2$. This, by the multiplication principle, is $5 \cdot 4 \cdot 3 = 60$. ∎

A convenient way to illustrate the multiplication principle is with the aid of a picture called a **tree diagram**. Consider the case where $X = \{x_1, x_2, x_3\}$ and $Y = \{y_1, y_2, y_3, y_4\}$. The tree diagram for $X \times Y$ is shown in Figure 1.2.1. Notice that there are three arcs leaving the root of the tree; these correspond to the elements of X and are labeled as such. Leaving each vertex at level 1 are four arcs corresponding to and labeled with the elements of Y. To obtain an element of $X \times Y$, we follow a path from the root, first to a vertex at level 1, and then to a leaf of the tree; if the first arc of this path is labeled x_i and the second arc is labeled y_j, then the associated element of $X \times Y$ is (x_i, y_j) and the leaf is given this label, as shown in the figure.

More generally, the arcs of a tree diagram may be labeled with sets. (The set labeling an arc might represent, for example, a set of choices that are equivalent, in some sense, at some stage in forming a combinatorial object.) If an arc is labeled with a set X of cardinality k, then we think of this arc as actually representing k multiple arcs, one corresponding to each element of X. In this case, if a path from the root to a leaf of the tree has arcs labeled with the sets $X_1, X_2, \ldots, X_n$, respectively, then the leaf is labeled with $X_1 \times X_2 \times \cdots \times X_n$. Alternately, we may label the leaf with $|X_1 \times X_2 \times \cdots \times X_n|$, which represents the total number of paths from the root to that leaf, in the sense that the arc labeled X_i really represents $|X_i|$ multiple arcs. The following example should help clarify these ideas.

■ *Example 1.2.7* Consider the state lottery described in Example 1.2.5. A tree diagram may be used to count the number of ways to play the lottery subject to the restriction of part (d), namely, that the last digit cannot be 0 nor can two 0s be consecutive. This tree diagram is shown in Figure 1.2.2; the rooted tree shown is $T = (V, A)$, where $V = \{a, b, c, d, e, f, u, v, w, x, y, z\}$ and $A = \{(a, b), (a, c), (b, d), (b, e), (c, f), (d, u), (e, v), (f, w), (u, x), (v, y), (w, z)\}$. We let $D = \{0, 1, \ldots, 9\}$ and let X be the set of ways to play the lottery subject to the given restriction.

Let $(x_1, x_2, x_3, x_4) \in X$. As noted in Example 1.2.5, the set of choices for x_2 depends on whether $x_1 \neq 0$ or $x_1 = 0$. Thus, our tree diagram has two arcs leaving the root a; arc (a, b) is labeled with the set $D - \{0\}$ and represents the choice $x_1 \neq 0$, whereas arc (a, c) is labeled with $\{0\}$, representing the choice $x_1 = 0$.

Next we focus our attention on vertex b. Since $x_1 \neq 0$ in this case, it is possible for x_2 to be 0, and the set of choices for x_3 depends on whether $x_2 \neq 0$ or $x_2 = 0$.

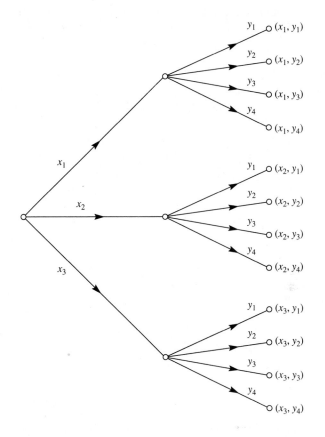

Figure 1.2.1 A tree diagram for $\{x_1, x_2, x_3\} \times \{y_1, y_2, y_3, y_4\}$.

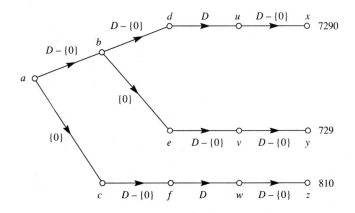

Figure 1.2.2 Tree diagram for Example 1.2.7.

Hence, the tree diagram has two arcs leaving vertex b; arc (b,d) is labeled $D - \{0\}$ and arc (b,e) is labeled $\{0\}$. Let us focus our attention next on vertex c. In this case, $x_1 = 0$, so that x_2 cannot be 0. So there is just one arc (c,f) leaving vertex c and this arc is labeled $D - \{0\}$.

At vertex d we have $x_2 \neq 0$, so that x_3 is allowed to be 0 but x_4 is not. Thus, from d there is a path of length two to the leaf x; arc (d,u) is labeled D and arc (u,x) is labeled $D - \{0\}$. The leaf x is labeled 7290; this is the cardinality of the set $(D - \{0\}) \times (D - \{0\}) \times D \times (D - \{0\})$ and is the number of sequences $(x_1, x_2, x_3, x_4) \in X$ with $x_1 \neq 0$ and $x_2 \neq 0$. Notice that $7290 = 9 \cdot 9 \cdot 10 \cdot 9$ is the product of the cardinalities of the sets labeling the arcs along the path from the root a to the leaf x.

At vertex e we have $x_2 = 0$, so that neither x_3 nor x_4 is allowed to be 0. This is the reason for the path of length two from e to the leaf y, with both arcs (e,v) and (v,y) labeled $D - \{0\}$. The label $729 = 9 \cdot 1 \cdot 9 \cdot 9$ on the leaf y is $|(D - \{0\}) \times \{0\} \times (D - \{0\}) \times (D - \{0\})|$ and is the number of sequences (x_1, x_2, x_3, x_4) in X with $x_1 \neq 0$ and $x_2 = 0$.

The analysis at vertex f is done in a similar fashion.

The cardinality of X is the total number of paths from the root a to the leaves of the tree. By our analysis, this is just the sum of the labels on the leaves, so that we obtain

$$|X| = 7290 + 729 + 810 = 8829 \qquad \blacksquare$$

EXERCISE SET 1.2

1. Robert, a mathematician, has seven shirts and ten ties. If he is of the opinion that any shirt goes with any tie, how many shirt-tie combinations does he have?

2. A professor has five books on algebra, six books on geometry, and seven books on number theory. How many ways are there for a student to choose two books not both on the same subject?

3. A Social Security number is a sequence of nine digits, for example, 080-55-1617.

   (a) How many possible Social Security numbers are there?

   How many possible Social Security numbers are there satisfying the following restrictions?

   (b) The digit 0 is not used.
   (c) The sequence neither begins nor ends with the digit 0.
   (d) No digit is repeated (used more than once).

4. How many different collections of soda pop cans can be formed from three (identical) Pepsi-Cola cans, three (identical) Coca-Cola cans, and six (identical) Seven-Up cans?

5. An identifier in the programming language Ada consists of letters, digits, and underscore characters. (Ada uses the underscore character "_" as a separation

there are n choices for each element. Hence, the following result follows from multiplication principle:

THEOREM 1.3.1 For positive integers m and n, let A be a set of cardinality m and let B be a set of cardinality n. Then the number of functions from A to B is n^m. □

Some books use the notation B^A to denote the set of functions from the set A to set B. Theorem 1.3.1 can then be stated as follows: For finite, nonempty sets A B,

$$|B^A| = |B|^{|A|}$$

Next, let us restrict the set of functions from A to B. Recall that $f : A \to B$ is **-to-one** provided the condition

$$a_1 \neq a_2 \to f(a_1) \neq f(a_2)$$

s for all $a_1, a_2 \in A$. For $A = \{1, 2, \ldots, m\}$, this condition is that the m images $, f(2), \ldots, f(m)$ are distinct or, in other words, that

$$|\text{im } f| = |\{f(1), f(2), \ldots, f(m)\}| = m$$

we ask, For A of cardinality m and B of cardinality n, how many one-to-one tions are there from A to B?

Again, we let $A = \{1, 2, \ldots, m\}$ and $B = \{1, 2, \ldots, n\}$. As before, a function $A \to B$ is determined by its sequence of images $(f(1), f(2), \ldots, f(m))$. If f is -to-one, then $(f(1), f(2), \ldots, f(m))$ is a sequence of m distinct elements from B.

DEFINITION 1.3.1 For a nonempty set S and a positive integer r, an **r-permutation** of S is an r-tuple $(x_1, x_2, \ldots, x_r)$, where $x_1, x_2, \ldots, x_r$ are distinct elements of S. If S is finite and $|S| = n \geq r$, then the number of r-permutations of S is denoted by $P(n, r)$ or $(n)_r$; also, an n-permutation of S is simply called a **permutation** of S. ■

As indicated by the definition, we can form an r-permutation of a nonempty set S naking an ordered selection of r distinct elements from S. For instance, suppose $\{1, 2, 3, 4\}$ and $r = 3$, so that we seek the 3-permutations (x_1, x_2, x_3) of S. If we ct, in order, $x_1 = 2$, $x_2 = 1$, and $x_3 = 4$, then we obtain the 3-permutation $(2, 1, 4)$. many 3-permutations of $\{1, 2, 3, 4\}$ are there? Well, there are four choices for irst element x_1, then three choices for x_2, and then two choices for x_3. By the iplication principle, there are $4 \cdot 3 \cdot 2 = 24$ choices in all; that is, the number of mutations of $\{1, 2, 3, 4\}$ is 24:

(1,2,3),	(1,3,2),	(2,1,3),	(2,3,1),	(3,1,2),	(3,2,1)
(1,2,4),	(1,4,2),	(2,1,4),	(2,4,1),	(4,1,2),	(4,2,1)
(1,3,4),	(1,4,3),	(3,1,4),	(3,4,1),	(4,1,3),	(4,3,1)
(2,3,4),	(2,4,3),	(3,2,4),	(3,4,2),	(4,2,3),	(4,3,2)

character, to make identifiers such as UNIT_COST easier to read.) An identifier must begin with a letter, may not end with an underscore character, and may not contain two consecutive underscore characters. (Also, Ada makes no distinction between uppercase and lowercase letters.)

(a) How many three-character Ada identifiers are there?

(b) How many four-character Ada identifiers are there?

(c) How many five-character Ada identifiers contain at most one underscore character?

(d) How many five-character Ada identifiers contain exactly two underscore characters?

(e) How many five-character Ada identifiers are there?

6. How many different five-digit integers can be formed using the five digits 1, 2, 2, 2, 3?

7. Consider the experiment of choosing, at random, a sequence of 4 cards from a standard deck of 52 playing cards. The cards are chosen one at a time *with replacement*, meaning that after a card is chosen it is replaced and the deck is reshuffled. We are interested in the sequence of cards that is obtained, for example (two of clubs, ace of hearts, king of spades, two of clubs).

(a) Determine the number of sequences.

Determine the number of sequences satisfying the following restrictions:

(b) None of the cards are spades.

(c) All 4 cards are spades.

(d) All 4 cards are the same suit.

(e) The first card is a king and the third card is not an ace.

(f) At least one of the cards is a spade.

8. How many five-letter strings can be formed using the letters a, b, c, d (with repeated letters allowed)? How many of these do not contain the substring "bad"?

9. A box contains three balls colored red, blue, and green. An experiment consists of choosing at random a sequence of three balls, one at a time, with replacement. An outcome for this experiment is written as an ordered triple, for example, (B, R, G) denotes that the first ball chosen is blue, the second is red, and the third is green.

(a) How many outcomes are there for this experiment?

(b) Draw a tree diagram that helps answer the question, How many outcomes for the experiment have the property that no two balls chosen consecutively have the same color?

(c) Draw a tree diagram that helps answer the question, How many outcomes for the experiment have the property that at least two of the balls chosen are blue?

10. Refer to Example 1.2.3 and use a tree diagram to help determine the number of outcomes in each of the following cases:

(a) The National League team wins the series in four games.

(b) The American League team wins the series in five games.

(c) The National League team wins the series in six games.

(d) The American League team wins the series in seven games.

11. Three different mathematics final examinations and two different computer science final examinations are to be scheduled during a five-day period. Suppose that each exam is to be scheduled from 1 P.M. to 4 P.M. on one of the days.

(a) In how many ways can these examinations be scheduled if there are no restrictions?

In how many ways can these examinations be scheduled under the following restrictions?

(b) No two exams can be scheduled for the same day.

(c) No two of the mathematics exams can be scheduled for the same day, nor can the two computer science exams be scheduled for the same day.

(d) Each mathematics exam must be the only exam scheduled for the day on which it is scheduled.

(Hint: Denote the mathematics exams A, B, and C, the computer science exams D and E, and the days 1, 2, 3, 4, 5; a complete schedule for the exams can be expressed as a 5-tuple, for example, $(1, 3, 5, 3, 2)$ indicates that exams A, B, C, D, and E are scheduled for days 1, 3, 5, 3, and 2, respectively.)

12. Prove Theorem 1.2.1 by using induction on $|S_2|$.

13. Consider the set $S = \{1000, 1001, 1002, \ldots, 9999\}$ of integers between 1000 and 9999, inclusive.

(a) Find $|S|$.

(b) How many elements of S are even?

(c) How many elements of S contain the digit 8 exactly once?

(d) How many elements of S contain the digit 8 (at least once)?

(e) How many elements of S are palindromes? (A positive integer is a **palindrome** if the integer remains the same when its digits are reversed, as, for example, in 1221 or 3773.)

14. Prove Theorem 1.2.4 by using induction on n.

15. In the game of Mastermind, one player, the "codemaker," selects a sequence of of four colors; this is the "code." The colors are chosen from the set {red, blue, green, white, black, yellow}. For example, (green, red, blue, red) and (white, black, yellow, red) are possible codes.

(a) How many codes are there?

(b) How many codes use four colors?

16. A box contains 12 distinct colored balls, numbered 1 through 12. Balls 1, 2, and 3 are red; balls 4, 5, 6, and 7 are blue; and balls 8, 9, 10, 11, and 12 are green. An experiment consists of choosing three balls at random from the box, one at a time, with replacement; the outcome is the sequence of three balls chosen: $(7, 4, 9)$ or $(3, 8, 3)$, for example.

(a) How many outcomes are there?

How many outcomes satisfy the following conditions?

(b) The first ball is red, the second is blue, and the third is green.

(c) The first and second balls are green and the third ball is blue.

(d) Exactly two balls are green.

(e) All three balls are red.

(f) All three balls are the same color.

(g) At least one of the three balls is red.

17. A certain make of lock has the integers 0 through 49 arranged in a for opening the lock consists of a sequence of three of these number for instance.

(a) How many codes are there?

(b) How many codes (x_1, x_2, x_3) have the property that $x_1 \neq x_2$ an

18. Generalizing Example 1.2.6, let $p_1, p_2, \ldots, p_k$ be distinct primes, let be positive integers, and let

$$n = p_1^{a_1} \cdot p_2^{a_2} \cdot \cdots \cdot p_k^{a_k}$$

Determine the number of positive divisors of n.

19. Let $X = \{1, 2, 3, 4, 5\}$ and let $Y = \{1, 2, 3, 4, 5, 6, 7\}$.

(a) Find the number of functions from X to Y. (Hint: Think o $f : X \to Y$ as the sequence $(f(1), f(2), f(3), f(4), f(5))$.)

(b) Find the number of one-to-one functions from X to Y.

(c) Find the number of functions from X onto Y.

(d) Find the number of functions f from X to Y with the property th for at least one $x \in X$.

(e) Find the number of functions f from X to Y with the propert odd or $f(5) = 5$. (Hint: Use the addition principle.)

(f) Find the number of functions f from X to Y with the property th {5}.

1.3   PERMUTATIONS AND COMBINATIONS

For positive integers m and n, let A be a set of cardinality m and let cardinality n. We ask, How many functions are there from A to B? In p let $A = \{1, 2, \ldots, m\}$, $B = \{1, 2, \ldots, n\}$, and $f : A \to B$. Then f is d its m images $f(1), f(2), \ldots, f(m)$. In fact, f can be specified by giving in a sequence: $(f(1), f(2), \ldots, f(m))$. For instance, if $m = 3$, $n = 4$, the sequence $(2, 4, 4)$, then the function $f : A \to B$ is defined by $f(1)$ 4, and $f(3) = 4$. Thus, the number of functions from A to B is the sequences $(f(1), f(2), \ldots, f(m))$ with each $f(k) \in B$. Such a sequence has

Note that each 3-permutation of $\{1, 2, 3, 4\}$ can be viewed as a function $f : \{1, 2, 3\} \to \{1, 2, 3, 4\}$ such that f is one-to-one. For instance, $(3, 4, 1)$ can be interpreted as the function f defined by $f(1) = 3$, $f(2) = 4$, and $f(3) = 1$. Conversely, any given one-to-one function $f : \{1, 2, 3\} \to \{1, 2, 3, 4\}$ determines a 3-permutation of $\{1, 2, 3, 4\}$, namely, $(f(1), f(2), f(3))$. Thus, the number of one-to-one functions from $\{1, 2, 3\}$ to $\{1, 2, 3, 4\}$ is $P(4, 3) = 24$.

In general, we let $A = \{1, 2, \ldots, m\}$ and $B = \{1, 2, \ldots, n\}$, where $1 \le m \le n$. Given an m-permutation of B, say, $(b_1, b_2, \ldots, b_m)$, this determines a one-to-one function $f : A \to B$, namely, $f(1) = b_1$, $f(2) = b_2, \ldots, f(m) = b_m$. Conversely, each one-to-one function $f : A \to B$ determines an m-permutation of B, namely, $(f(1), f(2), \ldots, f(m))$. Thus, we obtain the following result.

THEOREM 1.3.2 For integers m and n with $1 \le m \le n$, let A be a set of cardinality m and let B be a set of cardinality n. Then the number of one-to-one functions from A to B is $P(n, m)$. $\square$

In the context of this result, note that it is impossible to have one-to-one functions from an m-element set to an n-element set if $m > n$. This is the so-called *pigeonhole principle* discussed in the next section.

In light of the preceding discussion, it would be useful to have a formula for computing the number $P(n, r)$ of r-permutations of an n-element set. Such a formula is easily obtained by applying the multiplication principle.

THEOREM 1.3.3 For integers r and n with $1 \le r \le n$,

$$P(n, r) = \prod_{k=1}^{r} (n - k + 1) = n(n - 1) \cdots (n - r + 1)$$

Proof Let S be an n-element set. As noted, an r-permutation of S is a sequence $(x_1, x_2, \ldots, x_r)$ of distinct elements of S. There are n choices for x_1; thus, $P(n, 1) = n$ and the result of the theorem holds for the case $r = 1$. So we may assume $2 \le r \le n$. Since x_2 must be selected from the set $S - \{x_1\}$, there are $n - 1$ choices for x_2. In general, having already selected the first $k - 1$ elements $x_1, x_2, \ldots, x_{k-1}$, $k \ge 2$, the next element x_k must be selected from the set $S - \{x_1, x_2, \ldots, x_{k-1}\}$, so there are $n - (k - 1) = n - k + 1$ choices for x_k. The result of the theorem then follows from the multiplication principle. $\square$

It is easy to remember the formula for $P(n, r)$ given by Theorem 1.3.3, namely, $P(n, r)$ is a product of r factors, the first factor is n, and each factor thereafter is one less than the preceding factor.

Theorems 1.3.1 and 1.3.2 can be restated explicitly in the context of choosing a sequence $(x_1, x_2, \ldots, x_r)$ of r elements from an n-element set S. If repetition is allowed, then the number of ways to choose such a sequence is, by the multiplication principle, n^r. On the other hand, if repetition is not allowed, then necessarily $r \leq n$ and $(x_1, x_2, \ldots, x_r)$ is an r-permutation of S; thus, the number of ways to choose such a sequence is $P(n, r)$.

For a nonnegative integer n, recall that n factorial is written $n!$ and is defined recursively by

$$0! = 1$$
$$n! = n \cdot (n - 1)! \qquad n \geq 1$$

Alternately, for $n \geq 1$,

$$n! = \prod_{k=1}^{n} k$$

Thus, we can remark that

$$P(n, r) = \frac{n!}{(n - r)!}$$

We also see that $P(n, n) = n!$, yielding the following corollaries:

COROLLARY 1.3.4   For a positive integer n, the number of permutations of an n-element set is n factorial.   □

COROLLARY 1.3.5   For a positive integer n, let A and B be sets of cardinality n. Then the number of bijections from A to B is n factorial.   □

■ *Example 1.3.1*   To play the state lottery of Example 1.2.5, a person chooses a sequence (x_1, x_2, x_3, x_4) of four digits. As noted, there are 10^4 ways to play this lottery. If someone wishes to play subject to the restriction that no digit is repeated, then (x_1, x_2, x_3, x_4) is a 4-permutation of $D = \{0, 1, \ldots, 9\}$. Hence, there are

$$P(10, 4) = 10 \cdot 9 \cdot 8 \cdot 7 = 5040$$

ways to play subject to this restriction.   ■

■ *Example 1.3.2*   Consider forming five-letter code words using the letters a, b, c, d, e, f, g, and h, for example, fabda or aghcd. By the multiplication principle, there are 8^5 such code words.

(a) How many code words have no repeated letter?

(b) How many code words have no repeated letter and include the letter a?

(c) How many code words contain at least one repeated letter?

Solution

Think of a code word as a sequence $(x_1, x_2, x_3, x_4, x_5)$, where each $x_i \in \{$a, b, c, d, e, f, g, h$\}$.

(a) A code word with no repeated letter is a 5-permutation of $\{$a, b, c, d, e, f, g, h$\}$, so that the number of such words is

$$P(8,5) = 8 \cdot 7 \cdot 6 \cdot 5 \cdot 4 = 6720$$

(b) A code word with no repeated letter that includes the letter a can be formed in a unique way by first choosing a 4-permutation of $\{$b, c, d, e, f, g, h$\}$ and then choosing the position of the letter a relative to the other four letters. For instance, if we first choose the 4-permutation (e, b, c, d) and then choose to put the letter a in position four, then we obtain the code word ebcad. Since there are $P(7,4)$ 4-permutations of $\{$b, c, d, e, f, g, h$\}$ and five possible positions for the letter a, the number of code words that have no repeated letter and include the letter a is

$$P(7,4) \cdot 5 = 7 \cdot 6 \cdot 5 \cdot 4 \cdot 5 = 4200$$

(Exercise 33 suggests an alternate solution using the addition principle.)

(c) By using the addition principle, we can find the answer to this part by subtracting the number of code words with no repeated letters from the total number of code words. Thus, the number of code words containing at least one repeated letter is

$$8^5 - P(8,5) = 32,768 - 6720 = 26,048 \qquad \blacksquare$$

In choosing an r-permutation of a set S, the order in which the elements are chosen is important. In playing the lottery of Example 1.3.1, for instance, the sequences 7391 and 1379 are different, even though they both involve the same four elements 1, 3, 7, and 9. However, consider a lottery such as New York's Lotto 54. To play, someone chooses a *subset* of six numbers from the set $S = \{1, 2, \ldots, 54\}$, so that what matters is the set of six numbers selected and not the order in which they are selected. Suppose we wish to determine the number of ways to play Lotto 54, or the number of ways to play subject to the restriction that exactly four of the numbers chosen are prime. Since someone plays Lotto 54 by choosing a 6-element subset of $\{1, 2, \ldots, 54\}$, we wish to determine the number of 6-element subsets of a 54-element set. Similarly, finding the number of ways to play subject to the restriction that four of the numbers chosen are prime involves finding the number of 4-element subsets of a 16-element set, namely, $P = \{2, 3, 5, 7, 11, 13, 17, 19, 23, 29, 31, 37, 41, 43, 47, 53\}$, as well as the number of 2-element subsets of a 38-element set, namely, $S - P$.

Now we introduce some general terminology and notation. We let n be a positive integer, S be an n-element set, and r be an integer with $0 \le r \le n$. Then any r-element subset of S is called an **r-combination** of S. We denote the number of r-combinations of an n-element set by

$$C(n, r) \qquad \text{or} \qquad \binom{n}{r}$$

$C(n, r)$ is read "n choose r" and is sometimes referred to as the **number of combinations of n things taken r at a time**. Note that an r-combination of S is a subset of S and hence is not allowed to contain repeated elements.

To help us determine a formula for $C(n,r)$, we consider the following exercise: List the 3-combinations of $S = \{1, 2, 3, 4\}$, and next to each subset list the 3-permutations of S using the elements from that subset. Here goes:

$\{1, 2, 3\}$:   (1, 2, 3),   (1, 3, 2),   (2, 1, 3),   (2, 3, 1),   (3, 1, 2),   (3, 2, 1)
$\{1, 2, 4\}$:   (1, 2, 4),   (1, 4, 2),   (2, 1, 4),   (2, 4, 1),   (4, 1, 2),   (4, 2, 1)
$\{1, 3, 4\}$:   (1, 3, 4),   (1, 4, 3),   (3, 1, 4),   (3, 4, 1),   (4, 1, 3),   (4, 3, 1)
$\{2, 3, 4\}$:   (2, 3, 4),   (2, 4, 3),   (3, 2, 4),   (3, 4, 2),   (4, 2, 3),   (4, 3, 2)

Note that each of the four 3-combinations of S yields six 3-permutations of S. Thus, we see that

$$P(4, 3) = 24 = 6 \cdot 4 = P(3, 3) \cdot C(4, 3) = 3! \cdot C(4, 3)$$

This example suggests the following general relationship between $P(n, r)$ and $C(n, r)$:

THEOREM 1.3.6   For integers r and n with $1 \le r \le n$, we have

$$P(n, r) = r! \cdot C(n, r)$$

Proof   Consider any r-element subset of an n-element set S. Such a subset gives rise to r-factorial r-permutations of S, because the r elements in the subset can be ordered in $P(r, r) = r!$ ways. Moreover, every r-permutation of S is uniquely determined in this way. Thus, we obtain the relation

$$P(n, r) = r! \cdot C(n, r) \qquad\qquad\qquad \square$$

By Theorem 1.3.3,

$$P(n, r) = \frac{n!}{(n - r)!}$$

for $1 \le r \le n$. Applying Theorem 1.3.6 then yields

$$C(n, r) = \frac{P(n, r)}{r!} = \frac{n!}{r!(n - r)!}$$

for $1 \le r \le n$. This formula for $C(n, r)$ in terms of factorials is also seen to hold in the case $r = 0$.

COROLLARY 1.3.7   Let n be a positive integer and let r be an integer such that $0 \le r \le n$. Then

$$C(n, r) = \frac{n!}{r!(n - r)!} \qquad\qquad\qquad \square$$

character, to make identifiers such as UNIT_COST easier to read.) An identifier must begin with a letter, may not end with an underscore character, and may not contain two consecutive underscore characters. (Also, Ada makes no distinction between uppercase and lowercase letters.)

(a) How many three-character Ada identifiers are there?

(b) How many four-character Ada identifiers are there?

(c) How many five-character Ada identifiers contain at most one underscore character?

(d) How many five-character Ada identifiers contain exactly two underscore characters?

(e) How many five-character Ada identifiers are there?

6. How many different five-digit integers can be formed using the five digits 1, 2, 2, 2, 3?

7. Consider the experiment of choosing, at random, a sequence of 4 cards from a standard deck of 52 playing cards. The cards are chosen one at a time *with replacement*, meaning that after a card is chosen it is replaced and the deck is reshuffled. We are interested in the sequence of cards that is obtained, for example (two of clubs, ace of hearts, king of spades, two of clubs).

(a) Determine the number of sequences.

Determine the number of sequences satisfying the following restrictions:

(b) None of the cards are spades.

(c) All 4 cards are spades.

(d) All 4 cards are the same suit.

(e) The first card is a king and the third card is not an ace.

(f) At least one of the cards is a spade.

8. How many five-letter strings can be formed using the letters a, b, c, d (with repeated letters allowed)? How many of these do not contain the substring "bad"?

9. A box contains three balls colored red, blue, and green. An experiment consists of choosing at random a sequence of three balls, one at a time, with replacement. An outcome for this experiment is written as an ordered triple, for example, (B, R, G) denotes that the first ball chosen is blue, the second is red, and the third is green.

(a) How many outcomes are there for this experiment?

(b) Draw a tree diagram that helps answer the question, How many outcomes for the experiment have the property that no two balls chosen consecutively have the same color?

(c) Draw a tree diagram that helps answer the question, How many outcomes for the experiment have the property that at least two of the balls chosen are blue?

10. Refer to Example 1.2.3 and use a tree diagram to help determine the number of outcomes in each of the following cases:

(a) The National League team wins the series in four games.

(b) The American League team wins the series in five games.

(c) The National League team wins the series in six games.

(d) The American League team wins the series in seven games.

11. Three different mathematics final examinations and two different computer science final examinations are to be scheduled during a five-day period. Suppose that each exam is to be scheduled from 1 P.M. to 4 P.M. on one of the days.

(a) In how many ways can these examinations be scheduled if there are no restrictions?

In how many ways can these examinations be scheduled under the following restrictions?

(b) No two exams can be scheduled for the same day.

(c) No two of the mathematics exams can be scheduled for the same day, nor can the two computer science exams be scheduled for the same day.

(d) Each mathematics exam must be the only exam scheduled for the day on which it is scheduled.

(Hint: Denote the mathematics exams A, B, and C, the computer science exams D and E, and the days 1, 2, 3, 4, 5; a complete schedule for the exams can be expressed as a 5-tuple, for example, $(1, 3, 5, 3, 2)$ indicates that exams A, B, C, D, and E are scheduled for days 1, 3, 5, 3, and 2, respectively.)

12. Prove Theorem 1.2.1 by using induction on $|S_2|$.

13. Consider the set $S = \{1000, 1001, 1002, \ldots, 9999\}$ of integers between 1000 and 9999, inclusive.

(a) Find $|S|$.

(b) How many elements of S are even?

(c) How many elements of S contain the digit 8 exactly once?

(d) How many elements of S contain the digit 8 (at least once)?

(e) How many elements of S are palindromes? (A positive integer is a **palindrome** if the integer remains the same when its digits are reversed, as, for example, in 1221 or 3773.)

14. Prove Theorem 1.2.4 by using induction on n.

15. In the game of Mastermind, one player, the "codemaker," selects a sequence of of four colors; this is the "code." The colors are chosen from the set {red, blue, green, white, black, yellow}. For example, (green, red, blue, red) and (white, black, yellow, red) are possible codes.

(a) How many codes are there?

(b) How many codes use four colors?

16. A box contains 12 distinct colored balls, numbered 1 through 12. Balls 1, 2, and 3 are red; balls 4, 5, 6, and 7 are blue; and balls 8, 9, 10, 11, and 12 are green. An experiment consists of choosing three balls at random from the box, one at a time, with replacement; the outcome is the sequence of three balls chosen: $(7, 4, 9)$ or $(3, 8, 3)$, for example.

(a) How many outcomes are there?

How many outcomes satisfy the following conditions?

(b) The first ball is red, the second is blue, and the third is green.
(c) The first and second balls are green and the third ball is blue.
(d) Exactly two balls are green.
(e) All three balls are red.
(f) All three balls are the same color.
(g) At least one of the three balls is red.

17. A certain make of lock has the integers 0 through 49 arranged in a circle. A code for opening the lock consists of a sequence of three of these numbers—$(7, 32, 18)$, for instance.

    (a) How many codes are there?
    (b) How many codes (x_1, x_2, x_3) have the property that $x_1 \neq x_2$ and $x_2 \neq x_3$?

18. Generalizing Example 1.2.6, let $p_1, p_2, \ldots, p_k$ be distinct primes, let $a_1, a_2, \ldots, a_k$ be positive integers, and let

$$n = p_1^{a_1} \cdot p_2^{a_2} \cdot \cdots \cdot p_k^{a_k}$$

    Determine the number of positive divisors of n.

19. Let $X = \{1, 2, 3, 4, 5\}$ and let $Y = \{1, 2, 3, 4, 5, 6, 7\}$.

    (a) Find the number of functions from X to Y. (Hint: Think of the function $f : X \rightarrow Y$ as the sequence $(f(1), f(2), f(3), f(4), f(5))$.)
    (b) Find the number of one-to-one functions from X to Y.
    (c) Find the number of functions from X onto Y.
    (d) Find the number of functions f from X to Y with the property that $f(x)$ is odd for at least one $x \in X$.
    (e) Find the number of functions f from X to Y with the property that $f(1)$ is odd or $f(5) = 5$. (Hint: Use the addition principle.)
    (f) Find the number of functions f from X to Y with the property that $f^{-1}(\{5\}) = \{5\}$.

1.3 PERMUTATIONS AND COMBINATIONS

For positive integers m and n, let A be a set of cardinality m and let B be a set of cardinality n. We ask, How many functions are there from A to B? In particular, we let $A = \{1, 2, \ldots, m\}$, $B = \{1, 2, \ldots, n\}$, and $f : A \rightarrow B$. Then f is determined by its m images $f(1), f(2), \ldots, f(m)$. In fact, f can be specified by giving these images in a sequence: $(f(1), f(2), \ldots, f(m))$. For instance, if $m = 3$, $n = 4$, and we give the sequence $(2, 4, 4)$, then the function $f : A \rightarrow B$ is defined by $f(1) = 2$, $f(2) = 4$, and $f(3) = 4$. Thus, the number of functions from A to B is the number of sequences $(f(1), f(2), \ldots, f(m))$ with each $f(k) \in B$. Such a sequence has m elements and there are n choices for each element. Hence, the following result follows from the multiplication principle:

THEOREM 1.3.1   For positive integers m and n, let A be a set of cardinality m and let B be a set of cardinality n. Then the number of functions from A to B is n^m.   □

Some books use the notation B^A to denote the set of functions from the set A to the set B. Theorem 1.3.1 can then be stated as follows: For finite, nonempty sets A and B,

$$|B^A| = |B|^{|A|}$$

Next, let us restrict the set of functions from A to B. Recall that $f : A \rightarrow B$ is **one-to-one** provided the condition

$$a_1 \neq a_2 \rightarrow f(a_1) \neq f(a_2)$$

holds for all $a_1, a_2 \in A$. For $A = \{1, 2, \ldots, m\}$, this condition is that the m images $f(1), f(2), \ldots, f(m)$ are distinct or, in other words, that

$$|\text{im } f| = |\{f(1), f(2), \ldots, f(m)\}| = m$$

So we ask, For A of cardinality m and B of cardinality n, how many one-to-one functions are there from A to B?

Again, we let $A = \{1, 2, \ldots, m\}$ and $B = \{1, 2, \ldots, n\}$. As before, a function $f : A \rightarrow B$ is determined by its sequence of images $(f(1), f(2), \ldots, f(m))$. If f is one-to-one, then $(f(1), f(2), \ldots, f(m))$ is a sequence of m distinct elements from B.

DEFINITION 1.3.1   For a nonempty set S and a positive integer r, an **r-permutation** of S is an r-tuple $(x_1, x_2, \ldots, x_r)$, where $x_1, x_2, \ldots, x_r$ are distinct elements of S. If S is finite and $|S| = n \geq r$, then the number of r-permutations of S is denoted by $P(n, r)$ or $(n)_r$; also, an n-permutation of S is simply called a **permutation** of S.   ■

As indicated by the defintion, we can form an r-permutation of a nonempty set S by making an ordered selection of r distinct elements from S. For instance, suppose $S = \{1, 2, 3, 4\}$ and $r = 3$, so that we seek the 3-permutations (x_1, x_2, x_3) of S. If we select, in order, $x_1 = 2$, $x_2 = 1$, and $x_3 = 4$, then we obtain the 3-permutation $(2, 1, 4)$. How many 3-permutations of $\{1, 2, 3, 4\}$ are there? Well, there are four choices for the first element x_1, then three choices for x_2, and then two choices for x_3. By the multiplication principle, there are $4 \cdot 3 \cdot 2 = 24$ choices in all; that is, the number of 3-permutations of $\{1, 2, 3, 4\}$ is 24:

(1,2,3),	(1,3,2),	(2,1,3),	(2,3,1),	(3,1,2),	(3,2,1)
(1,2,4),	(1,4,2),	(2,1,4),	(2,4,1),	(4,1,2),	(4,2,1)
(1,3,4),	(1,4,3),	(3,1,4),	(3,4,1),	(4,1,3),	(4,3,1)
(2,3,4),	(2,4,3),	(3,2,4),	(3,4,2),	(4,2,3),	(4,3,2)

Note that each 3-permutation of $\{1,2,3,4\}$ can be viewed as a function $f : \{1,2,3\} \rightarrow \{1,2,3,4\}$ such that f is one-to-one. For instance, $(3,4,1)$ can be interpreted as the function f defined by $f(1) = 3$, $f(2) = 4$, and $f(3) = 1$. Conversely, any given one-to-one function $f : \{1,2,3\} \rightarrow \{1,2,3,4\}$ determines a 3-permutation of $\{1,2,3,4\}$, namely, $(f(1),f(2),f(3))$. Thus, the number of one-to-one functions from $\{1,2,3\}$ to $\{1,2,3,4\}$ is $P(4,3) = 24$.

In general, we let $A = \{1,2,\ldots,m\}$ and $B = \{1,2,\ldots,n\}$, where $1 \le m \le n$. Given an m-permutation of B, say, $(b_1,b_2,\ldots,b_m)$, this determines a one-to-one function $f : A \rightarrow B$, namely, $f(1) = b_1$, $f(2) = b_2,\ldots,f(m) = b_m$. Conversely, each one-to-one function $f : A \rightarrow B$ determines an m-permutation of B, namely, $(f(1),f(2),\ldots,f(m))$. Thus, we obtain the following result.

THEOREM 1.3.2 For integers m and n with $1 \le m \le n$, let A be a set of cardinality m and let B be a set of cardinality n. Then the number of one-to-one functions from A to B is $P(n,m)$. $\square$

In the context of this result, note that it is impossible to have one-to-one functions from an m-element set to an n-element set if $m > n$. This is the so-called *pigeonhole principle* discussed in the next section.

In light of the preceding discussion, it would be useful to have a formula for computing the number $P(n,r)$ of r-permutations of an n-element set. Such a formula is easily obtained by applying the multiplication principle.

THEOREM 1.3.3 For integers r and n with $1 \le r \le n$,

$$P(n,r) = \prod_{k=1}^{r}(n - k + 1) = n(n - 1)\cdots(n - r + 1)$$

Proof Let S be an n-element set. As noted, an r-permutation of S is a sequence $(x_1,x_2,\ldots,x_r)$ of distinct elements of S. There are n choices for x_1; thus, $P(n,1) = n$ and the result of the theorem holds for the case $r = 1$. So we may assume $2 \le r \le n$. Since x_2 must be selected from the set $S - \{x_1\}$, there are $n - 1$ choices for x_2. In general, having already selected the first $k - 1$ elements $x_1,x_2,\ldots,x_{k-1}$, $k \ge 2$, the next element x_k must be selected from the set $S - \{x_1,x_2,\ldots,x_{k-1}\}$, so there are $n - (k - 1) = n - k + 1$ choices for x_k. The result of the theorem then follows from the multiplication principle. $\square$

It is easy to remember the formula for $P(n,r)$ given by Theorem 1.3.3, namely, $P(n,r)$ is a product of r factors, the first factor is n, and each factor thereafter is one less than the preceding factor.

Theorems 1.3.1 and 1.3.2 can be restated explicitly in the context of choosing a sequence $(x_1,x_2,\ldots,x_r)$ of r elements from an n-element set S. If repetition is allowed, then the number of ways to choose such a sequence is, by the multiplication principle, n^r. On the other hand, if repetition is not allowed, then necessarily $r \leq n$ and $(x_1, x_2, \ldots, x_r)$ is an r-permutation of S; thus, the number of ways to choose such a sequence is $P(n, r)$.

For a nonnegative integer n, recall that n factorial is written $n!$ and is defined recursively by

$$0! = 1$$
$$n! = n \cdot (n-1)! \qquad n \geq 1$$

Alternately, for $n \geq 1$,

$$n! = \prod_{k=1}^{n} k$$

Thus, we can remark that

$$P(n, r) = \frac{n!}{(n-r)!}$$

We also see that $P(n, n) = n!$, yielding the following corollaries:

COROLLARY 1.3.4   For a positive integer n, the number of permutations of an n-element set is n factorial.   □

COROLLARY 1.3.5   For a positive integer n, let A and B be sets of cardinality n. Then the number of bijections from A to B is n factorial.   □

■ *Example 1.3.1*   To play the state lottery of Example 1.2.5, a person chooses a sequence (x_1, x_2, x_3, x_4) of four digits. As noted, there are 10^4 ways to play this lottery. If someone wishes to play subject to the restriction that no digit is repeated, then (x_1, x_2, x_3, x_4) is a 4-permutation of $D = \{0, 1, \ldots, 9\}$. Hence, there are

$$P(10, 4) = 10 \cdot 9 \cdot 8 \cdot 7 = 5040$$

ways to play subject to this restriction.   ■

■ *Example 1.3.2*   Consider forming five-letter code words using the letters a, b, c, d, e, f, g, and h, for example, fabda or aghcd. By the multiplication principle, there are 8^5 such code words.

(a) How many code words have no repeated letter?

(b) How many code words have no repeated letter and include the letter a?

(c) How many code words contain at least one repeated letter?

Solution

Think of a code word as a sequence $(x_1, x_2, x_3, x_4, x_5)$, where each $x_i \in \{$a, b, c, d, e, f, g, h$\}$.

(a) A code word with no repeated letter is a 5-permutation of $\{$a, b, c, d, e, f, g, h$\}$, so that the number of such words is

$$P(8,5) = 8 \cdot 7 \cdot 6 \cdot 5 \cdot 4 = 6720$$

(b) A code word with no repeated letter that includes the letter a can be formed in a unique way by first choosing a 4-permutation of $\{$b, c, d, e, f, g, h$\}$ and then choosing the position of the letter a relative to the other four letters. For instance, if we first choose the 4-permutation (e, b, c, d) and then choose to put the letter a in position four, then we obtain the code word ebcad. Since there are $P(7,4)$ 4-permutations of $\{$b, c, d, e, f, g, h$\}$ and five possible positions for the letter a, the number of code words that have no repeated letter and include the letter a is

$$P(7,4) \cdot 5 = 7 \cdot 6 \cdot 5 \cdot 4 \cdot 5 = 4200$$

(Exercise 33 suggests an alternate solution using the addition principle.)

(c) By using the addition principle, we can find the answer to this part by subtracting the number of code words with no repeated letters from the total number of code words. Thus, the number of code words containing at least one repeated letter is

$$8^5 - P(8,5) = 32,768 - 6720 = 26,048 \qquad \blacksquare$$

In choosing an r-permutation of a set S, the order in which the elements are chosen is important. In playing the lottery of Example 1.3.1, for instance, the sequences 7391 and 1379 are different, even though they both involve the same four elements 1, 3, 7, and 9. However, consider a lottery such as New York's Lotto 54. To play, someone chooses a *subset* of six numbers from the set $S = \{1, 2, \ldots, 54\}$, so that what matters is the set of six numbers selected and not the order in which they are selected. Suppose we wish to determine the number of ways to play Lotto 54, or the number of ways to play subject to the restriction that exactly four of the numbers chosen are prime. Since someone plays Lotto 54 by choosing a 6-element subset of $\{1, 2, \ldots, 54\}$, we wish to determine the number of 6-element subsets of a 54-element set. Similarly, finding the number of ways to play subject to the restriction that four of the numbers chosen are prime involves finding the number of 4-element subsets of a 16-element set, namely, $P = \{2, 3, 5, 7, 11, 13, 17, 19, 23, 29, 31, 37, 41, 43, 47, 53\}$, as well as the number of 2-element subsets of a 38-element set, namely, $S - P$.

Now we introduce some general terminology and notation. We let n be a positive integer, S be an n-element set, and r be an integer with $0 \le r \le n$. Then any r-element subset of S is called an **r-combination** of S. We denote the number of r-combinations of an n-element set by

$$C(n,r) \qquad \text{or} \qquad \binom{n}{r}$$

$C(n,r)$ is read "n choose r" and is sometimes referred to as the **number of combinations of n things taken r at a time**. Note that an r-combination of S is a subset of S and hence is not allowed to contain repeated elements.

To help us determine a formula for $C(n, r)$, we consider the following exercise: List the 3-combinations of $S = \{1, 2, 3, 4\}$, and next to each subset list the 3-permutations of S using the elements from that subset. Here goes:

$\{1, 2, 3\}$:	(1, 2, 3),	(1, 3, 2),	(2, 1, 3),	(2, 3, 1),	(3, 1, 2),	(3, 2, 1)
$\{1, 2, 4\}$:	(1, 2, 4),	(1, 4, 2),	(2, 1, 4),	(2, 4, 1),	(4, 1, 2),	(4, 2, 1)
$\{1, 3, 4\}$:	(1, 3, 4),	(1, 4, 3),	(3, 1, 4),	(3, 4, 1),	(4, 1, 3),	(4, 3, 1)
$\{2, 3, 4\}$:	(2, 3, 4),	(2, 4, 3),	(3, 2, 4),	(3, 4, 2),	(4, 2, 3),	(4, 3, 2)

Note that each of the four 3-combinations of S yields six 3-permutations of S. Thus, we see that

$$P(4, 3) = 24 = 6 \cdot 4 = P(3, 3) \cdot C(4, 3) = 3! \cdot C(4, 3)$$

This example suggests the following general relationship between $P(n, r)$ and $C(n, r)$:

THEOREM 1.3.6 For integers r and n with $1 \le r \le n$, we have

$$P(n, r) = r! \cdot C(n, r)$$

Proof Consider any r-element subset of an n-element set S. Such a subset gives rise to r-factorial r-permutations of S, because the r elements in the subset can be ordered in $P(r, r) = r!$ ways. Moreover, every r-permutation of S is uniquely determined in this way. Thus, we obtain the relation

$$P(n, r) = r! \cdot C(n, r) \qquad \qquad \square$$

By Theorem 1.3.3,

$$P(n, r) = \frac{n!}{(n - r)!}$$

for $1 \le r \le n$. Applying Theorem 1.3.6 then yields

$$C(n, r) = \frac{P(n, r)}{r!} = \frac{n!}{r!(n - r)!}$$

for $1 \le r \le n$. This formula for $C(n, r)$ in terms of factorials is also seen to hold in the case $r = 0$.

COROLLARY 1.3.7 Let n be a positive integer and let r be an integer such that $0 \le r \le n$. Then

$$C(n, r) = \frac{n!}{r!(n - r)!} \qquad \qquad \square$$

The above formula for $C(n,r)$ has theoretical value but should not be used to compute $C(n,r)$. Instead, notice that

$$C(n,0) = C(n,n) = 1$$

since a set $S = \{x_1, x_2, \ldots, x_n\}$ has one 0-element subset (the empty set) and one n-element subset (itself), and

$$C(n,1) = C(n, n-1) = n$$

since S has n one-element subsets, namely, $\{x_1\}, \{x_2\}, \ldots, \{x_n\}$, and n subsets of cardinality $n-1$, namely, $S - \{x_1\}, S - \{x_2\}, \ldots, S - \{x_n\}$. For $2 \le r \le n - 2$, we use the formula

$$C(n,r) = \frac{P(n,r)}{r!} = \frac{n(n-1)\cdots(n-r+1)}{r(r-1)\cdots 1}$$

which is easy to remember since both the numerator and denominator are products containing r factors.

■ *Example 1.3.3*   The number of ways to play Lotto 54 is

$$C(54,6) = \frac{P(54,6)}{6!} = \frac{54 \cdot 53 \cdot 52 \cdot 51 \cdot 50 \cdot 49}{6 \cdot 5 \cdot 4 \cdot 3 \cdot 2 \cdot 1} = 25,827,165$$

Find the number of ways to play Lotto 54 subject to the following restrictions:

(a) 19 is one of the numbers chosen.

(b) Exactly four of the numbers chosen are prime.

(c) At least one of the numbers chosen is prime.

(d) The difference between the largest number chosen and smallest number chosen is 30.

Solution
Let $S = \{1, 2, \ldots, 54\}$ and let

$$P = \{2, 3, 5, 7, 11, 13, 17, 19, 23, 29, 31, 37, 41, 43, 47, 53\}$$

(a) To play subject to the restriction that 19 is one of the numbers chosen, we must choose five numbers from the set $S - \{19\}$. The number of ways to do this is

$$C(53,5) = \frac{P(53,5)}{5!} = \frac{53 \cdot 52 \cdot 51 \cdot 50 \cdot 49}{5 \cdot 4 \cdot 3 \cdot 2 \cdot 1} = 2,869,685$$

(b) To play subject to the restriction that four of the numbers chosen are prime, we could first choose four numbers from the set P and then choose two numbers from the set $S - P$. By the multiplication principle, the number of ways to do this is

$$C(16,4) \cdot C(38,2) = \frac{P(16,4)}{4!} \cdot \frac{P(38,2)}{2!} = \frac{16 \cdot 15 \cdot 14 \cdot 13}{4 \cdot 3 \cdot 2 \cdot 1} \cdot \frac{38 \cdot 37}{2 \cdot 1}$$

$$= 1,279,460$$

(c) Let X be the set of ways to play Lotto 54 and let Y be the set of ways to play subject to the restriction that at least one of the six numbers chosen is prime. Then $X - Y$ is the set of ways to play without choosing any prime numbers, that is, all six numbers must be chosen from $S - P$. Thus, $|X - Y| = C(38, 6)$. Then, using part (a) and the addition principle,

$$|Y| = |X| - |X - Y| = C(54, 6) - C(38, 6) = 23,066,484$$

(d) To play subject to the restriction that the difference between the largest number and the smallest number chosen is 30, we could first decide on the smallest number to choose; call it k. Then the largest number to choose is $k + 30$, and the other four numbers must be chosen from the 29-element set $\{k + 1, k + 2, \ldots, k + 29\}$. Since $1 \leq k$ and $k + 30 \leq 54$, we have $1 \leq k \leq 24$. Thus, there are 24 choices for k and $C(29, 4)$ ways to choose the four numbers other than k and $k + 30$. Hence, by the multiplication principle, the answer to this part is

$$24 \cdot C(29, 4) = 570,024 \qquad \blacksquare$$

■ ***Example 1.3.4***   Let $A = \{1, 2, 3, 4, 5\}$ and $B = \{1, 2, 3\}$. Find the number of functions from A onto B.

Solution

Let S be the set of functions from A onto B and let $f \in S$. Then im $f = B$; in fact, $\{f^{-1}(\{1\}), f^{-1}(\{2\}), f^{-1}(\{3\})\}$ is a partition of A of degree 3. Since $|A| = 5$, there are two possibilities for the cardinalities of the sets in this partition: (1) $|f^{-1}(\{x\})| = 3$ and $|f^{-1}(\{y\})| = 1 = |f^{-1}(\{z\})|$, or (2) $|f^{-1}(\{x\})| = 2 = |f^{-1}(\{y\})|$ and $|f^{-1}(\{z\})| = 1$, where $\{x, y, z\} = \{1, 2, 3\}$. In words, either one element of B has three preimages and the other two elements of B have one preimage each, or two elements of B have two preimages each and the other element of B has one preimage.

This suggests using the addition principle to find $|S|$; let us partition S as $\{S_1, S_2\}$, where

$$S_1 = \{f \in S \mid \text{some } x \text{ in } B \text{ has three preimages under } f\}$$

and

$$S_2 = \{f \in S \mid \text{some distinct } x \text{ and } y \text{ in } B \text{ have two preimages each under } f\}$$

To find $|S_1|$, we note that an $f \in S_1$ is uniquely determined by the following sequence of choices: first, choose $x \in B$; second, choose three preimages in A for x; third, complete f by choosing a bijection from $A - f^{-1}(\{x\})$ to $B - \{x\}$. For example, choosing $x = 2$, then choosing $f^{-1}(\{2\}) = \{1, 3, 5\}$, and then completing f by choosing $f(2) = 1$ and $f(4) = 3$ yields f from A onto B defined by $f(1) = 2, f(2) = 1, f(3) = 2$, $f(4) = 3$, and $f(5) = 2$. Now there are $C(3, 1) = 3$ choices for x, then $C(5, 3) = 10$ ways to choose $f^{-1}(\{x\})$ (since $f^{-1}(\{x\})$ is a 3-element subset of A), and, finally, $2! = 2$ bijections from $A - f^{-1}(\{x\})$ to $B - \{x\}$. Thus, by the multiplication principle,

$$|S_1| = 3 \cdot 10 \cdot 2 = 60$$

To find $|S_2|$, we note that $f \in S_2$ is uniquely determined by the following sequence of choices: first, choose a subset $\{x, y\}$ of B; second, choose two preimages in A for x; third, choose two preimages in $A - f^{-1}(\{x\})$ for y. By default, f maps the one element of $A - (f^{-1}(\{x\}) \cup f^{-1}(\{y\}))$ to the one element z of $B - \{x, y\}$. For example, choosing $\{x, y\} = \{1, 3\}$, then choosing $f^{-1}(\{1\}) = \{1, 5\}$, and then choosing $f^{-1}(\{3\}) = \{2, 4\}$ yields f from A onto B defined by $f(1) = 1$, $f(2) = 3$, $f(3) = 2$, $f(4) = 3$, and $f(5) = 1$. Now there are $C(3, 2) = 3$ ways to choose $\{x, y\}$, then $C(5, 2) = 10$ ways to choose $f^{-1}(\{x\})$, and, finally, $C(3, 2) = 3$ ways to choose $f^{-1}(\{y\})$. Thus, by the multiplication principle,

$$|S_2| = 3 \cdot 10 \cdot 3 = 90$$

Therefore, using the addition principle,

$$|S| = |S_1| + |S_2| = 60 + 90 = 150$$

(Note that the number of functions from A to B is $3^5 = 243$.) ∎

■ *Example 1.3.5*   The purpose of this example is to point out an error that is frequently made in counting problems. Consider a box containing seven distinct colored balls: three blue, two red, and two green. A subset of three of the balls is to be selected at random. How many such subsets contain at least two blue balls?

Solution   (Incorrect)
We begin with an incorrect analysis of this problem. We want a subset that contains at least two blue balls. We start by choosing two blue balls; since there are three blue balls to choose from, this may be done in $C(3, 2)$ ways. Now any one of the remaining five balls may be chosen as the third ball in the subset. Thus, by the multiplication principle, the number of subsets containing at least two blue balls is

$$C(3, 2) \cdot C(5, 1) = 3 \cdot 5 = 15$$

Solution
To see that the above answer is incorrect, we first solve the problem by brute force. The number of balls was purposely kept small in this problem to allow a listing of all the 3-combinations with at least two blue balls. We denote the set of balls by $\{b_1, b_2, b_3, r_1, r_2, g_1, g_2\}$, where b_1, b_2, and b_3 are the blue balls. Also, we let x denote any one of the four nonblue balls. Then the 3-combinations with at least two blue balls are: $\{b_1, b_2, b_3\}$, the four subsets of the form $\{b_1, b_2, x\}$, the four subsets of the

form $\{b_1, b_3, x\}$, and the four subsets of the form $\{b_2, b_3, x\}$. Hence, there are 13 such subsets, not 15.

We can, of course, obtain the correct answer by properly applying our counting techniques. If a 3-combination is to contain at least two blue balls, then either it contains exactly two blue balls or it contains exactly three blue balls. The number of 3-combinations with exactly two blue balls is $C(3, 2) \cdot C(4, 1) = 12$. And there is just one 3-combination with exactly three blue balls. Thus, by the addition principle, the number of 3-combinations that contain at least two blue balls is $12 + 1 = 13$. (You are encouraged to find for yourself the flaw in the incorrect analysis.) ■

Recall that the power set $\mathcal{P}(S)$ of a set S is the set of all subsets of S. For example,

$$\mathcal{P}(\{\ \}) = \{\{\ \}\}$$
$$\mathcal{P}(\{1\}) = \{\{\ \}, \{1\}\}$$
$$\mathcal{P}(\{1, 2\}) = \{\{\ \}, \{1\}, \{2\}, \{1, 2\}\}$$
$$\mathcal{P}(\{1, 2, 3\}) = \{\{\ \}, \{1\}, \{2\}, \{3\}, \{1, 2\}, \{1, 3\}, \{2, 3\}, \{1, 2, 3\}\}$$

and so on. Note that the result

$$|S| = n \rightarrow |\mathcal{P}(S)| = 2^n$$

holds in each of these cases. We next prove, in general, that a set with n elements has 2^n subsets. We give two proofs of this—first, by induction on n and, second, by a more "combinatorial" justification.

THEOREM 1.3.8   For a nonnegative integer n, if a set S has cardinality n, then its power set $\mathcal{P}(S)$ has cardinality 2^n.

Proof   We first prove this result by induction on n. Let T be the set of those nonnegative integers n for which the statement of the theorem is true. Then, as noted, $0 \in T$.

Now let k be an arbitrary nonnegative integer and assume that $k \in T$. More specifically, the induction hypothesis is that if S' is any set with cardinality k, then $\mathcal{P}(S')$ has cardinality 2^k.

To show that $k + 1 \in T$, let S be a set such that $|S| = k + 1$. Since S is nonempty, let $s \in S$. Then $\mathcal{P}(S)$ may be partitioned as $\{\mathcal{P}_1, \mathcal{P}_2\}$, where

$$\mathcal{P}_1 = \{X \in \mathcal{P}(S) \mid s \notin X\} \qquad \text{and} \qquad \mathcal{P}_2 = \{X \in \mathcal{P}(S) \mid s \in X\}$$

Note that $X \in \mathcal{P}_1$ if and only if $X \in \mathcal{P}(S - \{s\})$. Thus, by the induction hypothesis,

$$|\mathcal{P}_1| = |\mathcal{P}(S - \{s\})| = 2^k$$

In a similar fashion, note that $X \in \mathcal{P}_2$ if and only if $X - \{s\} \in \mathcal{P}_1$. Thus,

$$|\mathcal{P}_2| = |\mathcal{P}_1| = 2^k$$

Hence, by the addition principle,

$$|\mathcal{P}(S)| = |\mathcal{P}_1| + |\mathcal{P}_2| = 2^k + 2^k = 2^{k+1}$$

showing that $k + 1 \in S$. Therefore, $T = \{0, 1, 2, 3, \ldots\}$. $\square$

Alternate Proof   The result of the theorem is obvious for $n = 0$, so let S be a nonempty set with $n \geq 1$ elements. We are to prove that $|\mathcal{P}(S)| = 2^n$. The type of combinatorial proof that we wish to illustrate is what is called a "bijective proof." The strategy is to construct a bijection f from $\mathcal{P}(S)$ to a set $\mathcal{F}$ that is known to have 2^n elements. It then follows that $\mathcal{P}(S)$ has 2^n elements by Theorem 0.2.1, part 3. Before we can construct f we need a candidate for the set $\mathcal{F}$. What set associated with S is known to have cardinality 2^n? Well, by Theorem 1.3.1, the number of functions from S to the set $\{0, 1\}$ is 2^n, so we let $\mathcal{F}$ be the set of all functions from S to $\{0, 1\}$. To complete the proof, we need a bijection f from $\mathcal{P}(S)$ to $\mathcal{F}$. Note that the elements of the domain of f are subsets of S. Given a subset T of S, its image $f(T)$ is to be a function from S to $\{0, 1\}$; let's denote this function by f_T. Is there a natural choice for f_T?

To get a better feel for what is going on here, let's consider a specific example. Let $S = \{1, 2, 3, 4, 5\}$ and suppose that $T = \{1, 3, 5\}$. Given $x \in S$, note that either $x \notin T$ or $x \in T$. So there is a natural way to decide whether $f_T(x) = 0$ or $f_T(x) = 1$, namely, we let $f_T(x) = 1$ if and only if $x \in T$. Thus, f_T in this specific example is defined by

$$f_T(1) = 1 \qquad f_T(2) = 0 \qquad f_T(3) = 1 \qquad f_T(4) = 0 \qquad f_T(5) = 1$$

In the general case, we define $f : \mathcal{P}(S) \to \mathcal{F}$ as follows: $f(T) = f_T$, where $f_T : S \to \{0, 1\}$ is given by

$$f_T(x) = \begin{cases} 0 & \text{if } x \notin T \\ 1 & \text{if } x \in T \end{cases}$$

(The function f_T is called the **characteristic function** of T; see Exercise 34 in Exercise Set 0.2.) To show that f is one-to-one, we let T_1 and T_2 be different subsets of S; without loss of generality, assume that $x_0 \in T_1 - T_2$. Then $f_{T_1}(x_0) = 1$, whereas $f_{T_2}(x_0) = 0$; hence, $f_{T_1} \neq f_{T_2}$. To show that f is onto, let g be a function from S to $\{0, 1\}$; we must exhibit a subset T of S such that $f_T = g$. Let T consist of those $x \in S$ such that $g(x) = 1$; then

$$g(x) = 1 \leftrightarrow x \in T \leftrightarrow f_T(x) = 1$$

so that $f_T = g$. This proves that f is a bijection. $\square$

Again, we let n be a nonnegative integer and let S be a set with cardinality n. An alternate formula for the cardinality of $\mathcal{P}(S)$ can be obtained by using the addition principle as follows: Partition $\mathcal{P}(S)$ as $\{\mathcal{P}_0, \mathcal{P}_1, \ldots, \mathcal{P}_n\}$, where $\mathcal{P}_r$ is the set of r-combinations of S, $0 \leq r \leq n$. Then $|\mathcal{P}_r| = C(n, r)$, and we obtain, by the addition principle,

$$|\mathcal{P}(S)| = C(n, 0) + C(n, 1) + \cdots + C(n, n) = \sum_{r=0}^{n} C(n, r)$$

This result, together with Theorem 1.3.8, gives us the following interesting combinatorial identity:

COROLLARY 1.3.9   For any nonnegative integer n,

$$\sum_{r=0}^{n} C(n, r) = 2^n \qquad\qquad \square$$

We now let n be a positive integer, S be a set such that $|S| = n$, and r be an integer with $1 \leq r \leq n$. A **circular r-permutation** of S is a sequence of r distinct elements from S, written

$$[x_0, x_1, \ldots, x_{r-1}]$$

with the understanding that two circular r-permutations $[x_0, x_1, \ldots, x_{r-1}]$ and $[y_0, y_1, \ldots, y_{r-1}]$ are equal provided there is an integer k, $0 \leq k \leq r - 1$, such that for all i, $0 \leq i \leq r - 1$, we have

$$y_i = x_j$$

where $j = (i + k) \bmod r$. A **circular permutation** of S is simply a circular n-permutation of S.

For example, let $S = \{1, 2, 3, 4, 5\}$. Then the 3-permutations $(1, 2, 3)$, $(2, 3, 1)$, and $(3, 1, 2)$ are different, but they are the same as circular 3-permutations of S; that is,

$$[1, 2, 3] = [2, 3, 1] = [3, 1, 2]$$

(To see that $[1, 2, 3] = [2, 3, 1]$, apply the definition of equality with $k = 1$; to see that $[1, 2, 3] = [3, 1, 2]$, apply the definition with $k = 2$.) Similarly, $(1, 3, 2, 5)$, $(3, 2, 5, 1)$, $(2, 5, 1, 3)$, and $(5, 1, 3, 2)$ are distinct 4-permutations of S; however,

$$[1, 3, 2, 5] = [3, 2, 5, 1] = [2, 5, 1, 3] = [5, 1, 3, 2]$$

Given a circular r-permutation of an n-element set S, say, $[x_1, x_2, \ldots, x_r]$, note that

$$[x_1, x_2, \ldots, x_r] = [x_2, x_3, \ldots, x_r, x_1] = \cdots = [x_r, x_1, x_2, \ldots, x_{r-1}]$$

whereas, the r r-permutations

$$(x_1, x_2, \ldots, x_r), (x_2, x_3, \ldots, x_r, x_1), \ldots, (x_r, x_1, x_2, \ldots, x_{r-1})$$

are distinct. That is, the number of r-permutations of S corresponding to any fixed circular r-permutation of S is r. This yields the following result:

THEOREM 1.3.10 For integers r and n with $1 \leq r \leq n$, the number of circular r-permutations of an n-element set is

$$\frac{P(n, r)}{r} \qquad \qquad \square$$

We remark, as an immediate consequence of Theorem 1.3.10, that the number of circular permutations of an n-element set is $(n - 1)$ factorial.

■ *Example 1.3.6* Six dignitaries are to be seated at a banquet.

(a) Find the number of ways to seat them in a row of six chairs at the head table.

(b) Find the number of ways to seat them at a circular table with six chairs. Assume that the particular position a dignitary is seated at is unimportant; all that matters to any dignitary is which person is seated to his or her left and which person is seated to his or her right.

Solution
Let S be the set of dignitaries.

(a) Here each seating arrangement corresponds to a unique permutation of S, so the number of seating arrangements is 6 factorial.

(b) In this part, the number of different seating arrangements is the same as the number of circular permutations of S, so the answer is 5 factorial. To see this, take a given assignment of dignitaries to chairs at the table. If, for some k, $0 \leq k \leq 5$, each dignitary moves k chairs to the left, then we have the same seating arrangement since each person still has the same left neighbor and the same right neighbor. ■

EXERCISE SET 1.3

1. Seven runners are entered in the mile race at a track meet. Different trophies are awarded to the first, second, and third place finishers. In how many ways might the trophies be awarded?

2. If the Yankees have nine players to trade and the Red Sox have seven players to trade, in how many ways might the two teams trade four players for four players?

3. Consider the game of Mastermind (see Exercise 15 in Exercise Set 1.2). How many codes use

   (a) only one color?   (b) exactly two colors?
   (c) exactly three colors?   (d) exactly four colors?

4. Ron has n friends who enjoy playing bridge. Every Wednesday evening, Ron invites three of these friends to his home for a bridge game. Ron always sits in the south position at the bridge table; he decides which friends are to sit in the west, north, and east positions. How large (at least) must n be if Ron is able to do this for 210 weeks before repeating a seating arrangement?

5. Ron has n friends who enjoy playing bridge, and he is able to invite a different subset of three of them to his home every Wednesday night for 104 weeks. How large (at least) is n?

6. How many four-digit numbers can be formed from the digits 1, 2, 3, 4, 5, 6? How many of these have distinct digits?

7. A checkerboard has 64 distinct squares arranged into eight rows and eight columns.

   (a) In how many ways can eight identical checkers be placed on the board so that no two checkers are in the same row or in the same column?
   (b) In how many ways can two identical red checkers and two identical black checkers be placed on the board so that no two checkers of the opposite color are in the same row or in the same column?

8. For a positive integer n and an integer k, $0 \le k \le n$, consider the identity

   $$C(n, k) = C(n, n - k)$$

   (a) Verify this identity, algebraically, using Corollary 1.3.7.
   (b) Give a "combinatorial" justification for this identity, arguing that the number of k-combinations of $\{1, 2, \ldots, n\}$ is the same as the number of $(n - k)$-combinations of $\{1, 2, \ldots, n\}$. One way to do this is to give a bijective proof.

9. The Department of Mathematics has four full professors, eight associate professors, and three assistant professors. A four-person curriculum committee is to be selected; this can be done in $C(15, 4)$ ways. How many ways are there to select the committee under the following restrictions?

   (a) At least one full professor is chosen.
   (b) The committee contains at least one professor of each rank.

10. In how many ways can tickets to eight different football games be distributed among 14 football fans if no fan gets more than one ticket?

11. A university is to interview six candidates for the position of provost; Dr. Deming is a member of the search committee.

    (a) After interviewing the candidates, Dr. Deming is to rank the candidates from first to sixth choice. In how many ways can this be done?
    (b) Suppose that, rather than ranking all the candidates, Dr. Deming is to give a first choice, a second choice, and a third choice. In how many ways can this be done?
    (c) Suppose that, rather than ranking the candidates, Dr. Deming is to indicate a set of three acceptable candidates. In how many ways can this be done?

(d) Suppose that, rather than ranking the candidates, Dr. Deming is to indicate a set of acceptable candidates. (This is called "approval voting.") In how many ways can this be done?

12. How many permutations of $\{1, 2, 3, 4, 5\}$ satisfy the following conditions?

   (a) 1 is in the first position.
   (b) 1 is in the first position and 2 is in the second position.
   (c) 1 occurs before 2.

13. How many binary strings of length 16 contain exactly six 1s?

14. In how many ways can a nonempty subset of people be chosen from six men and four women if the subset has equal numbers of men and women?

15. A hand in the card game of (standard) poker can be regarded as a 5-card subset of the standard deck of 52 cards. How many poker hands contain at least one card from each suit?

16. How many five-digit strings, formed using the digits 1, 2, 3, 4, and 5, satisfy the following conditions?

   (a) The string contains the substring 12 or the substring 21.
   (b) The string contains one of the substrings 123, 132, 213, 231, 312, or 321.

17. In how many ways can seven distinct keys be put on a key ring? (This question has several possible answers, depending on how one decides whether two ways of placing the keys on the ring are the "same" or "different." For example, is the orientation of the individual keys important? Problems of this sort are not atypical in combinatorics.)

18. Four married couples are to be seated at a circular table with eight chairs. As in Example 1.3.6, the particular position a person is seated at is unimportant; all that matters to any person is which person is seated to her or his left and which person is seated to the right.

   (a) How many ways are there to seat these eight people?

   How many ways are there to seat them under the following restrictions?

   (b) Each man is seated next to two women.
   (c) Each husband must be seated next to his wife.
   (d) The four men are seated consecutively.

19. How many 5-card poker hands (see Exercise 15) satisfy the following conditions?

   (a) The hand contains exactly one pair (not two pairs or three of a kind).
   (b) The hand contains exactly three of a rank (not four of a rank or a full house).
   (c) The hand is a full house (three of one rank, a pair of another).
   (d) The hand is a straight flush (for example, 7, 8, 9, 10, jack, all the same suit).
   (e) The hand is a straight (for example, 7, 8, 9, 10, jack, not all the same suit).
   (f) The hand is a flush (all the same suit, but not a straight)?

20. If the United States Senate had 52 Democrats and 48 Republicans, in how many ways could a committee of 6 Democrats and 4 Republicans be formed? How many ways would there be to form this committee if one of the Democrats is to

be appointed chair and one of the Republicans is to be appointed vice chair of the committee?

21. A pizza automatically comes with cheese and sauce, but any of the following extra items may be ordered: extra cheese, pepperoni, mushrooms, sausage, green peppers, onions, or anchovies.

    (a) How many kinds of pizza may be ordered?
    (b) How many kinds of pizza with exactly three extra items may be ordered?
    (c) If a vegetarian pizza may not contain pepperoni, sausage, or anchovies, how many kinds of vegetarian pizza are there?

22. A member of the Association for Computing Machinery (ACM) may belong to any of 33 Special Interest Groups (SIGs). Suppose the ACM wishes to assign each member a code number indicating the SIGs to which that member belongs. How many code numbers are needed?

23. Let m and n be positive integers, let $A = \{1, 2, \ldots, m\}$, and let $B = \{1, 2, \ldots, n\}$.

    (a) How many relations are there from A to B?
    (b) How many relations are there on A?

24. With A and B as in the previous exercise, find the number of functions from A onto B in each case:

    (a) $m = n + 1$ (b) $m = n + 2$ (c) $m = n + 3$

    (Hint: First work out the special case $n = 3$; see Example 1.3.4.)

25. The vertices of an n-gon are labeled with the numbers $1, 2, \ldots, n$.

    (a) How many different labelings are there if two labelings are considered the same provided the sequence of numbers encountered, beginning with the vertex labeled 1 and proceeding clockwise, is the same?
    (b) How many different labelings are there if two labelings are considered the same provided they yield the same set of edges.

26. Mary likes to listen to compact discs each evening. How many CDs must she have if she is able to listen each evening for 100 consecutive evenings to

    (a) A different subset of CDs?
    (b) A different subset of three CDs?

27. Suppose in Exercise 9 that, in addition to the 4-person curriculum committee, a 3-person administrative committee is also to be selected from the 15 members of the department.

    (a) How many ways are there to select both commitees (if there are no restrictions)?

    How many ways are there to select both committees under the following restrictions?

    (b) The committees must be disjoint.
    (c) The committees can have at most one member in common.

28. For a positive integer n, we wish to define an ordering of the permutations of $\{1, 2, \ldots, n\}$; this ordering is called **lexicographic (or lex) ordering**. Given two permutatuons $\pi_1 = (x_1, x_2, \ldots, x_n)$ and $\pi_2 = (y_1, y_2, \ldots, y_n)$, we use the notation $\pi_1 < \pi_2$ to indicate that π_1 precedes π_2 in lex order and use the notation $\pi_1 \leq \pi_2$ to indicate that $\pi_1 = \pi_2$ or $\pi_1 < \pi_2$. We then define $\pi_1 < \pi_2$ provided $x_1 < y_1$ or $x_1 = y_1, x_2 = y_2, \ldots, x_{i-1} = y_{i-1}$, and $x_i < y_i$ for some i, $1 < i \leq n$.

(a) Show that lex order is a total ordering of the set of permutations of $\{1, 2, \ldots, n\}$.
(b) List the permutations of $\{1, 2, 3, 4\}$ in lex order.
(c) For $n = 7$, what permutation is the immediate successor of $(1, 2, 3, 4, 6, 7, 5)$ (in lex order)?
(d) For $n = 7$, what permutation is the immediate successor of $(1, 6, 2, 7, 5, 4, 3)$?

29. How many partitions $\{A, B, C\}$ of $\{0, 1, 2, \ldots, 9\}$ are there such that $|A| = 2$, $|B| = 3$, and $|C| = 5$?

30. Refer to Exercise 28 and give a recursive definition of $\pi_1 < \pi_2$.

31. In how many ways can 7 cards be chosen from a standard deck of 52 cards so that there are 2 cards from each of any three suits and 1 card from the other suit?

32. Let n be a positive integer and let k be an integer, $0 \leq k \leq n$. We define an ordering, also called **lexicographic (or lex) ordering**, on the k-element subsets of $\{1, 2, \ldots, n\}$. Again, given two k-element subsets S_1 and S_2, the notation $S_1 < S_2$ indicates that S_1 precedes S_2 in lex order, and $S_1 \leq S_2$ indicates that $S_1 = S_2$ or $S_1 < S_2$. Now the empty set is the only 0-element subset. For $k \geq 1$, let $S_1 = \{x_1, x_2, \ldots, x_k\}$ and $S_2 = \{y_1, y_2, \ldots, y_k\}$ be two k-element subsets of $\{1, 2, \ldots, n\}$, and assume the elements are listed so that $x_1 < x_2 < \cdots < x_k$ and $y_1 < y_2 < \cdots < y_k$; then $S_1 < S_2$ provided $x_1 < y_1$ or $x_1 = y_1, x_2 = y_2, \ldots, x_{i-1} = y_{i-1}$, and $x_i < y_i$ for some i, $1 < i \leq k$.

(a) Show that lex order is a total ordering of the k-element subsets of $\{1, 2, \ldots, n\}$.
(b) List the 3-element subsets of $\{1, 2, 3, 4, 5, 6\}$ in lex order.
(c) For $n = 7$, what 4-element subset is the immediate successor of $\{1, 3, 4, 5\}$ (in lex order)?
(d) For $n = 7$, what 4-element subset is the immediate successor of $\{2, 4, 6, 7\}$?

33. Use the addition principle to give an alternate solution to Example 1.3.2, part (b), resulting in the answer $P(8, 5) - P(7, 5)$.

34. Refer to Exercise 32 and give a recursive definition of $S_1 < S_2$.

35. How many ways are there to distribute m different baseball cards to n children if

(a) There are no restrictions?
(b) $m < n$ and no child gets more than one card?

36. Let us extend the definition of lex order given in Exercise 32 to define a lexicographic ordering of all the subsets of $\{1, 2, \ldots, n\}$, namely, to $\mathcal{P}(\{1, 2, \ldots, n\})$. First, we agree that the empty subset precedes any nonempty subset. Then, for two nonempty subsets $S_1 = \{x_1, x_2, \ldots, x_k\}$ and $S_2 = \{y_1, y_2, \ldots, y_m\}$ with

$x_1 < x_2 < \cdots < x_k$ and $y_1 < y_2 < \cdots < y_m$, we define $S_1 < S_2$ provided $S_1 \subset S_2$, or $x_1 < y_1$, or $x_1 = y_1, x_2 = y_2, \ldots, x_{i-1} = y_{i-1}$, and $x_i < y_i$ for some i, $1 < i \leq k$.

(a) Show that lex order is a total ordering of $\mathcal{P}(\{1, 2, \ldots, n\})$.
(b) List the subsets of $\{1, 2, 3, 4\}$ in lex order.
(c) For $n = 7$, what subset is the immediate successor of $\{1, 3, 4, 5\}$ (in lex order)?
(d) For $n = 7$, what subset is the immediate successor of $\{1, 2, 3, 4, 5, 6, 7\}$?
(e) For $n = 7$, what subset is the immediate successor of $\{1, 7\}$?

37. Try to answer Exercise 35 if the baseball cards are identical (one part can be solved using a result discussed in Section 1.5).

38. Refer to Exercise 36 and give a recursive definition of $S_1 < S_2$.

1.4 THE PIGEONHOLE PRINCIPLE

For positive integers m and n, let A be an m-element set, B be an n-element set, and f be a function from A to B. As observed in the last section, if $m > n$, then it is not possible for f to be one-to-one. Stated in a more colloquial fashion, if pigeons are to be assigned to pigeonholes and there are more pigeons than there are pigeonholes, then some two pigeons must be assigned to the same pigeonhole.

THEOREM 1.4.1 (Pigeonhole Principle)   For positive integers m and n, let A be an m-element set, B be an n-element set, and f be a function from A to B. If $m > n$, then f is not one-to-one. More generally, if $m > kn$, for some positive integer k, then there is some element $b \in B$ that has at least $k + 1$ preimages under f, that is,

$$|f^{-1}(\{b\})| \geq k + 1$$

Proof   Let m, n, A, B, and $f : A \to B$ be as in the statement of the theorem, with $B = \{b_1, b_2, \ldots, b_n\}$. To prove the general result, we proceed to show the contrapositive. Suppose, for some positive integer k, that every element of B has at most k preimages under f, that is, suppose that

$$|f^{-1}(\{b_i\})| \leq k$$

for each i, $1 \leq i \leq n$. Note that the sets $f^{-1}(\{b_1\}), f^{-1}(\{b_2\}), \ldots, f^{-1}(\{b_n\})$ are pairwise disjoint and that

$$A = f^{-1}(\{b_1\}) \cup f^{-1}(\{b_2\}) \cup \cdots \cup f^{-1}(\{b_n\})$$

Hence, by the addition principle,

$$m = |A| = |f^{-1}(\{b_1\})| + |f^{-1}(\{b_2\})| + \cdots + |f^{-1}(\{b_n\})| \le kn \qquad \square$$

Theorem 1.4.1 is also known as the Dirichlet drawer principle, named for the famous German mathematician Peter Gustav Lejeune Dirichlet (1805–1859). It says that if m objects are assigned to n boxes (drawers), and $m > kn$ for some positive integer k, then some box has at least $k + 1$ of the objects assigned to it.

Although Theorem 1.4.1 is quite simple to state, prove, and understand, it has a wide variety of applications, some quite unexpected and nontrivial. This is illustrated by the following several examples.

■ *Example 1.4.1*   A clothes dryer contains 14 socks in 7 matching, but unsorted, pairs. Robert, a mathematician, randomly selects socks from the dryer, one at a time (without replacement). How many socks must Robert select to guarantee that he has a matching pair?

Solution
Associate the 7 pairs of socks with 7 boxes so that, when a sock is selected from the dryer, it is assigned to the appropriate box. It is possible for Robert to select 7 socks, one from each pair, so that each box has 1 sock in it and no matching pair exists. However, should 8 socks be selected, then, by the pigeonhole principle, some box must have 2 socks in it. This means that a matching pair of socks has been found. So the answer to the question is 8 socks. ■

■ *Example 1.4.2*   Given any set $S = \{x_1, x_2, \ldots, x_{10}\}$ of 10 integers, show that there exist subscripts j and k, with $0 \le j < k \le 10$, such that the sum

$$\sum_{i=j+1}^{k} x_i$$

is a multiple of 10.

Solution
Consider the 10 sums

$$s_1 = x_1 \qquad s_2 = x_1 + x_2 \qquad \cdots \qquad s_{10} = x_1 + x_2 + \cdots + x_{10}$$

If $10 \mid s_k$ for some k, then the stated result holds with $j = 0$. Thus, we may assume that $s_i \bmod 10 \in \{1, 2, \ldots, 9\}$ for each i, $1 \le i \le 10$. So $s_1 \bmod 10$, $s_2 \bmod 10, \ldots, s_{10} \bmod 10$ are 10 integers between 1 and 9. Hence, by the pigeonhole principle, some two of these numbers are equal, say,

$$s_j \bmod 10 = s_k \bmod 10$$

for some $1 \leq j < k \leq 10$. Therefore,

$$s_k - s_j = \sum_{i=j+1}^{k} x_i$$

is a multiple of 10, as was to be shown. ∎

■ *Example 1.4.3* Sarah is campaigning for the state assembly, visiting 45 towns in 30 days. If the number of towns she visits on any day is a positive integer, show that there must exist some period of consecutive days during which she will visit exactly 14 towns.

Solution

Let n_i be the number of towns Sarah visits during the first i days of the campaign, $1 \leq i \leq 30$. Note that

$$1 \leq n_1 < n_2 < \cdots < n_{30} = 45$$

We wish to show that

$$n_j = n_i + 14$$

for some $1 \leq i < j \leq 30$, because this means that following the ith day, up to and including the jth day of the campaign, Sarah visits exactly 14 towns. Now

$$15 \leq n_1 + 14 < n_2 + 14 < \cdots < n_{30} + 14 = 59$$

Thus, the numbers $n_1, n_2, \ldots, n_{30}, n_1 + 14, n_2 + 14, \ldots, n_{30} + 14$ are 60 integers between 1 and 59. By the pigeonhole principle, some two of these numbers must be equal. This yields the desired result. ∎

The following result is a corollary of Theorem 1.4.1; its proof is left for you to develop in Exercise 10.

COROLLARY 1.4.2 Let n be a positive integer and let $m_1, m_2, \ldots, m_n$ be n integers whose average $(m_1 + m_2 + \cdots + m_n)/n$ is denoted by $\overline{m}$.

1. If $\overline{m} > k$ for some integer k, then $m_i \geq k + 1$ for some i, $1 \leq i \leq n$.
2. If $\overline{m} < k$ for some integer k, then $m_i \leq k - 1$ for some i, $1 \leq i \leq n$. □

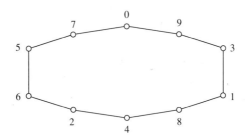

Figure 1.4.1

■ *Example 1.4.4*   A microcomputer laboratory plans to have six microcomputers and three printers. Each microcomputer is to be connected to some subset of the printers; it is required that, whenever x microcomputers require use of a printer and $x \leq 3$, then there exist x connections allowing these microcomputers to use x different printers. Show that at least 12 microcomputer-printer connections are required.

Solution
Denote the microcomputers by $c_1, c_2, c_3, c_4, c_5, c_6$ and the printers by p_1, p_2, p_3. For $1 \leq i \leq 3$, let m_i be the number of microcomputers connected to the ith printer, p_i, and let $\overline{m} = (m_1 + m_2 + m_3)/3$. We wish to show that $m_1 + m_2 + m_3 \geq 12$ or, equivalently, that $\overline{m} \geq 4$. Suppose, to the contrary, that $\overline{m} < 4$. Then, by Corollary 1.4.2, part 2, $m_i \leq 3$ for some i; suppose, without loss of generality, that $m_1 \leq 3$. Then, also without loss of generality, suppose that p_1 is not connected to any of c_4, c_5, or c_6. Now suppose that c_4, c_5, and c_6 simultaneously require printers. Since only the two printers p_2 and p_3 may be available for these three computers, we have a contradiction to the requirement that a request for simultaneous printer access by any three or fewer microcomputers can be accommodated.

   In fact, it can be shown that exactly 12 microcomputer-printer connections suffice to do the job; see Exercise 12.   ■

■ *Example 1.4.5*   The numbers 0, 1, 2, 3, 4, 5, 6, 7, 8, and 9 are randomly assigned to the vertices of a decagon (a polygon with 10 sides); Figure 1.4.1 shows one possible assignment. Show that the sum of the numbers at some three consecutive vertices is more than 13.

Solution
Label the vertices $v_0, v_1, \ldots, v_9$ and let x_i be the number assigned to vertex v_i, $0 \leq i \leq 9$. For each set of three consecutive vertices, we are interested in the sum of the numbers assigned to these vertices. Thus, there are 10 sums to consider, namely,

$$x_0 + x_1 + x_2 \qquad x_1 + x_2 + x_3 \qquad \cdots \qquad x_8 + x_9 + x_0 \qquad x_9 + x_0 + x_1$$

We denote these by $s_0, s_1, \ldots, s_8, s_9$. Note that each number x_i appears in exactly three of these sums. Thus, the average sum $\bar{s}$ is given by

$$\bar{s} = \frac{s_0 + s_1 + \cdots + s_8 + s_9}{10}$$

$$= \frac{1}{10} \sum_{i=1}^{9} 3x_i$$

$$= \frac{3}{10} \sum_{i=1}^{9} x_i$$

$$= \frac{3}{10}(0 + 1 + \cdots + 8 + 9)$$

$$= 13.5$$

Therefore, by Corollary 1.4.2, we have $s_i \geq 14$ for some i, $0 \leq i \leq 9$.   ∎

EXERCISE SET 1.4

1. Let five points be chosen on, or within, an equilateral triangle of side length 6 centimeters. Show that at least two of the chosen points are within 3 centimeters of each other. (Hint: Partition the triangle into four smaller triangles by joining the midpoints of its sides.)

2. Argue that, in a class of 20 students, there are 2 students who are acquainted with the same number of students in the class.

3. Argue that, in a round-robin tournament with 10 teams, either there is exactly 1 team with w wins for each w, $0 \leq w \leq 9$, or there are 2 teams with the same number of wins.

4. Generalize Example 1.4.2: Given any set $S = \{x_1, x_2, \ldots, x_n\}$ of n integers, show that there exist subscripts j and k, with $0 \leq j < k \leq n$, such that the sum

$$\sum_{i=j+1}^{k} x_i$$

   is a multiple of n.

5. Let $(x_0, x_1, \ldots, x_9)$ and $(y_0, y_1, \ldots, y_9)$ be binary strings such that

$$x_0 + x_1 + \cdots + x_9 = 5 = y_0 + y_1 + \cdots + y_9$$

   Show that some cyclic permutation of the string $(y_0, y_1, \ldots, y_9)$ matches the string $(x_0, x_1, \ldots, x_9)$ in at least five positions. The string $(z_0, z_1, \ldots, z_{n-1})$ is a "cyclic permutation" of the string $(y_0, y_1, \ldots, y_{n-1})$ provided there is an integer k, $0 \leq k \leq n - 1$ such that for all i, $0 \leq i \leq n - 1$,

$$z_i = y_j$$

where $j = (i + k) \bmod n$. (Hint: If the y string is cyclically permuted through k positions, then the number of matches with the x string is

$$m_k = f(x_0, y_k) + f(x_1, y_{k+1}) + \cdots + f(x_9, y_{k+9})$$

where the subscripts on the y's are taken mod 10 and $f(x, y) = xy + (1 - x)(1 - y)$. Find the average of $m_0, m_1, \ldots, m_9$ and apply Corollary 1.4.2.)

6. Each of 16 students has a positive whole number of dollars and together the students have 30 dollars. For each n, $1 \le n \le 30$, show that there is some subset S_n of the students such that the students in S_n together have exactly n dollars.

7. Let S be a subset of $\{1, 2, \ldots, 40\}$ such that $|S| = 21$. Show that there exist $m, n \in S$ such that n is a multiple of m.

8. A chess player has 77 days to prepare for a future tournament. She plans to practice by playing at least one game each day against a chess-playing computer program. However, so as not to acquire mental fatigue, she will play no more than 12 games in any 7-day period. Show that she will play exactly 21 games during some period of successive days.

9. Over a period of 10 days, Luis worked a total of 27 hours on calculus homework, working a whole number of hours each day.

   (a) Show that Luis worked at least 3 hours on calculus homework on some day.
   (b) Show that Luis worked at least 6 hours on calculus homework on some pair of consecutive days.

10. Prove Corollary 1.4.2:

    (a) First prove part 1 for the case that each $m_i > 0$. Let $m = m_1 + m_2 + \cdots + m_n$, let $A = \{1, 2, \ldots, m\}$, let $B = \{1, 2, \ldots, n\}$, and let $f : A \to B$ be such that $|f^{-1}(\{i\})| = m_i$, $1 \le i \le n$. Apply Theorem 1.4.1. Then, to prove part 1 in general, let d be a positive integer such that $m_i' = m_i + d > 0$ for each i, $1 \le i \le n$. Let $\overline{m}' = (m_1' + m_2' + \cdots + m_n')/n$; use the first part of the proof and the fact that $\overline{m}' = \overline{m} + d$.
    (b) To prove part 2, let $m_i' = -m_i$, $1 \le i \le n$, and let $\overline{m}' = (m_1' + m_2' + \cdots + m_n')/n$. If $\overline{m} < k$ for some integer k, then $\overline{m}' = -\overline{m} > -k$; now apply the result of part (a).

11. Any power of 2 is even and any power of 5 is a multiple of 5. Let p be a prime with $p \ne 2, 5$. Show that, among the numbers

$$p^1, p^2, p^3, \ldots, p^{40}$$

at least one of them has 01 as its last two digits.

12. In Example 1.4.4, show that 12 microcomputer-printer connections suffice.

13. Show that, for any positive integer n, some multiple of n has 0 and 1 as its only digits. (Hint: Consider the numbers 1 mod n, 11 mod n, 111 mod n, and so on.)

14. Generalize Example 1.4.4 by showing that, if there are m microcomputers and n printers, $m \ge n$, then at least $n(m - n + 1)$ microcomputer-printer connections are necessary.

1.5 ORDERED PARTITIONS AND DISTRIBUTIONS

In how many ways can the letters in GRAPE be permuted? Since we are dealing with five different letters, we see that the answer to this question is 5! = 120. What about the letters in APPLE? In how many ways can these letters be "permuted"? Suppose, for the moment, that we consider the two P's in APPLE to be different; let us subscript the P's as P_1 and P_2 so we can distinguish them. Then the number of ways to permute these five distinct letters is again 5!. Consider a particular permutation, say, AP_1LEP_2. If we take the natural approach and do not consider the two P's to be different, then this permutation and the permutation AP_2LEP_1 should be considered the same. Thus, the number of ways to permute the letters in APPLE is not 5!, but rather

$$\frac{5!}{2!} = 60$$

That is, we must reduce the number of permutations of five distinct objects by a factor of 2! = 2, since the two P's are the same.

How many ways are there to permute the letters in BANANA? As an exercise, you are encouraged to show that the answer is

$$\frac{6!}{3!2!} = 60$$

One of the counting techniques developed in this section allows us to answer questions of this sort.

Another counting result allows us to answer questions having the following general form:

Given positive integers m and n, in how many ways can m be expressed as a sum of n positive integers:

$$x_1 + x_2 + \cdots + x_n = m$$

where the order of the terms $x_1, x_2, \ldots, x_n$ in the sum is important?

For example, if $m = 5$ and $n = 3$, then we have

$$5 = 1 + 1 + 3 = 1 + 2 + 2 = 1 + 3 + 1 = 2 + 1 + 2 = 2 + 2 + 1 = 3 + 1 + 1$$

so the answer is six in this particular case.

We wish to place the preceding questions in the context of finding the number of functions from an m-element set A to an n-element set B, subject to various restrictions. In particular, for the sake of concreteness, let $A = \{1, 2, \ldots, m\}$ and $B = \{1, 2, \ldots, n\}$.

One fact we know already is that the number of functions from A to B is n^m. This is a direct consequence of the multiplication principle, because a function $f : A \to B$ is determined by the images $f(1), f(2), \ldots, f(m)$, and there are n choices for each of these images.

How many functions from A to B are one-to-one? We have also answered this question, the answer being $P(n, m)$, if $m \leq n$, and 0 otherwise. For if $m \leq n$ and $f : A \to B$ is one-to-one, then the sequence of images $((f(1), f(2), \ldots, f(m))$ is an m-permutation of B.

A tougher question is the following:

How many functions from *A* onto *B* are there?

Notice that the answer is 0 if $m < n$, so we assume that $m \geq n$. In this case, if $f : A \to B$ is onto, then the preimages $f^{-1}(\{1\}), f^{-1}(\{2\}), \ldots, f^{-1}(\{n\})$ are nonempty, pairwise-disjoint, and

$$A = \{1, 2, \ldots, m\} = f^{-1}(\{1\}) \cup f^{-1}(\{2\}) \cup \cdots \cup f^{-1}(\{n\})$$

namely, $\{f^{-1}(\{1\}), f^{-1}(\{2\}), \ldots, f^{-1}(\{n\})\}$ is a partition of *A*.

> **DEFINITION 1.5.1**   Let *S* be a nonempty set and let $\{S_1, S_2, \ldots, S_n\}$ be a partition of *S*. Then the *n*-tuple $(S_1, S_2, \ldots, S_n)$ is called an **ordered partition** of *S* **into *n* parts**. The number *n* is called the **degree** of the ordered partition. Two ordered partitions $(U_1, U_2, \ldots, U_n)$ and $(V_1, V_2, \ldots, V_k)$ of *S* are considered equal if and only if $n = k$ and $U_i = V_i$ for each i, $1 \leq i \leq n$.   ■

■ *Example 1.5.1*   Let $S = \{0, 1, 2, 3, 4, 5, 6, 7, 8, 9\}$. As pointed out in Example 1.2.1,

$$\{\{0, 1, 2, 9\}, \{3, 4, 8\}, \{5, 6\}, \{7\}\} \quad \text{and} \quad \{\{7\}, \{5, 6\}, \{3, 4, 8\}, \{0, 1, 2, 9\}\}$$

are equal partitions of *S* into 4 parts. However, the ordered partitions

$$(\{0, 1, 2, 9\}, \{3, 4, 8\}, \{5, 6\}, \{7\}) \quad \text{and} \quad (\{7\}, \{5, 6\}, \{3, 4, 8\}, \{0, 1, 2, 9\})$$

are different ordered partitions of *S* into 4 parts.   ■

A function *f* from *A* onto *B* uniquely determines an ordered partition of *A* into *n* parts, namely, $(f^{-1}(\{1\}), f^{-1}(\{2\}), \ldots, f^{-1}(\{n\}))$. Conversely, an ordered partition of *A* into *n* parts, say, $(A_1, A_2, \ldots, A_n)$, uniquely determines a function *f* from *A* onto *B*, namely, *f* is defined by

$$f(a) = b \leftrightarrow a \in A_b$$

■ *Example 1.5.2*   To make the preceding ideas more concrete, let's illustrate the particular case $m = 8$, $n = 4$. Consider the function $f : \{1, 2, \ldots, 8\} \to \{1, 2, 3, 4\}$ defined by $f(1) = 2$, $f(2) = 1$, $f(3) = 1$, $f(4) = 2$, $f(5) = 1$, $f(6) = 4$, $f(7) = 3$, and $f(8) = 4$. Then *f* is onto with $f^{-1}(\{1\}) = \{2, 3, 5\}$, $f^{-1}(\{2\}) = \{1, 4\}$, $f^{-1}(\{3\}) = \{7\}$, and $f^{-1}(\{4\}) = \{6, 8\}$. Notice that $(\{2, 3, 5\}, \{1, 4\}, \{7\}, \{6, 8\})$ is an ordered partition of $\{1, 2, \ldots, 8\}$ into 4 parts. Conversely, suppose we are given some ordered partition of $\{1, 2, \ldots, 8\}$ into 4 parts, say, $(\{1, 8\}, \{2, 7\}, \{3, 6\}, \{4, 5\})$. Then we may define a function *f* from $\{1, 2, \ldots, 8\}$ onto $\{1, 2, 3, 4\}$ such that $f^{-1}(\{1\}) = \{1, 8\}$, $f^{-1}(\{2\}) = \{2, 7\}$, $f^{-1}(\{3\}) = \{3, 6\}$, and $f^{-1}(\{4\}) = \{4, 5\}$, namely, we let

$$f(1) = 1, f(2) = 2, f(3) = 3, f(4) = 4, f(5) = 4, f(6) = 3, f(7) = 2, f(8) = 1 \quad ■$$

At this point, we see that the number of functions from $\{1, 2, \ldots, m\}$ onto $\{1, 2, \ldots, n\}$ is the same as the number of ordered partitions of $\{1, 2, \ldots, m\}$ into n parts, so we would like to know this latter number. Let us attack an easier problem first. Let $m_1, m_2, \ldots, m_n$ be n positive integers whose sum is m. We ask for the number of ordered partitions $(A_1, A_2, \ldots, A_n)$ of $A = \{1, 2, \ldots, m\}$ into n parts with $|A_i| = m_i$, $1 \leq i \leq n$.

■ *Example 1.5.3*   Determine the number of ordered partitions $(A_1, A_2, \ldots, A_n)$ of $A = \{1, 2, \ldots, 8\}$ with:

(a) $n = 2$, $|A_1| = 3$, $|A_2| = 5$
(b) $n = 4$, $|A_1| = 3$, $|A_2| = |A_3| = 2$, $|A_4| = 1$

Solution

(a) We may form an ordered partition (A_1, A_2) of A with $|A_1| = 3$ and $|A_2| = 5$ simply by choosing the three elements of A_1, because once this is done, the remaining five elements of A automatically belong to A_2. So the answer is

$$C(8, 3) = \frac{8!}{3!5!} = 56$$

(b) We may form an ordered partition (A_1, A_2, A_3, A_4) of A with $|A_1| = 3$, $|A_2| = 2$, $|A_3| = 2$, and $|A_4| = 1$ by first choosing three elements of A for A_1, then choosing two elements of $A - A_1$ for A_2, and then choosing two elements of $A - (A_1 \cup A_2)$ for A_3, with $A_4 = A - (A_1 \cup A_2 \cup A_3)$. The three elements for A_1 can be chosen in $C(8, 3)$ ways. This done, the two elements for A_2 can be chosen in $C(5, 2)$ ways. Finally, the two elements for A_3 can be chosen in $C(3, 2)$ ways. Thus, by the multiplication principle, the answer is

$$C(8, 3)C(5, 2)C(3, 2) = \frac{8!}{3!5!} \cdot \frac{5!}{2!3!} \cdot \frac{3!}{2!1!} = \frac{8!}{3!2!2!1!} = 1680 \qquad ■$$

As indicated by part (a) of Example 1.5.3, we already know a formula for the number of ordered partitions (A_1, A_2) of an m-element set A into 2 parts with $|A_1| = m_1$ and $|A_2| = m_2 = m - m_1$, namely, $C(m, m_1)$. For this reason, it makes sense to generalize the notation $C(m, m_1)$ by letting $C(m; m_1, m_2, \ldots, m_n)$ denote the number of ordered partitions $(A_1, A_2, \ldots, A_n)$ of an m-element set A into n parts with $|A_i| = m_i$, $1 \leq i \leq n$. Here, of course, $m_1, m_2, \ldots, m_n$ are positive integers satisfying $m_1 + m_2 + \cdots + m_n = m$. A common alternate notation for $C(m; m_1, m_2, \ldots, m_n)$ is

$$\binom{m}{m_1, m_2, \ldots, m_n}$$

We can now give a formula for $C(m; m_1, m_2, \ldots, m_n)$.

THEOREM 1.5.1 $C(m; m_1, m_2, \ldots, m_n) = \dfrac{m!}{m_1! m_2! \cdots m_n!}$

Proof   We can uniquely form an ordered partition $(A_1, A_2, \ldots, A_n)$ of an m-element set A into n parts with $|A_i| = m_i$, $1 \leq i \leq n$, in $n - 1$ steps. The first step is to choose m_1 elements of A for A_1. This can be done in $C(m, m_1)$ ways. For the ith step, $2 \leq i \leq n - 1$, we choose m_i elements of $A - (A_1 \cup \cdots \cup A_{i-1})$ for A_i. This can be done in

$$C(m - (m_1 + \cdots + m_{i-1}), m_i)$$

ways. Thus, by the multiplication principle, we have

$$C(m; m_1, m_2, \ldots, m_n) = C(m, m_1) C(m - m_1, m_2) \cdots$$
$$C(m - (m_1 + \cdots + m_{n-2}), m_{n-1})$$
$$= \frac{m!}{m_1! m_2! \cdots m_n!}$$

Verification of the last equality is left for you to develop in Exercise 16.   □

The next example shows how Theorem 1.5.1 can be applied to solve problems of the type mentioned at the beginning of this section.

■   ***Example 1.5.4***   Determine the number of different ways to form a string of letters using the letters in (a) BANANA and (b) MISSISSIPPI.

Solution

(a) A given string $c_1 c_2 c_3 c_4 c_5 c_6$ of the six letters in BANANA determines an ordered partition (A_1, A_2, A_3) of $\{1, 2, 3, 4, 5, 6\}$ with $|A_1| = 3$, $|A_2| = 1$, and $|A_3| = 2$; namely, let $A_1 = \{i \mid c_i = \text{A}\}$, $A_2 = \{i \mid c_i = \text{B}\}$, and $A_3 = \{i \mid c_i = \text{N}\}$. For instance, the string BANANA itself determines the ordered partition $(\{2, 4, 6\}, \{1\}, \{3, 5\})$. Conversely, each ordered partition (A_1, A_2, A_3) of $\{1, 2, 3, 4, 5, 6\}$ with $|A_1| = 3$, $|A_2| = 1$, and $|A_3| = 2$ determines a string $c_1 c_2 c_3 c_4 c_5 c_6$ of the letters in BANANA, namely, $c_i = \text{A}$, B, or C according to whether $i \in A_1$, A_2, or A_3, respectively. For instance, the ordered partition $(\{1, 4, 6\}, \{5\}, \{2, 3\})$ determines the string ANNABA. Thus, the number of different strings using the letters in BANANA is the same as the number of ordered partitions (A_1, A_2, A_3) of $\{1, 2, 3, 4, 5, 6\}$ with $|A_1| = 3$, $|A_2| = 1$, and $|A_3| = 2$. By Theorem 1.5.1, this number is

$$C(6; 3, 1, 2) = \frac{6!}{3! 1! 2!} = 60$$

(b) Using the same line of reasoning as in part (a), we see that the number of different sequences using the letters in MISSISSIPPI is the same as the number of ordered partitions (A_1, A_2, A_3, A_4) of $\{1, 2, \ldots, 11\}$ with $|A_1| = 4$, $|A_2| = 1$, $|A_3| = 2$, and $|A_4| = 4$. By Theorem 1.5.1, the number is

$$C(11; 4, 1, 2, 4) = \frac{11!}{4!1!2!4!} = 34,650$$ ∎

Now let's generalize the preceding example. Suppose we have n different types of objects, with m_i identical objects of type i, $1 \leq i \leq n$, where $m_1 + m_2 + \cdots + m_n = m$. How many different ways are there to list, or permute, these m objects? The number of such "permutations with repetition allowed" is given by the following result, whose formal proof is left for you to work out in Exercise 18.

THEOREM 1.5.2   Given a total of m objects of n different types, with m_i identical objects of type i, $1 \leq i \leq n$, and $m = m_1 + m_2 + \cdots + m_n$, the number of permutations of these m objects is given by $C(m; m_1, m_2, \ldots, m_n)$. □

At this point, we see that the number of functions from $\{1, 2, \ldots, m\}$ onto $\{1, 2, \ldots, n\}$, where $m \geq n$, is given by

$$\sum C(m; m_1, m_2, \ldots, m_n)$$

where the sum is over all n-tuples $(m_1, m_2, \ldots, m_n)$ of positive integers such that $m_1 + m_2 + \cdots + m_n = m$. How many terms are there in such a sum? This brings us to the second question mentioned at the beginning of this section, namely, How many ways are there to express a given positive integer m as a sum of n positive integers $(m \geq n)$ where the order of the terms in the sum is important?

Looking at the problem for $m = 5$ and $n = 3$, let us write 5 as $1 + 1 + 1 + 1 + 1$. Now, from among the four plus signs, we choose two of them, and combine the terms between the chosen plus signs. This uniquely yields an expression of 5 as a sum of 3 positive integers. For example, if we choose the first and fourth plus signs, then we obtain

$$5 = 1 + (1 + 1 + 1) + 1 = 1 + 3 + 1$$

Thus, the number of ways to express 5 as a sum of 3 positive integers is the same as the number of ways to choose two of four plus signs, namely, $C(4, 2) = 6$.

THEOREM 1.5.3   Let m and n be positive integers with $m \geq n$. The number of ways to express m as a sum of n positive integers, where the order of the terms in the sum is important, is $C(m - 1, n - 1)$.

Proof   We write m as $1 + 1 + \cdots + 1$. Then choose $n - 1$ of the $m - 1$ plus signs and combine any 1s before the first plus sign chosen, between two consecutive chosen plus signs, and after the last plus sign chosen. Note that this uniquely expresses m as a sum of n positive integers. Therefore, the number of ways to express m as a sum of n positive integers is the number of ways to choose $n - 1$ of $m - 1$ plus signs, that is, $C(m - 1, n - 1)$. □

Given $m = x_1 + x_2 + \cdots + x_n$, with each x_i a positive integer, we call the n-tuple $(x_1, x_2, \ldots, x_n)$ an **ordered partition of m into n parts**. So Theorem 1.5.3 states that the number of ordered partitions of m into n parts is $C(m-1, n-1)$. A related result is the following theorem:

THEOREM 1.5.4   The number of ways to express the positive integer m as a sum of n nonnegative integers, where the order of the terms in the sum is important, is $C(m+n-1, n-1)$.

Proof   Suppose m is written as

$$m = y_1 + y_2 + \cdots + y_n$$

where each y_i is a nonnegative integer. For $1 \le i \le n$, let $x_i = y_i + 1$. Then we have

$$m + n = x_1 + x_2 + \cdots + x_n$$

Conversely, if $m + n$ is written as above, where each x_i is a positive integer, let $y_i = x_i - 1$, for $1 \le i \le n$. Then $m = y_1 + y_2 + \cdots + y_n$. Thus, the number of ways to express m as a sum of n nonnegative integers is the same as the number of ways to express $m + n$ as a sum of n positive integers. This, by Theorem 1.5.2, is $C(m+n-1, n-1)$.   □

■ **Example 1.5.5**   A bakery sells seven kinds of donuts.

(a) How many ways are there to choose 12 donuts?

(b) How many ways are there to choose 12 donuts if there must be at least 1 donut of each kind?

Solution

What matters here is the number of donuts chosen of each kind. Thus, let x_i be the number of donuts of kind i chosen, $1 \le i \le 7$. Then

$$x_1 + x_2 + \cdots + x_7 = 12$$

(a) Here each x_i is a nonnegative integer, so we apply Theorem 1.5.4. Hence, the answer is $C(12 + 7 - 1, 7 - 1) = C(18, 6)$.

(b) In this part, each x_i is restricted to be positive, so we apply Theorem 1.5.3. This yields the answer $C(12 - 1, 7 - 1) = C(11, 6)$.   ■

The general type of problem suggested by the preceding example occurs frequently in combinatorics. Namely, suppose we have n different types of objects, with a sufficiently large number of identical objects of type i, $1 \le i \le n$. How many

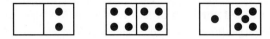

Figure 1.5.1   Several dominoes.

different ways are there to choose a collection of m of these objects, with repetition allowed? Such a collection is termed an ***m*-combination of *n* types, with repetition allowed**. (Some books use the term *multiset* to refer to a collection of elements with repeated elements allowed.) In forming such a collection, what matters is the number m_i of objects of type i selected, $1 \leq i \leq n$. Notice that $m_1, m_2, \ldots, m_n$ are nonnegative integers such that $m_1 + m_2 + \cdots + m_n = m$. Thus, we obtain the following result as an immediate corollary to Theorem 1.5.4.

> ***COROLLARY 1.5.5***   Given objects of n different types, with at least m identical objects of each type, then the number of ways to choose m of these objects is $C(m + n - 1, n - 1)$.   □

■   ***Example 1.5.6***   Each of the two squares of a domino contains between zero and six dots, as illustrated in Figure 1.5.1. How many different dominoes are there?

Solution
What matters in this problem is the number of dots on each of two squares of a domino. Thus, for $0 \leq i \leq 6$, let x_i be the number of squares with i dots. Then the number of different dominoes is equal to the number of solutions to

$$x_0 + x_1 + x_2 + x_3 + x_4 + x_5 + x_6 = 2$$

in nonnegative integers. In other words, we can think of a domino as a 2-combination of seven types, with repetition allowed. Thus, by Corollary 1.5.5, the number of different dominoes is $C(8, 6) = 28$.

We can also answer the question as follows. Either the numbers of dots on the two squares of a domino are equal or are different. There are seven different dominoes with equal numbers of dots on their two squares, one for each of the numbers $0, 1, \ldots, 6$. Also, the number of distinct dominoes with different numbers of dots on their two squares is $C(7, 2) = 21$. Thus, the total number of different dominoes is 28.   ■

To summarize then, the number of functions from $\{1, 2, \ldots, m\}$ onto $\{1, 2, \ldots, n\}$, where $m \geq n$, is given by

$$\sum C(m; m_1, m_2, \ldots, m_n)$$

where the sum is over all n-tuples $(m_1, m_2, \ldots, m_n)$ of positive integers such that $m_1 + m_2 + \cdots + m_n = m$. By Theorem 1.5.3, we know that there are $C(m - 1, n - 1)$ terms in this sum.

For example, the number of functions from $\{1, 2, 3, 4, 5\}$ onto $\{1, 2, 3\}$ is given by

$$C(5; 1, 1, 3) + C(5; 1, 2, 2) + C(5; 1, 3, 1) + C(5; 2, 1, 2) + C(5; 2, 2, 1) + C(5; 3, 1, 1)$$

$$= \frac{5!}{1!1!3!} + \frac{5!}{1!2!2!} + \frac{5!}{1!3!1!} + \frac{5!}{2!1!2!} + \frac{5!}{2!2!1!} + \frac{5!}{3!1!1!}$$

$$= 20 + 30 + 20 + 30 + 30 + 20$$

$$= 150$$

(Notice that the total number of functions from $\{1, 2, 3, 4, 5\}$ to $\{1, 2, 3\}$ is $3^5 = 243$.)

The problem of determining the number of functions from $\{1, 2, \ldots, m\}$ onto $\{1, 2, \ldots, n\}$ is considered from a different angle in Section 2.1, as an application of the general principle of inclusion/exclusion. There we see that the number of functions from $\{1, 2, \ldots, m\}$ onto $\{1, 2, \ldots, n\}$ is given by

$$\sum_{k=0}^{n-1} (-1)^k C(n, k)(n - k)^m$$

This same problem is met again in Section 2.2; there we find that the number of functions from $\{1, 2, \ldots, m\}$ onto $\{1, 2, \ldots, n\}$ is equal to

$$n!S(m, n)$$

where $S(m, n)$ is a special function denoting what are called **Stirling numbers of the second kind**.

Most of the results of this and the preceding section can be placed in the general context of what are termed **ordered distributions**—distributing a given number of objects to a given number of different (or "ordered") recipients under certain specified conditions. It is convenient here to think of the objects as "balls" and the recipients as "boxes" that are numbered $1, 2, \ldots, n$. Now, we may wish to think of the balls as being different from one another; for example, we may wish to distribute m golf balls, no two of which have exactly the same brand name, color, and number. In this case we can think of the distribution of balls to boxes as a function from the set of balls, say, $\{1, 2, \ldots, m\}$, to the set $\{1, 2, \ldots, n\}$. On the other hand, if the balls are to be considered all the same, then what matters is not which balls are placed in which box, but rather how many balls are placed in each box.

■ *Example 1.5.7*   In each of the following parts, find the number of ways to distribute the m balls to the n different boxes.

(a) $m = 8$, $n = 4$, and the balls are different

(b) $m = 4$, $n = 8$, the balls are different, and at most one ball is to be put in any box

(c) $m = 5$, $n = 3$, the balls are different, and at least one ball is to be placed in each box

(d) $m = 8$, $n = 4$, the balls are different, three balls are to be put in box 1, two balls are to be put in each of boxes 2 and 3, and one ball is to be put in box 4

(e) $m = 8$, $n = 4$, and the balls are identical

(f) $m = 4$, $n = 8$, the balls are identical, and at most one ball is to be put in any box

(g) $m = 8$, $n = 4$, the balls are identical, and at least one ball is to be placed in each box

Solution

(a) The number of ways to distribute 8 different balls to 4 different boxes is the same as the number of functions from $\{1, 2, \ldots, 8\}$ to $\{1, 2, 3, 4\}$, which is $4^8 = 4096$ by Theorem 1.3.1.

(b) The number of ways to distribute 4 different balls to 8 different boxes with at most one ball per box is the same as the number of one-to-one functions from $\{1, 2, 3, 4\}$ to $\{1, 2, \ldots, 8\}$; by Theorem 1.3.2, this is $P(8, 4) = 1680$.

(c) The number of ways to distribute 5 different balls to 3 different boxes with at least one ball per box is the same as the number of functions from $\{1, 2, 3, 4, 5\}$ onto $\{1, 2, 3\}$, which was found to be 150 prior to this example.

(d) Note that an assignment of 8 different balls to 4 different boxes (with at least one ball per box) is equivalent to an ordered partition (B_1, B_2, B_3, B_4) of the set of balls; namely, B_i is the set of balls assigned to box i. Here we require that $|B_1| = 3$, $|B_2| = |B_3| = 2$, and $|B_4| = 1$. Thus, by Theorem 1.5.1, the answer is $C(8; 3, 2, 2, 1) = 1680$.

(e) The number of ways to distribute 8 identical balls to 4 different boxes is the same as the number of solutions to $x_1 + x_2 + x_3 + x_4 = 8$, with each x_i a nonnegative integer (x_i is the number of balls placed in box i). By Theorem 1.5.4, this is $C(8 + 4 - 1, 4 - 1) = C(11, 3) = 165$.

(f) To determine an assignment of 4 identical balls to 8 different boxes with at most one ball in any box, we simply must decide which 4 boxes are to get a ball. Hence, the answer is $C(8, 4) = 56$.

(g) The number of ways to distribute 8 identical balls to 4 different boxes with at least one ball per box is the same as the number of solutions to $x_1 + x_2 + x_3 + x_4 = 8$, with each x_i a positive integer; by Theorem 1.5.3, this is $C(8 - 1, 4 - 1) = C(7, 3) = 35$.

∎

Taking our lead from the preceding example, the next theorem summarizes results concerning ordered distributions; the proof is left for you to work out in Exercise 20. In Section 2.4, we discuss "unordered distributions," namely, what happens in the problem when the boxes are considered to be identical or indistinguishable.

THEOREM 1.5.6   Consider the number of ways to distribute m balls to n different boxes.

1. If the balls are different, then:

(a) The number of ways to distribute the balls is n^m.

(b) If $m \leq n$, then the number of ways to distribute the balls with at most one ball per box is $P(n, m)$.

(c) If $m \geq n$, then the number of ways to distribute the balls with at least one ball per box is the same as the number of functions from $\{1, 2, \ldots, m\}$ onto $\{1, 2, \ldots, n\}$.

(d) If $m = m_1 + m_2 + \cdots + m_n$, with each $m_i \in \mathbb{Z}^+$, then the number of ways to distribute the balls with m_i balls placed in box i is $C(m; m_1, m_2, \ldots m_n)$.

2. If the balls are identical, then:

(a) The number of ways to distribute the balls is $C(m + n - 1, n - 1)$.

(b) If $m \leq n$, then the number of ways to distribute the balls with at most one ball per box is $C(n, m)$.

(c) If $m \geq n$, then the number of ways to distribute the balls with at least one ball per box is $C(m - 1, n - 1)$. □

EXERCISE SET 1.5

1. A store sells six different brands of chewing gum. In how many ways can someone choose to buy 12 packs of gum?

2. Given n men and their wives ($n \geq 3$), how many ways are there to select a subset of cardinality x from these $2n$ people such that the subset does not contain a married couple? Answer for:

(a) $x = 1$ (b) $x = 2$ (c) $x = 3$ (d) A general $x \leq n$

3. The clubhouse at a certain golf course sells five different types of sandwiches. A golfer orders eight sandwiches for herself and her seven friends. From the point of view of the person taking the order, how many orders of eight sandwiches are possible?

4. Consider the set S of positive integers between 100000 and 999999, inclusive.

(a) How many contain the digit 8 exactly twice?

(b) How many contain three 7s, two 8s, and a 9?

5. A regular six-sided die is rolled seven times and the sequence of results is recorded; for example: $(3, 1, 3, 6, 4, 4, 2)$.

(a) How many such sequences are there?

Determine the number of sequences with the following properties:

(b) Each of the numbers 1, 2, 3, 4, 5, 6 occurs.

(c) The sequence has three 2s, two 4s, and two 6s.

(d) Only two of the numbers 1, 2, 3, 4, 5, 6 occur.

6. Every Saturday night for 8 weeks, Sue invites one of her four boyfriends to a movie. How many ways can she do this subject to the following restrictions?

(a) The order of invitations is not important and each boyfriend receives at least one invitation.

(b) The order of invitations is important; Sue's favorite boyfriend, Andy, receives three invitations, Bob receives one invitation, and Carl and Dave receive two invitations each.

(c) The order of invitations is not important and there are no restrictions.

7. In how many ways can 12 golf balls be distributed to 4 golfers in each of the following cases?

(a) The balls are different.
(b) The balls are identical.
(c) The balls are different and each golfer gets at least one ball.
(d) The balls are identical and each golfer gets at least one ball.
(e) The balls are different and each golfer gets three balls.
(f) The balls are identical and each golfer gets three balls.

8. Ten students enter a local pub one at a time, and each orders a beer. Four of these students always drink Stroh's, three always drink Budweiser, two always drink Coors, and one student always drinks Molson's.

(a) How many sequences of ten orders are possible.
★(b) Six of the students decide to have another round. From the bartender's viewpoint, how many orders of six beers are possible?

9. How many ways are there to deal a 5-card poker hand (from a standard deck of 52 cards) to each of four players, Curly, Larry, Moe, and Joe, under the following conditions?

(a) There are no restrictions.
(b) Each player gets an ace.
(c) Curly gets a flush in spades, Larry a flush in hearts, Moe a flush in diamonds, and Joe a flush in clubs?

10. How many nine-letter strings can be formed from two A's, five C's, and two E's under the following conditions?

(a) There are no restrictions.
(b) No two vowels are consecutive.
(c) No two C's are consecutive.

11. In how many ways can a student distribute a good McIntosh apple, a good Golden Delicious apple, a rotten apple, and nine identical oranges to his four professors under the following conditions?

(a) There are no restrictions.
(b) Each professor gets at most one apple and at least one orange.
(c) Each professor gets (exactly) three pieces of fruit.

12. In how many ways can six coins be selected from six pennies, six nickels, six dimes, and six quarters? Assume that coins of the same value are identical.

13. In the game of Yahtzee, five dice are rolled. Assuming the dice are identical, how many outcomes are there?

14. Given n integers $b_1, b_2, \ldots, b_n$, find a formula for the number of ways to express the integer m as a sum of n integers:

$$m = m_1 + m_2 + \cdots + m_n$$

where the order of the terms is important and $m_i \geq b_i$, $1 \leq i \leq n$.

15. In how many ways can $2n$ persons be paired to form n bridge teams? (Answer for $n \in \{1, 2, 3, 4\}$ and then for a general n.)

16. Complete the proof of Theorem 1.5.1 by verifying the identity

$$C(m, m_1)C(m - m_1, m_2) \cdots C(m - (m_1 + \cdots + m_{n-2}), m_{n-1}) = \frac{m!}{m_1! m_2! \cdots m_n!}$$

17. In how many ways can 10 (single-scoop) ice cream cones be purchased from an ice cream parlor that offers 26 flavors under the following conditions?

(a) There are no restrictions.

(b) Exactly 3 flavors are used.

(c) At most one cone of any flavor is purchased.

18. Prove Theorem 1.5.2; in particular:

(a) Give a proof based on the solution of Example 1.5.4.

(b) Give a proof based on the ideas given at the beginning of this section.

19. How many ways are there to distribute 30 identical slices of pizza among five fraternity brothers

(a) with no restrictions?

How many ways are there to distribute the slices under the following restrictions?

(b) Each brother gets at least 1 slice.

(c) Each brother gets at least 3 slices.

20. Prove the following parts of Theorem 1.5.6:

(a) part 1(a)   (b) part 1(b)   (c) part 1(c)   (d) part 1(d)

(e) part 2(a)   (f) part 2(b)   (g) part 2(c)

21. If m_1 identical dice and m_2 identical coins are tossed, how many results are there?

22. Consider the equation

$$x_1 + x_2 + x_3 + x_4 = 8$$

where each x_i is a nonnegative integer. One possible solution to this equation is given by $x_1 = 1$, $x_2 = 2$, $x_3 = 3$, and $x_4 = 2$. This solution can be conveniently displayed as the string 1&11&111&11 of 1s and &'s. The solution $x_1 = 0$, $x_2 = 4$, $x_3 = 0$, and $x_4 = 4$, corresponds to the string &1111&&1111. In fact, any solution to the given equation determines a unique string of eight 1s and three &'s. Conversely, any string of eight 1s and three &'s determines a unique solution of the equation. Hence, the number of solutions to the equation in nonnegative integers is equal to the number of different strings of eight 1s and three &'s.

(a) Use this idea to give an alternate proof of Theorem 1.5.4.

(b) Prove Theorem 1.5.3 as a corollary to Theorem 1.5.4.

23. Let's play two modified versions of Yahtzee. In version 1, you roll five dice and win $1 if you get a pair or better; otherwise, you lose $1. In version 2, you roll six dice and win $1 if you get two pairs or better; otherwise, you lose $1. Which version would you prefer to play?

1.6 BINOMIAL AND MULTINOMIAL THEOREMS

In this section we prove a very important result called the binomial theorem. The binomial theorem involves the numbers $C(n, k)$ and is used to derive several interesting identities relating these numbers. We also consider the multinomial theorem, an extension of the binomial theorem involving the numbers $C(m; m_1, m_2, \ldots, m_n)$.

To begin, consider expanding the third power of the binomial $x + y$. We write

$$(x + y)^3 = (x_1 + y_1)(x_2 + y_2)(x_3 + y_3)$$

where $x = x_1 = x_2 = x_3$ and $y = y_1 = y_2 = y_3$. Here the subscripts are used to help us keep track of which factor an x or y comes from. For example, x_2 simply indicates the x from the second factor of $(x + y)^3$. Thus,

$$(x + y)^3 = (x_1 + y_1)(x_2 + y_2)(x_3 + y_3)$$
$$= x_1 x_2 x_3 + x_1 x_2 y_3 + x_1 y_2 x_3 + x_1 y_2 y_3$$
$$+ y_1 x_2 x_3 + y_1 x_2 y_3 + y_1 y_2 x_3 + y_1 y_2 y_3$$
$$= x^3 + 3x^2 y + 3xy^2 + y^3$$

We notice that the initial expansion of $(x + y)^3$ has $8 = 2^3$ terms. This is a result of the multiplication principle; in forming a term, we must decide, for each of the three factors, whether to choose the x or the y from that factor. Next, combining like terms, we obtain one x^3 term, three $x^2 y$ terms, three xy^2 terms, and one y^3 term. Note that each such term has the form $c_k x^{3-k} y^k$, where $0 \le k \le 3$, and that the coefficients c_k satisfy

$$c_0 = 1 = C(3, 0) \qquad c_1 = 3 = C(3, 1) \qquad c_2 = 3 = C(3, 2) \qquad c_3 = 1 = C(3, 3)$$

that is, the coefficient c_k of $x^{3-k} y^k$ is $C(3, k)$, for $0 \le k \le 3$.

In general, we desire a formula for expanding $(x + y)^n$, where n is a nonnegative integer. In addition to the expansion of $(x+y)^3$, let us work out expansions for several other small values of n:

$$(x + y)^0 = 1 = C(0, 0)$$
$$(x + y)^1 = x + y = C(1, 0)x + C(1, 1)y$$
$$(x + y)^2 = x^2 + 2xy + y^2 = C(2, 0)x^2 + C(2, 1)xy + C(2, 2)y^2$$
$$(x + y)^3 = x^3 + 3x^2 y + 3xy^2 + y^3$$
$$= C(3, 0)x^3 + C(3, 1)x^2 y + C(3, 2)xy^2 + C(3, 3)y^3$$

and

$$(x + y)^4 = x^4 + 4x^3 y + 6x^2 y^2 + 4xy^3 + y^4$$
$$= C(4, 0)x^4 + C(4, 1)x^3 y + C(4, 2)x^2 y^2 + C(4, 3)xy^3 + C(4, 4)y^4$$

In general, by expanding the binomial $(x + y)^n$ as a sum of terms of the form $c_k x^{n-k} y^k$, we can conjecture that

$$c_k = C(n, k)$$

for $0 \leq k \leq n$. This is, in fact, the binomial theorem, and because the numbers $C(n, k)$ occur in this context, they are often referred to as **binomial coefficients**.

THEOREM 1.6.1 (Binomial Theorem) For every nonnegative integer n and any numbers x and y,

$$(x + y)^n = \sum_{k=0}^{n} C(n, k) x^{n-k} y^k$$

Proof A combinatorial argument is given. As before, we write $(x + y)^n$ in the form

$$(x + y)^n = (x_1 + y_1)(x_2 + y_2) \cdots (x_n + y_n)$$

where $x_i = x$ and $y_i = y$ from the ith factor of $x + y$ on the right-hand side. Before suppressing subscripts and combining like terms, this expression expands as a sum of 2^n terms, because an arbitrary term is obtained by choosing, for each $i \in \{1, 2, \ldots, n\}$, either x_i or y_i from the ith factor. If we choose x from $n - k$ of the factors and y from the other k factors, where $0 \leq k \leq n$, then such a term has the form $x^{n-k} y^k$ when the subscripts are suppressed. The number of ways to obtain such a term is the number of ways to select the k factors, out of n factors, from which a y is chosen. This is precisely $C(n, k)$. Thus, when like terms are combined, the coefficent of $x^{n-k} y^k$ is $C(n, k)$. This proves the result. □

The binomial theorem can also be proved by induction on n; see Exercise 2.

■ *Example 1.6.1* Find the coefficient of:

(a) $x^7 y^3$ in the expansion of $(x + y)^{10}$.
(b) $x^4 y^7$ in the expansion of $(2x - y)^{11}$.

Solution

(a) By the binomial theorem,

$$(x + y)^{10} = \sum_{k=0}^{10} C(10, k) x^{10-k} y^k$$

Thus, the coefficient of $x^7 y^3$ is $C(10, 3) = 120$.

(b) In order to apply the binomial theorem, let $x' = 2x$ and $y' = -y$. Then

$$(2x - y)^{11} = (x' + y')^{11} = \sum_{k=0}^{11} C(11, k)(x')^{11-k}(y')^k$$

When $k = 7$, this expansion yields the term

$$C(11, 7)(x')^4(y')^7 = C(11, 7)(2x)^4(-y)^7 = -C(11, 7)(2^4)x^4y^7$$

Thus, the coefficient of x^4y^7 is $-C(11, 7)(2^4) = -5280$. ■

The binomial theorem is true for any numbers x and y; in particular, if we replace x by 1 and y by x, then we obtain the following corollary:

COROLLARY 1.6.2 For every nonnegative integer n and any number x,

$$(1 + x)^n = \sum_{k=0}^{n} C(n, k)x^k$$ □

We can go a step further and replace x by 1 in Corollary 1.6.2; this yields the identity

$$2^n = \sum_{k=0}^{n} C(n, k)$$

which is Corollary 1.3.9. If we let $x = -1$ in Corollary 1.6.2, we obtain the identity

$$0 = C(n, 0) - C(n, 1) + \cdots + (-1)^k C(n, k) + \cdots + (-1)^n C(n, n)$$

or

$$0 = \sum_{k=0}^{n} (-1)^k C(n, k)$$

This can be interpreted as saying that, for an n-element set S, the number of subsets of S of even cardinality is the same as the number of subsets of S of odd cardinality. See Exercise 16.

Several other interesting identities can be obtained from Corollary 1.6.2.

■ ***Example 1.6.2*** Show that

$$n \cdot 2^{n-1} = \sum_{k=1}^{n} kC(n, k)$$

for every positive integer n.

Solution
Our first method is to employ Corollary 1.6.2 and some calculus. We start with

$$(1 + x)^n = \sum_{k=0}^{n} C(n, k) x^k$$

This can be interpreted as stating that two (real-valued) functions are identical. Thus, we may differentiate both sides of this identity with respect to x:

$$n(1 + x)^{n-1} = \sum_{k=1}^{n} k C(n, k) x^{k-1}$$

(Recall that the derivative of a sum of functions is the sum of the derivatives of those functions.) Setting $x = 1$, we see that

$$n 2^{n-1} = \sum_{k=1}^{n} k C(n, k)$$

Alternate Solution
Our second method is combinatorial; one way to verify an identity is to show that the left-hand side and right-hand side count the same thing.
     Consider the left-hand side of this identity:

$$n 2^{n-1}$$

This has a rather obvious combinatorial interpretation; it is, by the multiplication principle, the number of ways to first choose an element x from $S = \{1, 2, \ldots, n\}$ and then choose a subset A of $S - \{x\}$. Now our job is to show that the right-hand side counts the same thing.
     Since the right-hand side

$$\sum_{k=1}^{n} k C(n, k)$$

is a sum, we should think of applying the addition principle. Note that for a fixed k, $1 \le k \le n$, the term $k C(n, k)$ counts the number of ways to choose a k-element subset B of S and then choose $x \in B$. So we have the following recipe for uniquely obtaining an ordered pair (x, A), where $x \in S$ and $A \subseteq S - \{x\}$: Step 1: Choose k, $1 \le k \le n$. Step 2: Choose a k-element subset B of S. Step 3: Choose x in B and let $A = B - \{x\}$. By the addition principle, the number of ways to follow this recipe, and hence the number of ordered pairs (x, A), is

$$\sum_{k=1}^{n} k C(n, k)$$

Therefore, it follows that

$$n2^{n-1} = \sum_{k=1}^{n} kC(n,k)$$

This result can also be shown by induction on n.   ■

The binomial coefficients $C(n,k)$, $0 \leq k \leq n$, are listed for successive values of n in the following triangular array:

$$C(0,0)$$
$$C(1,0) \quad C(1,1)$$
$$C(2,0) \quad C(2,1) \quad C(2,2)$$
$$C(3,0) \quad C(3,1) \quad C(3,2) \quad C(3,3)$$
$$C(4,0) \quad C(4,1) \quad C(4,2) \quad C(4,3) \quad C(4,4)$$
$$\vdots$$

Or,

$$1$$
$$1 \quad 1$$
$$1 \quad 2 \quad 1$$
$$1 \quad 3 \quad 3 \quad 1$$
$$1 \quad 4 \quad 6 \quad 4 \quad 1$$
$$\vdots$$

This array is called **Pascal's triangle**, after the seventeenth-century French mathematician Blaise Pascal. Although he did not discover the triangle, he did derive several of its properties. A careful look at Pascal's triangle reveals some interesting patterns. For example, notice that the array is symmetric with respect to the vertical column containing $C(0,0)$, $C(2,1)$, $C(4,2)$, and so on; that is, for $0 \leq k \leq n$, it appears that

$$C(n,k) = C(n,n-k)$$

This result is addressed in Exercise 8 of Exercise Set 1.3. Also, for $n \geq 2$ and $0 < k < n$, notice that $C(n,k)$ is the sum of the two numbers to its immediate left and right in the preceding row. In other words, it appears to be the case that

$$C(n,k) = C(n-1,k-1) + C(n-1,k)$$

This result is known, in fact, as Pascal's identity. It is very useful, deserving special mention and a combinatorial proof.

THEOREM 1.6.3 (Pascal's Identity) For every integer $n \geq 2$ and for each integer k, $0 < k < n$,

$$C(n, k) = C(n - 1, k - 1) + C(n - 1, k)$$

Proof Let $S = \{1, 2, \ldots, n\}$, where n is an integer, $n \geq 2$. For k an integer, $0 < k < n$, let $\mathcal{P}_k$ denote the set of k-element subsets of S. We partition $\mathcal{P}_k$ as $\{\mathcal{P}'_k, \mathcal{P}''_k\}$, where

$$\mathcal{P}'_k = \{T \in \mathcal{P}_k \mid n \in T\}$$
$$\mathcal{P}''_k = \{T \in \mathcal{P}_k \mid n \notin T\}$$

For $T \in \mathcal{P}_k$, note that

$$T \in \mathcal{P}'_k \leftrightarrow T - \{n\} \subseteq S - \{n\}$$

Thus, there is a one-to-one correspondence between the elements of $\mathcal{P}'_k$ and the $(k - 1)$-element subsets of $S - \{n\}$. In a similar fashion, note that

$$T \in \mathcal{P}''_k \leftrightarrow T \subseteq S - \{n\}$$

Hence, there is a one-to-one correspondence between the elements of $\mathcal{P}''_k$ and the k-element subsets of $S - \{n\}$. Therefore,

$$|\mathcal{P}'_k| = C(n - 1, k - 1) \quad \text{and} \quad |\mathcal{P}''_k| = C(n - 1, k)$$

Thus, by using the addition principle,

$$C(n, k) = |\mathcal{P}_k| = |\mathcal{P}'_k| + |\mathcal{P}''_k| = C(n - 1, k - 1) + C(n - 1, k) \qquad \square$$

Another interesting identity comes from looking at the sum of the elements on a diagonal that slopes upward from left to right (starting in a given row); for example, starting with row 4, observe that

$$C(0, 0) + C(1, 0) + C(2, 0) + C(3, 0) + C(4, 0) = 5 = C(5, 1)$$
$$C(1, 1) + C(2, 1) + C(3, 1) + C(4, 1) = 10 = C(5, 2)$$
$$C(2, 2) + C(3, 2) + C(4, 2) = 10 = C(5, 3)$$
$$C(3, 3) + C(4, 3) = 5 = C(5, 4)$$
$$C(4, 4) = 1 = C(5, 5)$$

Note that each left-hand side gives a sum of binomial coefficients $C(m, r)$, where m ranges from r to n. We can compute such a sum by repeated application of Pascal's identity; for example,

$$
\begin{aligned}
C(4,4) + C(5,4) + C(6,4) + C(7,4) &= C(5,5) + C(5,4) + C(6,4) + C(7,4) \\
&= C(6,5) + C(6,4) + C(7,4) \\
&= C(7,5) + C(7,4) \\
&= C(8,5)
\end{aligned}
$$

This suggests a proof by induction of the following result:

THEOREM 1.6.4 For every positive integer n and for each integer r, $0 \leq r \leq n$,

$$
\sum_{m=r}^{n} C(m, r) = C(n + 1, r + 1)
$$

Proof We proceed by induction on n. Let T be the set of positive integers n for which the result holds. To see that $1 \in T$, we must test the result for $r = 0$ and $r = 1$. For $r = 0$, we see that $C(0, 0) + C(1, 0) = C(2, 1)$ holds, whereas for $r = 1$, note that $C(1, 1) = C(2, 2)$. So $1 \in T$.

Now let k be an arbitrary positive integer and assume that $k \in T$. Our induction hypothesis is that

$$
C(r, r) + C(r + 1, r) + \cdots + C(k, r) = C(k + 1, r + 1) \tag{1.6.1}
$$

for $0 \leq r \leq k$. To show that $k + 1 \in T$, we must show that

$$
C(r, r) + C(r + 1, r) + \cdots + C(k, r) + C(k + 1, r) = C(k + 2, r + 1)
$$

for $0 \leq r \leq k + 1$. This statement for $r = k + 1$ is that

$$
C(k + 1, k + 1) = C(k + 2, k + 2)
$$

and is clearly true. To see that it holds for $0 \leq r \leq k$, we add $C(k + 1, r)$ to both sides of identity (1.6.1) and apply Pascal's identity:

$$
\begin{aligned}
C(r, r) + C(r + 1, r) + \cdots + C(k, r) + C(k + 1, r) &= C(k + 1, r + 1) + C(k + 1, r) \\
&= C(k + 2, r + 1)
\end{aligned}
$$

Hence, $k + 1 \in T$. Therefore, $T = \{1, 2, 3, \ldots\}$. $\square$

■ ***Example 1.6.3***   Find a (nice, compact) formula for the sum

$$\sum_{k=1}^{n} k(k+1)(k+2) = (1)(2)(3) + (2)(3)(4) + \cdots + n(n+1)(n+2)$$

Solution

We apply Theorem 1.6.4 to arrive at the value of this sum. The key is to realize that the kth term can be expressed in terms of a binomial coefficient:

$$k(k+1)(k+2) = P(k+2,3) = 3!C(k+2,3)$$

Thus,

$$\sum_{k=1}^{n} k(k+1)(k+2) = \sum_{k=1}^{n} 3!C(k+2,3)$$

$$= 3! \sum_{k=1}^{n} C(k+2,3)$$

$$= 3! \sum_{m=3}^{n+2} C(m,3)$$

$$= 3!C(n+3,4)$$

by Theorem 1.6.4. ■

Further identities and applications of the binomial theorem are explored in the exercises.

We next move on to the multinomial theorem. Just as the binomial theorem deals with expanding the nth power of a binomial, the multinomial theorem deals with expanding the nth power of a general multinomial:

$$(x_1 + x_2 + \cdots + x_k)^n$$

The terms in such an expansion have the general form

$$cx_1^{n_1} x_2^{n_2} \cdots x_k^{n_k}$$

where $n_1, n_2, \ldots, n_k$ are nonnegative integers such that $n_1 + n_2 + \cdots + n_k = n$. We wish to determine the coefficient c of this general term.

Let's look at the specific case $k = 3, n = 4$; we can verify that

$$(x_1 + x_2 + x_3)^4 = x_1^4 + 8x_1^3 x_2 + 6x_1^2 x_2^2 + 12x_1^2 x_2 x_3 + 6x_1^2 x_3^2 + 8x_1 x_2^3 + 12x_1 x_2^2 x_3$$
$$+ 12x_1 x_2 x_3^2 + 8x_1 x_3^3 + x_2^4 + 8x_2^3 x_3 + 6x_2^2 x_3^2 + 8x_2 x_3^3 + x_3^4$$

We note that the expansion of $(x_1 + x_2 + x_3)^4$ has 15 terms. This is because each term has the form

$$cx_1^{m_1} x_2^{m_2} x_3^{m_3}$$

where $m_1, m_2, m_3 \in \mathbb{N}$ and $m_1 + m_2 + m_3 = 4$, so, by Theorem 1.5.4, the number of such terms is $C(4 + 3 - 1, 3 - 1) = C(6, 2) = 15$. Now let's focus on the coefficients; for instance, note that 12 is the coefficient of $x_1^2 x_2 x_3$. Why is this so? Well, $(x_1 + x_2 + x_3)^4$ is the product of four factors. In the initial expansion of this product, before like terms are combined, we obtain a term of the form $x_1^2 x_2 x_3$ by choosing x_1 from exactly two of the four factors, then choosing x_2 from one of the two remaining factors, and then choosing x_3 from the remaining factor. So, by the multiplication principle, the number of such terms is

$$C(4, 2)C(2, 1)C(1, 1) = C(4; 2, 2, 1) = 12$$

Let's now consider the initial expansion of

$$(x_1 + x_2 + \cdots + x_k)^n$$

Before like terms are combined, the number of terms of the form $x_1^{n_1} x_2^{n_2} \cdots x_k^{n_k}$ is the number of ways to choose x_1 from exactly n_1 of the n factors, then choose x_2 from exactly n_2 of the remaining $n - n_1$ factors, and so on, until finally exactly n_k factors remain and x_k is chosen from each of these. By the multiplication principle, this number is

$$C(n, n_1)C(n - n_1, n_2) \cdots C(n - (n_1 + n_2 + \cdots + n_{k-2}), n_{k-1})C(n_k, n_k)$$

$$= \frac{n!}{n_1! n_2! \cdots n_k!} \quad (1.6.2)$$

When each n_i is positive, this number is written $C(n; n_1, n_2, \ldots, n_k)$ and is the number of ordered partitions $(S_1, S_2, \ldots, S_k)$ of an n-element set with $|S_i| = n_i$, $1 \leq i \leq k$. However, formula (1.6.2) holds even when some of the n_i's are 0. Thus, it makes sense to call $C(n; n_1, n_2, \ldots, n_k)$ a **multinomial coefficient** and to define it, for any nonnegative integers n and $n_1, n_2, \ldots, n_k$ such that $n = n_1 + n_2 + \cdots + n_k$, by the formula

$$C(n; n_1, n_2, \ldots, n_k) = \frac{n!}{n_1! n_2! \cdots n_k!}$$

With this understanding, we can state the multinomial theorem.

THEOREM 1.6.5 (Multinomial Theorem)   For every nonnegative integer n, every positive integer k, and any numbers $x_1, x_2, \ldots, x_k$,

$$(x_1 + x_2 + \cdots + x_k)^n = \sum C(n; n_1, n_2, \ldots, n_k) x_1^{n_1} x_2^{n_2} \cdots x_k^{n_k}$$

where the sum is over all k-tuples $(n_1, n_2, \ldots, n_k)$ of nonnegative integers such that $n_1 + n_2 + \cdots + n_k = n$. $\square$

■ ***Example 1.6.4*** Find the coefficient of:

(a) $x_1^3 x_2^2 x_3^1$ in the expansion of $(x_1 + x_2 + x_3)^6$.
(b) $x^4 y^3$ in the expansion of $(w + x + y + z)^7$.
(c) $x^2 y^3$ in the expansion of $(x + 2y + 3)^8$.
(d) x^{30} in the expansion of $(x^2 + x^3 + x^5)^{11}$.

Solution

(a) We apply the multinomial theorem with $n = 6$ and $k = 3$ to find that the answer is $C(6; 3, 2, 1) = 60$.

(b) We apply the multinomial theorem with $n = 7$, $k = 4$, $x_1 = w$, $x_2 = x$, $x_3 = y$, and $x_4 = z$. Then the coefficient of $x^4 y^3 = w^0 x^4 y^3 z^0$ is $C(7; 0, 4, 3, 0) = 35$.

(c) We apply the multinomial theorem with $n = 8$, $k = 3$, $x_1 = x$, $x_2 = 2y$, and $x_3 = 3$. Thus,

$$(x + 2y + 3)^8 = \sum C(8; n_1, n_2, n_3) x^{n_1} (2y)^{n_2} 3^{n_3}$$

So the coefficient of $x^2 y^3$ is $C(8; 2, 3, 3) 2^3 3^3 = 120{,}960$.

(d) First, we apply the multinomial theorem with $n = 11$, $k = 3$, $x_1 = x^2$, $x_2 = x^3$, and $x_3 = x^5$. This yields

$$(x^2 + x^3 + x^5)^{11} = \sum C(11; n_1, n_2, n_3)(x^2)^{n_1}(x^3)^{n_2}(x^5)^{n_3}$$
$$= \sum C(11; n_1, n_2, n_3) x^{2n_1 + 3n_2 + 5n_3}$$

Thus, the coefficient of x^{30} is

$$\sum C(11; n_1, n_2, n_3)$$

where the sum is over all triples (n_1, n_2, n_3) of nonnegative integers with $n_1 + n_2 + n_3 = 11$ and $2n_1 + 3n_2 + 5n_3 = 30$. Note that this system of two equations in three unknowns is equivalent to $n_1 + n_2 = 11 - 3n_3$ and $3n_1 + 2n_2 = 25$. This has three solutions: $(7, 2, 2)$, $(5, 5, 1)$, and $(3, 8, 0)$. Hence, the coefficient we are seeking is $C(11; 7, 2, 2) + C(11; 5, 5, 1) + C(11; 3, 8, 0)$. ■

EXERCISE SET 1.6

1. Use Pascal's identity (Theorem 1.6.3) to evaluate the following binomial coefficients:
   (a) $C(4, 2)$ (b) $C(6, 3)$

2. Prove the binomial theorem (Theorem 1.6.1) by induction on n.

3. Use the binomial theorem to expand each of the following binomials:
   (a) $(x + y)^5$ (b) $(x + y)^6$ (c) $(x + 3y)^7$

4. Prove Pascal's identity (Theorem 1.6.3) algebraically, using Corollary 1.3.7.

5. Give a combinatorial proof of Theorem 1.6.4. (Hint: Consider choosing an $(r + 1)$-element subset of $\{1, 2, \ldots, n + 1\}$; partition these subsets into $n - r + 1$ classes according to the largest element chosen, and apply the addition principle.)

6. Prove the following property, called the **unimodal property** of the binomial coefficients:

   (a) $C(n, k) \leq C(n, k + 1)$ for $0 \leq 2k < n$
   (b) $C(n, k) > C(n, k + 1)$ for $n \leq 2k \leq 2n - 2$
   (c) $C(n, k) = C(n, k + 1)$ if and only if $2k = n - 1$

7. Find the coefficient of:

   (a) $x^{11}y^4$ in the expansion of $(x + y)^{15}$
   (b) x^6y^4 in the expansion of $(2x + y)^{10}$
   (c) x^3y^5 in the expansion of $(3x - 2y)^8$
   (d) x^5 in the expansion of $(1 + x + x^2)(1 + x)^6$

8. For a positive integer n, prove that

$$\sum_{k=0}^{n} C(n, k)^2 = C(2n, n)$$

   (a) Use the binomial theorem and the fact that $(1 + x)^{2n} = (1 + x)^n(1 + x)^n$.
   (b) Give a combinatorial proof based on the fact that each n-element subset of $\{1, 2, \ldots, 2n\}$ is the union of a k-element subset of $\{1, 2, \ldots, n\}$ and an $(n-k)$-element subset of $\{n+1, n+2, \ldots, 2n\}$ for some k, $0 \leq k \leq n$; partition according to the value of k and apply the addition principle.

9. Use an appropriate identity to simplify each of the following expressions:

   (a) $C(12, 4) + C(12, 5)$
   (b) $C(5, 5) + C(6, 5) + C(7, 5) + \cdots + C(11, 5)$
   (c) $C(10, 0) + C(10, 1) + C(10, 2) + \cdots + C(10, 10)$
   (d) $1 \cdot C(9, 1) + 2 \cdot C(9, 2) + 3 \cdot C(9, 3) + \cdots + 9 \cdot C(9, 9)$

10. Consider Pascal's triangle with $n + 1$ rows; let $f(n)$ be the sum of the numbers that lie on the median from the lower left-hand corner to the opposite side. For instance, $f(5) = 1 + 4 + 3 = 8$ and $f(6) = 1 + 5 + 6 + 1 = 13$.

    (a) Express $f(n)$ as a sum of binomial coefficients.
    (b) Do you recognize the sequence $(f(0), f(1), f(2), \ldots)$?

11. Let n and k be integers with $0 \leq k < n$.

    (a) Give an algebraic proof (using Corollary 1.3.7) of the identity

    $$(n - k)C(n, k) = nC(n - 1, k)$$

(b) Give a combinatorial proof of the same identity. (Hint: Let $A = \{1, 2, \ldots, n\}$; how many ways are there to select a pair (x, B), where $B \subseteq A$, $|B| = k < n$, and $x \in A - B$?)

12. For n an integer, $n > 1$, prove the identity

$$\sum_{k=1}^{n} (-1)^{k-1} k C(n, k) = 0$$

13. Use the identity of Exercise 11 (and the fact that $C(k, k) = 1$) to compute the following binomial coefficients.

(a) $C(4, 2)$ (b) $C(6, 3)$

14. Let n and k be positive integers.

(a) Find integers a, b, and c such that

$$k^3 = aC(k, 3) + bC(k, 2) + cC(k, 1)$$

(b) Use part (a) and appropriate identities to find a formula for

$$1^3 + 2^3 + \cdots + n^3$$

15. For $n, r \in \mathbb{N}$, consider the identity

$$\sum_{k=0}^{r} C(n + k, k) = C(n + r + 1, r)$$

(a) Give a combinatorial proof of this identity. (Hint: Consider choosing an r-element subset of $\{1, 2, \ldots, n + r + 1\}$; partition these subsets according to the largest element not chosen, and apply the addition principle.)

(b) Prove the identity by induction on r.

⋆(c) Prove the identity by induction on n.

16. Let S be an n-element set, $n > 0$. Show that the number of subsets of S with even cardinality is the same as the number of subsets of S with odd cardinality. What is this number?

17. Let n be a positive integer and let r be a real number.

(a) Use Corollary 1.6.2 to show that

$$3^n = C(n, 0)2^0 + C(n, 1)2^1 + \cdots + C(n, n)2^n$$

(b) Generalize part (a) to give a formula for the sum

$$\sum_{k=0}^{n} C(n, k) r^k$$

18. Here is an outline of an alternate proof Corollary 1.6.2. We know that $f(x) = (1 + x)^n$ is a polynomial of degree n, say, $g(x) = c_0 + c_1 x + c_2 x^2 + \cdots + c_n x^n$, and we suppose that $f(x) = g(x)$ for all $x \in \mathbb{R}$. We wish to determine the coefficients $c_0, c_1, c_2, \ldots, c_n$. To determine c_0, use the fact that $f(0) = g(0)$; to determine c_1, use the fact that $f'(0) = g'(0)$; to determine c_2, use the fact that $f''(0) = g''(0)$, and so on.

19. Find a formula for

$$\sum_{k=0}^{n} (2k + 1)C(n, k)$$

20. For a given nonnegative integer n, the directed graph $G_n = (V_n, A_n)$ is defined as follows:

$$V_n = \{(x, y) \mid x, y \in \mathbb{N} \text{ and } n \geq x \geq y\}$$
$$A_n = \{((x_1, y_1), (x_2, y_2)) \mid x_2 = x_1 + 1 \text{ and } (y_2 = y_1 \text{ or } y_2 = y_1 + 1)\}$$

   (a) Draw the directed graph G_4.
   (b) In G_n, show that the number of paths from $(0, 0)$ to (n, k) is $C(n, k)$.

21. For $n \in \mathbb{Z}^+$, prove that $C(2n - 1, n - 1) < 4^n$.

22. Extend the result of Example 1.6.3 by showing that, for a positive integer m,

$$\sum_{k=1}^{n} P(m + k - 1, m) = m! C(n + m, m + 1)$$

23. Let n and k be integers with $0 \leq k < n$. Give a combinatorial proof of the identity

$$nC(n - 1, k) = (k + 1)C(n, k + 1)$$

24. For a positive integer n, prove that

$$\sum_{k=0}^{n} \frac{C(n, k)}{k + 1} = \frac{2^{n+1} - 1}{n + 1}$$

   (a) Use Corollary 1.6.2 and some integral calculus.
   (b) Use the identity of Exercise 23, which can be rewritten as

$$\frac{C(n, k)}{k + 1} = \frac{C(n + 1, k + 1)}{n + 1}$$

25. Find the coefficient of:

(a) $x_1^4 x_2^2 x_3^1 x_4^3$ in the expansion of $(x_1 + x_2 + x_3 + x_4)^{10}$
(b) $xy^3 z^5$ in the expansion of $(v + w + x + y + z)^9$
(c) $x^2 y^3$ in the expansion of $(x - 3y + 4)^8$
(d) x^{18} in the expansion of $(1 + x + x^2 + x^3)^7$

CHAPTER 1 PROBLEMS

1. A city police department has 10 detectives: 7 of whom are males and 3 of whom are females. In how many ways can a team of 3 detectives be chosen to work on a case under the following conditions?

(a) There are no restrictions.
(b) The team must have at least 1 male and at least 1 female.

2. A hand in a game of bridge consists of 13 of the 52 cards in a standard deck.

(a) How many bridge hands are there?

How many bridge hands satisfy the following properties?

(b) The hand has exactly two aces.
(c) The hand has exactly three aces, two kings, one queen, and zero jacks.
(d) The hand has at least one heart.
(e) The hand has four spades, three hearts, three diamonds, and three clubs.
(f) The hand has four cards in one suit and three cards in each of the other three suits.

3. Given 11 different combinatorics books and 7 different number theory books, how many ways are there for Curly to choose a combinatorics book, and then for Larry to choose a combinatorics or a number theory book, and then for Moe to choose both a combinatorics and a number theory book? (Assume that the books are chosen without replacement.)

4. If the numbers from 1 to 100,000 are listed, how many times does the digit 5 appear?

5. How many 8-letter sequences, constructed using the 26 (lowercase) letters of the alphabet, satisfy the following conditions?

(a) The sequence contains exactly three a's.
(b) The sequence contains three or four vowels (a, e, i, o, u).
(c) The sequence has no repeated letters.
(d) The sequence contains an even number of e's.

6. A certain make of keyless door lock has five buttons numbered $1, 2, 3, 4, 5$. The lock is set by choosing a code consisting of an ordered pair (A, B) of nonempty subsets of $\{1, 2, 3, 4, 5\}$. To open the door, one first pushes (simultaneously) the buttons corresponding to the elements of A and then pushes the buttons corresponding to the elements of B. How many ways are there to set the lock?

7. The mathematics club at a small college has six senior, five junior, four sophomore, and three freshmen members. Five members are to be chosen to represent the club on the mathematics department curriculum committee. In how many ways can these five students be chosen under the following conditions?

   (a) There are no restrictions.
   (b) At least two seniors are chosen.
   (c) At most one freshman is chosen.
   (d) Exactly two sophomores are chosen.

8. Let $P_{m,n}$ denote the set of nonzero polynomials of degee at most n with coefficients chosen from the set $\{0, 1, \ldots, m\}$. Find $|P_{m,n}|$.

9. A jar contains three red balls, four orange balls, five blue balls, and six green balls, numbered 1 through 18, respectively. A subset of four balls is to be selected from the jar at random. How many subsets satisfy the following conditions?

   (a) The subset has exactly two orange balls and one green ball.
   (b) The subset contains one ball of each color.
   (c) It is not the case that all four balls are the same color.
   (d) At least two balls in the subset are not blue.

10. How many relations are there on an n-element set? Determine the number of relations on $\{1, 2, \ldots, n\}$ that are:
    (a) Reflexive.   (b) Irreflexive.

    (c) Symmetric.   (d) Antisymmetric.

11. Determine the number of ways to arrange the 26 letters A, B, $\ldots$, Z:

    (a) In a row.
    (b) In a row so that no two vowels are consecutive.
    (c) In a circle.
    (d) In a circle so that no two vowels are consecutive.

12. Eight students and four faculty members must be seated at two (identical) six-person tables for lunch; what matters is how this set of twelve people is partitioned into two 6-element subsets.

    (a) In how many ways can this be done?
    (b) In how many ways can it be done so that each table has at least one faculty member?

13. The twelve face cards from a standard deck are arranged in a circle. Four cards are to be chosen.

    (a) In how many ways can this be done?
    (b) In how many ways can this be done if no two cards that are consecutive (in the circle) are chosen?

14. Let m and n be positive integers with $m < n$. We are interested in the following question: How many permutations of $\{1, 2, \ldots, n\}$ do not contain a permutation of $\{1, 2, \ldots, m\}$ as a "subpermutation"?

   (a) First, consider a specific case: How many permutations of $\{1, 2, 3, 4\}$ contain neither the permutation $(1, 2)$ nor the permutation $(2, 1)$?
   (b) Try another specific case: How many permutations of $\{1, 2, 3, 4, 5, 6, 7\}$ contain none of the permutations $(1, 2, 3)$, $(1, 3, 2)$, $(2, 1, 3)$, $(2, 3, 1)$, $(3, 1, 2)$, or $(3, 2, 1)$?
   (c) Formally define the term *subpermutation* and answer the general question.

15. To guarantee that there are 10 diplomats from the same continent at a party, how many diplomats must be invited if they are chosen from the following groups?

   (a) 12 Australasian, 14 African, 15 Asian, 16 European, 18 South American, and 20 North American diplomats
   (b) 7 Australasian, 14 African, 8 Asian, 16 European, 18 South American, and 20 North American diplomats

16. Show that, given any set of seven integers, there exist two integers in the set whose sum or difference is a multiple of 10.

17. A sequence $(x_1, x_2, \ldots, x_m)$ is called **monotonic** if it is increasing or decreasing, that is, if either $x_1 < x_2 < \cdots < x_m$ or $x_1 > x_2 > \cdots > x_m$. In this problem, we explore the following theorem of Erdös and Szekeres: Given a sequence $(x_1, x_2, \ldots, x_t)$ of $t = n^2 + 1$ distinct integers, it contains a monotonic subsequence of $n + 1$ integers.

   (a) Give an example of a sequence (x_1, x_2, x_3, x_4) of four distinct integers that does not contain a monotonic subsequence of length 3.
   (b) Give an example of a sequence $(x_1, x_2, \ldots, x_9)$ of nine distinct integers that does not contain a monotonic subsequence of length 4.
   (c) For an arbitrary integer $n > 1$, give an example of a sequence $(x_1, x_2, \ldots, x_s)$ of $s = n^2$ distinct integers that does not contain a monotonic subsequence of length $n + 1$.
   (d) Prove the result of Erdös and Szekeres. (Hint: For $1 \le i \le n^2 + 1$, let k_i be the maximum length of an increasing subsequence that begins with x_i. If $k_i > n$ for some i, then the result is established, so suppose that $1 \le k_i \le n$ for each i. Apply the pigeonhole principle to show that

$$k_{i_1} = k_{i_2} = \cdots = k_{i_{n+1}}$$

for some $1 \le i_1 < i_2 < \cdots < i_{n+1} \le n^2 + 1$, and then argue that $(x_{i_1}, x_{i_2}, \ldots, x_{i_{n+1}})$ is a decreasing subsequence of the original sequence.)

18. How many ways are there to distribute 20 different homework problems to 10 students under the following conditions?

   (a) There are no restrictions.
   (b) Each student gets 2 problems.

19. How many different positive integers can be formed using the following nine digits?
    (a) $1, 2, 3, 4, 5, 6, 7, 8, 9$ (b) $1, 1, 3, 3, 5, 5, 7, 7, 9$
    (c) $3, 3, 3, 6, 6, 6, 9, 9, 9$ (d) $8, 8, 9, 9, 9, 9, 9, 9, 9$

20. A local dairy offers 16 flavors of ice cream.

    (a) How many ways are there to choose eight scoops of ice cream?
    (b) How many ways are there to choose eight scoops of ice cream of eight different flavors?
    (c) How many ways are there to distribute the eight scoops of part (b) to four students so that each student gets two scoops?
    (d) How many ways are there to distribute eight (identical) scoops of chocolate ice cream to four students so that each student gets at least one scoop?
    (e) If a professor wants to purchase a triple-scoop ice cream cone, how many choices does she have? (Does the order of flavors on the cone matter? Answer the question for both cases.)

21. Consider choosing a subset of three digits from $\{1, 2, \ldots, 9\}$.

    (a) In how many ways can this be done?
    (b) In how many ways can this be done if no two of the digits chosen are consecutive?

22. Given five calculus books, three linear algebra books, and two number theory books (all distinct), how many ways are there to perform the following tasks?

    (a) Select three books, one on each subject.
    (b) Make a row of three books.
    (c) Make a row of three books, with one book on each subject.
    (d) Make a row of three books with exactly two of the subjects represented.
    (e) Make a row of three books all on the same subject.
    (f) Place the books on three different shelves, where the order of the books on a shelf matters.

23. A rumor is spread among m college presidents as follows: The person who starts the rumor telephones someone, and then that person telephones someone else, and so on. Assume any person can pass along the rumor to anyone except the person from whom the rumor was heard. How many different paths through the group can the rumor follow in n calls?

24. Given the books in Problem 22, in how many ways can they be arranged on a shelf under the following conditions?

    (a) There are no restrictions.
    (b) Books on the same subject must be placed together.

25. For $1 \leq k \leq n$, consider the identity

    $$kC(n, k) = (n - k + 1)C(n, k - 1)$$

    (a) Verify this identity algebraically, using Corollary 1.3.7.

(b) Give a combinatorial proof of this identity. (Hint: Argue that both sides of the identity count the number of ways to choose a k-combination of $\{1, 2, \ldots, n\}$, where one of the elements chosen is designated as "special.")

26. Let m, n_1, and n_2 be integers with $0 \le m \le n_1 \le n_2$. Prove **Vandermonde's identity**:

$$C(n_1 + n_2, m) = \sum_{k=0}^{m} C(n_1, k) C(n_2, m - k)$$

(a) Use the binomial theorem and the fact that $(1 + x)^{m+n} = (1 + x)^m (1 + x)^n$.
(b) Give a combinatorial proof based on the fact that each m-combination of $\{1, 2, \ldots, n_1, n_1+1, \ldots, n_1+n_2\}$ is the union of a k-combination of $\{1, 2, \ldots, n_1\}$ and an $(m - k)$-combination of $\{n_1 + 1, n_1 + 2, \ldots, n_1 + n_2\}$ for some k, $0 \le k \le m$; partition according to the value of k and apply the addition principle.
(c) Prove the result by induction on m.

27. Find the coefficient of x^4 in the expansion of $(10x + \frac{1}{2})^8$.

28. Use the binomial theorem to rewrite $(1 + \sqrt{3})^6$ in the form $a + b\sqrt{3}$.

29. Determine the coefficient of $x^7 y^5$ in the expansion of $(2x - y/4)^{12}$.

30. Prove the identity

$$1 \cdot C(n, 0) + 2 \cdot C(n, 1) + \cdots + (n + 1) \cdot C(n, n) = (n + 2) \cdot 2^{n-1}$$

31. The row of Pascal's triangle corresponding to $n = 6$ is

$$1 \quad 6 \quad 15 \quad 20 \quad 15 \quad 6 \quad 1$$

Use Pascal's identity to find the rows of Pascal's triangle corresponding to the following values of n:
(a) 7   (b) 8

32. Use binomial coefficents to give a formula for the sum

$$1 \cdot 2 + 2 \cdot 3 + \cdots + (n - 1) \cdot n$$

33. Compute $C(6, 3)$.

(a) Use Pascal's identity.
(b) Use the identity from Problem 25: $C(n, k) = (n - k + 1)C(n, k - 1)/k$.

34. Let n be a positive integer. Prove that

$$\left(\sum_{i=0}^{n} C(n, i) \right)^2 = \sum_{j=0}^{2n} C(2n, j)$$

35. Let m and n be integers with $0 \le m \le n$. Prove the identity

$$C(n, 0)C(n, m) + C(n, 1)C(n, m - 1) + \cdots + C(n, m)C(n, 0) = C(2n, m)$$

36. Find a formula for the sum

$$\sum_{k=0}^{n} \frac{(-1)^k C(n,k)}{k+1}$$

37. Let m, n_1, and n_2 be positive integers with $m \leq n_2$. Prove that

$$C(n_1 + n_2, n_1 + m) = \sum_{k=0}^{n_2} C(n_1, k) C(n_2, m+k)$$

(Note: By definition, $C(n, i) = 0$ when $n < i$.)

38. Find a formula for

$$\sum_{k=1}^{n} C(n, k-1) C(n, k)$$

39. For a nonnegative integer n, show that

$$\sum_{k=0}^{n} (-1)^k C(n, k) 3^{n-k} = 2^n$$

40. For positive integers n and k, show that

$$k^n = \sum C(n, n_1, n_2, \ldots, n_k)$$

where the sum is over all k-tuples $(n_1, n_2, \ldots, n_k)$ of nonnegative integers such that $n_1 + n_2 + \cdots + n_k = ns$.

41. Let n be a positive integer. Show that

(a) $\displaystyle\sum_{k=0}^{n} (2k+1) C(2n+1, 2k+1) = (2n+1)2^{2n-1}$

(b) $\displaystyle\sum_{k=1}^{n} 2k C(2n, 2k) = n2^{2n-1}$

42. Let k be a fixed positive integer. Show that any nonnegative integer r is uniquely represented as

$$r = C(m_1, 1) + C(m_2, 2) + \cdots + C(m_k, k)$$

for integers $0 \leq m_1 < m_2 < \cdots < m_k$. (See also Exercise 10 in Exercise Set 2.6.)

43. Let r and n be positive integers with $2 \leq r \leq n - 2$. Give a combinatorial argument to show that

$$C(n, r) = C(n-2, r) + 2C(n-2, r-1) + C(n-2, r-2)$$

44. Given a positive integer n and an integer r, $0 \leq r \leq n! - 1$, show that r is uniquely expressed as

$$\sum_{k=1}^{n-1} d_k k!$$

where each d_k is an integer, $0 \le d_k \le k$. This representation is called the **factorial representation** of r. For instance,

$$85 = 3 \cdot 4! + 2 \cdot 3! + 0 \cdot 2! + 1 \cdot 1!$$

so $d_4 = 3$, $d_3 = 2$, $d_2 = 0$, and $d_1 = 1$ in the factorial representation of 85. (See also Exercise 4 in Exercise Set 2.5.)

45. For positive integers m and n, determine the number of integer solutions to

$$x_1 + x_2 + \cdots + x_n + x_{n+1} + \cdots + x_{2n} = m$$

such that $x_1, \ldots, x_n \in \mathbb{N}$ and $x_{n+1}, \ldots, x_{2n} \in \mathbb{Z}^+$.

46. Let $n_1, n_2, \ldots, n_k$, and n be positive integers with $n = n_1 + n_2 + \cdots + n_k$. Show that the multinomial coefficient $C(n; n_1, n_2, \ldots, n_k)$ is equal to

$$C(n-1; n_1 - 1, n_2, \ldots, n_k) + C(n-1; n_1, n_2 - 1, n_3, \ldots, n_k) + \cdots$$
$$+ C(n-1; n_1, n_2, \ldots, n_k - 1)$$

47. Determine the constant term of $(x + x^{-1} + 2x^{-4})^{17}$.

48. Let m, k, and n be integers with $0 \le m \le k \le n$.

(a) Show that $C(k, m)C(n, k) = C(n, m)C(n - m, k - m) = C(n, m)C(n - m, n - k)$.
(b) Show that

$$\sum_{k=m}^{n} C(k, m)C(n, k) = C(n, m)2^{n-m}$$

49. Use the multinomial theorem to simplify $(1 + \sqrt{2} + \sqrt{5})^4$.

50. Eight scientists work at a secret research building. The building is to have D doors, each with four locks. Each scientist is to be issued a set of K keys. For reasons of security, it is required that at least four scientists be present together in order that, among them, they possess a subset of four keys that fit the four locks on one of the D doors. How many doors must the building have and how many keys must each scientist be issued?

51. Use the binomial theorem to determine whether $(1.01)^{100}$ is greater than 2.

52. If, in Problem 50, there is to be only one door with L locks, what should the values be for L and K if it is required that at least four scientists be present together in order to enter the building?

53. Use the binomial theorem to determine whether $(0.99)^{100}$ is less than 0.4.

54. How many ways are there to distribute 12 different books to three different bookshelves if the order of the books on a shelf is important? (Note: Some shelves may have no books assigned to them.)

55. How many ways are there to place eight rings on four fingers of your left hand? Answer assuming that

(a) The rings are identical.
(b) The rings are identical and at least one ring must be placed on each finger.
(c) The rings are different and the order of rings on a finger is not important.

    (d) The rings are different and the order of rings on a finger is important.

    (e) The rings are different, the order of rings on a finger is important, and at least one ring must be placed on each finger.

    (f) The rings are different, the order of rings on a finger is important, and exactly two rings must be placed on each finger.

56. Generalize Problems 54 and 55(d) to determine a formula for the number of ways to distribute m different balls to n different boxes if the order in which the balls are distributed to a box is important.

57. How many ways are there to distribute 12 different books to three different bookshelves if the order of the books on a shelf is important and at least 1 book must be placed on each shelf?

58. Generalize Problems 55(e) and 57 to determine a formula for the number of ways to distribute m different balls to n different boxes ($m \geq n$) if the order in which the balls are distributed to a box is important and at least one ball must be placed in each box.

59. Suppose in Problems 54 and 57 that we have a top shelf, a middle shelf, and a bottom shelf. How many ways are there to distribute the 12 books to these three shelves if the order of the books on a shelf is important, 3 books are placed on the top shelf, 4 books are placed on the middle shelf, and 5 books are placed on the bottom shelf.

60. Generalize Problems 55(f) and 59 to determine a formula for the number of ways to distribute m different balls to n different boxes if the order in which the balls are distributed to a box is important and exactly m_i balls are placed in box i, $1 \leq i \leq n$, where each $m_i \in \mathbb{N}$ and $m_1 + m_2 + \cdots + m_n = m$.

Intermediate Combinatorics

2.1 *PRINCIPLE OF INCLUSION–EXCLUSION*

In Chapter 1, Theorem 1.2.1 stated the addition principle for two sets. For convenience, we restate the theorem here.

> **THEOREM 1.2.1** If S_1 and S_2 are nonempty, disjoint finite sets, then
>
> $$|S_1 \cup S_2| = |S_1| + |S_2|$$ □

Now let us consider the more general situation, where S_1 and S_2 are not disjoint. Can we obtain a formula for $|S_1 \cup S_2|$? Indeed, for n finite sets $S_1, S_2, \ldots, S_n$, can we obtain a formula for $|S_1 \cup S_2 \cup \cdots \cup S_n|$?

We begin by letting $S = S_1 \cup S_2$ and drawing the Venn diagram in Figure 2.1.1(a). Suppose it is given that $|S_1| = 5$, $|S_2| = 8$, and $|S_1 \cap S_2| = 2$, and the problem is to determine $|S|$. We would like to use the addition principle; however, $\{S_1, S_2\}$ is not a partition of S. Is there a natural partition of S in this case? By looking at the Venn diagram, we can readily conjecture that $\{S_1, S_2 - S_1\}$ is a partition of S, as suggested by Figure 2.1.1(b). This result is easy to prove. Thus, by the addition principle,

$$|S| = |S_1| + |S_2 - S_1| \tag{2.1.1}$$

We know that $|S_1| = 5$, so now we need to figure out $|S_2 - S_1|$. Here the key is to recall that $\{S_1 \cap S_2, S_2 - S_1\}$ is a partition of S_2 (Exercise 8 in Exercise Set 0.1), so that

$$|S_2 - S_1| = |S_2| - |S_1 \cap S_2|$$
$$= 8 - 2 = 6 \tag{2.1.2}$$

Thus,

$$|S| = |S_1| + |S_2 - S_1| = 5 + 6 = 11$$

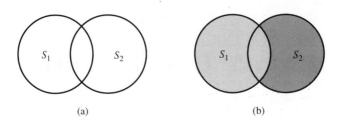

(a)   (b)

Figure 2.1.1

Note that identities (2.1.1) and (2.1.2) hold for any finite sets S_1 and S_2, where $S = S_1 \cup S_2$. Putting these together yields the following formula for $|S_1 \cup S_2|$:

$$|S_1 \cup S_2| = |S_1| + |S_2 - S_1|$$
$$= |S_1| + (|S_2| - |S_1 \cap S_2|)$$
$$= |S_1| + |S_2| - |S_1 \cap S_2|$$

This proves the following result:

THEOREM 2.1.1   Let S_1 and S_2 be finite sets. Then $S_1 \cup S_2$ is finite and

$$|S_1 \cup S_2| = |S_1| + |S_2| - |S_1 \cap S_2| \tag{2.1.3}$$

□

Theorem 2.1.1 is a special case of the general **principle of inclusion–exclusion**. Identity (2.1.3) is stated for two sets S_1 and S_2; the general principle will be stated for $n \geq 2$ finite sets $S_1, S_2, \dots, S_n$. The reason for the name "inclusion–exclusion principle" comes from a particular argument that can be given to justify its correctness. In the case of two sets, this argument goes as follows: Suppose that S_1 and S_2 are subsets of a universal set U, and let $S = S_1 \cup S_2$. To show that (2.1.3) is correct, we must show that each element $x \in U$ is accounted for the same number of times on both the left- and right-hand sides of the identity. First, suppose that $x \in \bar{S} = \bar{S}_1 \cap \bar{S}_2$. Note that such an x belongs to none of the sets S_1, S_2, and $S_1 \cap S_2$, so that it contributes 0 to both sides of (2.1.3). Next, the case $x \in S$ is divided into three subcases, depending on whether $x \in S_1 - S_2$ or $x \in S_1 \cap S_2$ or $x \in S_2 - S_1$. For the first subcase, note that each $x \in S_1 - S_2$ contributes 1 to $|S_1|$, 0 to $|S_2|$, and 0 to $|S_1 \cap S_2|$. Thus, such an x contributes 1 to both sides of (2.1.3). The third subcase is similar to the first, so let us consider the subcase $x \in S_1 \cap S_2$. Such an x contributes 1 to each of $|S_1|$, $|S_2|$, and $|S_1 \cap S_2|$. In other words, it is "included" once in S_1, is included again in S_2, but then is "excluded" once for belonging to $S_1 \cap S_2$. Thus, such an x also contributes 1 to both sides of (2.1.3), and so identity (2.1.3) is correct.

As an application of the principle of inclusion–exclusion, consider the following situation. Suppose that n people who work together in an office decide to have a party. They decide it might be fun to have each person buy a gift for some other person chosen at random from the group. To set this up, the n names are put into a hat, and then each person in turn picks a name at random (without replacement). We ask, What is the probability that no person draws her or his own name?

Suppose the people are numbered $1, 2, \ldots, n$ in the order they draw from the hat; that is, person 1 picks a number, then person 2 picks a number, and so on. The result can then be represented as a permutation $(x_1, x_2, \ldots, x_n)$ of $X = \{1, 2, \ldots, n\}$, where x_i is the number of the person picked by person number i. Since there are n-factorial permutations of X and they are equally likely to occur, the probability that no person draws his or her own name is $d(n)/n!$, where $d(n)$ is the number of permutations $(x_1, x_2, \ldots, x_n)$ of X such that $x_i \neq i$ for each i, $1 \leq i \leq n$. This leads to the following definition:

> **DEFINITION 2.1.1**   A permutation $(x_1, x_2, \ldots, x_n)$ of $\{1, 2, \ldots, n\}$ is called a **derangement** provided $x_i \neq i$ for each i, $1 \leq i \leq n$. More generally, let $Y = \{y_1, y_2, \ldots, y_n\}$, where $y_1 < y_2 < \cdots < y_n$. A permutation $(x_1, x_2, \ldots, x_n)$ of Y is called a **derangement of Y** provided $x_i \neq y_i$ for each i, $1 \leq i \leq n$. The number of derangements of an n-element (ordered) set is written $d(n)$.   ∎

Let $(x_1, x_2, \ldots, x_n)$ be a permutation of $\{1, 2, \ldots, n\}$. If $x_i = i$, then we say that the element i is in its "natural position" in the permutation. So the identity permutation $(1, 2, \ldots, n)$ has every element in its natural position, and a derangement has no element in its natural position. For example, $(4, 3, 2, 1)$ is a derangement of $\{1, 2, 3, 4\}$, but $(2, 1, 3, 4)$ is not, since 3 is in its natural position.

We wish to determine the number $d(n)$ of derangements of $\{1, 2, \ldots, n\}$. As a start, let's consider the problem for some small values of n.

The only permutation of $\{1\}$ is (1), so $d(1) = 0$. The only derangement of $\{1, 2\}$ is $(2, 1)$, so $d(2) = 1$. The six permutations of $\{1, 2, 3\}$ are $(1, 2, 3)$, $(1, 3, 2)$, $(2, 1, 3)$, $(2, 3, 1)$, $(3, 1, 2)$, and $(3, 2, 1)$. Of these, only $(2, 3, 1)$ and $(3, 1, 2)$ are derangements, so $d(3) = 2$.

For the case $n = 4$, rather than listing all 24 permutations of $\{1, 2, 3, 4\}$ and counting the derangements, let's use a tree diagram to systematically enumerate the derangements; such a diagram is shown in Figure 2.1.2. In forming a derangement (x_1, x_2, x_3, x_4) of $\{1, 2, 3, 4\}$, there are three choices for x_1: $x_1 = 2$, $x_1 = 3$, or $x_1 = 4$. So the tree diagram has three arcs leading from the root labeled with these three choices. Suppose, for example, that $x_1 = 3$; then there are two choices for x_2: $x_2 = 1$ or $x_2 = 4$. If $x_1 = 3$ and $x_2 = 1$, then we are forced to have $x_3 = 4$ and $x_4 = 2$, leading to the derangement $(3, 1, 4, 2)$. On the other hand, if $x_1 = 3$ and $x_2 = 4$, then (x_3, x_4) can be either permutation of $\{1, 2\}$, and we obtain the two derangements $(3, 4, 1, 2)$ and $(3, 4, 2, 1)$. The analysis for the other two choices for x_1 is similar and is left for you to provide. In all, we obtain the nine derangements labeling the leaves of the tree as shown. Thus, $d(4) = 9$.

The kind of analysis used above to find $d(n)$ for small n becomes more and more difficult as the value of n increases, so we require a different strategy. Approaching the problem of finding $d(n)$ from another direction, let us consider the easier problem of finding the number of permutations of $\{1, 2, \ldots, n\}$ with certain specified elements "deranged."

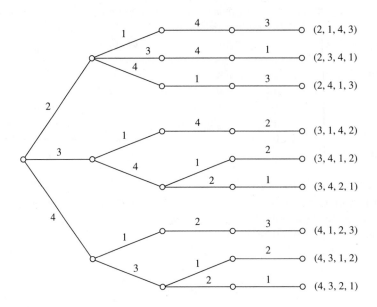

Figure 2.1.2   Tree diagram enumerating the nine derangements of $\{1,2,3,4\}$.

■ ***Example 2.1.1***   For $n \geq 2$, find the number of permutations $(x_1, x_2, \ldots, x_n)$ of $X = \{1, 2, \ldots, n\}$ such that $x_1 \neq 1$ and $x_2 \neq 2$.

Solution
Let U be the set of all n-factorial permutations of X. Let

$$S_1 = \{(x_1, x_2, \ldots, x_n) \in U \mid x_1 = 1\}$$

and

$$S_2 = \{(x_1, x_2, \ldots, x_n) \in U \mid x_2 = 2\}$$

We wish to determine $|\bar{S}_1 \cap \bar{S}_2|$. Since $\bar{S}_1 \cap \bar{S}_2 = \overline{S_1 \cup S_2}$, we first apply Theorem 2.1.1 to find $|S_1 \cup S_2|$. Note that

$$|S_1| = |S_2| = (n - 1)!$$

Also,

$$S_1 \cap S_2 = \{(x_1, x_2, \ldots, x_n) \in U \mid x_1 = 1 \text{ and } x_2 = 2\}$$

so that

$$|S_1 \cap S_2| = (n-2)!$$

Thus,

$$\begin{aligned}
|S_1 \cup S_2| &= |S_1| + |S_2| - |S_1 \cap S_2| \\
&= (n-1)! + (n-1)! - (n-2)! \\
&= [(n-1) + (n-1) - 1](n-2)! \\
&= (2n-3)(n-2)!
\end{aligned}$$

Hence, by the addition principle,

$$\begin{aligned}
|\bar{S}_1 \cap \bar{S}_2| &= |U| - |S_1 \cup S_2| \\
&= n! - (2n-3)(n-2)! \\
&= [n(n-1) - (2n-3)](n-2)! \\
&= (n^2 - 3n + 3)(n-2)!
\end{aligned}$$ ∎

Next, let S_1, S_2, and S_3 be finite sets; let us see if we can develop a formula for $|S_1 \cup S_2 \cup S_3|$. One idea is to first think of $S_1 \cup S_2 \cup S_3$ as the union of the two sets S_1 and $S_2 \cup S_3$ and apply Theorem 2.1.1:

$$\begin{aligned}
|S_1 \cup (S_2 \cup S_3)| &= |S_1| + |S_2 \cup S_3| - |S_1 \cap (S_2 \cup S_3)| \\
&= |S_1| + (|S_2| + |S_3| - |S_2 \cap S_3|) - |(S_1 \cap S_2) \cup (S_1 \cap S_3)| \\
&= |S_1| + |S_2| + |S_3| - |S_2 \cap S_3| \\
&\quad - (|S_1 \cap S_2| + |S_1 \cap S_3| - |(S_1 \cap S_2) \cap (S_1 \cap S_3)|) \\
&= |S_1| + |S_2| + |S_3| - |S_1 \cap S_2| - |S_1 \cap S_3| - |S_2 \cap S_3| \\
&\quad + |S_1 \cap S_2 \cap S_3|
\end{aligned}$$

This proves the following result, known as the principle of inclusion–exclusion for the case of three sets:

THEOREM 2.1.2 Let S_1, S_2, and S_3 be finite sets. Then $S = S_1 \cup S_2 \cup S_3$ is finite and

$$|S| = |S_1| + |S_2| + |S_3| - |S_1 \cap S_2| - |S_1 \cap S_3| - |S_2 \cap S_3| + |S_1 \cap S_2 \cap S_3|$$ □

■ *Example 2.1.2* For $n \geq 3$, find the number of permutations $(x_1, x_2, \ldots, x_n)$ of $X = \{1, 2, \ldots, n\}$ such that $x_1 \neq 1$, $x_2 \neq 2$, and $x_3 \neq 3$.

Solution

We define U as in Example 2.1.1 and define subsets S_1, S_2, and S_3 by:

$$S_1 = \{(x_1, x_2, \ldots, x_n) \in U \mid x_1 = 1\}$$
$$S_2 = \{(x_1, x_2, \ldots, x_n) \in U \mid x_2 = 2\}$$
$$S_3 = \{(x_1, x_2, \ldots, x_n) \in U \mid x_3 = 3\}$$

We wish to determine $|\bar{S}_1 \cap \bar{S}_2 \cap \bar{S}_3| = |U| - |S_1 \cup S_2 \cup S_3|$. As in Example 2.1.1,

$$|S_1| = |S_2| = |S_3| = (n-1)!$$

and

$$|S_1 \cap S_2| = |S_1 \cap S_3| = |S_2 \cap S_3| = (n-2)!$$

Moreover, since

$$S_1 \cap S_2 \cap S_3 = \{(x_1, x_2, \ldots, x_n) \in U \mid x_1 = 1 \text{ and } x_2 = 2 \text{ and } x_3 = 3\}$$

we have

$$|S_1 \cap S_2 \cap S_3| = (n-3)!$$

Now, by using Theorem 2.1.2,

$$\begin{aligned}
|S_1 \cup S_2 \cup S_3| &= |S_1| + |S_2| + |S_3| \\
&\quad - |S_1 \cap S_2| - |S_1 \cap S_3| - |S_2 \cap S_3| \\
&\quad + |S_1 \cap S_2 \cap S_3| \\
&= 3(n-1)! - 3(n-2)! + (n-3)! \\
&= [3(n-1)(n-2) - 3(n-2) + 1](n-3)! \\
&= (3n^2 - 12n + 13)(n-3)!
\end{aligned}$$

Thus,

$$\begin{aligned}
|\bar{S}_1 \cap \bar{S}_2 \cap \bar{S}_3| &= |U| - |S_1 \cup S_2 \cup S_3| \\
&= n! - (3n^2 - 12n + 13)(n-3)! \\
&= [n(n-1)(n-2) - (3n^2 - 12n + 13)](n-3)! \\
&= (n^3 - 6n^2 + 14n - 13)(n-3)!
\end{aligned}$$ ∎

Theorems 2.1.1 and 2.1.2 point to the form of the general principle of inclusion–exclusion. Let $S_1, S_2, \ldots, S_n$ be finite sets. To find the cardinality of $S_1 \cup S_2 \cup \cdots \cup S_n$, we first add together the cardinalities of the sets individually, then subtract the cardinalities of the sets intersected in pairs, then add back the cardinalities of the sets intersected three at a time, and so on.

THEOREM 2.1.3 (Principle of Inclusion–Exclusion)   For $n \geq 2$, let $S_1, S_2, \ldots, S_n$ be finite subsets of a universal set U. Then $S = S_1 \cup S_2 \cup \cdots \cup S_n$ is finite and

$$|S| = \sum_{k=1}^{n} (-1)^{k-1} \sum |S_{i_1} \cap S_{i_2} \cap \cdots \cap S_{i_k}| \tag{2.1.4}$$

where for a fixed value of k, $1 \leq k \leq n$, the inner sum

$$\sum |S_{i_1} \cap S_{i_2} \cap \cdots \cap S_{i_k}|$$

is taken over all k-combinations $\{i_1, i_2, \ldots, i_k\}$ of $\{1, 2, \ldots, n\}$.

Proof   One way to prove the theorem is by induction on n; this is left for you to work on in Exercise 20. Instead, we show here that each $x \in U$ contributes the same amount to both sides of identity (2.1.4). We distinguish the following cases:

Case 1. $x \in \overline{S}$. In this case x contributes 0 to the left-hand side of (2.1.4). Since $\overline{S} = \overline{S}_1 \cap \overline{S}_2 \cap \cdots \cap \overline{S}_n$, we have for any k, $1 \leq k \leq n$, and any k-combination $\{i_1, i_2, \ldots, i_k\}$ of $\{1, 2, \ldots, n\}$, that $x \notin S_{i_1} \cap S_{i_2} \cap \cdots \cap S_{i_k}$. Hence, x also contributes 0 to the right-hand side of (2.1.4).

Case 2.   Suppose $x \in S$ and x belongs to exactly m of the sets $S_1, S_2, \ldots, S_n$ for some m, $1 \leq m \leq n$. Note that x contributes 1 to the left-hand side of (2.1.4). To see the contribution of x to the right-hand side, we examine the inner sum for each value of k, $1 \leq k \leq n$. First, we consider $k = 1$. Since x belongs to exactly m of the sets $S_1, S_2, \ldots, S_n$, the contribution of x to

$$\sum |S_{i_1}|$$

must be $C(m, 1) = m$. We consider next the inner sum for $k = 2$, namely,

$$\sum |S_{i_1} \cap S_{i_2}|$$

where the sum is over all 2-combinations $\{i_1, i_2\}$ of $\{1, 2, \ldots, n\}$. The number of times x is counted in this sum is precisely the number of ways to choose two of the m subsets to which x belongs; this is just $C(m, 2)$. More generally, we consider the inner sum

$$\sum |S_{i_1} \cap S_{i_2} \cap \cdots \cap S_{i_k}|$$

corresponding to a fixed value of k, $1 \le k \le m$, where the sum is over all k-combinations $\{i_1, i_2, \ldots, i_k\}$ of $\{1, 2, \ldots, n\}$. The number of times x is counted in this sum is precisely the number of ways of choosing k of the m subsets that contain x, and this is $C(m, k)$. Moreover, for $k > m$ and any k-combination $\{i_1, i_2, \ldots, i_k\}$ of $\{1, 2, \ldots, n\}$, we have

$$x \notin S_{i_1} \cap S_{i_2} \cap \cdots \cap S_{i_k}$$

Hence, x contributes 0 to those terms on the right-hand side corresponding to values of k greater than m. Thus, the total contribution of x to the right-hand side of (2.1.4) is

$$C(m, 1) - C(m, 2) + \cdots + (-1)^m C(m, m) = \sum_{k=1}^{m} (-1)^{k-1} C(m, k)$$

In Section 1.6, as an immediate application of Corollary 1.6.2, we found that

$$\sum_{k=0}^{m} (-1)^k C(m, k) = 0$$

It follows that

$$\sum_{k=1}^{m} (-1)^k C(m, k) = -1$$

so that

$$\sum_{k=1}^{m} (-1)^{k-1} C(m, k) = 1$$

So the contribution of x to the right-hand side of (2.1.4) is 1, the same as its contribution to the left-hand side.

This completes the argument for Case 2 and, as a consequence, establishes the theorem. $\square$

Sometimes the principle of inclusion–exclusion is stated as a formula for finding

$$|\overline{S}_1 \cap \overline{S}_2 \cap \cdots \cap \overline{S}_n|$$

(where it is assumed that U is finite and $|U|$ is known). Since

$$\overline{S}_1 \cap \overline{S}_2 \cap \cdots \cap \overline{S}_n = \overline{S_1 \cup S_2 \cup \cdots \cup S_n}$$

such a result is obtained by applying the addition principle:

$$|\overline{S}_1 \cap \overline{S}_2 \cap \cdots \cap \overline{S}_n| = |U| - |S_1 \cup S_2 \cup \cdots \cup S_n|$$

and then using Theorem 2.1.3 to find $|S_1 \cup S_2 \cup \cdots \cup S_n|$.

To apply the principle of inclusion–exclusion to determine the cardinality of a set S, we must think of S as $S_1 \cup S_2 \cup \cdots \cup S_n$ (or $\overline{S}_1 \cap \overline{S}_2 \cap \cdots \cap \overline{S}_n$) for appropriately defined sets $S_1, S_2, \ldots, S_n$. In addition, it must be possible to compute

$$|S_{i_1} \cap S_{i_2} \cap \cdots \cap S_{i_k}|$$

for any k-combination $\{i_1, i_2, \ldots, i_k\}$ of $\{1, 2, \ldots, n\}$. It often helps to think of S_i as the set of elements (in the universal set) with some special property p_i. With this interpretation, $S_1 \cup S_2 \cup \cdots \cup S_n$ is the set of elements having at least one of the properties $p_1, p_2, \ldots, p_n$, whereas $\overline{S}_1 \cap \overline{S}_2 \cap \cdots \cap \overline{S}_n$ is the set of elements having none of the properties $p_1, p_2, \ldots, p_n$.

To illustrate the principle of inclusion–exclusion, we return to the problem of computing $d(n)$, the number of derangements of $X = \{1, 2, \ldots, n\}$. Following the lead of Examples 2.1.1 and 2.1.2, we let U be the set of all n-factorial permutations of X and, for $1 \le i \le n$, let

$$S_i = \{(x_1, x_2, \ldots, x_n) \in U \mid x_i = i\}$$

In other words, for $1 \le i \le n$, we let p_i be the property that a given permutation of X has i in its natural position; then S_i is the set of permutations with this property. Since a derangement of X has no number in its natural position, we wish to find the cardinality of the set of permutations of X having none of the properties $p_1, p_2, \ldots, p_n$, namely, we wish to find $|\overline{S}_1 \cap \overline{S}_2 \cap \cdots \cap \overline{S}_n|$. We can apply Theorem 2.1.3 provided that, for an arbitrary k, $1 \le k \le n$, and any k-combination $\{i_1, i_2, \ldots, i_k\}$ of $\{1, 2, \ldots, n\}$, we can compute

$$|S_{i_1} \cap S_{i_2} \cap \cdots \cap S_{i_k}|$$

Note that

$$S_{i_1} \cap S_{i_2} \cap \cdots \cap S_{i_k} =$$
$$\{(x_1, x_2, \ldots, x_n) \in U \mid x_{i_1} = i_1 \text{ and } x_{i_2} = i_2 \text{ and } \cdots \text{ and } x_{i_k} = i_k\}$$

This is just a set of permutations having some specified k numbers from $\{1, 2, \ldots, n\}$ occurring in their natural positions. In forming such a permutation, we only need to decide where to position the other $n - k$ numbers; thus,

$$|S_{i_1} \cap S_{i_2} \cap \cdots \cap S_{i_k}| = (n - k)!$$

Then, by applying Theorem 2.1.3:

$$|S_1 \cup S_2 \cup \cdots \cup S_n| = \sum_{k=1}^{n} (-1)^{k-1} \sum |S_{i_1} \cap S_{i_2} \cap \cdots \cap S_{i_k}|$$

$$= \sum_{k=1}^{n} (-1)^{k-1} \sum (n-k)!$$

$$= \sum_{k=1}^{n} (-1)^{k-1} C(n,k)(n-k)!$$

(since the inner sum is over all $C(n,k)$ k-combinations of X)

$$= \sum_{k=1}^{n} (-1)^{k-1} \frac{n!}{k!}$$

$$= n! \sum_{k=1}^{n} \frac{(-1)^{k-1}}{k!}$$

Thus,

$$|\bar{S}_1 \cap \bar{S}_2 \cap \cdots \cap \bar{S}_n| = |U| - |S_1 \cup S_2 \cup \cdots \cup S_n| = n! - n! \sum_{k=1}^{n} \frac{(-1)^{k-1}}{k!}$$

Therefore, we have the following result:

THEOREM 2.1.4 $d(n) = n! \left(\displaystyle\sum_{k=0}^{n} \frac{(-1)^k}{k!} \right)$ $\square$

As an example, let's use Theorem 2.1.4 to compute $d(5)$:

$$d(5) = 5! \left(\sum_{k=0}^{5} \frac{(-1)^k}{k!} \right) = 120 \left(1 - 1 + \frac{1}{2} - \frac{1}{6} + \frac{1}{24} - \frac{1}{120} \right)$$

$$= 60 - 20 + 5 - 1 = 44$$

Dividing $d(n)$ by $n!$ gives us the ratio $d(n)/n!$ of the number of derangements to the total number of permutations of $\{1, 2, \ldots, n\}$:

$$\frac{d(n)}{n!} = \sum_{k=0}^{n} \frac{(-1)^k}{k!}$$

This is also the probability that no person draws her or his own name from the hat in our office party example. The following table gives values of $d(n)$ and $d(n)/n!$ for $n \in \{1, 2, 3, 4, 5\}$:

$n =$	1	2	3	4	5
$d(n) =$	0	1	2	9	44
$\dfrac{d(n)}{n!} =$	0	$\dfrac{1}{2}$	$\dfrac{1}{3}$	$\dfrac{3}{8}$	$\dfrac{11}{30}$

Note that for a fixed value of i, $1 \leq i \leq n$, the probability that the ith person draws his or her own name from the hat is $(n-1)!/n! = 1/n$. Thus, as n gets larger and larger, the probability that any given person gets his or her own name gets smaller and smaller. We might then conjecture that the probability that some person gets her or his own name should tend to zero as n tends to infinity. On the other hand, as n increases, there are more people drawing from the hat, thus more chances for someone to get his or her own name. So perhaps the probability that someone will get her or his own name should increase as n increases. Somewhat surprisingly, however, computation of $d(n)/n!$ for small values of n leads to the conjecture that it is converging to some number greater than $1/3$ as n tends to infinity.

To obtain this limit, recall the Maclaurin series for e^x:

$$e^x = 1 + x + \frac{x^2}{2} + \frac{x^3}{6} + \cdots = \sum_{k=0}^{\infty} \frac{x^k}{k!}$$

(which is valid for any real number x). By letting $x = -1$, we obtain

$$e^{-1} = 1 - 1 + \frac{1}{2} - \frac{1}{6} + \cdots = \sum_{k=0}^{\infty} \frac{(-1)^k}{k!}$$

namely, the infinite series

$$1 - 1 + \frac{1}{2} - \frac{1}{6} + \frac{1}{24} - \frac{1}{120} + \cdots$$

converges to $1/e$. Since $d(n)/n!$ is just the nth partial sum of this series, it follows that $d(n)/n!$ approaches $1/e$ as n approaches infinity. In fact, considering the rapid rate of convergence of the series for $1/e$, we can say that $d(n)/n!$ is closely approximated by $1/e$, and $d(n)$ is closely approximated by $n!/e$. In fact, it can be shown that $d(n)$ is equal to the integer nearest $n!/e$. For example, $6!/e \simeq 264.87$, so that $d(6) = 265$.

COROLLARY 2.1.5   The number of derangements $d(n)$ of an n-element (ordered) set is equal to the integer nearest $n!/e$.   □

■ *Example 2.1.3*   Use the principle of inclusion–exclusion to determine the number of ways to select a five-card poker hand from a standard deck such that the hand contains at least one card from each suit.

Solution
Let U be the set of all five-card poker hands, so that $|U| = C(52, 5)$. Let S be the set of hands containing at least one card from each suit; we are asked to find $|S|$. Note that the complement of S is the set of hands that are void in some suit. (If a hand has

no cards from a given suit, then we say that the hand is "void" in that suit.) Thus, it makes sense to define the following subsets of U:

S_1 = the set of hands void in spades

S_2 = the set of hands void in hearts

S_3 = the set of hands void in diamonds

S_4 = the set of hands void in clubs

Then $S = \overline{S}_1 \cap \overline{S}_2 \cap \overline{S}_3 \cap \overline{S}_4$. In order to apply Theorem 2.1.3, note that

$$|S_1| = |S_2| = |S_3| = |S_4| = C(39, 5)$$
$$|S_1 \cap S_2| = |S_1 \cap S_3| = |S_1 \cap S_4| = |S_2 \cap S_3| = |S_2 \cap S_4| = |S_3 \cap S_4| = C(26, 5)$$
$$|S_1 \cap S_2 \cap S_3| = |S_1 \cap S_2 \cap S_4| = |S_1 \cap S_3 \cap S_4| = |S_2 \cap S_3 \cap S_4| = C(13, 5)$$
$$|S_1 \cap S_2 \cap S_3 \cap S_4| = 0$$

Thus,

$$|S_1 \cup S_2 \cup S_3 \cup S_4| = \sum_{k=1}^{4} (-1)^{k-1} \sum |S_{i_1} \cap \cdots \cap S_{i_k}|$$

$$= \sum_{k=1}^{4} (-1)^{k-1} C(4, k) C(52 - 13k, 5)$$

$$= C(4, 1)C(39, 5) - C(4, 2)C(26, 5) + C(4, 3)C(13, 5) - C(4, 4) \cdot 0$$

$$= 2,303,028 - 394,680 + 5148 - 0$$

$$= 1,913,496$$

Finally,

$$|\overline{S}_1 \cap \overline{S}_2 \cap \overline{S}_3 \cap \overline{S}_4| = C(52, 5) - 1,913,496$$
$$= 685,464 \qquad \blacksquare$$

In the statement of the principle of inclusion–exclusion, the inner sum

$$\sum |S_{i_1} \cap S_{i_2} \cap \cdots \cap S_{i_k}|$$

is taken over all k-combinations $\{i_1, i_2, \ldots, i_k\}$ of $\{1, 2, \ldots, n\}$. As in the preceding example, it often happens that

$$|S_{i_1} \cap S_{i_2} \cap \cdots \cap S_{i_k}|$$

has the same value for all $C(n, k)$ k-combinations $\{i_1, i_2, \ldots, i_k\}$ of $\{1, 2, \ldots, n\}$. In this case we obtain the following corollary of Theorem 2.1.3:

COROLLARY 2.1.6 For $n \geq 2$, let $S_1, S_2, \ldots, S_n$ be finite subsets of a universal set U, and suppose that for any k, $1 \leq k \leq n$, and any k-combination $\{i_1, i_2, \ldots, i_k\}$ of $\{1, 2, \ldots, n\}$,

$$|S_{i_1} \cap \cdots \cap S_{i_k}| = |S_1 \cap \cdots \cap S_k|$$

Then $S = S_1 \cup S_2 \cup \cdots \cup S_n$ is finite and

$$|S| = \sum_{k=1}^{n} (-1)^{k-1} C(n, k) |S_1 \cap \cdots \cap S_k| \qquad \square$$

■ *Example 2.1.4* In the next theorem, we use the principle of inclusion–exclusion to determine the number of functions from $\{1, 2, \ldots, m\}$ onto $\{1, 2, \ldots, n\}$, where m and n are positive integers and $m \geq n$. In order to get a feel for the general result, work out the special case $n = 3$.

Solution
Let U be the set of all functions from $A = \{1, 2, \ldots, m\}$ to $B = \{1, 2, 3\}$ and let S be the set of functions from A onto B. Then $|U| = 3^m$, and we want to find $|S|$. Note that $f \in S$ if and only if f maps at least one element of A to each element of B; in other words, $f \in S$ if and only if the preimages $f^{-1}(\{1\})$, $f^{-1}(\{2\})$, and $f^{-1}(\{3\})$ are all nonempty sets. Taking the complementary view, $f \notin S$ if and only if at least one of the preimages $f^{-1}(\{1\}), f^{-1}(\{2\}), f^{-1}(\{3\})$ is empty. Thus, for $i \in \{1, 2, 3\}$, we let

$$S_i = \{f \in U \mid f^{-1}(\{i\}) = \{ \, \}\}$$

Then $S = \overline{S_1} \cap \overline{S_2} \cap \overline{S_3}$.

Alternately, thinking in terms of properties, let us say that "f misses i" when $f^{-1}(\{i\}) = \{ \, \}$, and let p_i be the property that a given function $f \in U$ misses i. Then $S_i = \{f \in U \mid f \text{ has property } p_i\}$; more explicitly,

$$S_1 = \{f \in U \mid f \text{ misses } 1\} = \{f \in U \mid \text{im } f \subseteq \{2, 3\}\}$$
$$S_2 = \{f \in U \mid f \text{ misses } 2\} = \{f \in U \mid \text{im } f \subseteq \{1, 3\}\}$$
$$S_3 = \{f \in U \mid f \text{ misses } 3\} = \{f \in U \mid \text{im } f \subseteq \{1, 2\}\}$$

and our problem is to find $|\overline{S_1} \cap \overline{S_2} \cap \overline{S_3}|$, namely, the number of functions having none of the properties p_1, p_2, p_3.

In order to apply the principle of inclusion/exclusion, note that

$$|S_1| = |S_2| = |S_3| = 2^m$$
$$|S_1 \cap S_2| = |S_1 \cap S_3| = |S_2 \cap S_3| = 1^m$$
$$|S_1 \cap S_2 \cap S_3| = 0$$

Thus, by Corollary 2.1.4,

$$|S_1 \cup S_2 \cup S_3| = \sum_{k=1}^{3} (-1)^{k-1} C(3,k) |S_1 \cap \cdots \cap S_k|$$

$$= C(3,1)|S_1| - C(3,2)|S_1 \cap S_2| + C(3,3)|S_1 \cap S_2 \cap S_3|$$

$$= 3 \cdot 2^m - 3 \cdot 1^m + 1 \cdot 0$$

so that

$$|S| = |\overline{S}_1 \cap \overline{S}_2 \cap \overline{S}_3| = |U| - |S_1 \cup S_2 \cup S_3| = 3^m - 3 \cdot 2^m + 3$$

(Note that in the particular case $m = 5$, we obtain

$$|S| = 3^5 - 3 \cdot 2^5 + 3 = 243 - 3 \cdot 32 + 3 = 150$$

Compare this with the solution of the same problem given in Example 1.3.4, and in Section 1.5, following Example 1.5.6.)   ■

THEOREM 2.1.7   Let m and n be positive integers with $m \geq n$. Then the number of functions from $\{1, 2, \ldots, m\}$ onto $\{1, 2, \ldots, n\}$ is

$$\sum_{k=0}^{n-1} (-1)^k C(n,k)(n-k)^m$$

Proof   Let U be the set of all functions from $A = \{1, 2, \ldots, m\}$ to $B = \{1, 2, \ldots, n\}$ and let S be the set of functions from A onto B. Then $|U| = n^m$, and we want to find $|S|$. Taking our lead from the preceding example, for $1 \leq i \leq n$, we let

$$S_i = \{f \in U \mid f \text{ misses } i\} = \{f \in U \mid f^{-1}(\{i\}) = \{\ \}\}$$

Then $S = \overline{S}_1 \cap \overline{S}_2 \cap \cdots \cap \overline{S}_n$.

Let k be an integer, $1 \leq k \leq n$, and let $\{i_1, i_2, \ldots, i_k\}$ be any k-combination of $\{1, 2, \ldots, n\}$. To determine the value of the inner sum corresponding to k, we need to determine

$$|S_{i_1} \cap S_{i_2} \cap \cdots \cap S_{i_k}|$$

This is simply the number of functions f from $\{1, 2, \ldots, m\}$ to $\{1, 2, \ldots, n\}$ such that f misses each of $i_1, i_2, \ldots, i_k$, that is, $f^{-1}(\{i_1, i_2, \ldots, i_k\}) = \{\ \}$. For such an f, note that there are $n - k$ choices for each image $f(x)$. Thus,

$$|S_{i_1} \cap S_{i_2} \cap \cdots \cap S_{i_k}| = (n - k)^m$$

This value is independent of $\{i_1, i_2, \ldots, i_k\}$, so that we may employ Corollary 2.1.4. Hence,

$$|S_1 \cup S_2 \cup \cdots \cup S_n| = \sum_{k=1}^{n} (-1)^{k-1} C(n, k) |S_1 \cap \cdots \cap S_k|$$

$$= \sum_{k=1}^{n} (-1)^{k-1} C(n, k)(n - k)^m$$

And,

$$|\bar{S}_1 \cap \bar{S}_2 \cap \cdots \cap \bar{S}_n| = n^m - \sum_{k=1}^{n} (-1)^{k-1} C(n, k)(n - k)^m$$

$$= \sum_{k=0}^{n-1} (-1)^k C(n, k)(n - k)^m \qquad \square$$

EXERCISE SET 2.1

1. An insurance company classifies its policyholders according to age and marital status. Of 500 policyholders surveyed, 350 are married, 110 are married and less than 25 years of age, and 60 are not married and are 25 years of age or older. How many of the 500 policyholders are less than 25 years of age?

2. There are 30 students in a discrete mathematics class. Of these, 10 failed the midterm exam, 5 failed the final exam, and 3 failed both the midterm and the final.

    (a) How many students passed both the midterm and the final?
    (b) How many students passed the midterm or the final?

3. Refer to Exercise 7 in Exercise Set 1.2. Find the number of sequences such that the first card is a king or the third card is not an ace.

4. There are 225 computer science majors at a certain university. This semester, 100 of these students are taking the data structures course, 75 are taking the database management course, 50 are taking the artificial intelligence course, 7 are taking both data structures and database management, 4 are taking both data

structures and artificial intelligence, 31 are taking both database management and artificial intelligence, and 1 is taking all three of these courses.

(a) Find the number of computer science majors who are taking at least one of these three courses.

(b) How many computer science majors are taking none of these three courses?

5. Refer to Exercise 9 in Exercise Set 1.2 and find the number of outcomes for the experiment such that the first ball is red or the third ball is green.

6. Refer to Exercise 9 in Exercise Set 1.3 and find the number of ways to select the committee under the restriction that it contain exactly one full professor or exactly two associate professors or exactly one assistant professor.

7. Refer to Exercise 16 in Exercise Set 1.3. How many five-digit strings, formed from the digits 1, 2, 3, 4, and 5, satisfy the following conditions?

(a) The string contains the substring 12 or the substring 34.

(b) The string does not contain the substring 12 and does not contain the substring 23.

8. A bakery sells seven kinds of donuts. How many ways are there to choose one dozen donuts if no more than three donuts of any kind are used? (Hint: Let x_i be the number of donuts chosen of kind i, $1 \leq i \leq 7$; then find the number of 7-tuples $(x_1, x_2, \ldots, x_7)$ of integers such that $x_1 + x_2 + \cdots + x_7 = 12$ and $0 \leq x_i \leq 3$. Let $U = \{(x_1, x_2, \ldots, x_7) \mid x_1 + x_2 + \cdots + x_7 = 12$ and $x_i \in \mathbb{N}\}$ and, for $1 \leq i \leq 7$, let $S_i = \{(x_1, x_2, \ldots, x_7) \in U \mid x_i > 3\}$. Apply the principle of inclusion–exclusion to find $|\overline{S}_1 \cap \overline{S}_2 \cap \cdots \cap \overline{S}_7|$.)

9. Find the number of integer solutions to $x_1 + x_2 + x_3 + x_4 + x_5 + x_6 = 20$, where $1 \leq x_i \leq 4$ for $i \in \{1, 2, 3, 4\}$.

10. Find the number of nine-digit strings that contain each of the odd digits 1, 3, 5, 7, and 9.

11. How many five-card poker hands contain a jack, a queen, and a king?

12. Given five different pairs of gloves, in how many ways can each of five students select two gloves at random so that no student gets a matching pair? (Hint: Label the gloves g_{ij}, $1 \leq i \leq 5$, $j = 1, 2$, where g_{i1} and g_{i2} match, and label the students s_1, s_2, s_3, s_4, s_5. Then each of the five students is to select two of the gloves (a 2-combination) and we may assume that s_1 selects first, s_2 second, and so on. Let U be the set of all such ordered selections and, for each i, $1 \leq i \leq 5$, let S_i be the set of those elements in U such that s_i gets a matching pair of gloves. Find $|\overline{S}_1 \cap \overline{S}_2 \cap \overline{S}_3 \cap \overline{S}_4 \cap \overline{S}_5|$.)

13. How many nine-letter strings using the letters A, A, A, B, B, B, C, C, C have the three A's, or the three B's, or the three C's consecutive?

14. How many ways are there to distribute 30 identical slices of pizza among five fraternity brothers such that no brother gets more than 7 slices?

15. Given n identical objects of type 1, n identical objects of type 2, $\ldots$, n identical objects of type t, how many sequences of these nt objects have all the objects of some type appearing consecutively?

16. How many permutations of $\{0, 1, 2, \ldots, 9\}$, written as a 10-digit string, contain at least one of the substrings 012 or 34 or 56 or 789?

17. Let n be an odd positive integer.

    (a) How many functions f on $\{1, 2, \ldots, n\}$ have the property that $f(2k-1) \neq 2k-1$ for each k, $2 \leq 2k \leq n+1$?

    (b) How many permutations f on $\{1, 2, \ldots, n\}$ have the property that $f(2k-1) \neq 2k-1$ for each k, $2 \leq 2k \leq n+1$?

18. Refer to Exercise 18 in Exercise Set 1.3. In how many ways can n married couples be seated at a circular dinner table so that no husband is seated next to his wife? Answer for:

    (a) $n = 2$ (b) $n = 3$

    (c) $n = 4$ (d) An arbitrary $n \in \mathbb{Z}^+$

19. Find the number of integers between 1 and 10^{12} that are neither perfect squares nor perfect cubes nor perfect fourth powers.

★20. Prove Theorem 2.1.3 by induction on n.

21. Roberto has eight friends who enjoy playing bridge. He wishes to invite a different subset of three of them over to his home for a game every Wednesday evening for 5 consecutive weeks. In how many ways can this be done so that each friend receives at least one invitation?

22. Find the number of permutations $(x_1, x_2, \ldots, x_7)$ of $\{1, 2, \ldots, 7\}$ such that $x_i = i$ for exactly three values of i, $1 \leq i \leq 7$.

23. Find the number of permutations $(x_1, x_2, \ldots, x_n)$ of $\{1, 2, \ldots, n\}$ such that $x_i = i$ for exactly k values of i, $0 \leq k \leq n$.

24. For n a positive integer, prove the identity

$$n! = \sum_{k=0}^{n} C(n, k) d(n - k)$$

where $d(0)$ is defined to be 1.

25. Refer to Exercise 18. In how many ways can n married couples be seated at a circular dinner table so that exactly k husbands are seated next to their wives?

26. A group of m students is taking a computer class that meets in a classroom with n workstations (with at most one student per workstation). After taking a short break, in how many ways can the students return to the classroom so that no student uses the same workstation as before?

27. Refer to Exercise 12. Assume that each student owns exactly one pair of gloves (and no pair of gloves is owned by more than one student). In how many ways can each student select two gloves at random so that the following conditions are satisfied?

    (a) Each student gets a glove belonging to someone else.

    (b) Each student gets a pair of gloves belonging to someone else.

28. Refer to Exercise 18. In how many ways can n married couples be seated at a circular dinner table so that no husband is seated to the left of his wife and each man is seated next to two women?

29. A panel with five members is seated in a row at a table in front of an audience. After a break in their discussion, they sit down again at the same table. In how many ways can it happen that no panel member is to the right of the same member as before the break? (For a generalization of this exercise, see Problem 3 at the end of this chapter.)

30. Generalize the result of Exercise 12 to the case in which there are n pairs of gloves and n students.

31. A committee with five members is seated at a circular table. After a break in the meeting, they return to the same table. In how many ways can they be seated so that no committee member is to the left of the same member as before the break? (For a generalization, see Problem 5 at the end of this chapter.)

32. Generalize the result of Exercise 27 to the case in which there are n pairs of gloves and n students.

2.2 RECURRENCE EQUATIONS

The most famous mathematician of the Middle Ages was Leonardo Pisano (Leonardo of Pisa), who lived from approximately 1170 to 1250. As the son of an Italian businessman, Leonardo was taught early to figure and calculate, and he had the opportunity to travel widely throughout the Mediterranean region. This brought him in contact with Greek mathematics, such as Euclid's *Elements*, and the algebras of such Arab writers as al-Khowarizmi. At that time, Europe was still using the Roman system of numeration. Leonardo saw the advantages of the Hindu-Arabic system, the system we use today, and did much to advocate its adoption. His important work, the *Liber abaci*, introduced the Hindu-Arabic system to Europe and served to educate many to its uses.

A second edition of *Liber abaci* was published in 1228 and contained a large collection of problems, many of which appear (in some form) in present-day textbooks. One of these problems has led to what is without doubt the most famous example of a recurrence equation:

> How many pairs of rabbits are produced from a single pair in one year if every month each pair of rabbits more than 1 month old begets a new pair?

Let $f(n)$ be the number of pairs of rabbits present at the beginning of month n, $n \geq 1$. Also, we assume, of course, that a "pair" of rabbits consists of a male and a female, that the initial pair of rabbits is exactly 1 month old at the beginning of month 1, and that rabbits are born at the beginning of a month. Thus, $f(1) = 1$. At the beginning of month 2, the initial pair of rabbits begets a new pair, so that $f(2) = 2$.

Similarly, $f(3) = 3$. At the beginning of month 4, both the initial pair of rabbits and the pair born at the beginning of the second month beget new pairs, so that $f(4) = 5$, and so on. At the beginning of month n, $n \geq 3$, each of the $f(n-2)$ pairs of rabbits in existence at the beginning of month $n-2$ begets a new pair, which are then added to the population of $f(n-1)$ pairs present at month $n-1$. Thus, we obtain the recurrence equation

$$f(n) = f(n-2) + f(n-1)$$

which holds for $n \geq 3$. Indeed, the function f is recursively defined on the set of positive integers by this recurrence equation and the initial values $f(1) = 1$ and $f(2) = 2$.

You may recognize this recurrence equation. If not, perhaps the following table of values of $f(n)$ will jog some memory cells. These values are computed by using the recurrence equation and the initial values.

$n =$	1	2	3	4	5	6	7	8	9	10	11	12
$f(n) =$	1	2	3	5	8	13	21	34	55	89	144	233

You now no doubt recognize the sequence $(f(1), f(2), f(3), \ldots)$ as the Fibonacci sequence. (Often, f is defined on the set of nonnegative integers, with $f(0) = 1$.) Indeed, Leonardo Pisano is better known to us by the name Leonardo Fibonacci (Leonardo, son of Bonacci).

This section emphasizes the process of discovering recurrence equations for combinatorial functions. We also discuss several methods for "solving" recurrence equations; that is, given a recurrence equation and initial values defining a function, we often want to obtain an explicit formula for the function.

■ **Example 2.2.1**   Let $p(n)$ denote the number of permutations of an n-element set, say, $\{1, 2, \ldots, n\}$. From our work in Section 1.3, we know that $p(n) = n!$ and, since $n! = n \cdot (n-1)!$, we obtain the following recursive definition for $p : \mathbb{Z}^+ \to \mathbb{Z}^+$:

$$p(1) = 1$$
$$p(n) = np(n-1) \qquad n \geq 2$$

It is instructive to give a combinatorial justification for the above recurrence equation for the function p. In words, it says that the number of permutations of $\{1, 2, \ldots, n\}$ is equal to n times the number of permutations of $\{1, 2, \ldots, n-1\}$. To see that this is indeed the case, observe that the following two-step recursive "recipe" uniquely determines a permutation $(y_1, y_2, \ldots, y_n)$ of $S = \{1, 2, \ldots, n\}$, $n \geq 2$:

1.   Choose a permutation $(x_1, x_2, \ldots, x_{n-1})$ of $S - \{n\}$.

2.   Choose one of the n positions relative to $x_1, x_2, \ldots, x_{n-1}$ and insert n in this position to obtain $(y_1, y_2, \ldots, y_n)$.

For example, if $n = 5$, $(x_1, x_2, x_3, x_4) = (3, 2, 4, 1)$, and 5 is inserted between 2 and 4, then we obtain the permutation $(3, 2, 5, 4, 1)$. Now, since there are $p(n - 1)$ ways to perform step 1 and n ways to perform step 2, we have

$$p(n) = np(n - 1) \qquad n \geq 2$$

by the multiplication principle. ∎

■ **Example 2.2.2**   Let $g(n)$ be the number of k-element subsets of $\{1, 2, \ldots, n\}$, where k is a fixed nonnegative integer and $n \geq k$. (So $g(n) = C(n, k)$.) We note that $g(k) = 1$ and wish to obtain a recurrence equation for $g(n)$, $n > k$.

You may recall the result of Exercise 11 in Exercise Set 1.6 where, for $n > k$, the problem was to show that

$$(n - k)C(n, k) = nC(n - 1, k)$$

namely, that

$$g(n) = \frac{ng(n - 1)}{n - k}$$

This recurrence equation may be obtained algebraically by using Corollary 1.3.7:

$$g(n) = \frac{n!}{k!(n - k)!} = \frac{n(n - 1)!}{k!(n - k)(n - k - 1)!} = \frac{n}{n - k} \cdot \frac{(n - 1)!}{k!(n - 1 - k)!} = \frac{ng(n - 1)}{n - k}$$

Alternately, several combinatorial justifications may be given for this recurrence equation. One method is to observe that the following two-step recursive recipe produces a k-combination of $S = \{1, 2, \ldots, n\}$, where $n > k$:

1. Choose an element $x \in S$.
2. Choose a k-combination V of $S - \{x\}$.

Since there are n choices for x in step 1 and $g(n - 1)$ choices for V in step 2, this recipe leads to $ng(n - 1)$ subsets of cardinality k. However, these $ng(n - 1)$ subsets are not distinct. So we must determine the number of times each k-element subset of S is produced. A given k-element subset V may be chosen in step 2 whenever $x \in S - V$. This happens for $n - k$ choices of x in step 1. (The case for $n = 5$ and $k = 3$ is illustrated below; notice that $V = \{1, 2, 3\}$ is listed twice, for $x = 4$ and $x = 5$.)

Therefore, the number of distinct k-element subsets of S is $ng(n-1)/(n-k)$, which yields the recurrence equation

$$g(n) = \frac{ng(n-1)}{n-k}$$

x=1	x=2	x=3	x=4	x=5
{2,3,4}	{1,3,4}	{1,2,4}	{1,2,3}	{1,2,3}
{2,3,5}	{1,3,5}	{1,2,5}	{1,2,5}	{1,2,4}
{2,4,5}	{1,4,5}	{1,4,5}	{1,3,5}	{1,3,4}
{3,4,5}	{3,4,5}	{2,4,5}	{2,3,5}	{2,3,4}

■

■ *Example 2.2.3*   Let $d(n)$ denote the number of derangements of $S = \{1, 2, \ldots, n\}$. Then $d(1) = 0$ and $d(2) = 1$; show that $d(n)$ satisfies the recurrence equation

$$d(n) = (n-1)[d(n-2) + d(n-1)]$$

for $n \geq 3$.

Solution

We could proceed by using the result of Theorem 2.1.4:

$$d(n) = n! \left(\frac{1}{0!} - \frac{1}{1!} + \cdots + \frac{(-1)^n}{n!} \right)$$

However, we leave this route to Exercise 14 and use a combinatorial argument instead.

We claim that the following two-step recursive recipe uniquely determines a derangement $(x_1, x_2, \ldots, x_n)$ of S; verification of this claim is left for you to develop in Exercise 24.

1.   Choose $x_n = k \in S - \{n\}$.
2.   Perform either step (a) or step (b):

   (a) Let $(x_1, \ldots, x_{n-2})$ be a derangement of $S - \{n-1, n\}$. If $k = n-1$, then let $x_{n-1} = n$; else let $x_{n-1} = x_k$, replace x_k with n, and replace x_i with $n-1$ for that value of i such that $x_i = k$.
   (b) Let $(x_1, x_2, \ldots, x_{n-1})$ be a derangement of $S - \{n\}$. Replace x_i with n for that value of i such that $x_i = k$.

   There are $n-1$ choices for $x_n = k$ in step 1. If we choose to perform step 2(a), then there are $d(n-2)$ choices for the derangement. Hence, steps 1 and 2(a) produce $(n-1)d(n-2)$ derangements of S, namely, those with $x_n = k$ and $x_k = n$ for some k, $1 \leq k \leq n-1$. If, on the other hand, we choose to perform step 2(b), then there are $d(n-1)$ choices for the derangement. Hence, steps 1 and 2(b) produce $(n-1)d(n-1)$ derangements of S, namely, those with $x_n = k$ and $x_k \neq n$ for some k, $1 \leq k \leq n-1$.

Since each of the $d(n)$ derangements of S is uniquely determined by this process, we have, by the addition principle,

$$d(n) = (n - 1)[d(n - 2) + d(n - 1)]$$ ∎

■ ***Example 2.2.4***   Consider character strings consisting entirely of the letters A and B. Call those strings that do not contain two consecutive A's "good." Let $s(n)$ be the number of good strings of length n. Then $s(0) = 1$ (the empty string is good), $s(1) = 2$ (A and B), $s(2) = 3$ (AB, BA, and BB), and so on. Find a recurrence equation for $s(n)$, $n \geq 2$.

Solution
Let x denote a string of length $n \geq 2$ having no two consecutive A's. If the last character of x is B, then the first $n - 1$ characters of x form one of the good strings of length $n - 1$. On the other hand, if the last character of x is A, then necessarily the next to last character of x is B, and the first $n - 2$ characters form one of the $s(n - 2)$ good strings of length $n - 2$. It follows by the addition principle that

$$s(n) = s(n - 1) + s(n - 2)$$

Note that the recurrence equation for $s(n)$ is the same as that for the Fibonacci numbers; only the initial values differ.   ∎

■ ***Example 2.2.5***   Let $r(n)$ denote the number of "comparisons" required to sort a list of 2^n numbers using the "merge sort" technique. Assume that $r(0) = 0$. For $n > 0$, a merge sort is performed recursively by partitioning the list of 2^n numbers into two sublists of 2^{n-1} numbers each, sorting each of these sublists, and then "merging" them into a single sorted list. Assume that it takes k comparisons to merge two sorted sublists into a single sorted list of k numbers. Find a recurrence relation for $r(n)$, $n \geq 1$.

Solution
It takes $r(n-1)$ comparisons to sort each sublist of 2^{n-1} numbers, then 2^n comparisons to merge these sublists into a single sorted list. Thus, a recurrence relation for $r(n)$ is given by

$$r(n) = 2r(n - 1) + 2^n$$

for $n \geq 1$.   ∎

■ ***Example 2.2.6***   (**Towers of Hanoi**) We are given three pegs, labeled peg 1, peg 2, and peg 3, planted on a board. In addition, we are given n circular disks of different diameters, each with a hole drilled through its center. The disks are labeled

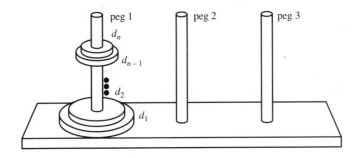

Figure 2.2.1 The "Towers of Hanoi."

$d_1, d_2, \ldots, d_n$, in order of decreasing size (d_1 is the largest and d_n the smallest), and are then placed on peg 1 in the order $d_1, d_2, \ldots, d_n$ (see Figure 2.2.1). The n disks are to be transferred from peg 1 to one of the other pegs, say, peg 3, subject to the following rules:

1. Each move consists of transferring the top disk from one peg to another peg.
2. At no time can a larger disk be placed on top of a smaller one.

Find the initial value and a recurrence equation for the number of moves $h(n)$ required to accomplish this.

Solution
Clearly, $h(1) = 1$. Suppose $n > 1$; then in order to transfer d_1 to a disk-free peg 3, we must first transfer $d_2, d_3, \ldots, d_n$ from peg 1 to peg 2. Notice that this is just a smaller case of the problem and, according to our notational convention, requires $h(n - 1)$ moves. Once this is accomplished, d_1 can be transferred from peg 1 to peg 3 in one move. At this point, the situation is as depicted in Figure 2.2.2, with $h(n-1)+1$ moves having taken place. Finally, $h(n - 1)$ moves are required to transfer $d_2, d_3, \ldots, d_n$ from peg 2 to peg 3. So the desired transfer of all n disks can be accomplished in

$$[h(n - 1) + 1] + h(n - 1) = 2h(n - 1) + 1$$

moves. Thus, the initial value and recurrence equation are given by

$$h(1) = 1$$
$$h(n) = 2h(n - 1) + 1 \qquad n \geq 2$$

∎

Suppose we wish to find an explicit formula for $h(n)$ in the problem of the Towers of Hanoi. Because we know the initial value and the recurrence equation for $h(n)$,

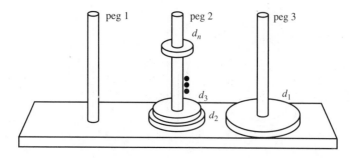

Figure 2.2.2

successive values of the function may be calculated as follows:

$h(1) = 1$
$h(2) = 2h(1) + 1 = 3$
$h(3) = 2h(2) + 1 = 7$
$h(4) = 2h(3) + 1 = 15$
$h(5) = 2h(4) + 1 = 31$

and so on.

This method of obtaining successive values of a recursively defined function is called **iteration**. Iteration is easily implemented in most programming languages by using a for loop; with a computer, values of a function such as h can be found for many values of n.

Sometimes, however, it is desirable to have an explicit formula for a function that is defined recursively. There exist general techniques for doing this, and we discuss some of these at this point. The general process of finding explicit formulas for recursively defined functions is referred to as "solving recurrence equations" (even though the initial values, as well as the recurrence equations, are used to obtain the solutions).

Perhaps the simplest method for solving a recurrence equation is that of "guess and check." Suppose the function g is defined recursively. Using the given initial values and recurrence equation, several values of $g(n)$ may be computed and, based on these values, we might conjecture an explicit formula for $g(n)$. We then check this formula to see whether it gives the correct initial values and satisfies the recurrence equation for g. For example, looking at the values of the function $h(n)$ shown above, it is reasonable to conjecture that

$h(n) = 2^n - 1$

Checking this, note that

$h(1) = 2^1 - 1 = 1$

and that

$$2h(n-1) + 1 = 2(2^{n-1} - 1) + 1 = 2^n - 2 + 1 = 2^n - 1 = h(n)$$

so that our guess has the correct initial value and satisfies the recurrence equation. Thus, we have our solution.

Another elementary technique is that of "repeated substitution." To illustrate this technique, let's use again the function $h : \mathbb{Z}^+ \to \mathbb{Z}^+$ defined recursively by

$$h(1) = 1$$
$$h(n) = 2h(n-1) + 1 \qquad n \geq 2 \qquad\qquad (2.2.1)$$

Now by using the recurrence equation (2.2.1), note that, for $n \geq 3$,

$$h(n-1) = 2h(n-2) + 1$$

Thus, substituting for $h(n-1)$ in (2.2.1) yields, for $n \geq 3$,

$$\begin{aligned} h(n) &= 2[2h(n-2) + 1] + 1 \\ &= 2^2 h(n-2) + 2 + 1 \end{aligned} \qquad\qquad (2.2.2)$$

Again using (2.2.1), we have, for $n \geq 4$,

$$h(n-2) = 2h(n-3) + 1$$

Substitution into (2.2.2) then yields

$$\begin{aligned} h(n) &= 2^2[2h(n-3) + 1] + 2 + 1 \\ &= 2^3 h(n-3) + 2^2 + 2 + 1 \end{aligned} \qquad\qquad (2.2.3)$$

Continuing in this fashion, we obtain the general relation

$$h(n) = 2^k h(n-k) + 2^{k-1} + \cdots + 2 + 1 \qquad\qquad (2.2.4)$$

for $n \geq k+1$, where k is a positive integer. (This can be proven rigorously by induction on k, but our aim is rather to obtain a formula for $h(n)$, as in the method of guess and check.) By now letting $k = n - 1$, we have

$$\begin{aligned} h(n) &= 2^{n-1} h(1) + 2^{n-2} + \cdots + 2 + 1 \\ &= \sum_{i=0}^{n-1} 2^i \\ &= 2^n - 1 \end{aligned}$$

using the well-known formula for the sum of a (finite) geometric series.

■   ***Example 2.2.7***   Use the technique of repeated substitution to find an explicit formula for the function $g : \mathbb{N} \to \mathbb{N}$ defined recursively by

$$g(0) = 0$$
$$g(n) = g(n-1) + 2n - 1 \qquad n \geq 1$$

Solution
Let k be a positive integer. Then, for $n \geq k$,

$$
\begin{aligned}
g(n) &= g(n-1) + 2n - 1 \\
&= [g(n-2) + 2(n-1) - 1] + 2n - 1 \\
&= g(n-2) + 4n - (1+3) \\
&= [g(n-3) + 2(n-2) - 1] + 4n - (1+3) \\
&= g(n-3) + 6n - (1+3+5) \\
&\;\;\vdots \\
&= g(n-k) + 2kn - \sum_{i=1}^{k}(2i-1)
\end{aligned}
$$

Letting $k = n$, we obtain

$$
\begin{aligned}
g(n) &= g(0) + 2n^2 - \sum_{i=1}^{n}(2i-1) \\
&= 0 + 2n^2 - n^2 \\
&= n^2
\end{aligned}
$$

(Note the use of the fact that the sum of the first n odd positive integers is n^2.) Checking the formula $g(n) = n^2$, we see that it does satisfy the initial condition $g(0) = 0$ and the recurrence equation $g(n) = g(n-1) + 2n - 1$, $n \geq 1$.   ■

Let us now consider a method for solving the special type of recurrence equation of the form

$$h(n) - a_1 h(n-1) - a_2 h(n-2) - \cdots - a_r h(n-r) = g(n) \tag{2.2.5}$$

where $n \geq r$ and $a_1, a_2, \ldots, a_r$ are constants with $a_r \neq 0$. This is called a **linear recurrence equation with constant coefficients**. If $g(n) = 0$ for all n, then this equation is called **homogeneous**; otherwise, it is called **nonhomogeneous**. For example, the recurrence equation

$$h(n) = 3h(n-1) + 6h(n-2) - 4h(n-3)$$

is a linear homogeneous recurrence equation, with $r = 3$, $a_1 = 3$, $a_2 = 6$, and $a_3 = -4$.

Another example is the recurrence equation for the Fibonacci sequence,

$$f(n) = f(n-1) + f(n-2) \tag{2.2.6}$$

which has $r = 2$, $a_1 = a_2 = 1$. Along with the initial values $f(0) = f(1) = 1$, this recurrence equation defines a function $f : \mathbb{N} \to \mathbb{Z}^+$. Let's attempt to find an explicit formula for $f(n)$. An examination of a table of values for $f(n)$ leads to the conjecture that $f(n)$ increases exponentially with n. In fact, Example 0.4.8 and Exercise 16 in Exercise Set 0.4 show that

$$(4/3)^n < f(n) < 2^n$$

for $n \geq 2$. Thus, suppose we conjecture a solution of the form

$$f(n) = b^n$$

for some $b > 0$. By substituting into the recurrence equation (2.2.6), we obtain

$$b^n = b^{n-1} + b^{n-2}$$

leading to

$$b^2 - b - 1 = 0$$

upon division by b^{n-2}. This is a quadratic equation in b and has solutions

$$b_1 = \frac{1 - \sqrt{5}}{2} \quad \text{and} \quad b_2 = \frac{1 + \sqrt{5}}{2}$$

This means that $f_1(n) = b_1^n$ and $f_2(n) = b_2^n$ are both solutions to the recurrence equation (2.2.6); in fact, if c_1 and c_2 are any constants, then

$$f(n) = c_1 b_1^n + c_2 b_2^n$$

is also a solution. This is shown as follows:

$$
\begin{aligned}
f(n-1) + f(n-2) &= (c_1 b_1^{n-1} + c_2 b_2^{n-1}) + (c_1 b_1^{n-2} + c_2 b_2^{n-2}) \\
&= c_1(b_1^{n-1} + b_1^{n-2}) + c_2(b_2^{n-1} + b_2^{n-2}) \\
&= c_1 b_1^n + c_2 b_2^n \\
&= f(n)
\end{aligned}
$$

The question now becomes this. Can the constants c_1 and c_2 be chosen so that $f(n)$ also gives the correct initial values $f(0) = f(1) = 1$? This requires that

$$f(0) = c_1 + c_2 = 1$$

$$f(1) = c_1 b_1 + c_2 b_2 = 1$$

This is a simple system of two linear equations in the two unknowns c_1 and c_2 and has the following solution:

$$c_1 = \frac{b_2 - 1}{b_2 - b_1} = \frac{-b_1}{b_2 - b_1} \qquad \text{and} \qquad c_2 = \frac{1 - b_1}{b_2 - b_1} = \frac{b_2}{b_2 - b_1}$$

Hence, we obtain the following result:

THEOREM 2.2.1   The terms of the Fibonacci sequence satisfy

$$f(n) = \frac{1}{\sqrt{5}} \left[\left(\frac{1 + \sqrt{5}}{2} \right)^{n+1} - \left(\frac{1 - \sqrt{5}}{2} \right)^{n+1} \right]$$

for all $n \in \mathbb{N}$.   □

Let us return to the general case of a linear homogeneous recurrence equation with constant coefficients:

$$h(n) - a_1 h(n-1) - a_2 h(n-2) - \cdots - a_r h(n-r) = 0 \tag{2.2.7}$$

Following the lead suggested by our work with the Fibonacci recurrence equation, we define the **characteristic equation** of this equation to be

$$x^r - a_1 x^{r-1} - a_2 x^{r-2} - \cdots - a_{r-1} x - a_r = 0 \tag{2.2.8}$$

(Note that 0 is not a solution of the characteristic equation since $a_r \neq 0$.) The roots of (2.2.8) are called the **characteristic roots** of the associated recurrence equation (2.2.7). Now suppose that $h_1(n) = b_1^n$ is a solution to (2.2.7) for some $b_1 \neq 0$. Then

$$b_1^n - a_1 b_1^{n-1} - \cdots - a_r b_1^{n-r} = 0$$

Dividing by b_1^{n-r}, we obtain

$$b_1^r - a_1 b_1^{r-1} - \cdots - a_r = 0$$

so that b_1 is a characteristic root of (2.2.8). Conversely, if b_1 is a characteristic root of (2.2.8), then $h_1(n) = b_1^n$ is a solution to (2.2.7). This yields the following general result:

THEOREM 2.2.2   For a complex number b_1, the function $h_1(n) = b_1^n$ is a solution of the linear homogeneous recurrence equation (2.2.7) if and only if b_1 is a characteristic root of the equation.   □

The form of the general solution of equation (2.2.7) depends on whether the characteristic equation (2.2.8) has distinct or repeated roots.

Suppose that (2.2.8) has r distinct (nonzero) roots $b_1, b_2, \ldots, b_r$. Then, as worked out in the particular case of the Fibonacci equation,

$$h_0(n) = c_1 b_1^n + c_2 b_2^n + \cdots + c_r b_r^n$$

is a solution of (2.2.7) for any choice of the constants $c_1, c_2, \ldots, c_r$ and is called the **general solution** of (2.2.7). This result is stated as Theorem 2.2.3; its proof is left for you to work out in Exercise 16.

> **THEOREM 2.2.3** If the characteristic equation (2.2.8) has r distinct roots $b_1, b_2, \ldots, b_r$, then the associated linear homogeneous recurrence equation (2.2.7) has the general solution
>
> $$h_0(n) = c_1 b_1^n + c_2 b_2^n + \cdots + c_r b_r^n \qquad (2.2.9)$$
>
> □

Given r initial values for the function h, say the values of $h(0), h(1), \ldots, h(r-1)$, we may substitute these values into the general solution (2.2.9) to obtain a system of r linear equations in the r unknowns $c_1, c_2, \ldots, c_r$. This system can then be solved to obtain a **particular solution** of (2.2.7). Note that a different set of initial values will lead to a different particular solution; that is, the general solution depends on the form of the recurrence equation (2.2.7), whereas the particular solution depends on the initial values.

■ *Example 2.2.8* Define $g : \mathbb{N} \to \mathbb{Z}^+$ recursively by

$$g(0) = 4, \quad g(1) = 5, \quad g(2) = 19$$
$$g(n) = 2g(n-1) + g(n-2) - 2g(n-3) \qquad n \geq 3$$

Find an explicit formula for $g(n)$.

Solution
The recurrence equation for g is a linear homogeneous equation with $r = 3$, $a_1 = 2$, $a_2 = 1$, and $a_3 = -2$. Thus, the characteristic equation is

$$x^3 - 2x^2 - x + 2 = 0$$

or

$$(x + 1)(x - 1)(x - 2) = 0$$

Hence, the characteristic roots are -1, 1, and 2, so the general solution is

$$g(n) = c_1(-1)^n + c_2(1)^n + c_3(2)^n$$
$$= c_1(-1)^n + c_2 + c_3(2)^n$$

From the initial values $g(0) = 4$, $g(1) = 5$, $g(2) = 19$, we obtain the following system of linear equations:

$$c_1 + c_2 + c_3 = 4$$
$$-c_1 + c_2 + 2c_3 = 5$$
$$c_1 + c_2 + 4c_3 = 19$$

The solution of this system is easily obtained:

$$c_1 = 2, \quad c_2 = -3, \quad c_3 = 5$$

Thus, the particular solution is given by

$$g(n) = 2(-1)^n + 5(2)^n - 3 \qquad \blacksquare$$

Now let us consider the case where the characteristic equation (2.2.8) has repeated roots. A simple example of this is the characteristic equation

$$x^2 - 4x + 4 = (x - 2)^2 = 0$$

which has 2 as a repeated root, of multiplicity 2. The associated recurrence equation is

$$h(n) = 4h(n - 1) - 4h(n - 2) \qquad\qquad\qquad (2.2.10)$$

We know that $h_1(n) = 2^n$ is a solution of (2.2.10) and desire to find another solution. A possibility is a function $h_2(n)$ of the form $h_2(n) = g(n)2^n$, where $g(n)$ is not a constant function. The simplest thing to try is $h_2(n) = n2^n$:

$$4[h_2(n - 1) - h_2(n - 2)] = 4[(n - 1)2^{n-1} - (n - 2)2^{n-2}]$$
$$= 2^n[2(n - 1) - (n - 2)]$$
$$= n2^n = h_2(n)$$

It works! Thus, the general solution of (2.2.10) is

$$h_0(n) = c_1 2^n + c_2 n2^n = (c_1 + c_2 n)2^n$$

Taking our lead from this example, we can develop the following result, the proof of which is left for you to work out in Exercise 18.

THEOREM 2.2.4   Suppose that b_1 is a root of multiplicity $t \geq 2$ of the characteristic equation (2.2.8), that is, suppose that

$$x^r - a_1 x^{r-1} - \cdots - a_{r-1} x - a_r = (x - b_1)^t q(x)$$

where $q(b_1) \neq 0$. Then each of

$$b_1^n, nb_1^n, \ldots, n^{t-1}b_1^n$$

is a solution of the linear homogeneous recurrence equation (2.2.7), so that

$$(c_1 + c_2 n + \cdots + c_t n^{t-1})b_1^n$$

is also a solution and is that portion of the general solution corresponding to the root b_1. □

■ **Example 2.2.9**   Define $h : \mathbb{N} \to \mathbb{Z}^+$ recursively by

$$h(0) = 1, \quad h(1) = 6$$
$$h(n) = 4h(n-1) - 4h(n-2) \qquad n \geq 2$$

Find an explicit formula for $h(n)$.

Solution
Note that the recurrence equation for h is (2.2.10), considered before Theorem 2.2.4. We saw there that the general solution is

$$h(n) = (c_1 + c_2 n)2^n$$

Now by using the initial value $h(0) = 1$, we find that $c_1 = 1$. Then using the initial value $h(1) = 6$ gives $c_2 = 2$. Therefore,

$$h(n) = (2n+1)2^n$$ ■

■ **Example 2.2.10**   Define $z : \mathbb{N} \to \mathbb{Z}^+$ recursively by

$$z(0) = 5, \quad z(1) = 13, \quad z(2) = 76, \quad z(3) = 343$$

$$z(n) = 7z(n-1) - 9z(n-2) - 27z(n-3) + 54z(n-4) \qquad n \geq 4$$

Find an explicit formula for $z(n)$.

Solution
Note that the recurrence equation for $z(n)$ is linear and homogeneous, with $r = 4$, $a_1 = 7$, $a_2 = -9$, $a_3 = -27$, and $a_4 = 54$. The characteristic equation is

$$x^4 - 7x^3 + 9x^2 + 27x - 54 = (x+2)(x-3)^3 = 0$$

with roots -2 and 3, where 3 is repeated with multiplicity 3. The portion of the general solution corresponding to the root -2 is $c_1(-2)^n$, whereas that corresponding to the root 3 is $(c_2 + c_3 n + c_4 n^2) 3^n$. Hence, the general solution is

$$z(n) = c_1(-2)^n + (c_2 + c_3 n + c_4 n^2) 3^n$$

Making use of the initial values, we obtain the system

$$c_1 + c_2 = 5$$
$$-2c_1 + 3c_2 + 3c_3 + 3c_4 = 13$$
$$4c_1 + 9c_2 + 18c_3 + 36c_4 = 76$$
$$-8c_1 + 27c_2 + 81c_3 + 243c_4 = 343$$

This has the following solution:

$$c_1 = 1, \quad c_2 = 4, \quad c_3 = 0, \quad c_4 = 1$$

Therefore,

$$z(n) = (-2)^n + (n^2 + 4)3^n$$ ■

By now, if you have taken a course in differential equations, you have probably noticed a direct analogy between the methods here and those for solving linear homogeneous differential equations.

The next section introduces the topic of generating functions. These can be used to solve a variety of recurrence equations, including nonhomogeneous linear recurrence equations.

Before ending this section, let's briefly look at recurrence equations for functions of more than one variable. We have seen at least one example of such a function already, namely, the function $C(n, k)$. This function can be recursively defined on $\mathbb{N} \times \mathbb{N}$ by Pascal's identity

$$C(n, k) = C(n - 1, k) + C(n - 1, k - 1)$$

along with the initial values $C(n, 0) = C(n, n) = 1$ and $C(n, k) = 0$ if $k > n$.

■ **Example 2.2.11**   For positive integers m and n with $1 \leq n \leq m$, let $S(m, n)$ denote the number of partitions (unordered) of an m-element set into n parts. Note that $S(m, 1) = 1 = S(m, m)$. Also, it is convenient to have the function S defined for all positive integers m and n; to do this, define $S(m, n) = 0$ if $n > m$.

(a) Find $S(5, 3)$.

(b) Show that S satisfies the recurrence equation

$$S(m, n) = S(m - 1, n - 1) + nS(m - 1, n)$$

for $1 < n < m$.

(c) Let $S'(m, n)$ denote the number of ordered partitions of an m-element set into n parts. Recall, from Section 1.5, that $S'(m, n)$ is also the number of functions from an m-element set onto an n-element set; Theorem 2.1.7 gives a formula for this number. Show that

$$S'(m, n) = n! S(m, n)$$

Solution

Without loss of generality, we use the m-element set $\{1, 2, \ldots, m\}$.

(a) A partition of $\{1, 2, 3, 4, 5\}$ into 3 parts can be of type 1: $\{\{a, b, c\}, \{d\}, \{e\}\}$, or of type 2: $\{\{a, b\}, \{c, d\}, \{e\}\}$. There are $C(5, 3) = 10$ partitions of type 1 (the number of ways to choose the subset $\{a, b, c\}$) and $C(5, 1) \cdot 3 = 15$ partitions of type 2 (the number of ways to first choose the subset $\{e\}$ and then decide how to partition the remaining four elements into two subsets of two elements each). Hence, $S(5, 3) = 25$.

(b) Let $\mathcal{P}$ denote the collection of partitions $\{1, 2, \ldots, m\}$ into n parts. We partition $\mathcal{P}$ as $\{\mathcal{P}_1, \mathcal{P}_2\}$, where

$$\mathcal{P}_1 = \{P \in \mathcal{P} \mid \{m\} \in P\}$$
$$\mathcal{P}_2 = \{P \in \mathcal{P} \mid \{m\} \notin P\}$$

that is, $\mathcal{P}_1$ consists of those partitions in which the element m is in a subset by itself.

If $P \in \mathcal{P}_1$, then P is uniquely determined by the partition $P - \{\{m\}\}$, which is a partition of $\{1, 2, \ldots, m - 1\}$ into $n - 1$ parts. Thus, $|\mathcal{P}_1| = S(m - 1, n - 1)$. To obtain $P \in \mathcal{P}_2$, we let $P' = \{A_1, A_2, \ldots, A_n\}$ be a partition of $\{1, 2, \ldots, m - 1\}$ into n parts; we choose i, $1 \leq i \leq n$, and replace A_i by $A_i \cup \{m\}$, that is,

$$P = (P' - \{A_i\}) \cup (\{A_i \cup \{m\}\})$$

This uniquely determines $P \in \mathcal{P}_2$. Since there are $S(m - 1, n)$ choices for P' and n choices for A_i, we have $|\mathcal{P}_2| = nS(m - 1, n)$. Therefore, by the addition principle,

$$S(m, n) = |\mathcal{P}| = |\mathcal{P}_1| + |\mathcal{P}_2| = S(m - 1, n - 1) + nS(m - 1, n)$$

(c) To obtain an ordered partition of P' of $\{1, 2, \ldots, m\}$ into n parts, we first choose an unordered partition $P = \{A_1, A_2, \ldots, A_n\}$, then choose a permutation $(i_1, i_2, \ldots, i_n)$ of $\{1, 2, \ldots, n\}$; then, we set

$$P' = (A_{i_1}, A_{i_2}, \ldots, A_{i_n})$$

This process uniquely determines an ordered partition of $\{1, 2, \ldots, m\}$ into n parts. Since there are $S(m, n)$ choices for P and $n!$ choices for the permutation $(i_1, i_2, \ldots, i_n)$, we have, by the multiplication principle,

$$S'(m, n) = n! S(m, n)$$

For example, we saw in Example 2.1.4 that $S'(5, 3) = 150$. Thus,

$$S(5, 3) = S'(5, 3)/3! = 150/6 = 25$$

as was obtained directly in part (a). ∎

The numbers $S(m, n)$ are called **Stirling numbers of the second kind** and are discussed again in Section 2.4.

EXERCISE SET 2.2

1. For a fixed positive integer n and for $0 \leq m \leq n$, let $k(m)$ be the number of m-element subsets of an n-element set. Then $k(0) = 1$. Find a recurrence equation for $k(m)$ in terms of $k(m - 1)$, $1 \leq m \leq n$.

2. Consider character strings of length n over the alphabet $\{A, B, C\}$. Find initial values and a recurrence equation for each of the following functions:

   (a) $h_1(n)$ is the number of strings (of length n) not containing the substring AA.
   (b) $h_2(n)$ is the number of strings not containing either of the substrings AA or BA.
   (c) $h_3(n)$ is the number of strings not containing the substring BBB.
   ★(d) $h_4(n)$ is the number of strings not containing the substring AB.

3. Let $t(n)$ be the number of ways in which $2n$ tennis players can be paired to play n matches. (For example, if $n = 3$ and the players are numbered 1, 2, 3, 4, 5, and 6, then one possible pairing is to have 1 play 2, 3 play 4, and 5 play 6; another possibility is to have 1 play 6, 2 play 5, and 3 play 4.) Find $t(1)$ and a recurrence equation for $t(n)$, $n \geq 2$. (Hint: Number the players $1, 2, \ldots, 2n$; once the opponent for player 1 has been decided, in how many ways can the remaining $2n - 2$ players be paired?)

4. At the beginning of year 1, Moe invests \$1000 in a savings account that pays 10% annual interest, payable at the end of each year. At the beginning of each subsequent year, Moe deposits an additional \$100 in his account. Let $v(n)$ be the value of the account at the beginning of year n, $n \geq 1$. Then $v(1) = 1000$, $v(2) = 1000+100+100 = 1200$, $v(3) = 1200+120+100 = 1420$, $v(4) = 1420+142+100 = 1662$, etc. Find a recurrence equation for $v(n)$, $n \geq 2$.

5. A wizard must climb an infinite staircase whose steps are numbered with the positive integers. At each step, the wizard may go up one stair or two. Let $w(n)$ be the number of ways for the wizard to reach step number n; then $w(1) = 1$, $w(2) = 2$, $w(3) = 3$, etc. Find a recurrence equation for $w(n)$, $n \geq 3$. (Hint: The set of ways for the wizard to reach step n may be partitioned into two sets according to whether, at his last step, the wizard went up one stair or two.)

6. Let $r(n)$ be the number of regions into which a plane is divided by n lines in general position (each pair of lines intersects at a point and no point is on more than two lines). Then $r(1) = 2$, $r(2) = 4$, $r(3) = 7$, and so on. Find a recurrence equation for $r(n)$, $n \geq 2$.

7. For $n \geq 3$, label the vertices of an n-sided convex polygon with the integers $1, 2, \ldots, n$. Let $t(n)$ be the number of ways to partition the interior of the n-gon into $n - 2$ triangular regions by adding $n - 3$ diagonals that do not intersect in the interior. Then $t(3) = 1$, $t(4) = 2$, $t(5) = 5$ (see Figure 2.2.3), and so on. Find a recurrence equation for $t(n)$, $n \geq 4$. (Hint: There is a triangle with vertices 1, 2, and k, for some k, $3 \leq k \leq n$; partition according to the value of k.)

8. Let $b(n)$ be the number of bees in the nth previous generation of a male bee. Assume that a male bee is produced asexually from a single female parent, whereas

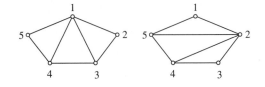

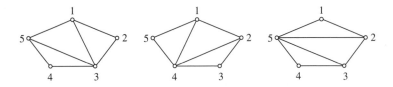

Figure 2.2.3

a female bee has both a male parent and a female parent. So $b(1) = 1$, $b(2) = 2$, $b(3) = 3$, and so on. Find a recurrence equation for $b(n)$, $n \geq 2$.

9. Moe obtains a home mortgage of $40,000 for which the annual interest rate is 10%. Moe wishes to pay off the mortgage in yearly installments of $6000 (except for the last payment, which may be less than $6000). Let $p(n)$ be the amount owed by Moe at the end of year n and, for convenience, define $p(0) = 40,000$. Then $p(1) = 40,000 + 4000 - 6000 = 38,000$, $p(2) = 38,000 + 3800 - 6000 = 35,800$, and so on. Find a recurrence equation for $p(n)$, $n \geq 1$.

10. Consider a single-elimination tournament involving 2^n players. Suppose that each player receives prize money in the amount of $100m$ dollars, where m is the number of players in the subtournament won by that player. (Thus, a player who loses in the first round gets $100, a player who loses in the second round gets $200, a player who loses in the third round gets $400, and so on.)

    (a) Let $r(n)$ be the number of rounds in the tournament; so $r(1) = 1$ and $r(2) = 2$. Find a recurrence equation for $r(n)$, $n \geq 2$.
    (b) Let $m(n)$ be the number of matches played in the tournament; so $m(1) = 1$ and $m(2) = 3$. Find a recurrence equation for $m(n)$, $n \geq 2$.
    (c) Let $p(n)$ be the total amount of prize money awarded; so $p(1) = 300$ and $p(2) = 800$. Find a recurrence equation for $p(n)$, $n \geq 2$.

11. Consider the following game played by a single player on a bit string $b_1 b_2 \ldots b_n$ of length n. At each move, the player switches the value of some bit b_i ($b_i := 1 - b_i$) subject to the following rules: (1) b_1 may be switched; (2) b_2 may be switched only if $b_1 = 1$; (3) for $2 < i \leq n$, b_i may be switched only if $b_{i-1} = 1$ and $b_k = 0$ for each k, $1 \leq k < i - 1$. Let $f(n)$ be the number of moves required to change $b_1 b_2 \ldots b_n$ from all 0s to the bit string with exactly one 1 in position n. Then $b(1) = 1$. Find a recurrence equation for $b(n)$, $n \geq 2$.

12. Consider the problem of seating n married couples at a round table for dinner. As in Example 1.3.6, the particular position in which a person is seated is

unimportant; all that matters to any person is which person is seated to her or his left and which person is seated to the right.

(a) Let $g_1(n)$ be the number of ways to seat these $2n$ people subject to the condition that every husband sits next to his wife. Then $g_1(1) = 1$, $g_1(2) = 4$, $g_1(3) = 16$, and so on. Find a recurrence equation for $g_1(n)$.

(b) Let $g_2(n)$ be the number of ways to seat these $2n$ people subject to the condition that no husband sits next to his wife. Then $g_2(1) = 0$, $g_2(2) = 2$, $g_2(3) = 32$, and so on. Try to find a recurrence relation for $g_2(n)$. Is there a difficulty? (See also Exercise 18 in Exercise Set 2.1.)

13. Use the method of repeated substitution to show that the function r of Example 2.2.5 has the explicit form $r(n) = n2^n$. (If $m = 2^n$ and $r'(m) = r(n)$, then

$$r'(m) = r(n) = n2^n = nm = mn = m \log_2 m$$

The interpretation of this is that the merge sort requires (approximately) $m \log_2 m$ comparisons to sort a list of m numbers.)

14. Prove the recurrence equation $d(n) = (n-1)[d(n-1)+d(n-2)]$ of Example 2.2.3 by using the result of Theorem 2.1.4.

15. Let $s(m)$ denote the number of comparisons needed to find the smallest number in a list of m (distinct) numbers. Assume that $s(1) = 0$.

(a) For $m > 1$, we could find the smallest number recursively as follows. First, we find the smallest of the first $m-1$ numbers in the list, then we compare this with the last number in the list. Find and solve a recurrence equation for $s(m)$ based on this method.

(b) Alternately, for $m = 2^n > 1$, we could find the smallest number recursively as follows: We partition the list of 2^n numbers into two sublists of 2^{n-1} numbers each, find the smallest number in each of these sublists, and then compare these two numbers. Find and solve a recurrence equation for $s(m)$ based on this method. (Hint: Let $s'(n) = s(2^n)$.)

16. Prove Theorem 2.2.3.

17. Let $t(m)$ denote the number of comparisons required to sort a list of m numbers using the "selection sort" technique. Assume that $t(1) = 0$. For $m \geq 2$, a selection sort of m numbers can be described recursively as follows: First, find the smallest of the m numbers and place it at the beginning of the list. Assume that this step requires $m-1$ comparisons. Then sort the remaining $m-1$ numbers.

(a) Find a recurrence equation for $t(m)$.

(b) Show that $t(m) = m(m-1)/2$. Compare this with the result of Exercise 13.

18. Prove Theorem 2.2.4.

19. Each of the following parts recursively defines a function from $\mathbb{N}$ to $\mathbb{Q}$. Find an explicit formula for the function.

(a) $h_1(0) = 0$, $h_1(1) = 1$, $h_1(n) = 3h_1(n-1) + 4h_1(n-2)$, $n \geq 2$

(b) $h_2(0) = h_2(1) = 2$, $h_2(n) = 2h_2(n-1) - h_2(n-2)$, $n \geq 2$

(c) $h_3(0) = 1$, $h_3(1) = 0$, $h_3(n) = h_3(n-2)$, $n \geq 2$

(d) $h_4(0) = h_4(1) = 1$, $h_4(2) = 5$, $h_4(n) = 3h_4(n-1) + 3h_4(n-2) - h_4(n-3)$, $n \geq 3$

(e) $h_5(0) = 0$, $h_5(1) = 1$, $h_5(n) = 4h_5(n-2)$, $n \geq 2$

(f) $h_6(0) = 1$, $h_6(n) = (n+1)h_6(n-1)$, $n \geq 1$

(g) $h_7(0) = 0$, $h_7(1) = 1$, $h_7(2) = 2$, $h_7(n) = h_7(n-1) + 9h_7(n-2) - 9h_7(n-3)$, $n \geq 3$

(h) $h_8(0) = -1/2$, $h_8(1) = -1$, $h_8(n) = 8h_8(n-1) - 16h_8(n-2)$, $n \geq 2$

(i) $h_9(0) = 1$, $h_9(1) = h_9(2) = 0$, $h_9(n) = 3h_9(n-2) - 2h_9(n-3)$, $n \geq 3$

(j) $h_0(0) = 0$, $h_0(1) = -3$, $h_0(2) = 5$, $h_0(3) = 39$, $h_0(n) = 5h_0(n-1) - 6h_0(n-2) - 4h_0(n-3) + 8h_0(n-4)$, $n \geq 4$

20. Consider the linear homogeneous recurrence equation (2.2.7) with constant coefficients:

$$h(n) - a_1 h(n-1) - a_2 h(n-2) - \cdots - a_r h(n-r) = 0 \qquad (2.2.7)$$

Suppose that $h_1(n)$ and $h_2(n)$ are both solutions of this recurrence equation; prove that

$$c_1 h_1(n) + c_2 h_2(n)$$

is also a solution for any constants c_1 and c_2.

21. Consider the function $g : \mathbb{N} \to \mathbb{Z}$ defined recursively by $g(0) = g(1) = -1$, and $g(n) = 5g(n-1) - 6g(n-2)$, $n \geq 2$.

(a) Compute $g(2)$, $g(3)$, and $g(4)$.

(b) Find an explicit formula for $g(n)$.

22. Consider the function $h : \mathbb{N} \to \mathbb{Z}$ defined recursively by $h(0) = 0$, $h(1) = 1$, and $h(n) = -h(n-2)$, $n \geq 2$.

(a) Compute the first several values of $h(n)$ and guess an explicit formula for $h(n)$. (Hint: The value of $h(n)$ depends on the value of n mod 4.) Check your guess.

(b) Use the theory of linear homogeneous recurrence relations to find an explicit formula for $h(n)$.

23. Each of the following parts recursively defines a function from $\mathbb{N}$ to $\mathbb{Z}$. Find an explicit formula for the function.

(a) $a_1(0) = 0$, $a_1(1) = 3$, $a_1(2) = 21$, $a_1(n) = 3a_1(n-1) - 4a_1(n-3)$, $n \geq 3$

(b) $a_2(0) = 1$, $a_2(1) = 7$, $a_2(2) = 65$, $a_2(n) = 5a_2(n-1) - 3a_2(n-2) - 9a_2(n-3)$, $n \geq 3$,

(c) $a_3(0) = 7$, $a_3(1) = a_3(2) = 3$, $a_3(n) = 3a_3(n-1) - a_3(n-2) + 3a_3(n-3)$, $n \geq 3$

24. Verify the claim that the two-step recipe given in Example 2.2.3 uniquely determines a derangement $(x_1, x_2, \ldots, x_n)$ of $S = \{1, 2, \ldots, n\}$.

25. For positive integers m and n, let $t(m, n)$ be the number of ordered partitions of m into n parts. Find initial values and a recurrence relation for $t(m, n)$.

26. Each of the following parts recursively defines a function from $\mathbb{N}$ to $\mathbb{Z}$ (a, b, and c are integer constants). Find an explicit formula for the function.

   (a) $f_1(0) = b, f_1(n) = f_1(n-1) + a, \quad n \geq 1$
   (b) $f_2(0) = b, f_2(n) = f_2(n-1) + a(2n-1), \quad n \geq 1$
   (c) $f_3(0) = c, f_3(n) = f_3(n-1) + 2an - a + b, \quad n \geq 1$
   (d) $f_4(0) = 1, f_4(n) = af_4(n-1), \quad n \geq 1$
   (e) $f_5(0) = 1 + b, f_5(n) = af_5(n-1) + b(1-a), \quad n \geq 1$
   (f) $f_6(0) = a, f_6(n) = f_6(n-1), \quad n \geq 1$

27. Let $f(n)$ denote the nth Fibonacci number.

   (a) Show that $f(0) + f(1) + \cdots + f(n) = f(n+2) - 1$.
   (b) Find and verify a formula for $f(1) + f(3) + \cdots + f(2n+1)$.
   (c) Find and verify a formula for $f(0) + f(2) + \cdots + f(2n)$.
   (d) Find and verify a formula for

$$\sum_{k=0}^{n} (-1)^k f(k)$$

28. Let $f(n)$ denote the nth Fibonacci number.

   (a) Show that $f(n)f(n+2) = f^2(n+1) + (-1)^n$.
   (b) Show that $f(n)$ and $f(n+1)$ are relatively prime.

29. Let $f(n)$ be the nth Fibonacci number. Show that $f(n)$ is equal to the integer nearest

$$\frac{1}{\sqrt{5}} \left(\frac{1 + \sqrt{5}}{2} \right)^{n+1}$$

30. Let $g(n)$ be the number of ways to express the positive integer n as a sum, where the order of terms in the sum is important and each term is 1, 2, or 5. For example, the ways to express 5 as such a sum are 5, $2+2+1$, $2+1+2$, $2+1+1+1$, $1+2+2$, $1+2+1+1$, $1+1+2+1$, $1+1+1+2$, and $1+1+1+1+1$, so $g(5) = 9$. Find initial values and a recurrence equation for $g(n)$.

31. The Lucas numbers (see Exercise 29 of Exercise Set 0.4) $L(n)$, $n \geq 0$, are defined by:

$$L(0) = 1, \quad L(1) = 3$$
$$L(n) = L(n-1) + L(n-2) \quad n \geq 2$$

   Find an explicit formula for $L(n)$.

32. Let $\lambda(m)$ denote the number of comparisons used to find both the smallest and largest numbers in a list of m (distinct) numbers.

   (a) Consider using a method analogous to that of Exercise 15(a). The initial values are $\lambda(1) = 0$ and $\lambda(2) = 1$; find and solve a recurrence equation for $\lambda(m)$, $m \geq 3$.

(b) Consider using a method analogous to that of Exercise 15(b). Let $\lambda'(n) = \lambda(2^n)$. The initial value is $\lambda'(0) = 0$; find and solve a recurrence equation for $\lambda'(n)$, $n \geq 1$. Compare with the result of part (a).

2.3 GENERATING FUNCTIONS

Let $g(x)$ be a real-valued function, and assume g has the following representation as a power series:

$$g(x) = c_0 + c_1 x + c_2 x^2 + \cdots = \sum_{n=0}^{\infty} c_n x^n$$

valid for all x such that $|x| < r$, where r is some positive real number. Then $g(x)$ is called a **generating function** for the sequence (of coefficients)

$$(c_0, c_1, c_2, \ldots)$$

In this section we see how generating functions provide an ingenious method for solving a variety of combinatorial problems. We will also see how generating functions can be used to solve recurrence equations.

Let $c(n) = c_n$ be some function of the nonnegative integer n that is of combinatorial interest; for example, c_n might be the number of ways to distribute n identical balls to three different boxes. If we can find a generating function $g(x)$ for $(c_0, c_1, \ldots, c_n, \ldots)$, and if we can find the representation of $g(x)$ as a power series, then, in theory, we know the value of c_n for any n, namely, c_n is the coefficient of x^n in the power series for g. For example, we will see that the generating function for the number c_n of ways to distribute n identical balls to three different boxes is $g(x) = 1/(1 - x)^3$, and that

$$g(x) = \frac{1}{(1 - x)^3} = 1 + 3x + 6x^2 + \cdots + C(n + 2, 2)x^n + \cdots$$

So $c_n = C(n + 2, 2)$. Of course, in this case we can find c_n without using generating functions—see Theorem 1.5.6, part 2(a)—but often generating functions provide the best, or perhaps the only, way to get the answer we are seeking.

Our first task is to develop a catalog of certain basic or standard generating functions.

■ *Example 2.3.1*

(a) The geometric series $1 + x + x^2 + \cdots + x^n + \cdots$ converges to $1/(1 - x)$ for all x such that $|x| < 1$. Hence, the function g defined on $(-1, 1)$ by

$$g(x) = \frac{1}{1 - x}$$

is a generating function for the constant sequence $(1, 1, 1, \ldots)$.

(b) Let m be a positive integer. By Corollary 1.6.2, we have

$$(1+x)^m = \sum_{n=0}^{m} C(m,n)x^n$$

Let us define $C(m,n)$ to be 0 for $n > m$. Thus, we may write

$$(1+x)^m = \sum_{n=0}^{\infty} C(m,n)x^n$$

and this is valid for any real number x. Hence, the function b defined on $\mathbb{R}$ by

$$b(x) = (1+x)^m$$

is a generating function for the sequence of binomial coefficients

$$(C(m,0), C(m,1), \ldots, C(m,n), \ldots)$$

(c) Let $f(x)$ be the polynomial defined on $\mathbb{R}$ by

$$f(x) = (1+x)(1+x+x^2)(x+x^2+x^3)$$

We have written f as a product of three factors for a reason. Consider expanding f to its standard polynomial form, as a sum of terms, each term being a monomial. The initial expansion of f, before like terms are combined, has $2 \cdot 3 \cdot 3 = 18$ terms; these are obtained by the 18 ways of multiplying together 3 terms, one chosen from each factor of f. If the term

$$x^{y_i}$$

is chosen from the ith factor, $1 \le i \le 3$, then, when these are multiplied, we obtain

$$x^{y_1} \cdot x^{y_2} \cdot x^{y_3} = x^{y_1+y_2+y_3}$$

For example, if we choose $1 = x^0$ from the first factor, x^1 from the second factor, and x^3 from the third factor, we obtain

$$x^0 \cdot x^1 \cdot x^3 = x^{0+1+3} = x^4$$

Thus, when like terms are combined, the coefficient a_n of x^n in the expansion of f is just the number of ways to choose

$$x^{y_i}$$

from the ith factor, $1 \leq i \leq 3$, so that $y_1 + y_2 + y_3 = n$. Note that the exponents y_1, y_2, and y_3 are integers satisfying $0 \leq y_1 \leq 1$, $0 \leq y_2 \leq 2$, and $1 \leq y_3 \leq 3$. Therefore, a_n is the number of ways to express n as a sum of three integers:

$$n = y_1 + y_2 + y_3$$

where $0 \leq y_1 \leq 1$, $0 \leq y_2 \leq 2$, and $1 \leq y_3 \leq 3$ (and order is important), and f is a generating function for the sequence $(a_0, a_1, \ldots, a_n, \ldots)$.

If we expand f (perhaps using our favorite computer algebra package), we obtain

$$f(x) = x + 3x^2 + 5x^3 + 5x^4 + 3x^5 + x^6$$

So we have $a_0 = 0$, $a_1 = 1$, $a_2 = 3$, $a_3 = 5$, $a_4 = 5$, $a_5 = 3$, $a_6 = 1$, and $a_n = 0$ for $n \geq 7$.

This specific example illustrates a general technique for finding a whole family of generating functions: Let k be a positive integer and, for each $i \in \{1, 2, \ldots, k\}$, let l_i and u_i be (nonnegative) integers with $l_i \leq u_i$. Let a_n be the number of solutions to the equation

$$y_1 + y_2 + \cdots + y_k = n$$

in integers $y_1, y_2, \ldots, y_k$ satisfying the constraints $l_i \leq y_i \leq u_i$ for $i \in \{1, 2, \ldots, k\}$. Then the generating function for the sequence $(a_0, a_1, \ldots, a_n, \ldots)$ is (the polynomial)

$$\prod_{i=1}^{k} \left(\sum_{j=l_i}^{u_i} x^j \right)$$

Note that the ith factor in the generating function contains a term x^j for each integer j that is a possible value for the variable y_i.   ■

Up to its domain, a generating function g for a sequence $(c_0, c_1, \ldots, c_n, \ldots)$ is unique. We shall therefore refer to "the" generating function for a given sequence. Often, generating functions are treated in a purely algebraic way, as "formal" power series, without regard to the issue of convergence. However, our approach is to ensure that all power series under discussion in a given problem do converge in some interval; this allows us to bring the full power of analysis to bear on the problem. Moreover, the study of generating functions provides an excellent opportunity for you to review and apply results about power series first encountered in calculus.

To continue, recall the following results about power series.

THEOREM 2.3.1   Let r be a positive real number. If the function g has the power series representation

$$g(x) = \sum_{n=0}^{\infty} c_n x^n$$

valid for $|x| < r$, then each of the following statements is true:

1. g is differentiable on $(-r, r)$, and its derivative g' is represented on $(-r, r)$ by the power series

$$c_1 + 2c_2 x + 3c_3 x^2 + \cdots = \sum_{n=0}^{\infty} (n+1)c_{n+1} x^n$$

2. For any constant a, the function ag is represented on $(-r, r)$ by the power series

$$\sum_{n=0}^{\infty} ac_n x^n$$

3. Let the function h be defined from $(-s, s)$ to $(-r, r)$, where s is a positive real number (or ∞). Then the composite function $g \circ h$ is represented on $(-r, r)$ by the power series

$$\sum_{n=0}^{\infty} c_n h^n(x)$$

4. The coefficient c_n is determined by **Taylor's formula**:

$$c_n = \frac{g^{(n)}(0)}{n!}$$

(Recall that $g^{(n)}(a)$ denotes the nth derivative of g at a.)   □

The next result follows immediately from parts 1 and 2 of Theorem 2.3.1.

COROLLARY 2.3.2   Let g be the generating function for the sequence $(c_0, c_1, \ldots, c_n, \ldots)$. Then g' is the generating function for the sequence $(c_1, 2c_2, \ldots, (n+1)c_{n+1}, \ldots)$ and, for any constant a, the function ag is the generating function for the sequence $(ac_0, ac_1, \ldots, ac_n, \ldots)$.   □

■ *Example 2.3.2*   We have seen that the function g defined on $(-1, 1)$ by $g(x) = 1/(1-x)$ is the generating function for the constant sequence $(1, 1, 1, \ldots)$. Differentiating and applying Corollary 2.3.2, we find that g', defined on $(-1, 1)$ by

$$g'(x) = \frac{1}{(1-x)^2}$$

is the generating function for the sequence of positive integers $(1, 2, \ldots, n+1, \ldots)$.

Let $b_n = n+1$, so that g' is the generating function for $(b_0, b_1, \ldots, b_n, \ldots)$. Taking the derivative of g' (the second derivative of g), and again applying Corollary 2.3.2, we find that g'', defined on $(-1, 1)$ by

$$g''(x) = \frac{2}{(1-x)^3}$$

generates the sequence $(1 \cdot 2, 2 \cdot 3, \ldots, (n+1)b_{n+1}, \ldots) = (2, 6, 12, \ldots, (n+1)(n+2), \ldots)$. Let

$$g_3(x) = \frac{g''(x)}{2} = \frac{1}{(1-x)^3}$$

Now apply Corollary 2.3.2 (the second part) with $a = 1/2$. We find that g_3 is the generating function for the sequence

$$(1, 3, 6, \ldots, C(n+2, 2), \ldots)$$

of "triangle numbers." These are so called because $C(n+2, 2)$ is the number of dots in an equilateral triangular array of dots with $n+1$ dots on a side. At the beginning of this section, we learned that g_3 is the generating function for the number of ways to distribute n identical balls to three different boxes. An alternate derivation of this result will be given shortly.

Continuing the above analysis, we let m be a positive integer and define the function g_m on $(-1, 1)$ by

$$g_m(x) = \frac{1}{(1-x)^m}$$

It is left for you in Exercise 2 to show that g_m is the generating function for the sequence

$$(1, m, \ldots, C(n+m-1, m-1), \ldots) = (1, m, \ldots, C(n+m-1, n), \ldots)$$

Recall, from Theorem 1.5.4 (with the roles played by m and n interchanged), that $C(n+m-1, m-1)$ is the number of ways to express n as a sum of m nonnegative integers, where the order of the terms is important. Alternately, we have, by Theorem 1.5.6, part 2(a), that $C(n+m-1, m-1)$ is the number of ways to distribute n identical balls to m different boxes. ∎

∎ *Example 2.3.3*   Again, let g be defined on $(-1, 1)$ by $g(x) = 1/(1-x)$. Then g generates $(1, 1, 1, \ldots)$. By using Theorem 2.3.1, part (c), we have

$$\frac{1}{1+x} = g(-x) = 1 - x + x^2 - x^3 + \cdots = \sum_{n=0}^{\infty} (-1)^n x^n$$

Thus, the function h, defined on $(-1, 1)$ by $h(x) = 1/(1 + x)$, generates the sequence

$$(1, -1, 1, -1, \ldots, (-1)^n, \ldots)$$

More generally, for any nonzero real number r,

$$\frac{r}{r - x} = \frac{1}{1 - \frac{x}{r}} = g\left(\frac{x}{r}\right) = 1 + \frac{x}{r} + \frac{x^2}{r^2} + \cdots = \sum_{n=0}^{\infty} r^{-n} x^n$$

The above power series is valid for $|x/r| < 1$, namely, for $|x| < |r|$. That is, the function h_r, defined on $(-|r|, |r|)$ by $h_r(x) = r/(r - x)$, is the generating function for the sequence

$$(1, r^{-1}, r^{-2}, \ldots, r^{-n}, \ldots)$$

Next, we start with $h(x) = 1/(1 + x)$ and replace x by x^2. We now obtain

$$\frac{1}{1 + x^2} = h(x^2) = 1 - x^2 + x^4 - x^6 + \cdots = \sum_{n=0}^{\infty} (-1)^n x^{2n}$$

for $x \in (-1, 1)$. Namely, the function p defined on $(-1, 1)$ by $p(x) = 1/(1 + x^2)$ generates the sequence $(1, 0, -1, 0, 1, 0, -1, 0, \ldots)$. ∎

When working with the more familiar generating functions, we sometimes allow ourselves to be imprecise and to say, for example, "$1/(1-x)^m$ is the generating function for the sequence $(1, m, \ldots, C(n + m - 1, n), \ldots)$," or "$1/(1 - x)^m$ is the generating function for $C(n + m - 1, n)$." Here it is understood that we are speaking of the function, defined on $(-1, 1)$, whose value at x is $1/(1 - x)^m$.

Recall another result from calculus, namely, that under suitable restrictions power series may be added and multiplied.

THEOREM 2.3.3 Let r_1 and r_2 be positive real numbers. Assume that the function g_1 has the power series representation

$$g_1(x) = \sum_{n=0}^{\infty} a_n x^n$$

on $(-r_1, r_1)$ and that the function g_2 has the power series representation

$$g_2(x) = \sum_{n=0}^{\infty} b_n x^n$$

on $(-r_2, r_2)$. Let r be the minimum of r_1 and r_2. Then, on $(-r, r)$:

1. The function $g_1 + g_2$ has the power series representation

$$(g_1 + g_2)(x) = \sum_{n=0}^{\infty}(a_n + b_n)x^n$$

2. The function $g_1 g_2$ has the power series representation

$$(g_1 g_2)(x) = \sum_{n=0}^{\infty} c_n x^n$$

where the coefficient c_n is determined as follows:

$$c_n = \sum_{k=0}^{n} a_k b_{n-k}$$

(so $c_0 = a_0 b_0$, $c_1 = a_0 b_1 + a_1 b_0$, $c_2 = a_0 b_2 + a_1 b_1 + a_2 b_0$, $c_3 = a_0 b_3 + a_1 b_2 + a_2 b_1 + a_3 b_0$, and so on). □

COROLLARY 2.3.4   If g_1 is the generating function for the sequence $(a_0, a_1, \ldots, a_n, \ldots)$ and g_2 is the generating function for the sequence $(b_0, b_1, \ldots, b_n, \ldots)$, then:

1. $g_1 + g_2$ is the generating function for the sequence $(a_0 + b_0, a_1 + b_1, \ldots, a_n + b_n, \ldots)$.
2. $g_1 g_2$ is the generating function for the sequence $(c_0, c_1, \ldots, c_n, \ldots)$, where

$$c_n = \sum_{k=0}^{n} a_k b_{n-k} \qquad\qquad □$$

Suppose $c(n) = c_n$ is some function of the nonnegative integer n that is of combinatorial interest. We may, for instance, know initial values and a recurrence equation for c_n and wish to solve the recurrence equation. One approach, to be illustrated shortly, is to assume that c_n has a generating function g:

$$g(x) = \sum_{n=0}^{\infty} c_n x^n$$

and then to attempt to use the given recurrence equation to find an explicit formula for $g(x)$. If this works, then we know $g(x)$, but usually g will not be one of our standard generating functions. We then try to build up g from standard generating functions, using the operations of Theorems 2.3.1 and 2.3.3, so that Corollaries 2.3.2 and 2.3.4 can be applied to determine c_n. This idea of "decomposing" a given function into standard generating functions is illustrated by the next example.

■ *Example 2.3.4* Find the sequence generated by each of the following generating functions:

(a) $f_1(x) = (8 - 5x)/(2 - 3x + x^2)$

(b) $f_2(x) = x/(1 + x)$

(c) $f_3(x) = (1 + x)^3/(1 - x)$

Solution

(a) Note that

$$f_1(x) = \frac{8 - 5x}{(1 - x)(2 - x)}$$

The method of partial fraction decomposition can be applied to express f_1 as a sum:

$$f_1(x) = \frac{3}{1 - x} + \frac{2}{2 - x}$$

By letting $g_1(x) = 3/(1 - x)$ and $g_2(x) = 2/(2 - x)$, we see that g_1 generates $(3, 3, 3, \ldots)$ and g_2 generates $(1, \frac{1}{2}, \ldots, 2^{-n}, \ldots)$. So by Corollary 2.3.4, part 1 (with $a_n = 3$ and $b_n = 2^{-n}$), we find that f_1 generates the sequence $(3, \frac{7}{2}, \ldots, (3 + 2^{-n}), \ldots)$.

(b) Recall that

$$\frac{1}{1 + x} = 1 - x + x^2 - x^3 + \cdots$$

namely, that $1/(1 + x)$ is the generating function for the sequence $(1, -1, 1, -1, \ldots)$. So

$$\frac{x}{1 + x} = x - x^2 + x^3 - x^4 + \cdots = \sum_{n=0}^{\infty} (-1)^n x^{n+1}$$

namely, $x/(1 + x)$ generates the sequence $(0, 1, -1, 1, -1, \ldots)$. More generally, if $g(x)$ is the generating function for the sequence $(c_0, c_1, \ldots, c_n, \ldots)$:

$$g(x) = \sum_{n=0}^{\infty} c_n x^n$$

then

$$xg(x) = \sum_{n=0}^{\infty} c_n x^{n+1}$$

namely, $xg(x)$ generates the sequence $(0, c_0, c_1, c_2, \ldots)$. Note that the effect of multiplication by x is to shift each number in the generated sequence one position to the right. For a generalization of this result, work Exercise 4.

(c) Here we apply Corollary 2.3.4, part 2, with $g_1(x) = (1 + x)^3$ and $g_2(x) = 1/(1 - x)$. Hence, $a_n = C(3, n)$ and $b_n = 1$. Note that $a_n = 0$ for $n > 3$, so that $c_0 = a_0 = 1$,

$c_1 = a_0 + a_1 = 1 + 3 = 4$, $c_2 = a_0 + a_1 + a_2 = 1 + 3 + 3 = 7$, and $c_n = a_0 + a_1 + a_2 + a_3 = 8$ for $n \geq 3$. Thus,

$$f_3(x) = (g_1 g_2)(x) = 1 + 4x + 7x^2 + 8(x^3 + x^4 + \cdots) = 1 + 4x + 7x^2 + 8\sum_{n=3}^{\infty} x^n$$

namely, f_3 generates the sequence $(1, 4, 7, 8, 8, 8, \ldots)$. The same result can be obtained by using division and Corollary 2.3.4, part 1:

$$f_3(x) = \frac{1 + 3x + 3x^2 + x^3}{1 - x} = -7 - 4x - x^2 + \frac{8}{1 - x}$$
$$= -7 - 4x - x^2 + 8(1 + x + x^2 + x^3 + \cdots)$$
$$= 1 + 4x + 7x^2 + 8(x^3 + x^4 + \cdots)$$
■

■ **Example 2.3.5** We can use Corollary 2.3.4, part 2, to derive the results of Example 2.3.2.

First, let's apply the corollary with $g_1(x) = g_2(x) = 1/(1 - x)$; so $a_n = b_n = 1$. Then

$$c_n = \sum_{k=0}^{n} a_k b_{n-k} = n + 1$$

so that $(g_1 g_2)(x) = 1/(1 - x)^2$ generates the sequence of positive integers: $(1, 2, \ldots, n + 1, \ldots)$.

Using this result, this time we let $g_1(x) = 1/(1 - x)^2$ and $g_2(x) = 1/(1 - x)$; so $a_n = n + 1$ and $b_n = 1$. Then

$$c_n = \sum_{k=0}^{n} a_k b_{n-k} = \sum_{k=0}^{n} (k + 1) = C(n + 2, 2)$$

and we find that $(g_1 g_2)(x) = 1/(1 - x)^3$ generates the following sequence: $(1, 3, 6, \ldots, C(n + 2, 2), \ldots)$.

Continuing this approach, we can show by induction on m that $1/(1 - x)^m$ is the generating function for the sequence $(1, m, \ldots, C(n + m - 1, n), \ldots)$; see Exercise 2. Another connection can be seen as follows:

$$\frac{1}{(1 - x)^m} = \left(\frac{1}{1 - x}\right)^m = (1 + x + x^2 + x^3 + \cdots)^m$$

When Corollary 2.3.4 is used to expand $(1 + x + x^2 + x^3 + \cdots)^m$, the coefficent of x^n will be the number of ways to choose

x^{y_i}

from the *i*th factor, $1 \leq i \leq m$, so that

$$x^{y_1} \cdot x^{y_2} \cdot \cdots \cdot x^{y_m} = x^{y_1 + y_2 + \cdots + y_m} = x^n$$

Note that each exponent y_i is a nonnegative integer. Thus, the coefficient of x^n in the expansion of $1/(1-x)^m$ is the number $C(n+m-1,n)$ of ways to express n as a sum of m nonnegative integers:

$$y_1 + y_2 + \cdots + y_m = n$$

where order is important. This again shows that $1/(1-x)^m$ generates $C(n+m-1,n)$.

■

The next several examples illustrate how generating functions are used to obtain explicit formulas for recursively defined functions.

■ ***Example 2.3.6*** The Fibonacci sequence is defined recursively by $f(0) = f(1) = 1$, $f(n) = f(n-2) + f(n-1)$, $n \geq 2$. Let g be the generating function for the sequence $(f(0), f(1), f(2), \ldots)$ of Fibonacci numbers; thus,

$$g(x) = \sum_{n=0}^{\infty} f(n)x^n$$

(Because $f(n) \leq 2^n$ for all n, it follows that g is defined on $(-1/2, 1/2)$.) Our goal is to find an explicit formula for $f(n)$. To do this, we first try to find an explicit formula for $g(x)$. We begin by moving those terms of the power series corresponding to the initial values of f to the left-hand side:

$$g(x) - f(0) - f(1)x = \sum_{n=2}^{\infty} f(n)x^n$$

The power series on the right-hand side now begins with $n = 2$, allowing us to use the recurrence equation for $f(n)$:

$$g(x) - f(0) - f(1)x = \sum_{n=2}^{\infty} [f(n-2) + f(n-1)]x^n$$

$$= \sum_{n=2}^{\infty} f(n-2)x^n + \sum_{n=2}^{\infty} f(n-1)x^n$$

We next manipulate each of the power series on the right-hand side to obtain an expression involving $g(x)$:

$$g(x) - 1 - x = x^2 \sum_{n=2}^{\infty} f(n-2)x^{n-2} + x \sum_{n=2}^{\infty} f(n-1)x^{n-1}$$

$$= x^2 \sum_{n=0}^{\infty} f(n)x^n + x \left(\sum_{n=0}^{\infty} f(n)x^n - f(0) \right)$$

$$= x^2 g(x) + x[g(x) - f(0)]$$

$$= x^2 g(x) + xg(x) - x$$

Thus,

$$g(x) - 1 = x^2 g(x) + xg(x)$$

This last equation can be solved for $g(x)$ to obtain $g(x) = 1/(1 - x - x^2)$. So we have found the generating function for the sequence of Fibonacci numbers. Now let

$$\alpha = \frac{1 + \sqrt{5}}{2} \qquad \text{and} \qquad \beta = \frac{1 - \sqrt{5}}{2}$$

and note that

$$\frac{1}{1 - x - x^2} = \frac{1}{(1 - \alpha x)(1 - \beta x)}$$

$$= \frac{1}{\alpha - \beta} \left(\frac{\alpha}{1 - \alpha x} - \frac{\beta}{1 - \beta x} \right)$$

Recall that $1/(1 - \alpha x)$ generates the sequence $(1, \alpha, \alpha^2, \ldots, \alpha^n, \ldots)$, so $\alpha/(1 - \alpha x)$ generates $(\alpha, \alpha^2, \alpha^3, \ldots \alpha^{n+1}, \ldots)$. Similarly, $\beta/(1 - \beta x)$ generates $(\beta, \beta^2, \beta^3, \ldots, \beta^{n+1}, \ldots)$. It follows by Corollaries 2.3.2 and 2.3.4 that $g(x)$ is the generating function for the sequence $(f(0), f(1), \ldots, f(n), \ldots)$, where

$$f(n) = \frac{1}{\sqrt{5}} \left(\alpha^{n+1} - \beta^{n+1} \right) = \frac{1}{\sqrt{5}} \left[\left(\frac{1 + \sqrt{5}}{2} \right)^{n+1} - \left(\frac{1 - \sqrt{5}}{2} \right)^{n+1} \right]$$

This is the same result as that of Theorem 2.2.1. ∎

■   **Example 2.3.7**   Use generating functions to find an explicit formula for the function r of Example 2.2.5, defined recursively on $\mathbb{N}$ by

$$r(0) = 0$$
$$r(n) = 2r(n-1) + 2^n \qquad n \geq 1$$

Note that the recurrence equation for r is linear but nonhomogeneous.

Solution
Let g be the generating function for $r(n)$, so that

$$g(x) = \sum_{n=0}^{\infty} r(n)x^n$$

(It can be shown, by induction, that $r(n) \leq 3^n$, so that g is defined on $(-1/3, 1/3)$.) Then

$$g(x) - r(0) = \sum_{n=1}^{\infty} r(n)x^n$$

$$= \sum_{n=1}^{\infty} [2r(n-1) + 2^n]x^n$$

$$= 2\sum_{n=1}^{\infty} r(n-1)x^n + \sum_{n=1}^{\infty} 2^n x^n$$

so that

$$g(x) = 2x\sum_{n=1}^{\infty} r(n-1)x^{n-1} + \sum_{n=0}^{\infty} 2^n x^n - 1$$

$$= 2x\sum_{n=0}^{\infty} r(n)x^n + \sum_{n=0}^{\infty} 2^n x^n - 1$$

$$= 2xg(x) + \frac{1}{1-2x} - 1$$

Thus,

$$(1 - 2x)g(x) = \frac{1}{1-2x} - 1$$

so that

$$g(x) = \frac{1}{1-2x}\left(\frac{1}{1-2x} - 1\right)$$

Now let $g_1(x) = 1/(1 - 2x)$ and $g_2(x) = g_1(x) - 1$. We know that g_1 generates the sequence $(1, 2, 4, \ldots, 2^n, \ldots)$; hence, g_2 generates $(0, 2, 4, \ldots, 2^n, \ldots)$. Let a_n and b_n denote the nth terms of these two sequences, respectively. Then, by Corollary 2.3.4, we see that $g = g_1 g_2$ generates the sequence $(c_0, c_1, \ldots, c_n, \ldots)$, where $c_0 = a_0 b_0 = 0$ and, for $n \geq 1$,

$$
\begin{aligned}
c_n &= \sum_{k=0}^{n} a_k b_{n-k} \\
&= \sum_{k=0}^{n-1} a_k b_{n-k} \qquad \text{(since } b_0 = 0\text{)} \\
&= \sum_{k=0}^{n-1} 2^k 2^{n-k} \\
&= \sum_{k=0}^{n-1} 2^n = n2^n
\end{aligned}
$$

Therefore, g generates the sequence $(0, 2, \ldots, n2^n, \ldots)$, so that

$$r(n) = n2^n \qquad \blacksquare$$

■ ***Example 2.3.8*** Consider an expression such as $4 - 3 - 2 - 1$. Such an expression is ambiguous; in fact, different answers may result depending on the order in which the subtractions are performed. For instance, $4 - (3 - (2 - 1)) = 2$, whereas $(4 - 3) - (2 - 1) = 0$. (Recall that subtraction is not an associative operation on $\mathbb{Z}$.) In general, let $*$ denote some abstract, nonassociative operation (on $\mathbb{Z}$, say) and consider the ambiguous expression

$$x_1 * x_2 * \cdots * x_n$$

with n operands. Let $c(n)$ be the number of ways to parenthesize this expression to make it unambiguous. We wish to find initial values and a recurrence equation for $c(n)$ and then apply generating functions to obtain an explicit formula for $c(n)$.

Let us work out a few values for $c(n)$. The expressions x_1 and $x_1 * x_2$ are unambiguous as they stand, so $c(1) = c(2) = 1$. For $n = 3$, the ways to parenthesize the expression are $(x_1 * x_2) * x_3$ and $x_1 * (x_2 * x_3)$, so that $c(3) = 2$. For $n = 4$, the ways to parenthesize the expression are as follows:

$$((x_1 * x_2) * x_3) * x_4 \qquad (x_1 * (x_2 * x_3)) * x_4 \qquad (x_1 * x_2) * (x_3 * x_4)$$
$$x_1 * ((x_2 * x_3) * x_4) \qquad x_1 * (x_2 * (x_3 * x_4))$$

Thus, $c(4) = 5$.

In general, we may partition the set of ways to parenthesize

$$x_1 * x_2 * \cdots * x_n$$

according to which of the $n - 1$ operations $*$ is to be performed last. If the kth $*$ is the last one to be applied, then our expression looks like

$$(x_1 * \cdots * x_k) * (x_{k+1} * \cdots * x_n)$$

To then completely parenthesize the expression, each of the subexpressions $x_1 * \cdots * x_k$ and $x_{k+1} * \cdots * x_n$ must be parenthesized. By the multiplication principle, this can be done in $c(k)c(n - k)$ ways. Hence, by the addition principle, we have

$$c(n) = \sum_{k=1}^{n-1} c(k)c(n - k)$$

for $n \geq 2$. It is convenient to define $c(0) = 0$. We may then define the function c recursively on $\mathbb{N}$ by

$$c(0) = 0, \quad c(1) = 1$$

$$c(n) = \sum_{k=0}^{n} c(k)c(n - k) \qquad n \geq 2$$

Note that the recurrence equation for $c(n)$ is nonlinear. As usual, we let g be the generating function for the sequence $(0, 1, \ldots, c(n), \ldots)$, so that

$$g(x) = \sum_{n=0}^{\infty} c(n)x^n$$

Proceeding as usual:

$$g(x) - c(0) - c(1)x = \sum_{n=2}^{\infty} c(n)x^n$$

or

$$g(x) - x = \sum_{n=2}^{\infty} \left(\sum_{k=0}^{n} c(k)c(n - k) \right) x^n$$

In this last equation, we may begin the outer summation with $n = 0$ since the inner sum has value 0 for $n = 0$ and for $n = 1$. Thus,

$$g(x) - x = \sum_{n=0}^{\infty} \left(\sum_{k=0}^{n} c(k)c(n - k) \right) x^n = g^2(x)$$

by Theorem 2.3.3. Let $y = g(x)$; then y is a solution of the quadratic equation

$$y^2 - y + x = 0$$

Solving for y, we obtain

$$y = \frac{1 - \sqrt{1 - 4x}}{2} \qquad \text{or} \qquad y = \frac{1 + \sqrt{1 - 4x}}{2}$$

Since $g(0) = c(0) = 0$, it must be that

$$g(x) = \frac{1 - \sqrt{1 - 4x}}{2}$$

(Note that g is defined on $(-1/4, 1/4)$.)

Not recognizing g as one of our standard generating functions, nor as a sum and/or product of standard generating functions, we instead apply Taylor's formula (part 4 of Theorem 2.3.1):

$$c(n) = \frac{g^{(n)}(0)}{n!}$$

Note that

$$g'(x) = (1 - 4x)^{-1/2}, \quad g''(x) = 2(1 - 4x)^{-3/2}, \quad g'''(x) = 3 \cdot 2^2 (1 - 4x)^{-5/2}$$
$$g^{(4)}(x) = 5 \cdot 3 \cdot 2^3 (1 - 4x)^{-7/2}$$

and so on. In Exercise 10 you are asked to show that, for $n \geq 2$,

$$g^{(n)}(0) = 2^{n-1} \prod_{k=1}^{n-1} (2k - 1)$$

Thus, for $n \geq 2$,

$$c(n) = \frac{2^{n-1} \prod_{k=1}^{n-1} (2k - 1)}{n!}$$

By doing some algebra with factorials, we can show that this last expression is equal to

$$\frac{(2n - 2)!}{n!(n - 1)!} = \frac{1}{n} \cdot \frac{(2n - 2)!}{(n - 1)!(n - 1)!}$$

so that

$$c(n) = \frac{C(2n - 2, n - 1)}{n}$$

for $n \geq 2$. Note that the above formula also works when $n = 1$. The numbers $c(n)$ are called the **Catalan numbers** and arise in a number of seemingly diverse combinatorial settings, some of which are explored in the exercises. ∎

The next several examples illustrate how generating functions are used to solve counting problems.

∎   ***Example 2.3.9***   Let $e(n)$ be the number of ways for a team in the National Football League to score n points. We develop the generating function for the sequence $(e(0), e(1), \ldots, e(n), \ldots)$.

There are four ways for a team to score: a safety is 2 points, a field goal is 3 points, a touchdown without conversion is 6 points, and a touchdown with conversion is 7 points. Thus, $e(n)$ is the number of solutions (n_1, n_2, n_3, n_4) in nonnegative integers of

$$2n_1 + 3n_2 + 6n_3 + 7n_4 = n$$

Let $y_1 = 2n_1$, $y_2 = 3n_2$, $y_3 = 6n_3$, and $y_4 = 7n_4$. Then $e(n)$ is equal to the number of solutions (y_1, y_2, y_3, y_4) of

$$y_1 + y_2 + y_3 + y_4 = n$$

where each y_i is a nonnegative integer, y_1 is a multiple of 2, y_2 is a multiple of 3, y_3 is a multiple of 6, and y_4 is a multiple of 7. Let

$$g_1(x) = \frac{1}{1 - x^2} = 1 + x^2 + x^4 + \cdots + x^{2n_1} + \cdots$$

$$g_2(x) = \frac{1}{1 - x^3} = 1 + x^3 + x^6 + \cdots + x^{3n_2} + \cdots$$

$$g_3(x) = \frac{1}{1 - x^6} = 1 + x^6 + x^{12} + \cdots + x^{6n_3} + \cdots$$

$$g_4(x) = \frac{1}{1 - x^7} = 1 + x^7 + x^{14} + \cdots + x^{7n_4} + \cdots$$

Note that the exponents of the (nonzero) terms in the power series for $g_i(x)$ yield the possible values for the variable y_i. Before combining like terms,

$$g(x) = g_1(x)g_2(x)g_3(x)g_4(x)$$

contains terms of the form

$$x^{2n_1} x^{3n_2} x^{6n_3} x^{7n_4} = x^{2n_1 + 3n_2 + 6n_3 + 7n_4}$$

For instance, the term

$$x^2 x^9 x^0 x^{14}$$

indicates one way for a team to score 25 points—one safety, three field goals, no touchdowns without conversion, and two touchdowns with conversion. Hence, when like terms are combined, the coefficient of x^n is the number of ways to express n as $y_1 + y_2 + y_3 + y_4$, that is, $e(n)$. Therefore,

$$g(x) = \frac{1}{(1-x^2)(1-x^3)(1-x^6)(1-x^7)}$$

is the generating function for the sequence $(e(0), e(1), \ldots, e(n), \ldots)$.

Suppose we want the value $e(24)$, the number of ways for a team to score 24 points. We let $e_1(n)$ be the number of ways for a team to score n points with at most twelve safeties, at most eight field goals, at most four touchdowns without conversion, and at most three touchdowns with conversion. Then $e(24) = e_1(24)$. The generating function for $e_1(n)$ is the polynomial

$$(1 + x^2 + \cdots + x^{2 \cdot 12})(1 + x^3 + \cdots + x^{3 \cdot 8})(1 + x^6 + \cdots + x^{4 \cdot 6})(1 + x^7 + \cdots + x^{7 \cdot 3})$$

$$= (1 + x^2 + \cdots + x^{24})(1 + x^3 + \cdots + x^{24})(1 + x^6 + \cdots + x^{24})(1 + x^7 + \cdots + x^{21})$$

so our answer is the coefficient of x^{24} in the expansion of this expression. ∎

■ **Example 2.3.10**   Let $c(n)$ be the number of ways to distribute n identical slices of pizza to five fraternity brothers if no brother gets more than seven slices. We seek the generating function for the sequence $(c(0), c(1), \ldots, c(n), \ldots)$. We let

$$g(x) = (1 + x + x^2 + \cdots + x^7)^5$$

Before combining like terms, the expansion of $g(x)$ contains terms such as

$$x^{y_1} x^{y_2} x^{y_3} x^{y_4} x^{y_5} = x^{y_1 + y_2 + y_3 + y_4 + y_5}$$

where $0 \le y_i \le 7$ for $1 \le i \le 5$. Thus, when like terms are combined, the coefficient of x^n is the number of integer solutions $(y_1, y_2, y_3, y_4, y_5)$ of

$$y_1 + y_2 + y_3 + y_4 + y_5 = n$$

where $0 \le y_i \le 7$ for $1 \le i \le 5$. Interpreting y_i as the number of slices of pizza received by brother i, we see that the coefficient of x^n is $c(n)$. Thus, g is the generating function for the sequence $(c(0), c(1), \ldots, c(n), \ldots)$. Note that $c(n) = 0$ for $n > 35$.

In Exercise 14 of Exercise Set 2.1, you were asked to compute the value of $c(30)$ using the principle of inclusion–exclusion. In theory, generating functions provide a more straightforward method of solving this and similar problems; once we have the generating function g for $c(n)$, we simply need to find the coefficient $c(30)$ of x^{30} in the power series for g. Let us illustrate a mathematical technique for computing such a coefficient.

First, we use the formula for a geometric sum (see Exercise 11(b) in Exercise Set 0.4) to write

$$g(x) = (1 + x + x^2 + \cdots + x^7)^5 = \left(\frac{1 - x^8}{1 - x} \right)^5$$

Then we can consider g to be the product $g_1 g_2$ of two generating functions, where $g_1(x) = 1/(1 - x)^5$ and $g_2(x) = (1 - x^8)^5$. We know that g_1 generates the sequence $(a_0, a_1, \ldots, a_n, \ldots)$, where $a_n = C(n + 4, 4)$. Also, by the binomial theorem,

$$g_2(x) = C(5,0) - C(5,1)x^8 + C(5,2)x^{16} - C(5,3)x^{24} + C(5,4)x^{32} - C(5,5)x^{40}$$

so g_2 generates the sequence $(b_0, b_1, \ldots, b_n, \ldots)$, where $b_0 = 1$, $b_8 = -5$, $b_{16} = 10$, $b_{24} = -10$, $b_{32} = 5$, $b_{40} = -1$, and all other values of b_n are 0. Hence, by Corollary 2.3.4, part 2, g generates $(c_0, c_1, \ldots, c_n, \ldots)$, where

$$c_{30} = \sum_{k=0}^{30} a_k b_{30-k} = a_6 b_{24} + a_{14} b_{16} + a_{22} b_8 + a_{30} b_0$$

$$= -10C(10,4) + 10C(18,4) - 5C(26,4) + C(34,4)$$

Note that this is exactly the same answer as that obtained by applying the principle of inclusion–exclusion; in a kind of magical way, using generating functions has done all the tricky analysis for us! (Of course, we have to become proficient with generating functions in order to use them.) ∎

■ *Example 2.3.11* Find the number of ways to distribute 30 identical balls to six different boxes under the restriction that each of the first four boxes must receive between 3 and 7 balls.

Solution
Let $t(n)$ be the number of ways to distribute n identical balls to six different boxes with the first four boxes each receiving between 3 and 7 balls. Note that $t(n)$ is also the number of integer solutions $(y_1, y_2, y_3, y_4, y_5, y_6)$ to

$$y_1 + y_2 + y_3 + y_4 + y_5 + y_6 = n$$

where $3 \leq y_k \leq 7$ for $k \in \{1, 2, 3, 4\}$, $y_5 \geq 0$, and $y_6 \geq 0$. (To see this, let y_i be the number of balls placed in box i, $1 \le i \le 6$.) Thus, the generating function for the sequence $(t(0), t(1), \dots, t(n), \dots)$ is

$$g(x) = \frac{(x^3 + x^4 + x^5 + x^6 + x^7)^4}{(1-x)^2}$$

The factor $(x^3 + x^4 + x^5 + x^6 + x^7)^4$ results from the restriction that each of the first four boxes must receive between 3 and 7 balls; the factor $1/(1-x)^2$ comes from the fact that any (nonnegative) number of balls may be placed in the fifth and sixth boxes.

We want to find the coefficient $t(30)$ of x^{30} in (the power series for) g. First, we factor the numerator of g and then use the formula for a geometric sum to write

$$g(x) = \frac{x^{12}(1 + x + x^2 + x^3 + x^4)^4}{(1-x)^2} = \frac{x^{12}(1 - x^5)^4}{(1-x)^6}$$

Note that the coefficient of x^{30} in g is the same as the coefficient c_{18} of x^{18} in

$$\frac{g(x)}{x^{12}} = \frac{(1 - x^5)^4}{(1-x)^6}$$

We consider this to be the product $g_1 g_2$ of two generating functions, where $g_1(x) = 1/(1-x)^6$ and $g_2(x) = (1 - x^5)^4$. We know that g_1 generates the sequence $(a_0, a_1, \dots, a_n, \dots)$, where $a_n = C(n+5, 5)$. Also, by the binomial theorem,

$$g_2(x) = C(4,0) - C(4,1)x^5 + C(4,2)x^{10} - C(4,3)x^{15} + C(4,4)x^{20}$$

so g_2 generates the sequence $(b_0, b_1, \dots, b_n, \dots)$, where $b_0 = 1$, $b_5 = -4$, $b_{10} = 6$, $b_{15} = -4$, $b_{20} = 1$, and all other values of b_n are 0. Hence, by Corollary 2.3.4, part 2, $g(x)/x^{12}$ generates $(c_0, c_1, \dots, c_n, \dots)$, where

$$c_{18} = \sum_{k=0}^{18} a_k b_{18-k} = a_3 b_{15} + a_8 b_{10} + a_{13} b_5 + a_{18} b_0$$

$$= -4C(8,5) + 6C(13,5) - 4C(18,5) + C(23,5) \qquad \blacksquare$$

EXERCISE SET 2.3

1. Find the generating function for $(a_0, a_1, a_2, \dots, a_n, \dots)$, where a_n is the number of ways to select n golf balls from three Titleist balls, three Top Flight balls, four Maxfli balls, and four Pro Staff balls. Assume that balls having the same brand name are identical.

2. Let m be a positive integer. Show that the function g_m, defined over $(-1, 1)$ by

$$g_m(x) = \frac{1}{(1-x)^m}$$

is the generating function for the sequence $(1, m, \ldots, C(n+m-1, n), \ldots)$.

(a) Use induction on m and Corollary 2.3.2.
(b) Use induction on m and Corollary 2.3.4.

3. Find the generating function for $(b_0, b_1, b_2, \ldots, b_n, \ldots)$, where b_n is the number of ways to distribute n identical golf balls to five different golfers under the restriction that the first three golfers each get an even number of balls between 2 and 8 and the other two golfers get between 1 and 4 balls each.

4. Suppose g is the generating function for the sequence $(c_0, c_1, \ldots, c_n, \ldots)$ and k is a fixed positive integer.

(a) Find the generating function for the sequence $(a_0, a_1, a_2, \ldots)$ defined by

$$a_n = \begin{cases} 0 & n < k \\ c_{n-k} & n \geq k \end{cases}$$

The sequence $(a_0, a_1, a_2, \ldots)$ is called a **right k-shift** of the sequence $(c_0, c_1, c_2, \ldots)$.

(b) Find the generating function for the sequence $(b_0, b_1, b_2, \ldots)$ defined by

$$b_n = c_{n+k}$$

The sequence $(b_0, b_1, b_2, \ldots)$ is called a **left k-shift** of the sequence $(c_0, c_1, c_2, \ldots)$.

5. Find the generating function for each of the following sequences:

(a) $(1, -m, \ldots, (-1)^n C(m, n), \ldots)$ (where m is a fixed positive integer)
(b) $(0, 0, 0, 0, 1, -1, 1, -1, \ldots)$
(c) $(0, 2, 4, \ldots, 2n, \ldots)$
(d) $(1, 3, 5, \ldots, 2n+1, \ldots)$
(e) $(0, 1, 4, \ldots, n^2, \ldots)$
(f) $(4, 5, 6, \ldots, n+4, \ldots)$
(g) $(1, 1, 1/2, \ldots, 1/n!, \ldots)$
(h) $(0, 1, 8, \ldots, n^3, \ldots)$
(i) $(0, 0, 0, 1, 4, C(n, 3), \ldots)$
(j) $(1, -1/2, 1/3, \ldots, (-1)^n/(n+1), \ldots)$

6. Let r be a nonzero real number and let m be a positive integer. Define the function g over $(-|r|, |r|)$ by

$$g(x) = \frac{1}{(1-rx)^m}$$

Show that g is the generating function for the sequence
$(1, mr, \ldots, C(m+n-1, n)r^n, \ldots)$.

7. Use generating functions to find an explicit formula for each of the following recursively defined functions:

   (a) $f_1(0) = 1, f_1(1) = -2, f_1(n) = 5f_1(n-1) - 6f_1(n-2), \quad n \geq 2$

   (b) $f_2(0) = 0, f_2(1) = 1, f_2(n) = 2f_2(n-2), \quad n \geq 2$

   (c) $f_3(0) = 0, f_3(1) = 1, f_3(2) = 2, f_3(n) = f_3(n-1) + 9f_3(n-2) - 9f_3(n-3),$
   $n \geq 3$

   (d) $f_4(0) = -1, f_4(1) = 0, f_4(n) = 8f_4(n-1) - 16f_4(n-2), \quad n \geq 2$

   (e) $f_5(0) = 1, f_5(1) = 0, f_5(2) = 0, f_5(n) = 3f_5(n-2) - 2f_5(n-3), \quad n \geq 3$

8. Find the generating function for the sequence $(a(0), a(1), a(2), \ldots, a(n), \ldots)$ where $a(n)$ is the number of solutions (x_1, x_2, x_3) of $x_1 + x_2 + x_3 = n$ such that each x_i is a nonnegative integer, x_2 is even, and x_3 is odd. Then apply Corollary 2.3.2 and/or Corollary 2.3.4 to show that $a(n) = C(\lceil n/2 \rceil + 1, 2)$.

9. For $n \geq 3$, label the vertices of a convex polygon with n sides with the integers $1, 2, \ldots, n$. Let $t(n)$ be the number of ways to partition the interior of the polygon into $n - 2$ triangular regions by adding $n - 3$ diagonals that do not intersect in the interior. (See Exercise 7 in Exercise Set 2.2.) Define $c : \mathbb{N} \to \mathbb{Z}$ by $c(0) = 0$, $c(1) = 1$, and $c(n) = t(n+1)$ for $n \geq 2$. Show that $c(n)$ is the Catalan number $C(2n-2, n-1)/n$.

10. For $n \geq 1$, label the vertices of a convex polygon with $n+2$ sides with the integers $1, 2, \ldots, n+2$. Let $b(n)$ be the number of regions into which the interior is divided by the $(n+2)(n-1)/2$ diagonals, assuming that no three diagonals intersect at a common point. Then $b(1) = 1$, $b(2) = 4$, $b(3) = 11$, and so on.

    (a) Show that b satisfies the recurrence equation
    $$b(n) = b(n-1) + n + \frac{n(n+1)(n-1)}{6}$$
    for $n \geq 2$. (Hint: Add the diagonals from vertex $n+2$ to the vertices $2, 3, \ldots, n$ last. Before these are added, there are $b(n-1) + 1$ regions; show that adding these diagonals produces $n - 1 + n(n+1)(n-1)/6$ additional regions.)

    (b) Define $b(0) = 0$ so that the recurrence equation of part (a) holds for $n \geq 1$. Find the generating function for the sequence $(b(0), b(1), \ldots, b(n), \ldots)$. Note that the recurrence equation for $b(n)$ can be written as
    $$b(n) = b(n-1) + n + C(n+1, 3)$$

    (c) Find an explicit formula for $b(n)$.

11. Find the generating function for the sequence $(c_0, c_1, \ldots, c_n, \ldots)$, where c_n is the number of ways to have a total of n cents using pennies, nickels, dimes, and quarters.

12. Find the generating function for the sequence $(c_1(0), c_1(1), \ldots, c_1(n), \ldots)$, where $c_1(n)$ is the number of ways to have a total of n cents using pennies, nickels, dimes, and quarters under the restriction that one uses at most four pennies, at most one nickel, and at most two dimes. Use the generating function to find an explicit formula for $c_1(n)$.

13. Find the generating function for the sequence $(e_0, e_1, \ldots, e_n, \ldots)$, where e_n is the number of integer solutions to

$$y_0 + y_1 + y_2 + \cdots + y_9 = n$$

where, for each $i \in \{0, 1, 2, \ldots, 9\}$, $0 \le y_i \le 9$ and y_i is even or odd according to whether i is even or odd.

14. Complete Example 2.3.8 by showing that, for $n \ge 2$,

$$g^{(n)}(0) = 2^{n-1} \prod_{k=1}^{n-1} (2k - 1)$$

15. Find the generating function for the sequence $(a_0, a_1, \ldots, a_n, \ldots)$, where a_n is the number of solutions (x_1, x_2, x_3) to $x_1 + x_2 + x_3 = n$, where each x_i is an integer and $0 \le x_1 \le x_2 \le x_3$. (Hint: Let $y_1 = x_1$, $y_2 = x_2 - x_1$, $y_3 = x_3 - x_2$ and consider $3y_1 + 2y_2 + y_3$.)

16. Show that

$$\frac{1}{1 - (x + x^2)}$$

is the generating function for the sequence $(b_2(0), b_2(1), \ldots, b_2(n), \ldots)$, where $b_2(0) = 1$ and, for $n \ge 1$, $b_2(n)$ is the number of ways to express n as a sum of terms such that each term is 1 or 2 and the order of the terms is important. As an example, $4 = 2 + 2 = 2 + 1 + 1 = 1 + 2 + 1 = 1 + 1 + 2 = 1 + 1 + 1 + 1$, so that $b_2(4) = 5$.

17. Find the generating function for the sequence $(c_0, c_1, \ldots, c_n, \ldots)$, where c_n is the number of n-combinations from m types (m a fixed positive integer) with repetition allowed.

18. Generalize Exercise 16. For a fixed positive integer m, show that

$$\frac{1}{1 - (x + x^2 + \cdots + x^m)}$$

is the generating function for the sequence $(b_m(0), b_m(1), \ldots, b_m(n), \ldots)$, where $b_m(0) = 1$ and, for $n \ge 1$, $b_m(n)$ is the number of ways to express n as a sum of terms such that each term is an integer between 1 and m and the order of the terms is important.

19. Find the generating function for the sequence $(c_0, c_1, \ldots, c_n, \ldots)$, where c_n is the number of ways to collect n dollars from 10 professors $p_1, p_2, \ldots, p_{10}$, if each of $p_1, p_2, \ldots, p_8$ may give no dollars or one dollar and each of p_9 and p_{10} may give any (nonnegative) whole number of dollars.

20. Let m be a fixed positive integer. Find the generating function for $(d_0, d_1, \ldots, d_n, \ldots)$, where d_n is the number of ways for m distinct dice to show a sum of n. Find d_{28} for the case $m = 7$.

21. Find the generating function for the sequence $(e_0, e_1, \ldots, e_n, \ldots)$, where e_n is the number of integer solutions to

$$y_1 + y_2 + y_3 + y_4 + y_5 = n$$

where, for each $i \in \{1, 2, 3, 4, 5\}$, $-2 \le y_i \le 2$.

22. Find the generating function for the number c_n of ways to select four numbers from $\{1, 2, \ldots, n\}$, no two of which are consecutive. (Hint: Suppose the set of four numbers selected is $\{x_1, x_2, x_3, x_4\}$, with $x_1 < x_2 < x_3 < x_4$. Let $y_1 = x_1 - 1$, $y_2 = x_2 - x_1 - 1$, $y_3 = x_3 - x_2 - 1$, $y_4 = x_4 - x_3 - 1$, and $y_5 = n - x_4$; then $y_1 + y_2 + y_3 + y_4 + y_5 = n - 4$.) Show that $c_n = C(n - 3, 4)$ for $n \ge 7$.

23. In Example 2.3.7, having found $g(x)$, proceed to find $r(n)$ as follows:

$$g(x) = \frac{1}{1 - 2x}\left(\frac{1}{1 - 2x} - 1\right) = \frac{2x}{(1 - 2x)^2} = \cdots = \sum_{n=0}^{\infty} n2^n x^n$$

Therefore, $r(n) = n2^n$.

24. Consider the sequence $(1, 2, 6, \ldots, C(2n, n), \ldots)$.

(a) Show that $1/\sqrt{1 - 4x}$ is the generating function for the sequence.

(b) Find

$$\sum_{k=0}^{n} C(2k, k)C(2n - 2k, n - k)$$

(c) For $m > 4$, find

$$\sum_{n=0}^{\infty} C(2n, n)m^{-n}$$

25. Use generating functions to prove Vandermonde's identity in Problem 26 of Chapter 1.

2.4 *UNORDERED PARTITIONS AND DISTRIBUTIONS*

We considered two general questions in Section 1.5:

1. Given positive integers m and n, how many *ordered* partitions of $\{1, 2, \ldots, m\}$ into n parts are there?

2. Given positive integers m and n, how many *ordered* partitions of m into n parts are there?

We saw in Section 1.5 that the number of ordered partitions of $\{1, 2, \ldots, m\}$ into n parts is the same as the number of functions from $\{1, 2, \ldots, m\}$ onto $\{1, 2, \ldots, n\}$; so, denoting this number by $S'(m, n)$, we have, by Theorem 2.1.7,

$$S'(m, n) = \sum_{k=0}^{n-1}(-1)^k C(n, k)(n - k)^m$$

The second question asks for the number of ways to express m as a sum of n positive integers:

$$x_1 + x_2 + \cdots + x_n = m$$

where the order of the terms $x_1, x_2, \ldots, x_n$ in the sum is important. In this context we call the n-tuple $(x_1, x_2, \ldots, x_n)$ an **ordered partition of m into n parts**. (Some books call this a **composition of m into n parts**.) Theorem 1.5.2 states that the number of ordered partitions of m into n parts is $C(m - 1, n - 1)$.

In this section we wish to explore questions such as those above, but here we are interested in the case when order is not important. The first question thus becomes the following:

Given positive integers m and n, how many partitions are there of $\{1, 2, \ldots, m\}$ into n parts?

Suppose the positive integer m is written as a sum of positive integers:

$$m = x_1 + x_2 + \cdots + x_n$$

and the order of the terms is not important; then we have what is called a **partition of m**. More particularly, if the number of terms is n, then we say we have a **partition of m into n parts**. So the analogue of the second question is:

Given positive integers m and n, how many partitions of m into n parts are there?

The number of partitions of an m-element set into n parts is written $S(m, n)$; these numbers are called **Stirling numbers (of the second kind)** and were introduced in Example 2.2.11. Note that $S(m, n) = 0$ if $n > m$. For $1 \leq n \leq m$, we have the following result:

THEOREM 2.4.1   The Stirling numbers satisfy the initial conditions

$$S(m, 1) = 1 = S(m, m)$$

and, for $1 < n < m$, the recurrence equation

$$S(m, n) = S(m - 1, n - 1) + nS(m - 1, n) \qquad (2.4.1)$$

Proof   There is a unique partition of $X = \{1, 2, \ldots, m\}$ with one part, namely, $\{X\}$, and exactly one partition of X with m parts, namely, $\{\{1\}, \{2\}, \ldots, \{m\}\}$. Thus, $S(m, 1) = 1 = S(m, m)$.

To verify the relation (2.4.1), consider that the following recursive recipe uniquely determines a partition P of X into n parts, $1 < n < m$: Either (1) choose a partition P_1 of $X - \{m\}$ into $n - 1$ parts and add to it the part $\{m\}$, or (2) choose a partition P_2 of $X - \{m\}$ into n parts and add m to one the parts. In option 1 there are $S(m - 1, n - 1)$ choices for the partition P_1. In option 2 there are $S(m - 1, n)$

				n			
	1	2	3	4	5	6	7
1	1						
2	1	1					
3	1	3	1				
m 4	1	7	6	1			
5	1	15	25	10	1		
6	1	31	90	65	15	1	
7	1	63	301	350	140	21	1

Figure 2.4.1 Stirling's triangle.

ways to choose the partition P_2 and then n choices for the part of P_2 into which m is to be inserted. So there are $S(m-1, n-1)$ partitions P resulting from option 1 (those with the element m in a part by itself) and $nS(m-1, n)$ partitions P resulting from option 2 (those with the element m not in a part by itself). Relation (2.4.1) follows by the addition principle.   ☐

Identity (2.4.1) may remind you of Pascal's identity for the binomial coefficients. Indeed, it can be used to form a triangular array of the Stirling numbers, similar to Pascal's triangle for the binomial coefficients; this array is called **Stirling's triangle**. With the rows of this array indexed by m and the columns indexed by n, we can interpret the recurrence equation (2.4.1) as saying that, for $1 < n < m$, the entry in row m, column n of Stirling's triangle is the sum of the entry in row $m-1$, column $n-1$, and n times the entry in row $m-1$, column n. Also, the initial conditions $S(m, 1) = 1 = S(m, m)$ tell us that all the entries in the first column and on the main diagonal of Stirling's triangle are 1. The first several rows of Stirling's triangle are shown in Figure 2.4.1.

If we start with a given partition of $\{1, 2, \ldots, m\}$ into n parts, then its parts can be ordered in $n!$ ways to produce $n!$ different ordered partitions of $\{1, 2, \ldots, m\}$ into n parts. Doing this for each of the $S(m, n)$ partitions of $\{1, 2, \ldots, m\}$ into n parts produces each of the $S'(m, n)$ ordered partitions of $\{1, 2, \ldots, m\}$ into n parts exactly once. Thus, we see that

$$S'(m, n) = n! S(m, n)$$

Putting this together with Theorem 2.1.7 yields the following result.

THEOREM 2.4.2 $S(m, n) = \dfrac{\sum_{k=0}^{n-1} (-1)^k C(n, k)(n-k)^m}{n!}$ ☐

Let $B(m)$ denote the total number of partitions of an m-element set. It then follows immediately from the addition principle that

$$B(m) = \sum_{n=1}^{m} S(m, n)$$

namely, $B(m)$ is the sum of the numbers in row m of Stirling's triangle. For example, $B(1) = 1$, $B(2) = 2$, $B(3) = 5$, $B(4) = 15$, and $B(5) = 52$. It also makes sense to let $B(0) = 1$, because $\{\{ \}\}$ is a partition of the empty set. (We could also say that $S(0,0) = 1$.) The numbers $B(m)$ are called **Bell numbers**, after the mathematician E. T. Bell who studied them.

Besides being able to compute $B(m)$ as a sum of Stirling numbers, the Bell numbers also satisfy a recurrence equation.

THEOREM 2.4.3 For $m \geq 1$,

$$B(m) = \sum_{i=0}^{m-1} C(m-1, i)B(i)$$

Proof Let $X = \{1, 2, \ldots, m\}$. We claim that the following recursive recipe uniquely determines a partition P of X: (1) Choose $i \in \{0, 1, \ldots, m-1\}$; (2) choose an i-element subset X' of $X - \{m\}$; (3) choose a partition P' of X'; (4) let $P = P' \cup \{X - X'\}$.

First, to see that each partition of X is produced in only one way by the recipe, suppose we are given a partition P of X. Then m is in some part Y of P and P is produced by the recipe only if i, X', and P' are chosen so that $X' = X - Y$, $i = |X'|$, and $P' = P - \{Y\}$.

Now, there are $C(m-1, i)$ ways to choose X' in step 2 and then $B(i)$ ways to choose P' in step 3. Thus, summing over all possible choices of i in step 1 gives the desired result. $\square$

As an example using Theorem 2.4.3, we have

$$B(6) = \sum_{i=0}^{5} C(5, i)B(i)$$

so

$$
\begin{aligned}
B(6) &= C(5,0)B(0) + C(5,1)B(1) + C(5,2)B(2) + C(5,3)B(3) \\
&\quad + C(5,4)B(4) + C(5,5)B(5) \\
&= 1 \cdot 1 + 5 \cdot 1 + 10 \cdot 2 + 10 \cdot 5 + 5 \cdot 15 + 1 \cdot 52 \\
&= 1 + 5 + 20 + 50 + 75 + 52 = 203
\end{aligned}
$$

We next turn to the subject of partitions of positive integers. Again, a **partition** of m is a representation of m as a sum of positive integers such that the order of the terms in the sum is not important. For example, the partitions of 5 are 5, $4+1$, $3+2$, $3+1+1$, $2+2+1$, $2+1+1+1$, and $1+1+1+1+1$. Letting $p(m)$ denote the number of partitions of m, we see that $p(5) = 7$.

A partition of m, say,

$$x_1 + x_2 + \cdots + x_n = m$$

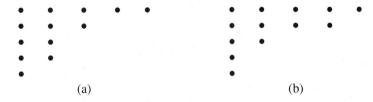

Figure 2.4.2   A Ferrers diagram (a) and its transpose (b).

is customarily written so that $x_1 \geq x_2 \geq \cdots \geq x_n$. Also, given such a partition, we can collect like terms and write

$$1y_1 + 2y_2 + \cdots + my_m = m$$

where each y_i is a nonnegative integer, namely, the number of parts of the partition equal to i. The list $(y_1, y_2, \ldots, y_m)$ is called the **type representation** of the partition, or just its **type** for short. For instance, consider the partition $5 + 2 + 2 + 2 + 1 + 1$ of 13. Its type is $(2, 3, 0, 0, 1, 0, \ldots, 0)$; usually, we shorten such a type representation to $(2, 3, 0, 0, 1)$ by deleting the trailing 0s. By considering types, we see immediately that $p(m)$ is the number of nonnegative integer solutions $(y_1, y_2, \ldots, y_m)$ to the equation

$$1y_1 + 2y_2 + \cdots + my_m = m$$

A **Ferrers diagram** such as Figure 2.4.2 represents the partition $x_1 + x_2 + \cdots + x_r$ of m $(x_1 \geq x_2 \geq \cdots \geq x_r)$ as an array of m dots (or squares) organized into r rows and x_1 columns, with x_i dots in the ith row located in columns 1 through x_i. Figure 2.4.2(a) shows the Ferrers diagram for the partition $5 + 3 + 2 + 2 + 1$ of 13. Let $\mathcal{F}$ be the Ferrers diagram representing the partition $x_1 + x_2 + \cdots + x_r$ of m. The **transpose** of $\mathcal{F}$ is the Ferrers diagram $\mathcal{F}'$ having x_1 rows and r columns, with x_i dots in the ith column located in rows 1 through x_i. Notice that $\mathcal{F}'$ also represents a partition of m, called the **transpose** of the partition represented by $\mathcal{F}$. For instance, the transpose of the Ferrers diagram of Figure 2.4.2(a) is shown in Figure 2.4.2(b); it represents the partition $5 + 4 + 2 + 1 + 1$ of 13.

For positive integers m and n, let $p(m, n)$ be the number of partitions of m with n parts. Note that $p(m, 1) = 1 = p(m, m)$ and $p(m, n) = 0$ if $n > m$. Also,

$$\sum_{n=1}^{m} p(m, n) = p(m)$$

THEOREM 2.4.4   The numbers $p(m, n)$ satisfy the following recurrence equations for $1 < n < m$:

1. $p(m, n) = p(m - 1, n - 1) + p(m - n, n)$
2. $p(m, n) = \sum_{k=1}^{n} p(m - n, k)$

Proof   Let $\mathcal{P}(m, n)$ denote the set of partitions of m into n parts.

To prove 1, we construct a bijection

$$f : \mathcal{P}(m, n) \rightarrow \mathcal{P}(m - 1, n - 1) \cup \mathcal{P}(m - n, n)$$

We let $x_1 + x_2 + \cdots + x_n$ be a partition of m (with $x_1 \geq x_2 \geq \cdots \geq x_n$). If $x_n = 1$, then f maps this partition to the partition $x_1 + x_2 + \cdots + x_{n-1}$ of $m - 1$; on the other hand, if $x_n > 1$, then f maps the given partition to the partition $(x_1 - 1) + (x_2 - 1) + \cdots + (x_n - 1)$ of $m - n$. It is clear that f is one-to-one. To see that f is onto, note that the preimage of the partition $y_1 + y_2 + \cdots + y_{n-1}$ of $m - 1$ is the partition $y_1 + y_2 + \cdots + y_{n-1} + 1$ of m, whereas the preimage of the partition $z_1 + z_2 + \cdots + z_n$ of $m - n$ is the partition

$$(z_1 + 1) + (z_2 + 1) + \cdots + (z_n + 1)$$

of m.

For 2, we implicitly construct a bijection from $\mathcal{P}(m, n)$ to

$$\mathcal{P}(m - n, 1) \cup \mathcal{P}(m - n, 2) \cup \cdots \cup \mathcal{P}(m - n, n)$$

Given a partition $x_1 + x_2 + \cdots + x_n$ of m, we subtract 1 from each part and delete any 0s that result; this yields a partition of $m - n$ into n or fewer parts. Conversely, given a partition $y_1 + y_2 + \cdots + y_k$ of $m - n$ into k parts, $k \leq n$, we add 1 to each part, and also add $n - k$ parts equal to 1; this gives a partition of m into n parts. □

■ *Example 2.4.1* We noted above that $p(5, 1) = p(5, 4) = p(5, 5) = 1$ and $p(5, 2) = p(5, 3) = 2$. Now we use these results and those of Theorem 2.4.4 to compute $p(9, 4)$. First, using part 1 we have

$$p(9, 4) = p(8, 3) + p(5, 4)$$
$$= p(7, 2) + p(5, 3) + p(5, 4)$$
$$= p(6, 1) + p(5, 2) + p(5, 3) + p(5, 4)$$
$$= 1 + 2 + 2 + 1 = 6$$

Using part 2 we obtain

$$p(9, 4) = \sum_{k=1}^{4} p(9 - 4, k) = p(5, 1) + p(5, 2) + p(5, 3) + p(5, 4) = 1 + 2 + 2 + 1 = 6$$

■

Consider the function, with domain the set of partitions of m into n parts, that maps each partition to its transpose. It is clear that this function is one-to-one and that its image is the set of partitions of m with largest part n. For instance, the partition $6 + 3 + 2 + 1 + 1$ of 13 has five parts, and its transpose $5 + 3 + 2 + 1 + 1 + 1$ has largest part 5. We thus have a bijective proof of the following result.

THEOREM 2.4.5 The number $p(m, n)$ of partitions of m with n parts is equal to the number of partitions of m with largest part n. ☐

Although many interesting facts are known concerning partitions, such as the result of the preceding theorem, no one has as yet discovered a formula for computing the numbers $p(m, n)$ directly from m and n, formulas such as those we have for the binomial coefficients $C(m, n)$ and for the Stirling numbers $S(m, n)$. We will explore these ideas further in the exercises and chapter problems and do some work with generating functions.

Just as results concerning ordered partitions can be applied in the context of ordered distributions—distributing a given number of objects to a given number of different (or "ordered") recipients (under certain specified conditions), so, too, can unordered partitions be studied in the context of **unordered distributions**—distributing a given number of objects to a given number of identical (or "unordered") recipients. As with ordered distributions, it is convenient to think of the objects as "balls" and the recipients as identical (or indistinguishable) "boxes." If the balls are to be considered all different—for example, we may wish to distribute m golf balls, no two of which have exactly the same brand name, color, and number—then we can think of the distribution of balls to boxes as a partition of the set of balls. On the other hand, if the balls are considered to be identical, then what matters is not how the balls are partitioned; rather, the distribution of balls to boxes determines a partition of the number of balls.

■ *Example 2.4.2* In each of the following parts, find the number of ways to distribute the m balls to n identical boxes.

(a) $m = 8$, $n = 4$, and the balls are different

(b) $m = 4$, $n = 8$, the balls are different, and at most one ball is to be put in any box

(c) $m = 5$, $n = 3$, the balls are different, and at least one ball is to be placed in each box

(d) $m = 8$, $n = 4$, the balls are different, three balls are to be put in one box, two balls are to be put in each of two other boxes, and one ball is to be put in another box

(e) $m = 8$, $n = 4$, and the balls are identical

(f) $m = 4$, $n = 8$, the balls are identical, and at most one ball is to be put in any box

(g) $m = 8$, $n = 4$, the balls are identical, and at least one ball is to be placed in each box

Solution

(a) The number of ways to distribute 8 different balls to 4 identical boxes is the same as the number of partitions of $\{1, 2, \ldots, 8\}$ with at most 4 parts. This is

$$S(8, 1) + S(8, 2) + S(8, 3) + S(8, 4)$$

(b) There is only one way to distribute 4 different balls to 8 identical boxes with at most one ball per box! For any such distribution, 4 boxes will be empty and 4 boxes will contain exactly one ball and, since the boxes are identical, we don't care which are which.

(c) The number of ways to distribute 5 different balls to 3 identical boxes with at least one ball per box is the same as the number of partitions of $\{1, 2, 3, 4, 5\}$ with exactly 3 parts, namely, $S(5, 3)$.

(d) Here the answer is the number of partitions of $\{1, 2, \ldots, 8\}$ of type $(1, 2, 1)$, that is, having 1 part of cardinality 1, 2 parts of cardinality 2, and 1 part of cardinality 3. The number of ordered partitions with this type is $12 \cdot C(8; 1, 2, 2, 3)$; here the factor of 12 comes from the fact that there are $C(4; 1, 2, 1) = 12$ ordered partitions of 4 of type $(1, 2, 1)$. Since each unordered partition with 4 parts determines $4! = 24$ ordered partitions with 4 parts, we find that the number of partitions of $\{1, 2, \ldots, 8\}$ of type $(1, 2, 1)$ is

$$\frac{12 \cdot C(8; 1, 2, 2, 3)}{24} = 840$$

(e) The number of ways to distribute 8 identical balls to 4 identical boxes is the same as the number of partitions of 8 with at most 4 parts, namely,

$$p(8, 1) + p(8, 2) + p(8, 3) + p(8, 4)$$

(f) Like part (b), this is a trick question—there is only one way to distribute 4 identical balls to 8 identical boxes with at most one ball in any box!

(g) The number of ways to distribute 8 identical balls to 4 identical boxes with at least one ball per box is the same as the number of partitions of 8 with exactly 4 parts, namely, $p(8, 4)$.   ∎

Taking our lead from the preceding example, the next theorem summarizes results concerning unordered distributions; the proof is left for you to develop in Exercise 20.

THEOREM 2.4.6   Consider the number of ways to distribute m balls to n identical boxes.

1.   If the balls are different, then:

(a) The number of ways to distribute the balls is the same as the number of partitions of $\{1, 2, \ldots, m\}$ into at most n parts, namely,

$$\sum_{k=1}^{n} S(m, k)$$

(b) If $m \geq n$, then the number of ways to distribute the balls with at least one ball per box is the same as the number of partitions of $\{1, 2, \ldots, m\}$ into n parts, namely, $S(m, n)$.

(c) Consider a partition $m_1 + m_2 + \cdots + m_n$ of m of type $(y_1, y_2, \ldots, y_m)$. Then the number of ways to distribute the balls so that y_i boxes contain i balls, $1 \leq i \leq m$, is

$$\frac{C(n; y_1, y_2, \ldots, y_m) C(m; m_1, m_2, \ldots, m_n)}{n!}$$

2. If the balls are identical, then:

(a) The number of ways to distribute the balls is the same as the number of partitions of m with at most n parts, namely,

$$\sum_{k=1}^{n} p(m, k)$$

(b) If $m \geq n$, then the number of ways to distribute the balls with at least one ball per box is the same as the number of partitions of m with n parts, namely, $p(m, n)$. □

EXERCISE SET 2.4

1. Extend Stirling's triangle as shown in Figure 2.4.1 to find the Stirling numbers $S(8, n)$ and $S(9, n)$.

2. Use Theorem 2.4.2 to compute $S(7, 4)$.

3. Use Theorem 2.4.3 to compute $B(7)$. Also, compute $B(8)$ and $B(9)$ (using any method you wish). Make a conjecture about the rate of growth of $B(m)$ as a function of m.

4. Prove that

$$S(m, n) = \sum_{i=0}^{m-1} C(m - 1, i) S(i, n - 1)$$

5. Give a table of values of $p(m, n)$, like that of Stirling's triangle, for $1 \leq n \leq m \leq 9$. Try to use Theorem 2.4.4 as much as possible. Then find $p(m)$ for $1 \leq m \leq 9$.

6. Recall that the number of permutations of n things taken k at a time is

$$P(n, k) = n(n - 1) \cdots (n - k + 1)$$

This number is also written $(n)_k$ and is called a **(falling) factorial power of n of degree k**; note that it is a polynomial of degree k in n. Thus, given positive integers m and n, there is reason to expect the existence of coefficients $S(m, k)$, $1 \leq k \leq n$, such that

$$n^m = \sum_{k=1}^{n} S(m, k)(n)_k$$

Indeed, show that the coefficients $S(m, k)$ are Stirling numbers of the second kind. (Hint: Show that both sides count the number of functions from $\{1, 2, \ldots, m\}$ to $\{1, 2, \ldots, n\}$.) The above identity was discovered by Stirling and it is for this reason that the numbers $S(m, k)$ are named for him.

7. Use the result of Exercise 6 to find coefficients a_1, a_2, a_3, a_4 such that
$$n^4 = a_1(n)_1 + a_2(n)_2 + a_3(n)_3 + a_4(n)_4$$

8. Use part 1 of Theorem 2.4.4 to prove part 2.

9. List the partitions of $\{1, 2, 3, 4, 5\}$ into three parts.

10. Give a formula for each of the following Stirling numbers (as a function of m).
   (a) $S(m, 2)$
   (b) $S(m, m - 1)$

11. Consider the partition $6 + 3 + 2 + 2 + 2$ of 15.
   (a) Give its type representation.
   (b) Give its Ferrers diagram.
   (c) Find its transpose.
   (d) Give the type of representation of the transpose.

12. Give an alternate proof of Theorem 2.4.3 using the result of Exercise 4. As a hint, the proof begins as follows:

$$B(m) = \sum_{n=1}^{m} S(m, n) = \sum_{n=1}^{m} \sum_{i=0}^{m-1} C(m - 1, i)S(i, n - 1)$$

$$= \sum_{i=0}^{m-1} \sum_{n=1}^{m} C(m - 1, i)S(i, n - 1)$$

$$= \sum_{i=0}^{m-1} C(m - 1, i) \sum_{n=1}^{m} S(i, n - 1) = \cdots$$

13. In how many ways can 12 golf balls be distributed to four identical golf bags in each of the following cases?
   (a) The balls are different.
   (b) The balls are identical.
   (c) The balls are different and each golf bags gets at least one ball.
   (d) The balls are identical and each golf bags gets at least one ball.
   (e) The balls are different and each golf bag gets three balls.
   (f) The balls are identical and each golf bag gets three balls.

14. Develop a necessary and sufficient condition for $p(m, n)$ and $p(m - 1, n - 1)$ to be equal.

15. A computer system with four identical processors has seven distinct tasks to perform. In how many ways can the tasks be distributed to the processors assuming each processor gets at least one task?

16. Prove that $p(m)$ is equal to the number of partitions of $m + 1$ with smallest part 1.

17. List the partitions of 6.

18. Prove that the number of partitions of m with at most n parts is equal to the number of partitions of m whose largest part is at most n. (Hint: Use the idea of transposes.)

19. In how many ways can 16 identical balls be distributed to seven identical boxes so that each box gets at least 2 balls?

20. Prove Theorem 2.4.6:
    (a) Part 1(a).   (b) Part 1(b).   (c) Part 1(c).
    (d) Part 2(a).   (e) Part 2(b).

21. Prove that $p(2m, m) = p(m)$.

22. Show that the number of partitions of m with each part greater than 1 is $p(m) - p(m - 1)$.

23. Give a formula for each of the following (as a function of m).
    (a) $p(m, 2)$ (b) $p(m, 3)$

24. Use the result of Exercise 22 to show that

$$p(m + 1) + p(m - 1) \geq 2p(m)$$

25. Give a formula for each of the following:
    (a) $p(m, m - 1), \quad m \geq 2$ (b) $p(m, m - 2), \quad m \geq 4$
    (c) $p(m, m - 3), \quad m \geq 6$ (d) $p(m, m - 4), \quad m \geq 8$

26. For a fixed positive integer m, let $q(n, m)$ be the number of partitions of n such that no term exceeds m. For example, $q(5, 3) = 5$. Find the generating function $g_m(x)$ for the sequence $(q(0, m), q(1, m), \ldots, q(n, m), \ldots)$.

27. Show that $n! \, p(m, n) \geq C(m - 1, n - 1)$.

28. Show that the generating function for the sequence $(p(0), p(1), \ldots, p(n), \ldots)$ is given by the infinite product

$$\prod_{k=1}^{\infty} \frac{1}{1 - x^k}$$

29. Call a partition of m **symmetric** if it is equal to its transpose. Prove that the number of symmetric partitions of m is equal to the number of partitions of m with distinct odd parts. (Hint: For example, consider the symmetric partition $6 + 4 + 4 + 3 + 1 + 1$ of 19. Draw its Ferrers diagram. Say that a dot has level i if i is the smallest value for which that dot is in the ith row or the ith column; if the dots are partitioned by level, we obtain the partition $11 + 5 + 3$ of 19, having distinct odd parts.)

★30. Prove that, among all the partitions of m, the total number of parts equal to 1 is the same as the total number of distinct parts.

2.5 *LISTING, RANKING, AND UNRANKING PERMUTATIONS*

In the next two sections we discuss algorithms for listing, ranking, and unranking combinatorial objects belonging to various families.

The type of problem we wish to consider is first introduced for the family of k-permutations of $\{1, 2, \ldots, n\}$, where k and n are fixed positive integers with $1 \le k \le n$.

First of all, we may desire to have a list of all these permutations in some definite order. Such a list could be used by a program that must process each of the permutations in turn, perhaps to test some conjecture about them. There are many possible ways to order the k-permutations of $\{1, 2, \ldots, n\}$. Perhaps the most natural one is the order called **lexicographic order**, or **lex order** for short.

This ordering of permutations is like the ordering of words in a dictionary, which is based on the underlying natural order of the letters of the alphabet. Thus, "apple" comes before "peach" because a comes before p; "peach" comes before "pear" since, comparing the fourth letters, c comes before r; "grape" comes before "grapefruit" because "grape" is a prefix of "grapefruit," and so forth.

Lexicographic ordering of permutations of the set $\mathbb{Z}^+$ is based on the underlying natural order of the positive integers:

$$1 < 2 < 3 < 4 < \cdots$$

Let $\pi_1 = (x_1, x_2, \ldots, x_k)$ and $\pi_2 = (y_1, y_2, \ldots, y_k)$ be two k-permutations of the set $\mathbb{Z}^+$ of positive integers. We use the notation $\pi_1 < \pi_2$ to indicate that π_1 precedes π_2 in lexicographic order, and $\pi_1 \le \pi_2$ to indicate that $\pi_1 = \pi_2$ or $\pi_1 < \pi_2$. We then define **lexicographic order** for k-permutations recursively as follows: $\pi_1 < \pi_2$ if $x_1 < y_1$ or $x_1 = y_1$, $k \ge 2$, and

$$(x_2, \ldots, x_k) < (y_2, \ldots, y_k)$$

In other words, given $\pi_1 \ne \pi_2$, there is a smallest subscript i, $1 \le i \le k$, such that $x_i \ne y_i$; then $\pi_1 < \pi_2$ if $x_i < y_i$ and $\pi_2 < \pi_1$ if $y_i < x_i$.

■ *Example 2.5.1*   The 60 3-permutations of $\{1, 2, 3, 4, 5\}$ listed in lex order are:

(1, 2, 3),	(1, 2, 4),	(1, 2, 5),	(1, 3, 2),	(1, 3, 4),	(1, 3, 5),
(1, 4, 2),	(1, 4, 3),	(1, 4, 5),	(1, 5, 2),	(1, 5, 3),	(1, 5, 4),
(2, 1, 3),	(2, 1, 4),	(2, 1, 5),	(2, 3, 1),	(2, 3, 4),	(2, 3, 5),
(2, 4, 1),	(2, 4, 3),	(2, 4, 5),	(2, 5, 1),	(2, 5, 3),	(2, 5, 4),
(3, 1, 2),	(3, 1, 4),	(3, 1, 5),	(3, 2, 1),	(3, 2, 4),	(3, 2, 5),
(3, 4, 1),	(3, 4, 2),	(3, 4, 5),	(3, 5, 1),	(3, 5, 2),	(3, 5, 4),
(4, 1, 2),	(4, 1, 3),	(4, 1, 5),	(4, 2, 1),	(4, 2, 3),	(4, 2, 5),
(4, 3, 1),	(4, 3, 2),	(4, 3, 5),	(4, 5, 1),	(4, 5, 2),	(4, 5, 3),
(5, 1, 2),	(5, 1, 3),	(5, 1, 4),	(5, 2, 1),	(5, 2, 3),	(5, 2, 4),
(5, 3, 1),	(5, 3, 2),	(5, 3, 4),	(5, 4, 1),	(5, 4, 2),	(5, 4, 3) ■

To solve the problem of generating the k-permutations of $\{1, 2, \ldots, n\}$ in lex order, it suffices to solve the problem of finding the (immediate) successor of any given k-permutation. That is, suppose we have a function NEXT_PERM(PI) that returns the successor of any given k-permutation PI in the lex order list. We can initialize PI to $(1, 2, \ldots, k)$, the first k-permutation in lex order; then, whenever we want the next permutation we can execute the statement

PI := NEXT_PERM(PI);

to update PI.

The lex order list of the k-permutations of $\{1, 2, \ldots, n\}$ contains $P(n, k)$ permutations. Given a permutation π, we would like to determine its **rank**, that is, its "position" on the list. For certain mathematical reasons, the first permutation on the list has rank 0, the second permutation on the list has rank 1, and so on, and the last permutation on the list has rank $P(n, k) - 1$. Thus, the **(lexicographic) rank** of a given k-permutation π of $\{1, 2, \ldots, n\}$ is an integer r, $0 \leq r \leq P(n, k) - 1$, where r is the number of k-permutations that precede π in the lex order list.

The ranking problem could be solved by generating the k-permutations of $\{1, 2, \ldots, n\}$ in lex order until the permutation π is obtained, keeping count of the number of permutations generated so far. However, if π has a rank close to $P(n, k)$, then such a method is far too inefficient. Instead, we desire an algorithm that takes no more than n steps to find the rank of π.

The inverse problem of the ranking problem is called "unranking": given an integer r and positive integers k and n with $0 \leq r \leq P(n, k) - 1$, determine the k-permutation π that has rank r among all the k-permutations of $\{1, 2, \ldots, n\}$.

■ *Example 2.5.2*   Let $n = 6$ and $k = 3$.

(a) Find the rank of the permutation $\alpha = (4, 2, 5)$.

(b) Find the permutation β with rank 58.

Solution

(a) To find the rank of $\alpha = (4, 2, 5)$, we must find the number of 3-permutations $\pi = (x_1, x_2, x_3)$ of $U = \{1, 2, \ldots, 6\}$ that precede α in lex order. If $x_1 \in \{1, 2, 3\}$, then $\pi < \alpha$. By the multiplication principle, the number of such permutations is

$$3P(5, 2) = 60$$

It is also possible to have $\pi < \alpha$ if $x_1 = 4$ and $x_2 = 1$. The number of 3-permutations of the form $(4, 1, x_3)$ is 4. Another possibility is that $\pi = (4, 2, x_3)$ with $x_3 \in \{1, 3\}$. Obviously, there are 2 such permutations. These are the only possibilities for π and thus, by the addition principle, the rank of $(4, 2, 3)$ is

$$60 + 4 + 2 = 66$$

(b) Let $\beta = (y_1, y_2, y_3)$. Of the 3-permutations (x_1, x_2, x_3) of U, there are $P(5, 2) = 20$ with $x_1 = 1$, 20 with $x_1 = 2$, 20 with $x_1 = 3$, and so on. Since

$$2 \cdot 20 = 40 \leq 58 < 60 = 3 \cdot 20$$

it follows that $y_1 = 3$. Now, of the 3-permutations $(3, x_2, x_3)$ of U, there are $P(4, 1) = 4$ of them for each value of $x_2 \in \{1, 2, 4, 5, 6\}$. Since

$$40 + 4 \cdot 4 = 56 \leq 58 < 40 + 5 \cdot 4 = 60$$

it follows that $y_2 = 6$. Finally, the 3-permutation $(3, 6, x_3)$ of U has rank 56, 57, 58, or 59 as $x_3 = 1$, 2, 4, or 5, respectively. Therefore, $\beta = (3, 6, 4)$. ∎

Another problem that we wish to consider is that of choosing a k-permutation of $\{1, 2, \ldots, n\}$ at random. This is to be done "uniformly," meaning that each of the $P(n, k)$ permutations has the same probability of being chosen.

To summarize, then, we desire algorithms to accomplish the following:

1. Given a k-permutation π of $\{1, 2, \ldots, n\}$, find its successor in the lex order list of all the k-permutations of $\{1, 2, \ldots, n\}$. For convenience, we wish to think of the lex order list as "circular," so that the successor of the last permutation $(n, n-1, \ldots, n-k+1)$ is the first permutation $(1, 2, \ldots, k)$.

2. Given a k-permutation π of $\{1, 2, \ldots, n\}$, find its rank.

3. Given integers r, k, and n with $0 \leq r \leq P(n, k) - 1$, find the k-permutation π of $\{1, 2, \ldots, n\}$ with rank r.

4. Given n and k, choose a k-permutation π of $\{1, 2, \ldots, n\}$ uniformly at random.

In designing these algorithms, we assume that we have at our disposal a package of basic operations for working with permutations. (In modern computer science terminology, we would implement the type PERMUTATION as an abstract data type.) This package provides (among others) the following basic operations:

1. The function LENGTH(π) returns the length of the permutation π.

2. The function IDENTITY(n) returns the identity permutation $(1, 2, \ldots, n)$ for $n > 0$; for $n = 0$, this function returns the "empty permutation" ( ).

3. The function ELEMENT(j, π) returns the jth element of the permutation π, where $1 \leq j \leq$ LENGTH(π).

4. The function POSITION(x, π) returns the position of the element x in the permutation π. If x is not an element of π, then $1 +$ LENGTH(π) is returned.

5. Given $1 \leq j \leq$ LENGTH(π), the procedure REPLACE(j, x, π) replaces the jth element of the permutation π with the element x.

6. Given $1 \leq j \leq$ LENGTH$(\pi)+1$, the procedure INSERT(j, x, π) inserts the element x into the permutation π at position j, thereby increasing the length of π by 1.

7. The function SUBPERM(π, i, j) returns the subpermutation

(ELEMENT$(i, \pi), \ldots,$ ELEMENT$(j, \pi))$

of the permutation π (assuming $1 \leq i \leq j \leq$ LENGTH(π)).

We also assume that we have at our disposal the function $P(n,k)$ that returns the number of k-permutations of an n-element set, and the function MINIMUM(a,b) that returns the minimum of the two integers a and b.

For the benefit of those of you with a computer science orientation, there is a method that can be used to implement the abstract data type PERMUTATION. The most obvious method is to think of a permutation as a list, implemented as a record type with two fields, such as:

> type ELEMENT_ARRAY_TYPE is array (1 .. MAX) of POSITIVE;
> type PERMUTATION is record
> > > LENGTH: NATURAL;
> > > ELEMENT: ELEMENT_ARRAY_TYPE;
> > end record;

For an object PI of type PERMUTATION, the idea here is to let PI.LENGTH be the length of PI and PI.ELEMENT(J) be the Jth element of PI.

With such a data structure, notice that each of the operations LENGTH, ELEMENT, and REPLACE can be implemented to execute in time $\Theta(1)$; however, the POSITION operation is $O(k)$, where k is the length of π, since finding the position of x in π involves searching. The INSERT operation is also $O(k)$ in general; however, the special case

> INSERT(LENGTH$(\pi) + 1, x, \pi$);

merely appends x to π and so is $\Theta(1)$. Such details are left to Exercise 2. We emphasize here that, as an abstract data type, the type PERMUTATION is assumed to be "encapsulated"—users of the type are not allowed direct access to the LENGTH and ELEMENT fields of an object PI. The only means by which PI can be accessed or modified is through use of the operations LENGTH, ELEMENT, POSITION, REPLACE, and so on, which are provided by the basic package.

We are now ready to discuss the successor algorithm. Let $\pi = (x_1, x_2, \ldots, x_k)$ be a k-permutation of $\{1, 2, \ldots, n\}$. We first consider the case $k = n$.

■ ***Example 2.5.3*** Suppose $k = n = 7$. Find the next four permutations π_1, π_2, π_3, and π_4 in lex order after $\pi_0 = (1, 6, 2, 7, 5, 3, 4)$.

Solution
The successor of π_0 is $\pi_1 = (1, 6, 2, 7, 5, 4, 3)$. To find the successor of π_1, note that π_1 is last in lex order among the $P(4) = 24$ permutations that begin with the prefix $(1, 6, 2)$. The next 24 permutations in lex order have prefix $(1, 6, 3)$, and the first of these is $\pi_2 = (1, 6, 3, 2, 4, 5, 7)$. The successor of π_2 is $\pi_3 = (1, 6, 3, 2, 4, 7, 5)$. To find the successor of π_3, note that π_3 is last in lex order among the $P(2) = 2$ permutations that begin with the prefix $(1, 6, 3, 2, 4)$. The next 2 permutations in lex order begin with the prefix $(1, 6, 3, 2, 5)$ and the first of these is $\pi_4 = (1, 6, 3, 2, 5, 4, 7)$. ■

Hopefully the preceding example gives an indication of how to find the successor of $\pi = (x_1, x_2, \ldots, x_n)$. First, we find that subscript i such that

$$x_i < x_{i+1} \qquad \text{and} \qquad x_{i+1} > x_{i+2} > \cdots > x_n$$

If such an i does not exist, then

$$x_1 > x_2 > \cdots > x_n$$

and π is the last permutation in lex order, in which case the successor of π is $(1, 2, \ldots, n)$. Having found i, we know that π is the last permutation in lex order among the $P(n-i)$ permutations having the prefix $(x_1, \ldots, x_{i-1}, x_i)$. Thus, the successor of π is the first permutation among those with the prefix $(x_1, \ldots, x_{i-1}, x_j)$, where x_j is the smallest element of $\{x_{i+1}, x_{i+2}, \ldots, x_n\}$ greater than x_i. Since $x_{i+1} > x_{i+2} > \cdots > x_n$, we know that j is the largest subscript among $i+1, i+2, \ldots, n$ for which $x_j > x_i$. Having found j, the successor of π is

$$(x_1, \ldots, x_{i-1}, y_i, y_{i+1}, \ldots, y_n)$$

where $y_i = x_j$ and $y_{i+1}, \ldots, y_n$ are the elements of $\{x_i, x_{i+1}, \ldots, x_n\} - \{x_j\}$ arranged in increasing order.

Note that, once i and j are found, we may obtain the successor of π by first swapping the values of x_i and x_j and then reversing the sublist $(x_{i+1}, \ldots, x_n)$. For example, if $\pi = (x_1, x_2, x_3, x_4, x_5, x_6, x_7) = (3, 6, 2, 7, 5, 4, 1)$, then $i = 3$ and $j = 6$. Swapping the values of x_3 and x_6 yields $(3, 6, 4, 7, 5, 2, 1)$, and then reversing the sublist (x_4, x_5, x_6, x_7) gives $(3, 6, 4, 1, 2, 5, 7)$ as π's successor.

Now, what happens if $k < n$? For instance, suppose that $k = 5$, $n = 7$, and $\pi = (3, 6, 2, 7, 5)$. The key is to observe that π is the prefix of $P(n-k)$ n-permutations, and the last of these in lex order, call it ρ, can be obtained by appending to π, in decreasing order, the "missing" elements—those elements of $\{1, 2, \ldots, n\}$ that are not elements of π. For our specific example, $\rho = (3, 6, 2, 7, 5, 4, 1)$. Then we can find the successor of ρ among the n-permutations—this is $(3, 6, 4, 1, 2, 5, 7)$ in our example. Finally, the successor of π is found by taking the k-element prefix of ρ, giving $(3, 6, 4, 1, 2)$ as the successor of $(3, 6, 2, 7, 5)$.

ALGORITHM 2.5.1   Given a k-permutation π of $\{1, 2, \ldots, n\}$, the function

NEXT_PERM(π, n)

returns the successor of π in the lex order list of all the k-permutations of $\{1, 2, \ldots, n\}$. For convenience, the successor of the last permutation $(n, n-1, \ldots, n-k+1)$ is defined to be the first permutation $(1, 2, \ldots, k)$.

NEXT_PERM uses the following local variables:

ρ is a permutation and is initialized to π;
k is a positive integer and is initialized to LENGTH(π);
i is a nonnegative integer and is initialized to MINIMUM$(k, n-1)$;
j and m are nonnegative integers and both are intialized to n;

NEXT_PERM also uses the following (local) functions and procedures:

The function RHO_MEMBER(x) determines whether the positive integer x is an element of ρ (i.e., RHO_MEMBER(x) is true if and only if x is an element of ρ) and is defined as follows:

```
begin
    return (POSITION(x, ρ) ≤ LENGTH(ρ));
end RHO_MEMBER;
```

The function RHO_ELEMENT(i) returns the ith element of ρ if $i > 0$, or it returns 0 if $i = 0$. It is defined as follows:

```
begin
    if i = 0 then
        return 0;
    else
        return ELEMENT(i, ρ);
    end if;
end RHO_ELEMENT;
```

The procedure RHO_SWAP(i, j) swaps the ith and jth elements of ρ and is defined as follows:

```
The local variable t is initialized to ELEMENT(i, ρ);
begin
    REPLACE(i, ELEMENT(j, ρ), ρ);
    REPLACE(j, t, ρ);
end RHO_SWAP;
```

The function NEXT_PERM is then defined as follows:

```
begin
    while LENGTH(ρ) < n loop
        if not RHO_MEMBER(m) then
            INSERT(LENGTH(ρ) + 1, m, ρ);
        end if;
        m := m − 1;
    end loop;
```

```
    while RHO_ELEMENT(i) > RHO_ELEMENT(i + 1) loop
        i := i - 1;
    end loop;
    while RHO_ELEMENT(i) > RHO_ELEMENT(j) loop
        j := j - 1;
    end loop;
    if i > 0 then
        RHO_SWAP(i, j);
    end if;
    for t in i + 1 .. (i + n) div 2 loop
        RHO_SWAP(t, i + n + 1 - t);
    end loop;
    return SUBPERM(ρ, 1, k);
end NEXT_PERM;
```

Now let $\pi = (x_1 x_2, \ldots, x_k)$ be a k-permutation of $U = \{1, 2, \ldots, n\}$. The next problem is to find the (lexicographic) rank r of π. Initially, set r to 0 and consider x_1. For any given value of x_1, there are $P(n - 1, k - 1)$ k-permutations of U that begin with that value. If $x_1 = 1$, then $r = 0$ may be the rank of π; on the other hand, if $x_1 > 1$, then there are least $(x_1 - 1)P(n - 1, k - 1)$ k-permutations of U that precede π in lex order, namely, those that begin with a value less than x_1. In any case, we may increment r by $(x_1 - 1)P(n - 1, k - 1)$. Here it is convenient to think of the number $P(n - 1, k - 1)$ as a "unit cost," because it is the amount by which r is incremented for each element of U that is less than x_1.

In general, at the ith step, $2 \leq i \leq \min\{k, n - 1\}$, we have taken into account $x_1, \ldots, x_{i-1}$. We assume that r is the rank of the permutation $(x_1, \ldots, x_{i-1}, y_i, \ldots, y_k)$, where $y_i, \ldots, y_k$ are the smallest k elements of $V_i = U - \{x_1, \ldots, x_{i-1}\}$ in increasing order; that is, r is the minimum rank among all k-permutations of U that have the prefix $(x_1, \ldots, x_{i-1})$. For a given value of x_i, there are $P(n - i, k - i)$ k-permutations that begin with the prefix $(x_1, \ldots, x_{i-1}, x_i)$. If $x_i = y_i$, then it is possible that r has the correct value; on the other hand, if $x_i > y_i$, then there are at least $w(x_i)P(n - i, k - i)$ additional permutations that precede π in lex order, where $w(x_i)$ is the number of elements of V_i that are less than x_i. In any case, we should increment r by the amount $w(x_i)P(n - i, k - i)$, and we think of the number $P(n - i, k - i)$ as being the unit cost at the ith step.

Thus, in our algorithm, we let the variable UNIT_COST hold the value $P(n - i, k - i)$ at step i. Initially, UNIT_COST$= P(n - 1, k - 1)$; it is then easily modified, at the end of step i, by the assignment statement

UNIT_COST := UNIT_COST div $(n - i)$;

Also, our algorithm uses an array of values $w(1), w(2), \ldots, w(n)$ such that $w(j)$ is, at step i, the number of elements of V_i that are less than j. Letting $V_1 = U$, note that $w(j) = j - 1$, initially. At the end of step i, the value of $w(j)$ is decremented for each j, $x_i < j \leq n$.

ALGORITHM 2.5.2 Given a k-permutation π of $\{1, 2, \ldots, n\}$, the function

PERM_TO_RANK(π, n)

returns its lexicographic rank among the $P(n, k)$ k-permutations of $\{1, 2, \ldots, n\}$. PERM_TO_RANK uses the following local variables (as described above):

r is a nonnegative integer and is initialized to 0;
k is a nonnegative integer and is initialized to
MINIMUM(LENGTH(π), $n - 1$);
UNIT_COST is a nonnegative integer and is initialized to
$P(n - 1, k - 1)$;
$(w(1), w(2), \ldots, w(n))$ is an array of nonnegative integers, initialized to
$(0, 1, \ldots, n - 1)$;

```
begin
    for i in 1 .. k loop
        r := r + (w(ELEMENT(i, π))*UNIT_COST);
        UNIT_COST:= UNIT_COST div (n − i);
        for j in ELEMENT(i, π) + 1 .. n loop
            w(j) := w(j) − 1;
        end loop;
    end loop;
    return r;
end PERM_TO_RANK;
```

■ **Example 2.5.4** Trace Algorithm 2.5.2 for $n = 7$ and $\pi = (3, 6, 2, 7, 5)$.

Solution
Initially, $r = 0$, $k = 5$, UNIT_COST $= P(6, 4) = 360$, and $w(j) = j - 1$, $1 \leq j \leq 7$. We now execute the main loop for $i \in \{1, 2, 3, 4, 5\}$.
   Step 1: $i = 1$. Since ELEMENT($1, \pi$) = 3, we have

$$r := r + (w(3) * \text{UNIT\_COST}) = 0 + (2 * 360) = 720$$

UNIT_COST := UNIT_COST div 6 = 60, and w is updated to $(0, 1, 2, 2, 3, 4, 5)$.
   Step 2: $i = 2$. Since ELEMENT($2, \pi$) = 6, we have

$$r := r + (w(6) * \text{UNIT\_COST}) = 720 + (4 * 60) = 960$$

UNIT_COST:= UNIT_COST div 5 = 12, and w is updated to $(0, 1, 2, 2, 3, 4, 4)$.

Step 3: $i = 3$. Since ELEMENT$(3, \pi) = 2$, we have

$$r := r + (w(2) * \text{UNIT\_COST}) = 960 + (1 * 12) = 972$$

UNIT_COST:= UNIT_COST div 4 = 3, and w is updated to $(0, 1, 1, 1, 2, 3, 3)$.
   Step 4: $i = 4$. Since ELEMENT$(4, \pi) = 7$, we have

$$r := r + (w(7) * \text{UNIT\_COST}) = 972 + (3 * 3) = 981$$

UNIT_COST:= UNIT_COST div 3 = 1, and w is not changed.
   Step 5: $i = 5$. Since ELEMENT$(5, \pi) = 5$, we have

$$r := r + (w(5) * \text{UNIT\_COST}) = 981 + (2 * 1) = 983$$

UNIT_COST:= UNIT_COST div 2 = 0, and w is updated to $(0, 1, 1, 1, 2, 2, 2)$, and the value $r = 983$ is returned.   ∎

Next we consider the unranking algorithm: Given integers r, k, and n with $0 \le r \le P(n, k) - 1$, we wish to find the k-permutation π of $U = \{1, 2, \ldots, n\}$ with lexicographic rank r. As will be our custom, we think of the given rank r as a budgeted amount of dollars used to purchase a permutation; the variable BUDGET holds the amount of dollars available to spend at a given step, so that the initial value of BUDGET is the given rank r. Let $\pi = (x_1, x_2, \ldots, x_k)$. To determine x_1, we divide BUDGET by UNIT_COST$= P(n - 1, k - 1)$, obtaining the quotient q. Then $x_1 = q + 1$ is the largest value of x_1 we can afford, because a permutation with $x_1 \ge q + 2$ has rank at least $(q + 1)P(n - 1, k - 1)$ and this exceeds BUDGET. So $x_1 = q + 1$ and the value of BUDGET is replaced by the remainder

BUDGET $- (\text{UNIT\_COST} * q)$

For the general step i, $2 \le i \le \min\{k, n-1\}$, we have determined $x_1, \ldots, x_{i-1}$ and now wish to determine x_i. We let $V_i = U - \{x_1, \ldots, x_{i-1}\} = \{e_0, e_1, \ldots, e_{n-i}\}$, where $e_0 < e_1 < \cdots < e_{n-i}$. To determine x_i, divide BUDGET by UNIT_COST$= P(n - i, k - i)$, obtaining quotient q. Then e_q is the largest value of x_i we can afford; so $x_i = e_q$, and the value of BUDGET is replaced by the remainder BUDGET$-(\text{UNIT\_COST} * q)$.

With $V_1 = U$ and $V_i = U - \{x_1, \ldots, x_{i-1}\}$ for $i > 1$, we see that our algorithm must maintain an ordered list $(e_0, e_1, \ldots, e_{n-i})$ of the elements of V_i in increasing order. Initially, $e_j = j + 1$, $0 \le j \le n - 1$. At the ith step, when it is determined that $x_i = e_q$, then the value e_q is deleted from the list. Also, if $k = n$, then $\pi = (x_1, x_2, \ldots, x_n)$ and x_n is the last value that remains on the list after completion of the first $n - 1$ steps.

ALGORITHM 2.5.3   Given integers r, k, and n with $0 \le r \le P(n, k) - 1$, the function

RANK_TO_PERM(r, k, n)

returns the k-permutation π of $\{1, 2, \ldots, n\}$ with lexicographic rank r.

RANK_TO_PERM uses the following local variables (as described above):

BUDGET is a nonnegative integer and is initialized to r;
q is a nonnegative integer;
UNIT_COST is a nonnegative integer and is initialized to $P(n-1,k-1)$;
$(e_0, e_1, \ldots, e_{n-1})$ is an array of positive integers, initialized to $(1, 2, \ldots, n)$;
π is a permutation and is initialized to (the empty permutation)
IDENTITY(0);

```
begin
    for i in 1 .. MINIMUM(k, n - 1) loop
        q := BUDGET div UNIT_COST;
        BUDGET := BUDGET - (UNIT_COST * q);
        UNIT_COST:= UNIT_COST div (n - i);
        INSERT(i, e_q, π);
        for j in q .. n - 1 - i loop
            e_j := e_{j+1};
        end loop;
    end loop;
    if k ≥ n then
        INSERT(n, e_0, π);
    end if;
    return π;
end RANK_TO_PERM;
```

■ *Example 2.5.5*   Trace Algorithm 2.5.3 for $n = 7$, $k = 5$, and $r = 983$.

Solution
Initially, BUDGET = 983, UNIT_COST= 360, $e_j = j + 1$, $0 \le j \le n - 1$, and π is the empty permutation. The main loop is now executed for $i \in \{1, 2, 3, 4, 5\}$.
Step 1: $i = 1$. We have

$q :=$ BUDGET div UNIT_COST $= 2$
BUDGET $:=$ BUDGET $-$ (UNIT_COST $* q$) $= 983 - (360 * 2) = 263$
UNIT_COST $:=$ UNIT_COST div $6 = 60$

and $e_2 = 3$ is appended to π so that $\pi = (3)$. The element 3 is then deleted from the list e, so that $e = (1, 2, 4, 5, 6, 7)$. (Actually, of course, $e = (1, 2, 4, 5, 6, 7, 7)$, but the last 7 is now ignored.)
Step 2: $i = 2$. We have

$q :=$ BUDGET div UNIT_COST $= 4$
BUDGET $:=$ BUDGET $-$ (UNIT_COST $* q$) $= 263 - (60 * 4) = 23$
UNIT_COST $:=$ UNIT_COST div $5 = 12$

and $e_4 = 6$ is appended to π so that $\pi = (3, 6)$. Then 6 is deleted from the list e, so that $e = (1, 2, 4, 5, 7)$.

Step 3: $i = 3$. We have

$q :=$ BUDGET div UNIT_COST $= 1$

BUDGET $:=$ BUDGET $-$ (UNIT_COST $* q$) $= 23 - (12 * 1) = 11$

UNIT_COST $:=$ UNIT_COST div $4 = 3$

and $e_1 = 2$ is appended to π so that $\pi = (3, 6, 2)$. The element 2 is then deleted from e, so that $e = (1, 4, 5, 7)$.

Step 4: $i = 4$. At this step,

$q :=$ BUDGET div UNIT_COST $= 3$

BUDGET $:=$ BUDGET $-$ (UNIT_COST $* q$) $= 11 - (3 * 3) = 2$

UNIT_COST $:=$ UNIT_COST div $3 = 1$

and $e_3 = 7$ is appended to π so that π becomes $(3, 6, 2, 7)$. The element 7 is then deleted from e, so that $e = (1, 4, 5)$.

Step 5: $i = 5$. At this, the final step,

$q :=$ BUDGET div UNIT_COST $= 2$

BUDGET $:=$ BUDGET $-$ (UNIT_COST $* q$) $= 2 - (1 * 2) = 0$

UNIT_COST $:=$ UNIT_COST div $2 = 0$

and $e_2 = 5$ is appended to π, giving $\pi = (3, 6, 2, 7, 5)$. Deleting 5 from the list e leaves $e = (1, 4)$.

Thus, $\pi = (3, 6, 2, 7, 5)$ is returned. ∎

The last algorithm we consider selects a k-permutation $\pi = (x_1, x_2, \ldots, x_k)$ of $U = \{1, 2, \ldots, n\}$ uniformly at random. Initially, we set π equal to the identity permutation of U, so that $\pi = (x_1, x_2, \ldots, x_n)$ has length n, and $x_i = i$, $1 \le i \le n$. We want x_1 to be selected uniformly at random from U. Equivalently, we can select a subscript i_1 uniformly at random, $1 \le i_1 \le n$, and then swap the values of x_1 and x_{i_1}. This places the value i in x_1 and the value 1 in x_{i_1}. Next we want x_2 to be selected uniformly at random from $V_2 = U - \{x_1\}$. This is accomplished, quite neatly, by selecting a subscript i_2 uniformly at random, $2 \le i_2 \le n$, and then swapping the values of x_2 and x_{i_2}. The algorithm continues in this fashion; at step j, $1 \le j \le k$, select a subscript i_j uniformly at random, $j \le i_j \le n$, and then swap the values of x_j and x_{i_j}. At the end, return the prefix $(x_1, x_2, \ldots, x_k)$. Our algorithm assumes the existence of a function

RANDOM_INTEGER(a, b)

Given integers a and b with $a \le b$, this function returns an integer x selected uniformly at random from $[a, b]$. (Thus, each integer in $[a, b]$ has probability $1/(b - a + 1)$ of being selected.)

ALGORITHM 2.5.4 Given integers n and k with $1 \leq k \leq n$, the function

RANDOM_PERM(k, n)

returns a k-permutation π of $\{1, 2, \ldots, n\}$ selected uniformly at random.
RANDOM_PERM uses the following local variables:

π is a permutation and is intialized to IDENTITY(n);
i is a positive integer;

RANDOM_PERM also uses the procedure PI_SWAP(i, j) to swap the ith and jth elements of π; this procedure is defined as follows:

The local variable t is initialized to ELEMENT(i, π).
begin
    REPLACE(i, ELEMENT(j, π), π);
    REPLACE(j, t, π);
end PI_SWAP;

RANDOM_PERM is then defined as follows:

begin
    for j in 1 .. k loop
        i := RANDOM_INTEGER(j, n);
        PI_SWAP(i, j);
    end loop;
    return SUBPERM($\pi, 1, k$);
end RANDOM_PERM;

EXERCISE SET 2.5

1. For $k = 5$ and $n = 8$, apply Algorithm 2.5.1 to find the immediate successor in lex order of the given 5-permutations.
   (a) $(1, 2, 4, 5, 7)$ (b) $(1, 5, 3, 6, 8)$
   (c) $(2, 4, 8, 7, 6)$ (d) $(8, 7, 6, 5, 4)$

2. As mentioned, a permutation can be thought of as a list of elements so that the

abstract data type PERMUTATION can be implemented as a record type with two fields, such as

> type ELEMENT_ARRAY_TYPE is array (1 .. MAX) of POSITIVE;
> type PERMUTATION is record
>
> > LENGTH: NATURAL;
> > ELEMENT: ELEMENT_ARRAY_TYPE;
> > end record;

The idea is that for an object PI of type PERMUTATION, PI.LENGTH is the length of PI and, for PI.LENGTH> 0, PI.ELEMENT(J) is the Jth element of PI, $1 \leq J \leq$ PI.LENGTH. Given this, discuss the implementation of the LENGTH, IDENTITY, ELEMENT, POSITION, REPLACE, INSERT, and SUBPERM operations.

3. For $k = 5$ and $n = 8$, apply Algorithm 2.5.2 to find the lexicographic rank of the given 5-permutations.

   (a) $(1, 2, 3, 4, 5)$ (b) $(1, 5, 3, 6, 8)$

   (c) $(2, 4, 8, 7, 6)$ (d) $(8, 7, 6, 5, 4)$

4. Given a positive integer n and an integer r, $0 \leq r \leq n! - 1$, show that r is uniquely expressed as

$$\sum_{k=1}^{n-1} d_k k!$$

where each d_k is an integer, $0 \leq d_k \leq k$. This representation is called the **factorial representation** of r. For instance,

$$85 = 3 \cdot 4! + 2 \cdot 3! + 0 \cdot 2! + 1 \cdot 1!$$

so $d_4 = 3$, $d_3 = 2$, $d_2 = 0$, and $d_1 = 1$ in the factorial representation of 85.) (Hint: Let the permutation $\pi = (x_1, x_2, \ldots, x_n)$ of $\{1, 2, \ldots, n\}$ have lexicographic rank r; then x_1 determines d_{n-1}, x_2 determines d_{n-2}, and so on, until finally x_{n-1} determines d_1.) See also Problem 44 in Chapter 1.

5. For $k = 5$ and $n = 8$, apply Algorithm 2.5.3 to find the 5-permutation π with the given lexicographic rank r.

   (a) 0   (b) 30   (c) 1199   (d) 6719

6. For a k-permutation $(x_1, x_2, \ldots, x_k)$ of $\{1, 2, \ldots, n\}$, its (lexicographic) **corank** is the number of k-permutations that succeed it in lexicographic order. Develop an algorithm to determine the corank of a given k-permutation of $\{1, 2, \ldots, n\}$.

7. Express the function NEXT_PERM of Algorithm 2.5.1 in terms of the PERM_TO_RANK and RANK_TO_PERM functions of Algorithms 2.5.2 and 2.5.3, respectively. Discuss the efficiency of this method of implementing NEXT_PERM.

8. Algorithms 2.5.1, 2.5.2, and 2.5.3 are based on the lex order list of the permutations of $\{1, 2, \ldots, n\}$. A different set of algorithms for listing, ranking, and

unranking permutations is due to H. Trotter [Algorithm 115 PERM, *Communications of the ACM* 5 (1962): 434–435] and S. Johnson [Generation of Permutations by Adjacent Transposition, *Mathematics of Computation* 17 (1963): 282–285]. Their method has the advantage that the successor of a given permutation π is obtained from π merely by a "transposition," or swapping, of two adjacent elements. For example, here are the lists for $n \in \{1, 2, 3\}$ of the permutations of $\{1, 2, \ldots, n\}$ in Trotter-Johnson order:

$$(1)$$

$$(1, 2), (2, 1)$$

$$(1, 2, 3), (1, 3, 2), (3, 1, 2), (3, 2, 1), (2, 3, 1), (2, 1, 3)$$

(a) To construct the list for $\{1, 2, \ldots, n, n + 1\}$ from the list for $\{1, 2, \ldots, n\}$, repeat each n-permutation $n + 1$ times, then insert the element $n + 1$ into each n-permutation so that its position follows the pattern

$$n + 1, n, \ldots, 1, 1, 2, \ldots, n, \ldots, n + 1$$

Use this rule to construct the list for $n = 4$ from the list for $n = 3$.

Given a permutation $\pi = (x_1, x_2, \ldots, x_n)$ of $\{1, 2, \ldots, n\}$, the **Trotter-Johnson rank** of π is an integer between 0 and $n! - 1$ giving the number of permutations that precede π on the list of the permutations of $\{1, 2, \ldots, n\}$ in Trotter-Johnson order.

⋆(b) Develop an algorithm that finds the Trotter-Johnson rank of a given permutation π of $\{1, 2, \ldots, n\}$. (Hint: Let j be the position of n in π, and let π' be the permutation of $\{1, \ldots, n - 1\}$ obtained by deleting n from π; develop a formula for rank of π in terms of the rank of π' and j.)

⋆(c) Develop an algorithm that, given a positive integer n and an integer r, $0 \le r < n!$, finds the permutation π of $\{1, 2, \ldots, n\}$ with Trotter-Johnson rank r.

⋆(d) Develop an algorithm that inputs n and outputs the list of the permutations of $\{1, 2, \ldots, n\}$ in Trotter-Johnson order.

9. Develop a recursive implementation of Algorithm 2.5.1.

10. Develop a recursive implementation of Algorithm 2.5.2.

11. Develop a recursive implementation of Algorithm 2.5.3.

2.6 *LISTING, RANKING, AND UNRANKING SUBSETS*

In this section we consider algorithms for listing, ranking, and unranking subsets. For a fixed positive integer n, we first consider the lex order list of all subsets of $U = \{1, 2, \ldots, n\}$; then, for k an integer, $0 \le k \le n$, we restrict our attention to just the k-element subsets of U.

Subsets of $\{1, 2, \ldots, n\}$

There are several well-known methods for ordering the subsets of $U = \{1, 2, \ldots, n\}$; perhaps the most natural one is the order called **lexicographic order**, or **lex order** for short, which is similar to lexicographic order as defined for permutations in the last section.

Lexicographic ordering of subsets of the set $\mathbb{Z}^+$ is based on the underlying natural order of the positive integers:

$$1 < 2 < 3 < 4 < \cdots$$

For two subsets A and B, let us use the notation $A < B$ to mean that A precedes B in lexicographic order, and $A \leq B$ to mean $A = B$ or $A < B$. We then define **lexicographic order** recursively using the following two rules:

1. For any nonempty set A, we have $\{\ \} < A$.
2. For nonempty sets A and B with $A \neq B$, let a and b be the smallest elements of A and B, respectively. Then $A < B$ provided either $a < b$ or $a = b$ and $A - \{a\} < B - \{b\}$.

In other words, for nonempty sets A and B with $A \neq B$, let x be the smallest element of the symmetric difference $(A - B) \cup (B - A)$. If $x \in A$, then $A < B$; otherwise, $B < A$.

■ *Example 2.6.1*   List the subsets of these sets in lex order: (a) $\{1\}$, (b) $\{1, 2\}$, (c) $\{1, 2, 3\}$.

Solution

(a) The list is $\{\ \}, \{1\}$. Notice that, by rule 1, the empty set is always the first subset listed.

(b) The list is $\{\ \}, \{1\}, \{1, 2\}, \{2\}$. Here we use rule 2 to see that $\{1\} < \{1, 2\}$ and $\{1, 2\} < \{2\}$.

(c) The list is $\{\ \}, \{1\}, \{1, 2\}, \{1, 2, 3\}, \{1, 3\}, \{2\}, \{2, 3\}, \{3\}$. Observe that, if we delete from this list every subset that contains 3, then we obtain the list of part (b).   ■

To solve the problem of generating the subsets of $\{1, 2, \ldots, n\}$ in lex order, it suffices to solve the problem of finding the (immediate) successor of any given subset. That is, suppose we have a function NEXT_SUBSET(S) that returns the successor of any given subset S in the lex order list. We can initialize S to the empty set; then, whenever we want the next subset we can execute the statement

$$S := \text{NEXT\_SUBSET}(S)$$

to update S.

The lex order list of the subsets of $\{1, 2, \ldots, n\}$ contains 2^n subsets. Given a subset S, we would like to determine its **rank**, that is, its "position" in the list. For certain mathematical reasons, the first element of the list, $\{\ \}$, has rank 0, the second element of the list, $\{1\}$, has rank 1, and so on, and the last element of the list, $\{n\}$, has rank $2^n - 1$. Thus, the **(lexicographic) rank** of the subset S may be defined as the number of subsets that precede it in lex order.

We note that the ranking problem could be solved by generating the subsets of $\{1, 2, \ldots, n\}$ in lex order until the subset S is obtained, keeping count of the number of subsets generated so far. However, if S has rank close to 2^n, then such a method is far too inefficient. Instead, we desire an algorithm that takes no more than n steps to find the rank of any subset.

The inverse problem of the ranking problem is called "unranking": Given integers r and n with $0 \leq r \leq 2^n - 1$, determine the subset S that has rank r among all the subsets of $\{1, 2, \ldots, n\}$.

■ **Example 2.6.2** The following table lists the subsets of $\{1, 2, 3, 4\}$ in lex order and gives the rank of each subset:

Subset	Rank
$\{\ \}$	0
$\{1\}$	1
$\{1, 2\}$	2
$\{1, 2, 3\}$	3
$\{1, 2, 3, 4\}$	4
$\{1, 2, 4\}$	5
$\{1, 3\}$	6
$\{1, 3, 4\}$	7
$\{1, 4\}$	8
$\{2\}$	9
$\{2, 3\}$	10
$\{2, 3, 4\}$	11
$\{2, 4\}$	12
$\{3\}$	13
$\{3, 4\}$	14
$\{4\}$	15

■

Another problem that we wish to consider is that of choosing a subset of $\{1, 2, \ldots, n\}$ at random. This is to be done "uniformly," meaning that each of the 2^n subsets has the same probability of being chosen.

To summarize, then, we desire algorithms to accomplish the following:

1. Given a subset S of $\{1, 2, \ldots, n\}$, find its successor in lexicographic order. For convenience, we wish to think of the lex order list as "circular," so that the successor of the last subset $\{n\}$ is the first subset $\{\ \}$.

2. Given a subset S of $\{1, 2, \ldots, n\}$, find its rank.

3. Given integers r and n with $0 \leq r \leq 2^n - 1$, find the subset S with rank r among the subsets of $\{1, 2, \ldots, n\}$.

4. Given n, choose a subset S of $\{1, 2, \ldots, n\}$ uniformly at random.

In designing such algorithms, we assume that we have at our disposal a package of basic operations for working with sets. (In modern computer science terminology, we would implement the type SET as an abstract data type.) This package provides (among others) the following basic operations:

1. The function CARDINALITY(S) returns the cardinality of the set S.

2. The function EMPTY_SET returns the empty set.

3. The function ELEMENT(h, S) returns the hth smallest element of S, given $1 \leq h \leq$ CARDINALITY(S). Thus, ELEMENT($1, S$) is the smallest element of S, and ELEMENT(CARDINALITY(S), S) is the largest element of S.

4. The procedure APPEND(m, S) updates S to the set $S \cup \{m\}$, provided that either S is empty or ELEMENT(CARDINALITY(S), S) $< m$.

5. The procedure DELETE_LARGEST(S) modifies S by deleting its largest element, that is, S is updated to $S - \{$ELEMENT(CARDINALITY(S), S)$\}$. (It is assumed that CARDINALITY(S) > 0 (S is nonempty).)

6. The procedure INCREMENT_LARGEST(S) modifies S by adding 1 to its largest element. (It is assumed that S is nonempty.)

One method that can be used to implement the abstract data type SET is to think of a set as an ordered list, implemented as a record type with two fields, such as

```
type ELEMENT_ARRAY_TYPE is array (1 .. MAX) of POSITIVE;
type SET is record
            CARDINALITY: NATURAL;
            ELEMENT: ELEMENT_ARRAY_TYPE;
          end record;
```

For an object S of type SET, the idea here is to let S.CARDINALITY be the cardinality of S. Also, if S.CARDINALITY > 0, then

$$\text{S.ELEMENT}(1) < \text{S.ELEMENT}(2) < \cdots < \text{S.ELEMENT}(\text{S.CARDINALITY})$$

are the elements of S (arranged in increasing order).

With such a data structure, notice that each of the operations CARDINALITY, ELEMENT, APPEND, DELETE_LARGEST, and so on, can be implemented to execute in time $\Theta(1)$ (see Exercise 6). However, as an abstract data type, the type SET is "encapsulated"—that is, users of the type are not allowed direct access to the CARDINALITY and ELEMENT fields of an object S. The only means by which S can be accessed or modified is through use of the operations CARDINALITY, ELEMENT, APPEND, DELETE_LARGEST, and so on, provided by the basic package.

We are now ready to apply the basic set operations to implement the successor algorithm. We let $S = \{x_1, x_2, \ldots, x_k\}$ be a subset of $\{1, 2, \ldots, n\}$, where $x_1 < x_2 < \cdots < x_k$. If $x_k < n$, then the successor of S is $\{x_1, x_2, \ldots, x_k, x_k + 1\}$. For example, in the case $n = 5$, the successor of $\{1,3\}$ is $\{1,3,4\}$, and the successor of $\{1,3,4\}$ is $\{1,3,4,5\}$. What if $x_k = n$? Well, if $k = 1$, so that $S = \{n\}$, then S is the last subset in lex order and, by our agreement, the successor of S is the empty set. However, if $k > 1$, then the successor of S is $\{x_1, \ldots, x_{k-1} + 1\}$. As an example in the case $n = 5$, the successor of $\{1,3,4,5\}$ is $\{1,3,5\}$, and the successor of $\{1,3,5\}$ is $\{1,4\}$. These observations yield the following algorithm:

ALGORITHM 2.6.1 Given a subset S of $\{1,2,\ldots,n\}$, the function

NEXT_SUBSET(S, n)

returns the successor of S in lexicographic order. For convenience, the successor of the last subset $\{n\}$ is defined to be the first subset $\{\ \}$.

NEXT_SUBSET uses the following local variables:
The nonnegative integer k is initialized to CARDINALITY(S);
The set RESULT is initialized to S;

```
begin
    if k = 0 then
        APPEND(1, RESULT);
    elsif ELEMENT(k, S) < n then
        APPEND(ELEMENT(k, S) + 1, RESULT);
    else
        DELETE_LARGEST(RESULT);
        if k > 1 then
            INCREMENT_LARGEST(RESULT);
        end if;
    end if;
    return RESULT;
end NEXT_SUBSET;
```

Next, we consider the ranking algorithm. To aid in understanding this and several other algorithms in this section, we introduce a family of labeled directed graphs.

Given a positive integer n, the digraph G_n has vertex set V_n, where

$$V_n = \{(x,y) \mid x \text{ and } y \text{ are nonnegative integers and } n \geq x \geq y\}$$

and arc set A_n, where

$$A_n = \{((x_1, y_1), (x_2, y_2)) \mid x_2 = x_1 + 1 \text{ and } (y_2 = y_1 \text{ or } y_2 = y_1 + 1)\}$$

(This family of digraphs was introduced in Exercise 20 of Exercise Set 1.6.) The digraph G_4 is shown in Figure 2.6.1. Note that the digraph G_n can be drawn so

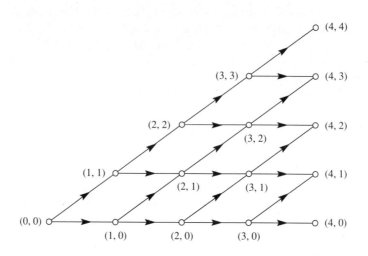

Figure 2.6.1   The digraph G_4.

that its vertices are points in a coordinate plane. We also want to label the arcs; each "horizontal" arc $((x_1, y_1), (x_1 + 1, y_1))$ is labeled 0, whereas each "diagonal" arc $((x_1, y_1), (x_1 + 1, y_1 + 1))$ is labeled 1.

In G_n, how many paths are there from $(0, 0)$ to (n, k), where $0 \leq k \leq n$? It is not difficult to see that the answer to this question is $C(n, k)$. This is because a path

$$(0, 0) = (0, y_0), (1, y_1), \ldots, (n, y_n) = (n, k)$$

from $(0, 0)$ to (n, k) can be encoded as a binary string $(b_1, b_2, \ldots, b_n)$ of length n, where $b_x = 1$ if and only if $y_x = y_{x-1} + 1$. In other words, b_x is just the label on the xth arc of the path. For example, the binary string 1010 encodes the path

$$(0, 0), (1, 1), (2, 1), (3, 2), (4, 2)$$

in G_4. Now $(b_1, b_2, \ldots, b_n)$ encodes a path from $(0, 0)$ to (n, k) if and only if it contains exactly k 1s, because the number of 1s is the number of times that an arc labeled 1 is used and, hence, is the number of times that y is incremented along the path. Thus, there is a one-to-one correspondence between paths in G_n from $(0, 0)$ to (n, k) and binary strings of length n with exactly k 1s, and this latter number is $C(n, k)$.

In fact, notice that a binary string $(b_1, b_2, \ldots, b_n)$ of length n encodes a subset S of $\{1, 2, \ldots, n\}$ in the obvious way: $b_x = 1$ if and only if $x \in S$. We shall often ignore the distinction between a subset S of $\{1, 2, \ldots, n\}$, its corresponding binary string $(b_1, b_2, \ldots, b_n)$, and the corresponding path in the digraph G_n.

Given the path $(b_1, b_2, \ldots, b_n)$, our problem is then to find the rank r of the corresponding subset S. Initially, we set r to 0 and let $k = |S|$. If $k = 0$, then $S = \{\ \}$ and r has the correct value. So suppose $k > 0$. If $b_1 = 1$ (so that $1 \in S$), then the empty set precedes S in lex order, so we increment r by 1. We also decrement k by 1 since, at step x, k is to be the cardinality of the subset represented by the path

$(0, \ldots, 0, b_x, \ldots, b_n)$. On the other hand, if $b_1 = 0$ (so that $1 \notin S$), then the 2^{n-1} subsets of $\{1, 2, \ldots, n\}$ that contain 1 all precede S in lex order, so we increment r by 2^{n-1}. In either case we think of the amount by which r is incremented at the first step as being the "cost" of using the arc labeled b_1. Thus, with $k > 0$, the cost of the arc from $(0, 0)$ to $(1, 1)$ is 1, whereas the cost of the arc from $(0, 0)$ to $(1, 0)$ is 2^{n-1}.

In general, at step $x \geq 2$, we have taken account of $b_1, b_2, \ldots, b_{x-1}$. We assume k is the cardinality of the subset $(0, \ldots, 0, b_x, \ldots, b_n)$ (that is, k is the number of 1s among $b_x, \ldots, b_n$), and r is the rank of the subset represented by $(b_1, \ldots, b_{x-1}, 0, \ldots, 0)$. Now, if $k = 0$, then r has the correct value and we are done. So suppose $k > 0$. If $b_x = 1$, then the subset represented by $(b_1, \ldots, b_{x-1}, 0, \ldots, 0)$ precedes S in lex order, so r should be incremented by 1 and k should be decremented by 1. On the other hand, if $b_x = 0$, then the 2^{n-x} subsets represented by $(b_1, \ldots, b_{x-1}, 1, a_{x+1}, \ldots, a_n)$, with $a_i = 0$ or 1 for $x + 1 \leq i \leq n$, all precede S in lex order, so r should be incremented by 2^{n-x}. Again, think of the amount by which r is incremented at step x as the cost of using the arc labeled b_x (when $k > 0$), so that any diagonal arc costs 1, whereas the horizontal arc from $(x - 1, y)$ to (x, y) costs 2^{n-x}.

We are now ready to present the ranking algorithm, except for two final observations.

The first is that we could just as well label the arcs of the digraph G_n with the Boolean values FALSE and TRUE instead of 0 and 1. Since the type BOOLEAN is usually provided by modern programming languages (such as Ada), this is what we will do now. Thus, our algorithm associates with a subset S of $\{1, 2, \ldots, n\}$ a Boolean array PATH, where PATH(x) is TRUE if and only if $x \in S$, $1 \leq x \leq n$.

The other observation concerns the horizontal arc cost. Let the variable HCOST hold this value. Since the horizontal arc cost at step x is 2^{n-x}, we initialize HCOST to 2^n and then divide HCOST by 2 at the beginning of each step.

ALGORITHM 2.6.2 Given a subset S of $\{1, 2, \ldots, n\}$, the function

SUBSET_TO_RANK(S, n)

determines its lexicographic rank.

> SUBSET_TO_RANK uses the following local variables (as discussed above):
> The nonnegative integer r is initialized to 0;
> The nonnegative integer k is initialized to CARDINALITY(S);
> The Boolean array (PATH(1), PATH(2), ..., PATH(n)) is initialized to
> (FALSE, FALSE, ..., FALSE);
> The positive integer x is initialized to 1;
> The positive integer HCOST is initialized to 2^n;
>
> begin
>   for j in 1 .. k loop
>   PATH(ELEMENT(j, S)) := TRUE;
>   end loop;

```
        while k > 0 loop
          HCOST := HCOST div 2;
          if PATH(x) then
            r := r + 1;
            k := k − 1;
          else
            r := r + HCOST;
          end if;
          x := x + 1;
        end loop;
        return r;
      end SUBSET_TO_RANK;
```

Once we understand the idea of the preceding algorithm, then it can be implemented without (explicitly) using the PATH array; this is left to Exercise 2.

■ **Example 2.6.3**   Trace Algorithm 2.6.2 for $n = 4$ and $S = \{2, 4\}$.

Solution
Initially, $r = 0$, $k = 2$, PATH = (FALSE, TRUE, FALSE, TRUE), $x = 1$, and HCOST = 16. We now execute the main loop while $k > 0$.
    Step 1: HCOST := HCOST div 2 = 8; since PATH(1) = FALSE, $r := r + $ HCOST = 8; $x := x + 1 = 2$.
    Step 2: HCOST := HCOST div 2 = 4; since PATH(2) = TRUE, $r := r + 1 = 9$ and $k := k − 1 = 1$. Also, $x := x + 1 = 3$.
    Step 3: HCOST := HCOST div 2 = 2; since PATH(3) = FALSE, $r := r + $ HCOST = 11; $x := x + 1 = 4$.
    Step 4: HCOST := HCOST div 2 = 1; since PATH(4) = TRUE, $r := r + 1 = 12$ and $k := k − 1 = 0$. Also, $x := x + 1 = 5$.
    At this point, $k = 0$, so the loop is exited. The value $r = 12$ is returned.   ■

In the ranking algorithm, the rank of a subset S is determined by summing the arc costs along the path in G_n determined by S. Let's use a dollar as our basic cost unit, so that, when $k > 0$, each diagonal arc costs one dollar to traverse whereas the horizontal arc from $(x − 1, y)$ to (x, y) costs $2^{n−x}$ dollars to traverse. Then, when ranking, we are in a "seller" position—we have a subset S to sell and we wish to know its value; this amount is the rank of S.

For the unranking algorithm, we simply turn the situation around, placing ourselves in a "buyer" position. Now, we are given the rank r—a budgeted amount to spend—and we want to know what subset S can be purchased for exactly r dollars.

Let the variable BUDGET hold the number of dollars we have available to spend at a given step, so that the initial value of BUDGET is r. Begin, at step 1, in the digraph G_n at $(0, 0)$ with $\{\ \}$ as the initial value of S. If BUDGET = 0, then the only subset we can afford is $\{\ \}$ (which is free!), so we are done. So suppose BUDGET > 0. If BUDGET ≤ HCOST = $2^{n−1}$, then we cannot afford to exclude the element 1 from S, for a nonempty subset that does not contain 1 has rank at least $2^{n-1} + 1$, and this exceeds our budget. So $1 \in S$ and we move to the vertex $(1, 1)$ along the diagonal arc from $(0, 0)$. Since it costs one dollar to use this arc, we reduce the value of BUDGET by 1. On the other hand, if BUDGET $>$ HCOST, then we can afford to exclude 1 from S. It turns out that our algorithm should be "greedy"—at any step, if we can afford to exclude an element, we will do so. Thus, $1 \notin S$ and we move to the vertex $(1, 0)$ along the horizontal arc from $(0, 0)$. Since this costs HCOST, we reduce the value of BUDGET by HCOST.

In general, at step $x > 1$, we are at some vertex $(x - 1, y)$ and we have determined for each i, $1 \le i < x$, whether $i \in S$ or $i \notin S$. If BUDGET $= 0$, then S has the correct value and we are done. If $0 <$ BUDGET $\le$ HCOST $= 2^{n-x}$, then we cannot afford to exclude x from S. So $x \in S$ and we move to vertex $(x, y + 1)$ along the diagonal arc. This costs one dollar, so BUDGET is reduced by 1. If BUDGET $>$ HCOST, then we can afford to exclude x from S. So $x \notin S$ and we move to vertex (x, y) along the horizontal arc. This costs HCOST dollars, so BUDGET is reduced by HCOST. The unranking algorithm continues in this manner until BUDGET $= 0$.

ALGORITHM 2.6.3 Given a positive integer n and an integer r, with $0 \le r \le 2^n - 1$, the function

RANK_TO_SUBSET(r, n)

returns the subset S with lexicographic r among the subsets of $\{1, 2, \ldots, n\}$.

RANK_TO_SUBSET uses the following local variables (as discussed above):

> The set S is initialized to EMPTY_SET;
> BUDGET is a nonnegative integer and is initialized to r;
> HCOST is a positive integer and is initialized to 2^n;
> The positive integer x is initialized to 1;

```
begin
   while BUDGET > 0 loop
      HCOST := HCOST div 2;
      if BUDGET ≤ HCOST then
         BUDGET := BUDGET − 1;
         APPEND(x, S);
      else
         BUDGET := BUDGET − HCOST;
      end if;
      x := x + 1;
   end loop;
   return S;
end RANK_TO_SUBSET;
```

■ ***Example 2.6.4***   Trace Algorithm 2.6.3 for $n = 4$ and $r = 12$.

Solution
Initially, $S = \{\ \}$, BUDGET = 12, HCOST = 16, and $x = 1$. We now execute the main loop while BUDGET > 0.

Step 1: HCOST := HCOST div 2 = 8. Since BUDGET $>$ HCOST, we have BUDGET := BUDGET − HCOST = 4 and $x := x + 1 = 2$.

Step 2: HCOST := HCOST div 2 = 4. Since BUDGET $\leq$ HCOST, we have BUDGET := BUDGET − 1 = 3 and x is appended to S so that $S = \{2\}$. Then $x := x + 1 = 3$.

Step 3: HCOST := HCOST div 2 = 2. Since BUDGET $>$ HCOST, we have BUDGET := BUDGET − HCOST = 1 and $x := x + 1 = 4$.

Step 4: HCOST := HCOST div 2 = 1. Since BUDGET $\leq$ HCOST, we have BUDGET := BUDGET − 1 = 0 and x is appended to S so that $S = \{2, 4\}$. Then $x := x + 1 = 5$.

At this point, BUDGET = 0, so the loop is exited. The set $S = \{2, 4\}$ is returned.
■

The final algorithm in this part selects a subset S of $\{1, 2, \ldots, n\}$ uniformly at random. Such an algorithm is straightforward; since any x, $1 \leq x \leq n$, is an element of exactly half of the subsets, x should be included in S with probability $1/2$. As in the preceding section, we assume the existence of a function

RANDOM_INTEGER(a, b)

Given integers a and b with $a \leq b$, this function returns an integer x selected uniformly at random from $[a, b]$.

ALGORITHM 2.6.4   Given a positive integer n, the function

RANDOM_SUBSET(n)

returns a subset S of $\{1, 2, \ldots, n\}$ selected uniformly at random.

The set S is initialized to EMPTY_SET;

```
begin
    for x in 1 .. n loop
        if RANDOM_INTEGER(0, 1) = 1 then
            APPEND(x, S);
        end if;
        end loop;
        return S;
    end RANDOM_SUBSET;
```

Another method for listing the subsets of $\{1, 2, \ldots, n\}$ is known as a **Gray code**. In such a list, each subset (after the first) differs in cardinality from its predecessor by exactly 1. In other words, each subset is obtained from its predecessor by the addition or deletion of a single element. For example, a Gray code for $n = 3$ is

$$\{\ \}, \{1\}, \{1, 2\}, \{2\}, \{2, 3\}, \{1, 2, 3\}, \{1, 3\}, \{3\}$$

Listing, ranking, and unranking algorithms connected with Gray codes are explored in the exercises.

k-Subsets of $\{1, 2, \ldots, n\}$

In this part we restrict the problem considered in the last part. Again, let n be a positive integer. Rather than being interested in the lex order list of all the subsets of $\{1, 2, \ldots, n\}$, we are interested here only in those subsets of cardinality k for some fixed k, $0 \le k \le n$. Such a subset is called a **k-subset** of $\{1, 2, \ldots, n\}$.

■ *Example 2.6.5* List the k-subsets of $\{1, 2, 3, 4, 5\}$ in lexicographic order for these values of k: (a) 0, (b) 1, (c) 2, (d) 3, (e) 4, (f) 5.

Solution

(a) The only 0-subset is $\{\ \}$.

(b) The list of 1-subsets in lex order is $\{1\}, \{2\}, \{3\}, \{4\}, \{5\}$.

(c) The list of 2-subsets in lex order is $\{1, 2\}, \{1, 3\}, \{1, 4\}, \{1, 5\}, \{2, 3\}, \{2, 4\}, \{2, 5\}, \{3, 4\}, \{3, 5\}, \{4, 5\}$.

(d) Here we obtain the list $\{1, 2, 3\}, \{1, 2, 4\}, \{1, 2, 5\}, \{1, 3, 4\}, \{1, 3, 5\}, \{1, 4, 5\}, \{2, 3, 4\}, \{2, 3, 5\}, \{2, 4, 5\}, \{3, 4, 5\}$ of 3-subsets in lex order.

(e) The 4-subsets in lex order are $\{1, 2, 3, 4\}, \{1, 2, 3, 5\}, \{1, 2, 4, 5\}, \{1, 3, 4, 5\}, \{2, 3, 4, 5\}$.

(f) The only 5-subset is $\{1, 2, 3, 4, 5\}$. ■

Recall that the number of k-subsets of $\{1, 2, \ldots, n\}$ is $C(n, k)$. The **(lexicographic) rank** of a k-subset S (with respect to the lex order list of all the k-subsets of $\{1, 2, \ldots, n\}$) is the number of k-subsets that precede S in lex order. Thus, the rank of S is an integer r, $0 \le r \le C(n, k) - 1$.

■ *Example 2.6.6* The following table lists the 3-subsets of $\{1, 2, 3, 4, 5, 6\}$ in lex order and gives the rank of each subset.

Subset	Rank
$\{1,2,3\}$	0
$\{1,2,4\}$	1
$\{1,2,5\}$	2
$\{1,2,6\}$	3
$\{1,3,4\}$	4
$\{1,3,5\}$	5
$\{1,3,6\}$	6
$\{1,4,5\}$	7
$\{1,4,6\}$	8
$\{1,5,6\}$	9
$\{2,3,4\}$	10
$\{2,3,5\}$	11
$\{2,3,6\}$	12
$\{2,4,5\}$	13
$\{2,4,6\}$	14
$\{2,5,6\}$	15
$\{3,4,5\}$	16
$\{3,4,6\}$	17
$\{3,5,6\}$	18
$\{4,5,6\}$	19

∎

Analogous to the previous part, we desire algorithms to accomplish the following:

1. Given a k-subset S of $\{1,2,\ldots,n\}$, find its successor in the lex order list of all the k-subsets of $\{1,2,\ldots,n\}$. For convenience, we wish to think of the lex order list as "circular," so that the successor of the last subset $\{n-k+1,\ldots,n\}$ is the first subset $\{1,\ldots,k\}$.

2. Given a k-subset S of $\{1,2,\ldots,n\}$, find its rank.

3. Given integers r, k, and n with $0 \leq r \leq C(n,k) - 1$, find the k-subset S of $\{1,2,\ldots,n\}$ with rank r.

4. Given n and k, choose a k-subset S of $\{1,2,\ldots,n\}$ uniformly at random.

The successor algorithm is fairly easy to understand. Let $S = \{x_1,x_2,\ldots,x_k\}$ be a k-subset of $\{1,2,\ldots,n\}$, and assume that $x_1 < x_2 < \cdots < x_k$. If $x_k < n$, the successor of S is $\{x_1,\ldots,x_{k-1},x_k+1\}$. Suppose that $x_k = n$. If $k = 1$, then the successor of S is $\{1\}$, so assume further that $k \geq 2$. Now consider x_{k-1}. If $x_{k-1} < n-1$, then the successor of S is $\{x_1,\ldots,x_{k-2},x_{k-1}+1,x_{k-1}+2\}$. On the other hand, if $x_{k-1} = n-1$ and $k = 2$, then the successor is $\{1,2\}$, whereas if $k \geq 3$, we next consider x_{k-2}, and so on.

In general, we must find the largest subscript h, $1 \leq h \leq k$, such that x_h is not at its maximum value $n-k+h$. If no such h exists (namely, $x_1 = n-k+1$), then S is the last k-subset in lex order and the successor of S is $\{1,2,\ldots,k\}$. Otherwise, the successor of S is the k-subset $\{y_1,y_2,\ldots,y_k\}$, where $y_i = x_i$, $1 \leq i < h$, and $y_i = x_h - h + 1 + i$, $h \leq i \leq k$.

ALGORITHM 2.6.5 Given a k-subset S of $\{1, 2, \ldots, n\}$, the function

NEXT_K_SUBSET(S, n)

returns the k-subset that is its successor in lexicographic order. For convenience, the successor of the last subset $\{n - k + 1, \ldots, n\}$ is defined to be the first subset $\{1, 2, \ldots, k\}$.

NEXT_K_SUBSET uses the following local variables:
RESULT is a set whose initial value is S;
k is a nonnegative integer initialized to CARDINALITY(S);
h is a nonnegative integer initialized to k;
m is a positive integer;

```
begin
    while h > 0 and then ELEMENT(h, RESULT) ≥ n − k + h loop
        h := h − 1;
        DELETE_LARGEST(RESULT);
    end loop;
    if h = 0 then
        for x in 1 .. k loop
            APPEND(x, RESULT);
        end loop;
    else
        INCREMENT_LARGEST(RESULT);
        m := ELEMENT(h, RESULT) − h;
        for j in h + 1 .. k loop
            APPEND(m + j, RESULT);
        end loop;
    end if;
    return RESULT;
end NEXT_K_SUBSET;
```

To aid our understanding of the ranking and unranking algorithms, we turn again to the digraph G_n introduced in the first part. In particular, for given values of n and k, we are interested in all the paths of G_n from $(0,0)$ to (n,k) since each such path corresponds to a k-subset of $\{1, 2, \ldots, n\}$. For example, Figure 2.6.2 shows that portion of the digraph G_6 consisting of all paths from $(0,0)$ to $(6,3)$.

Given a path $(b_1, b_2, \ldots, b_n)$ from $(0,0)$ to (n,k), our problem is to find the rank r of the corresponding k-subset S. Initially, we set r to 0 and let $k = |S|$. We assume $0 < k < n$, for otherwise the rank of S is 0 and the problem is trivial. So we begin at $(0,0)$ and consider b_1. If $b_1 = 1$ (so that $1 \in S$), then it is possible that S is the first k-subset in lex order, so we leave r at 0 and move along the diagonal arc of G_n from $(0,0)$ to $(1,1)$. On the other hand, if $b_1 = 0$ (so that $1 \notin S$), then the $C(n - 1, k - 1)$ k-subsets of $\{1, 2, \ldots, n\}$ that contain 1 all precede S in lex order, so we increment r

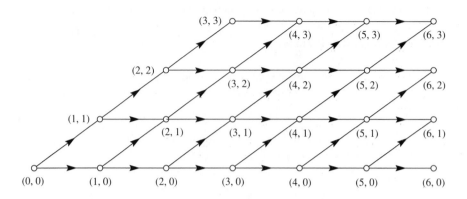

Figure 2.6.2

by $C(n - 1, k - 1)$ and move along the horizontal arc of G_n from $(0, 0)$ to $(1, 0)$. In either case we think of the amount by which r is incremented at the first step as being the "cost" of using the arc labeled b_1. Thus, the cost of the arc from $(0, 0)$ to $(1, 1)$ is 0, whereas the cost of the arc from $(0, 0)$ to $(1, 0)$ is $C(n - 1, k - 1)$.

In general, at step x, $2 \leq x < n$, we have taken into account $b_1, b_2, \ldots, b_{x-1}$ and are at the vertex $(x - 1, y - 1)$ of G_n, where $y - 1$ is equal to the number of 1s among $b_1, \ldots, b_{x-1}$. Assume that r is the rank of the subset represented by $(b_1, \ldots, b_{x-1}, c_x, \ldots, c_n)$, where $c_i = 1$ if $x \leq i \leq x + k - y$ and $c_i = 0$ if $i > x + k - y$. Now, if $y = k + 1$, then r has the correct value and we are done. So suppose $y \leq k$. If $b_x = 1$, then it is possible that r has the correct value, so r is not changed and we move along the diagonal arc from $(x - 1, y - 1)$ to (x, y). On the other hand, if $b_x = 0$, then the $C(n - x, k - y)$ subsets represented by $(b_1, \ldots, b_{x-1}, 1, a_{x+1}, \ldots, a_n)$, where $(a_{x+1}, \ldots, a_n)$ represents a path from (x, y) to (n, k), all precede S in lex order. So r should be incremented by $C(n - x, k - y)$, and we move along the horizontal arc from $(x - 1, y - 1)$ to $(x, y - 1)$. Again, we think of the amount by which r is incremented at step x as the cost of using the arc labeled b_x, so that any diagonal arc costs 0, whereas the horizontal arc from $(x - 1, y - 1)$ to $(x, y - 1)$ costs $C(n - x, k - y)$.

We assume the existence of the function $C(n, k)$ that returns the value of the binomial coefficient $C(n, k)$ if $0 \leq k \leq n$ and returns 0 if $k < 0$ or $n < k$. As with Algorithm 2.6.2, we let the variable HCOST hold the value of the horizontal arc cost at step x. Initially, HCOST $= C(n - 1, k - 1)$. In general, we are at some vertex $(x - 1, y - 1)$ with HCOST $= C(n - x, k - y)$. If we next move diagonally to (x, y), then HCOST should change to $C(n - x - 1, k - y - 1)$. Note that this is accomplished by the assignment statement

HCOST := $((k - y) * \text{HCOST})$ div $(n - x)$;

On the other hand, if we next move horizontally to $(x, y - 1)$, then HCOST should change to $C(n - x - 1, k - y)$; this is accomplished by the assignment statement

HCOST := $(((n - x) - (k - y)) * \text{HCOST})$ div $(n - x)$;

Also as in Algorithm 2.6.2, we use a Boolean array PATH to encode the subset S, where PATH(x) is TRUE if and only if $x \in S$, $1 \leq x \leq n$. Again, this is a convenience since the algorithm can be implemented without making explicit use of such an array; see Exercise 4.

ALGORITHM 2.6.6   Given a positive integer n and a k-subset S of $\{1, 2, \ldots, n\}$, the function

K_SUBSET_TO_RANK(S, n)

returns the lexicographic rank of S (among the k-subsets of $\{1, 2, \ldots, n\}$).

> K_SUBSET_TO_RANK uses the following local variables:
> RANK is a nonnegative integer and is initialized to 0;
> k is a nonnegative integer and is initialized to CARDINALITY(S);
> (PATH(1), PATH(2), ..., PATH(n)) is a Boolean array and is initially
>   (FALSE, FALSE, ..., FALSE);
> HCOST is a nonnegative integer and is initially $C(n-1, k-1)$;
> y is a positive integer, initialized to 1;

```
begin
    if 0 < k and k < n then
        for j in 1 .. k loop
            PATH(ELEMENT(j, S)) := TRUE;
        end loop;
        for x in 1 .. n − 1 loop
            if PATH(x) then
                HCOST := ((k − y) ∗ HCOST) div (n − x);
                y := y + 1;
            else
                RANK := RANK + HCOST;
                HCOST := (((n − x) − (k − y)) ∗ HCOST) div (n − x);
            end if;
        end loop;
    end if;
    return RANK;
end K_SUBSET_TO_RANK;
```

■ **Example 2.6.7**   Trace Algorithm 2.6.6 for $n = 6$ and $S = \{2, 4, 6\}$.

Solution
Before the main loop is executed, RANK = 0, $k = 3$, HCOST = 10, $y = 1$, and PATH = (FALSE, TRUE, FALSE, TRUE, FALSE, TRUE). We now execute the main loop for $x = 1, 2, 3, 4, 5$.

Step 1: $x = 1$. Since PATH(1) = FALSE, we have

RANK := RANK + HCOST = 10

HCOST := $(((n - x) - (k - y)) * \text{HCOST})$ div $(n - x) = ((5 - 2) * 10)$ div $5 = 6$

Step 2: $x = 2$. Since PATH(2) = TRUE, we have

HCOST := $((k - y) * \text{HCOST})$ div $(n - x) = (2 * 6)$ div $4 = 3$

$y := y + 1 = 2$

Step 3: $x = 3$. Since PATH(3) = FALSE, we have

RANK := RANK + HCOST = 10 + 3 = 13

HCOST := $(((n - x) - (k - y)) * \text{HCOST})$ div $(n - x) = ((3 - 1) * 3)$ div $3 = 2$

Step 4: $x = 4$. Since PATH(4) = TRUE, we have

HCOST := $((k - y) * \text{HCOST})$ div $(n - x) = (1 * 2)$ div $2 = 1$

$y := y + 1 = 3$

Step 5: $x = 5$. Since PATH(5) = FALSE, we have

RANK := RANK + HCOST = 13 + 1 = 14

HCOST := $(((n - x) - (k - y)) * \text{HCOST})$ div $(n - x) = ((1 - 0) * 1)$ div $1 = 1$

Thus, the value RANK = 14 is returned. ∎

The unranking algorithm is similar in spirit to Algorithm 2.6.3. As in that algorithm, we let BUDGET hold the amount of dollars we have to spend at a given step
so that the initial value of BUDGET is the given rank r. Also, S is initially empty,
and we begin in the digraph G_n at vertex $(0, 0)$.

If $k = 0$, then S has the correct value and we are done, so suppose $k > 0$. If
BUDGET $<$ HCOST $= C(n - 1, k - 1)$, then we cannot afford to exclude the element 1
from S, for a k-subset of $\{1, 2, \ldots, n\}$ that excludes 1 has rank at least $C(n - 1, k - 1)$,
and this exceeds our budget. So $1 \in S$ and we move to vertex $(1, 1)$ along the diagonal
arc from $(0, 0)$. Since it costs nothing to use this arc, the value of BUDGET does not
change. On the other hand, if BUDGET $\geq$ HCOST, then we can afford to exclude
1 from S and we do so, moving to vertex $(1, 0)$ along the horizontal arc from $(0, 0)$.
Since it costs HCOST dollars to use this arc, the value of BUDGET is reduced by
this amount.

For the general step, we are at some vertex $(x - 1, y - 1)$; we have determined
whether $i \in S$ for each i, $1 \leq i < x$, and the cardinality of S is $y - 1$. If $y - 1 = k$,
then S has been determined and we are done. So suppose $y - 1 < k$. If BUDGET $<$

HCOST = $C(n - x, k - y)$, then we cannot afford to exclude the element x from S. So $x \in S$ and we move to vertex (x, y) along the diagonal arc. If BUDGET $\geq$ HCOST, then we can afford to exclude x from S. So $x \notin S$ and we move to vertex $(x, y - 1)$ along the horizontal arc from $(x - 1, y - 1)$. This costs HCOST dollars, so BUDGET is reduced by HCOST.

ALGORITHM 2.6.7 Given a positive integer n, an integer k with $0 \leq k \leq n$, and an integer r with $0 \leq r \leq C(n, k) - 1$, the function

RANK_TO_K_SUBSET(r, k, n)

returns the k-subset S with lexicographic rank r among the k-subsets of $\{1, 2, \ldots, n\}$.

RANK_TO_K_SUBSET uses the following local variables:
The set S is initialized to EMPTY_SET;
The nonnegative integer BUDGET is initialized to r;
The nonnegative integer HCOST is initialized to $C(n - 1, k - 1)$;
The positive integer x is initialized to 1;
The positive integer y is initialized to 1;

```
begin
    if k > 0 then
        loop
            if BUDGET < HCOST then
                APPEND(x, S);
                exit when y = k;
                HCOST := ((k - y) * HCOST) div (n - x);
                y := y + 1;
            else
                BUDGET := BUDGET - HCOST;
                HCOST := (((n - x) - (k - y)) * HCOST) div (n - x);
            end if;
            x := x + 1;
        end loop;
    end if;
    return S;
end RANK_TO_K_SUBSET;
```

■ *Example 2.6.8* Trace Algorithm 2.6.7 for $n = 6$, $k = 3$, and $r = 14$.

Solution
Initially, $S = \{\ \}$, BUDGET = 14, HCOST = 10, $x = 1$, and $y = 1$. We now execute the main loop until CARDINALITY(S) = k.

Step 1: Since BUDGET $\geq$ HCOST, we have

BUDGET := BUDGET $-$ HCOST = 14 $-$ 10 = 4
HCOST := $(((n - x) - (k - y)) * \text{HCOST})$ div $(n - x) = ((5 - 2) * 10)$ div $5 = 6$

and $x := x + 1 = 2$.

Step 2: Since BUDGET $<$ HCOST, 2 is appended to S so that $S = \{2\}$,

HCOST := $((k - y) * \text{HCOST})$ div $(n - x) = (2 * 6)$ div $4 = 3$
$y := y + 1 = 2$

and $x := x + 1 = 3$.

Step 3: Since BUDGET $\geq$ HCOST, we have

BUDGET := BUDGET $-$ HCOST = 4 $-$ 3 = 1
HCOST := $(((n - x) - (k - y)) * \text{HCOST})$ div $(n - x) = ((3 - 1) * 3)$ div $3 = 2$

and $x := x + 1 = 4$.

Step 4: Since BUDGET $<$ HCOST, 4 is appended to S so that $S = \{2, 4\}$,

HCOST := $((k - y) * \text{HCOST})$ div $(n - x) = (1 * 2)$ div $2 = 1$
$y := y + 1 = 3$

and $x := x + 1 = 5$.

Step 5: Since BUDGET $\geq$ HCOST, we have

BUDGET := BUDGET $-$ HCOST = 1 $-$ 1 = 0
HCOST := $(((n - x) - (k - y)) * \text{HCOST})$ div $(n - x) = ((1 - 0) * 1)$ div $1 = 1$

and $x := x + 1 = 16$.

Step 6: Since BUDGET $<$ HCOST, 6 is appended to S so that $S = \{2, 4, 6\}$. At this point $y = 3 = k$, so the main loop is exited and $S = \{2, 4, 6\}$ is returned.   ∎

The last problem discussed in this section concerns the random selection of a k-subset S of $\{1, 2, \ldots, n\}$. We may assume that $2k \leq n$. (Why?) We discuss two different algorithms for solving this problem, with a third method left to the exercises.

Many states (provinces, countries) have a lottery in which a player essentially chooses a subset of k numbers from $\{1, 2, \ldots, n\}$ for some positive integers k and n. Some areas have automated systems, including computers, that allow players to enter the lottery. Then a player may either decide which subset of numbers to play or have the computer select the subset. The computer should then select a k-subset of $\{1, 2, \ldots, n\}$ uniformly at random.

Equivalently, we may consider the problem of choosing a path in the digraph G_n from $(0, 0)$ to (n, k) uniformly at random. We begin, in G_n, at the vertex $(0, 0)$. There are two possible ways to go—along the diagonal arc to $(1, 1)$ or along the horizontal arc to $(1, 0)$—and these correspond to the choice of whether to include the element 1 in S or not. The number of k-subsets of $\{1, 2, \ldots, n\}$ that include 1 is $C(n-1, k-1)$, whereas the total number of k-subsets is $C(n, k)$. Therefore, since each subset is to have the same chance of being selected, we have

$$\text{Prob}(1 \in S) = \text{the probability that } 1 \in S = \frac{C(n-1, k-1)}{C(n, k)} = \frac{k}{n}$$

Hence, the diagonal arc to $(1, 1)$ is selected with probability k/n (and the horizontal arc to $(1, 0)$ is selected with the complementary probability $(n-k)/n$).

In general, at step $x \geq 1$, we have decided whether to include the element i in S for $1 \leq i < x$, and we are at the vertex $(x-1, y-1)$ of G_n, where $y-1$ is the number of elements included in S so far. We let $k' = k - y + 1$ be the number of elements needed to complete S and let $n' = n - x + 1$ be the number of remaining elements from which these k' elements are to be chosen. Then the probability that $x \in S$ is the probability that a path in G_n from $(x-1, y-1)$ to (n, k) uses the diagonal arc from $(x-1, y-1)$ to (x, y). Since the number of paths from $(x-1, y-1)$ to (n, k) that use this arc is $C(n-x, k-y)$, and the total number of paths is $C(n-x+1, k-y+1)$, the probability we are seeking is

$$\frac{C(n-x, k-y)}{C(n-x+1, k-y+1)} = \frac{C(n'-1, k'-1)}{C(n', k')} = \frac{k'}{n'}$$

Thus, our algorithm has (potentially) n steps, where at step x, $1 \leq x \leq n$, the element x is included in S with probability k'/n'. Initially, $k' = k$ and $n' = n$; n' is decremented after each step, but k' is decremented only when it is determined that $x \in S$. Note that the algorithm can halt when $k' = 0$.

ALGORITHM 2.6.8 Given a positive integer n and an integer k, $0 \leq k \leq n$, the function

RANDOM_K_SUBSET1(k, n)

returns a k-subset of $\{1, 2, \ldots, n\}$ selected uniformly at random.

RANDOM_K_SUBSET1 uses the following local variables:

The set S is initialized to EMPTY_SET;
The nonnegative integer n' is initialized to n;
The nonnegative integer k' is initialized to k;

```
begin
    while k' > 0 loop
        if RANDOM_INTEGER(1, n') ≤ k' then
            APPEND(n + 1 − n', S);
            k' := k' − 1;
        end if;
        n' := n' − 1;
    end loop;
    return S;
end RANDOM_K_SUBSET1;
```

For given values of n and k, we can show that Algorithm 2.6.8 makes $k(n + 1)/(k + 1)$ calls to RANDOM_INTEGER, on average, so that Algorithm 2.6.8 executes in time $\Theta(n)$, on average. (See Exercise 14.)

Algorithm 2.6.8 considers the numbers $1, 2, 3, \ldots$ in turn, deciding whether each should be included in S and halting when $|S| = k$. Another approach is to proceed as follows.

As usual, S is initially empty. As a general step, we choose an integer x from $U = \{1, 2, \ldots, n\}$ uniformly at random and replace S with $S \cup \{x\}$. We continue in this way until $|S| = k$. It is obvious that every element of U has the same chance of being selected for S, so it is clear that this method selects S uniformly at random.

In order to implement this algorithm, we assume that our basic set package includes a procedure

INSERT(x, S)

This procedure updates S to $S \cup \{x\}$. (Thus, if $x \in S$ before the operation is applied, then S is unchanged by the operation.)

ALGORITHM 2.6.9   Given a positive integer n and an integer k, $0 \le k \le n$, the function

RANDOM_K_SUBSET2(k, n)

returns a k-subset of $\{1, 2, \ldots, n\}$ selected uniformly at random.

RANDOM_K_SUBSET2 uses the following local variables:

The set S is initialized to EMPTY_SET;
x is a positive integer;

```
begin
    while CARDINALITY(S) < k loop
        x := RANDOM_INTEGER(1, n);
        INSERT(x, S);
    end loop;
    return S;
end RANDOM_K_SUBSET2;
```

It can be shown that Algorithm 2.6.9 makes

$$n\left(\frac{1}{n-k+1} + \frac{1}{n-k+2} + \cdots + \frac{1}{n}\right)$$

calls to RANDOM_INTEGER, on average. For k fixed, $2k \leq n$, this expression is $\Theta(k)$. That's the good news. The bad news is that using the INSERT operation to build S causes the algorithm's execution time to grow to $\Theta(k^2)$, on average (assuming an ordered linear data structure is used to implement sets, as described earlier). See Exercise 15.

In conclusion, if k is small compared with n, so that k^2 is much less than n, then Algorithm 2.6.9 executes faster, on average, than Algorithm 2.6.8. On the other hand, if k^2 is much greater than n, then Algorithm 2.6.8 is preferable. For example, if $n = 1000$, then Algorithm 2.6.8 is better for $k \simeq 100$, whereas Algorithm 2.6.9 is better for $k \simeq 10$.

In Exercise 16 we explore a rather more efficient, but more complicated, algorithm for this problem that executes in average time $\Theta(k)$.

In this and the previous section you have been introduced to the problem of listing, ranking, and unranking combinatorial objects. Of course, similar algorithms can be considered for other families of combinatorial objects besides permutations and combinations, and for various ways of ordering the objects in a given family. For example, a "lexicographic order" for partitions of $\{1, 2, \ldots, n\}$ can be defined and then we can consider such operations as NEXT_PARTITION, PARTITION_TO_RANK, RANK_TO_PARTITION, and RANDOM_PARTITION relative to this lexicographic order. Some of these ideas are explored further in the exercises and chapter problems.

EXERCISE SET 2.6

1. For $n = 6$, apply Algorithm 2.6.1 to find the lex order successor of each of the given subsets.

   (a) $\{1, 3, 5\}$ (b) $\{1, 2, 4, 6\}$ (c) $\{2, 4, 5, 6\}$ (d) $\{6\}$

2. Implement Algorithm 2.6.2 without making (explicit) use of the PATH array.

3. For $n = 5$, apply Algorithm 2.6.2 to find the lexicographic rank of each of the given subsets.

(a) { }   (b) $\{1, 3\}$ (c) $\{2, 4, 5\}$ (d) $\{5\}$

4. Implement Algorithm 2.6.6 without (explicit) use of the PATH array.

5. For $n = 5$, apply Algorithm 2.6.3 to find the subset S with the given lexicographic rank r.

(a) 0   (b) 11   (c) 24   (d) 31

6. As mentioned, one way to implement the abstract data type SET is to think of a set as an ordered list, implemented as a record type with two fields, such as

> type ELEMENT_ARRAY_TYPE is array (1 .. MAX) of POSITIVE;
> type SET is record
>   CARDINALITY: NATURAL;
>   ELEMENT: ELEMENT_ARRAY_TYPE;
> end record;

The idea is that for an object S of type SET, S.CARDINALITY is the cardinality of S and, if S.CARDINALITY > 0, then

> S.ELEMENT(1) < S.ELEMENT(2) < $\cdots$
>   < S.ELEMENT(S.CARDINALITY)

are the elements of S (arranged in increasing order). Given this, discuss the implementation of the CARDINALITY, EMPTY_SET, ELEMENT, APPEND, DELETE_LARGEST, INCREMENT_LARGEST, and INSERT operations.

7. For $n = 8$, apply Algorithm 2.6.5 to find the 4-subset that is the lex order successor of each of the given 4-subsets.

(a) $\{1, 2, 3, 4\}$ (b) $\{1, 3, 4, 7\}$
(c) $\{2, 4, 7, 8\}$ (d) $\{5, 6, 7, 8\}$

8. For a k-subset $\{x_1, x_2, \ldots, x_k\}$ of $\{1, 2, \ldots, n\}$, its (lexicographic) **corank** is the number of k-subsets that succeed it in lexicographic order. Develop an algorithm to determine the corank of a given k-subset of $\{1, 2, \ldots, n\}$. Try to do this directly, without using (either implicitly or explicitly) K_SUBSET_TO_RANK.

9. For $n = 8$, apply Algorithm 2.6.6 to find the lexicographic rank of each of the given 4-subsets.

(a) $\{1, 2, 3, 4\}$ (b) $\{1, 3, 4, 7\}$
(c) $\{2, 4, 7, 8\}$ (d) $\{5, 6, 7, 8\}$

10. Given a positive integer n and an integer r, $0 \leq r \leq C(n, k) - 1$, show that r is uniquely expressed as

$$\sum_{i=1}^{k} C(m_i, i)$$

where $0 \leq m_1 < m_2 < \cdots < m_k \leq n - 1$. (Hint: Let r be the lexicographic corank of the k-subset $\{x_1, x_2, \ldots, x_k\}$ of $\{1, 2, \ldots, n\}$, and let $m_i = n - x_{k+1-i}$; see Exercise 8.) See also Problem 42 in Chapter 1.

11. For $n = 8$, apply Algorithm 2.6.7 to find the 4-subset S with the given lexicographic rank r.

(a) 0   (b) 27   (c) 63   (d) 69

12. Express the function NEXT_SUBSET of Algorithm 2.6.1 in terms of the SUBSET_TO_RANK and RANK_TO_SUBSET functions of Algorithms 2.6.2 and 2.6.3, respectively. Discuss the efficiency of this method.

13. Express the function NEXT_K_SUBSET of Algorithm 2.6.5 in terms of the K_SUBSET_TO_RANK and RANK_TO_K_SUBSET functions of Algorithms 2.6.6 and 2.6.7, respectively. Discuss the efficiency of this method.

⋆14. For given values of n and k, show that Algorithm 2.6.8 makes $k(n+1)/(k+1)$ calls to RANDOM_INTEGER, on average, so that Algorithm 2.6.8 executes in time $\Theta(n)$, on average. As a hint, consider the following equivalent problem. A box contains n balls; k are red and $n - k$ are blue. Balls are selected from the box at random, one at a time, without replacement. Let X be the number of selections required to obtain all k red balls. Find the expected value of X. (Note that, if $X = x$, then the last ball selected is red and the first $x - 1$ balls selected constitute a random sample from the box with $k - 1$ red balls and $x - k$ blue balls; thus, the probability that $X = x$ involves a hypergeometric probability.)

⋆15. Concerning Algorithm 2.6.9:

(a) Show that

$$n\left(\frac{1}{n-k+1} + \frac{1}{n-k+2} + \cdots + \frac{1}{n}\right)$$

calls are made to RANDOM_INTEGER, on average. As a hint, let $X_1 = 1$ be the number of calls to RANDOM_INTEGER required to obtain the first element for the subset, and for $1 < i \leq k$, assuming that $i - 1$ elements for the subset have been chosen, let the random variable X_i be the number of additional calls to RANDOM_INTEGER required to obtain the ith element for the subset. Note that X_i has a geometric distribution with probability of success $(n - i + 1)/n$; we want

$$E(X_1 + X_2 + \cdots + X_k) = E(X_1) + E(X_2) + \cdots + E(X_k)$$

(b) For k fixed, $2k \leq n$, show that the expression in part (a) is $\Theta(k)$.

⋆16. Here is another method for selecting a k-subset of $\{1, 2, \ldots, n\}$ at random. This method uses a "divide and conquer" approach. (This method is discussed in *Combinatorial Algorithms*, 2d ed., by Nijenhuis and Wilf.) It executes in time $\Theta(k)$ and uses an auxiliary array of k integers. We call this function

RANDOM_K_SUBSET3(k, n); it uses the following local variables:

c and LEFT are nonnegative integers initialized to 0;
$(\text{AUX}(1), \text{AUX}(2), \ldots, \text{AUX}(k))$ is an array of nonnegative integers;
i, j, and RIGHT are positive integers;
S and T are sets initialized to EMPTY_SET;

```
begin
      -- Partition {1, 2, ..., n} into k "bins" (of consecutive integers) of
      -- approximately equal size. AUX(j) - ((j - 1) * n)/k is the number of
      -- elements to be selected from the jth bin. Initially,
      -- AUX(j) = ((j - 1) * n) div k.
   for j in 1 .. k loop
      AUX(j) := ((j - 1) * n) div k;
   end loop;
      -- Decide how many elements to select from each bin.
   while c < k loop
      i := RANDOM_INTEGER(1, n);
      j := 1 + ((k * i) - 1) div n;
      if i > AUX(j) then
         c := c + 1;
         AUX(j) := AUX(j) + 1;
      end if;
   end loop;
      -- Select the elements for each bin using RANDOM_K_SUBSET2.
   for j in 1 .. k loop
      RIGHT := (j * n) div k;
      T := RANDOM_K_SUBSET2(AUX(j) - LEFT, RIGHT - LEFT);
      for i in 1 .. CARDINALITY(T) loop
         APPEND(LEFT + ELEMENT(i, T), S);
      end loop;
      LEFT := RIGHT;
   end loop;
   return S;
end RANDOM_K_SUBSET3;
```

(a) For a given $x \in \{1, 2, \ldots, n\}$, show that this method selects x for the subset with probability k/n. As a hint, let BIN(x) be the "bin" to which x belongs, let X be the event that x is selected for the subset, and let E_m be the event that m numbers are selected from BIN(x). Also, let t be the cardinality of BIN(x). Then the probability of X can be found using the law of total probability:

$$\text{Prob}(X) = \sum_{m=0}^{t} \text{Prob}(X \mid E_m) \, \text{Prob}(E_m)$$

Argue that $\text{Prob}(X \mid E_m) = m/t$, and employ the hypergeometric distribution to find $\text{Prob}(E_m)$.

(b) Argue that this method executes in time $\Theta(k)$.

17. Algorithms 2.6.1, 2.6.2, and 2.6.3 are based on the lex order list of the subsets of $\{1, 2, \ldots, n\}$. A different set of algorithms for listing, ranking, and unranking these subsets is based on the so-called Gray code. This method has the advantage that the successor of a given subset S is obtained from S by the insertion or deletion of a single element. For example, here are Gray code lists for $n = 1, 2, 3$:

$$\{\,\}, \{1\}$$
$$\{\,\}, \{1\}, \{1,2\}, \{2\}$$
$$\{\,\}, \{1\}, \{1,2\}, \{2\}, \{2,3\}, \{1,2,3\}, \{1,3\}, \{3\}$$

(a) A Gray code list L_{n+1} of subsets of $\{1, 2, \ldots, n+1\}$ is constructed recursively from the list L_n of subsets of $\{1, 2, \ldots, n\}$ as follows: Let L_n' be the reverse of L_n and let L_n'' be the list obtained by appending the element $n+1$ to each member of L_n'; then L_{n+1} is the catenation of L_n with L_n''. Use this rule to construct a Gray code list for $n = 4$ from the list for $n = 3$ given above.

Let the method given in part (a) for finding a Gray code list of the subsets of $\{1, 2, \ldots, n\}$ be fixed. Given a subset S of $\{1, 2, \ldots, n\}$, its **Gray rank** is an integer between 0 and $2^n - 1$ giving the number of subsets that precede S on the Gray code list of the subsets of $\{1, 2, \ldots, n\}$.

⋆(b) Develop an algorithm that determines the Gray rank of a given subset S of $\{1, 2, \ldots, n\}$.

⋆(c) Develop an algorithm that, given a positive integer n and an integer r, $0 \le r < 2^n$, finds the subset S of $\{1, 2, \ldots, n\}$ with Gray rank r.

⋆(d) Develop an algorithm that inputs n and outputs the Gray code list of the subsets of $\{1, 2, \ldots, n\}$.

18. Another way to select a k-subset of $\{1, 2, \ldots, n\}$ at random is to first select a random integer r between 0 and $C(n, k) - 1$ and then use RANK_TO_K_SUBSET to convert r to a k-subset. Discuss the advantages and/or disadvantages of this approach.

CHAPTER 2 PROBLEMS

1. How many permutations of the letters a, b, c, d, e contain neither the word "bad" nor the word "ade"?

2. Find the number of solutions of $x_1 + x_2 + x_3 = 11$ where each x_i is a nonnegative integer and $x_1 \le 3$, $x_2 \le 4$, and $x_3 \le 5$.

3. How many permutations $(x_1, x_2, \ldots, x_n)$ of $\{1, 2, \ldots, n\}$ are such that 2 does not immediately follow 1, 3 does not immediately follow 2, $\ldots$, and n does not immediately follow $n - 1$? Call this number $q(n)$.

   (a) Find $q(1)$, $q(2)$, and $q(3)$.
   (b) Use a tree diagram to compute $q(4)$.
   (c) Use the principle of inclusion–exclusion to develop a general formula for $q(n)$.
   (d) Show that $q(n) = d(n - 1) + d(n)$, where $d(n)$ is the number of derangements of $\{1, 2, \ldots, n\}$.

4. Find the number of permutations of $\{1, 2, \ldots, 8\}$ that satisfy the following conditions:

   (a) No element is mapped to itself.
   (b) Each odd element is mapped to an odd element and no even element is mapped to itself.
   (c) 1 is not mapped to itself and 8 is not mapped to itself.

5. How many circular permutations $[x_1, x_2, \ldots, x_n]$ of $\{1, 2, \ldots, n\}$ are such that 2 does not immediately follow 1, 3 does not immediately follow 2, $\ldots$, n does not immediately follow $n - 1$, and 1 does not immediately follow n? Call this number $q^*(n)$.

   (a) Find $q^*(1)$, $q^*(2)$, $q^*(3)$, and $q^*(4)$.
   (b) Use a tree diagram to compute $q^*(5)$.
   (c) Use the principle of inclusion–exclusion to develop a general formula for $q^*(n)$.

6. Given a permutation $\alpha = (x_1, x_2, \ldots, x_n)$ of $\{1, 2, \ldots, n\}$, if $x_i = j$ and $x_j = i$ for some $1 \le i < j \le n$, then α is said to have an i, j-**interchange** . Let $c(n)$ denote the number of derangements of $\{1, 2, \ldots, n\}$ with at least one interchange.

   (a) Find $c(n)$.
   (b) Show that

   $$\lim_{n \to \infty} \frac{c(n)}{d(n)} = 1$$

7. Let $d(n)$ be the number of derangements of $\{1, 2, \ldots, n\}$. Show that $d(n)$ satisfies the recurrence equation

   $$d(n) = nd(n - 1) + (-1)^n$$

8. For a positive integer n, let $d_k(n)$ be the number of permutations $(x_1, x_2, \ldots, x_n)$ of $\{1, 2, \ldots, n\}$ such that $x_i = i$ for exactly k values of i, $1 \le i \le n$. Exercise 23 in Exercise Set 2.1 asks for a formula for $d_k(n)$. Show that

   $$\lim_{n \to \infty} \frac{d_k(n)}{n!} = \frac{1}{ek!}$$

9. Use the principle of inclusion–exclusion to verify the identity

$$\sum_{k=0}^{m}(-1)^k C(n,k)C(n-k,m-k) = 0$$

where m and n are positive integers and $m \le n$.

10. For a positive integer n, let $\Phi(n)$ denote the number of positive integers less than n that are relatively prime to n; Φ is called the **Euler totient function**. For example, 1, 5, 7, and 11 are relatively prime to 12, so $\Phi(12) = 4$. Use the principle of inclusion–exclusion to compute

    (a) $\Phi(675)$ (b) $\Phi(756)$ (c) $\Phi(462)$

11. How many permutations of the letters in TORONTO have no two consecutive letters the same?

12. How many permutations of the letters in COMBINATORICS have both C's before both I's, both I's before both O's, and both O's before the T?

13. A mathematician has four colleagues. During a summer workshop, she had lunch alone 6 times; she lunched with any colleague 7 times, with any subset of two colleagues 4 times, with any subset of three colleagues 3 times, and with all four colleagues on 2 occasions. How many days did the workshop last (assuming that the mathematician had lunch each day)?

14. A simple coding scheme involves the replacement of each alphanumeric character by an alphanumeric character such that no two characters are encoded the same. In how many ways can this be done? In how many ways can it be done if no character is encoded as itself?

15. In how many ways can 30 slices of pizza be distributed to 5 fraternity brothers if Curly gets at least 3 slices, Larry gets at most 4 slices, Moe gets between 2 and 6 slices (inclusive), Joe gets between 4 and 7 slices, and Al gets between 1 and 8 slices?

16. Consider Pascal's triangle with $n + 1$ rows; let $g(n)$ denote the sum of the numbers that lie on the median from the lower left-hand corner to the opposite side. Then

$$g(n) = \sum_{k=0}^{n} C(n-k,k)$$

Show that $g(n) = f(n)$, where $f(n)$ is the nth Fibonacci number. (Hint: Note that $g(0) = g(1) = 1$; show that $g(n)$ satisfies the Fibonacci recurrence equation $g(n) = g(n-2) + g(n-1)$, $n \ge 2$.)

17. Let $f(n)$ denote the nth Fibonacci number. Show that

$$\sum_{k=0}^{n} f^2(k) = f(n)f(n+1)$$

18. Let $F(m,n)$ be the number of ways to distribute m identical slices of pizza to n fraternity brothers, with each brother getting 2, 3, or 4 slices. Find initial values and a recurrence equation for $F(m,n)$.

19. A wizard must climb an infinite staircase whose steps are numbered with the positive integers. At each step, the wizard may go up one stair or three. Let $w(n)$ be the number of ways for the wizard to reach step number n.

(a) Find $w(1)$, $w(2)$, and $w(3)$.
(b) Find a recurrence equation for $w(n)$, $n \geq 4$.

20. Find initial values and a recurrence relation for the number of ternary sequences (sequences of 0s, 1s, and 2s) of length n in which the first 0 (if any) occurs before the first 1 (if any).

21. Let $g_3(n)$ be the number of ways to partition a set having cardinality $3n$ into n subsets with 3 elements in each subset. Find initial values and a recurrence equation for $g_3(n)$.

22. Generalize Problem 21: let $g_k(n)$ be the number of ways to partition a set having cardinality kn into n subsets with k elements in each subset. Find initial values and a recurrence equation for $g_k(n)$.

23. Each of the following parts recursively defines a function from $\mathbb{N}$ to $\mathbb{Z}$. Find an explicit formula for the function.

(a) $h_1(0) = 1$, $h_1(1) = -2$, $h_1(n) = 5h_1(n-1) - 6h_1(n-2)$, $n \geq 2$
(b) $h_2(0) = 0$, $h_2(1) = 1$, $h_2(n) = 4h_2(n-2)$, $n \geq 2$
(c) $h_3(0) = 0$, $h_3(1) = 1$, $h_3(2) = 2$, $h_3(n) = h_3(n-1) + 9h_3(n-2) - 9h_3(n-3)$, $n \geq 3$
(d) $h_4(0) = -1$, $h_4(1) = 0$, $h_4(n) = 8h_4(n-1) - 16h_4(n-2)$, $n \geq 2$
(e) $h_5(0) = 1$, $h_5(1) = 0$, $h_5(2) = 0$, $h_5(n) = 3h_5(n-2) - 2h_5(n-3)$, $n \geq 3$
(f) $h_6(0) = 1$, $h_6(1) = -3$, $h_6(2) = 25$, $h_6(n) = -h_6(n-1) + 5h_6(n-2) - 3h_6(n-3)$, $n \geq 3$
(g) $h_7(0) = 1$, $h_7(n) = 2nh_7(n-1)$, $n \geq 1$
(h) $h_8(0) = 1$, $h_8(1) = 7$, $h_8(2) = 65$, $h_8(n) = 5h_8(n-1) - 3h_8(n-2) - 9h_8(n-3)$, $n \geq 3$

24. Let us consider a general **linear nonhomogeneous recurrence equation** with constant coefficients:

$$h(n) - a_1 h(n-1) - a_2 h(n-2) - \cdots - a_r h(n-r) = g(n) \qquad (*)$$

Suppose that $h_0(n)$ is the general solution of the corresponding homogeneous recurrence equation and that $p(n)$ is any **particular solution** of the nonhomogeneous recurrence equation $(*)$. Show that $h_0(n) + p(n)$ is a solution of $(*)$; this is called a **general solution** of $(*)$.

25. Each of the following parts recursively defines a function from $\mathbb{N}$ to $\mathbb{Z}$ and gives a particular solution $p(n)$ to the recurrence equation. Use the result of Problem 24 to find an explicit formula for the function.

(a) $f_7(0) = 5$, $f_7(n) = 3f_7(n-1) - 8$, $p(n) = 4$, $n \geq 1$
(b) $f_8(0) = 1$, $f_8(n) = 2f_8(n-1) - n$, $p(n) = n + 2$, $n \geq 1$
(c) $f_9(0) = 9$, $f_9(n) = -2f_9(n-1) + 27(n^2 + n)$, $p(n) = 9n^2 + 21n + 8$, $n \geq 1$
(d) $f_0(0) = 2$, $f_0(n) = -3f_0(n-1) + 4^n$, $p(n) = 4^{n+1}/7$, $n \geq 1$

26. Suppose that $f : \mathbb{N} \to \mathbb{Z}$ and $g : \mathbb{N} \to \mathbb{Z}$ are functions and, for all $n \geq 1$,

$$f(n) - f(n-1) = g(n) - g(n-1)$$

Prove that $f(n) = g(n) + c$ for some constant $c \in \mathbb{Z}$. (Hint: Consider $h(n) = f(n) - g(n)$.)

27. Find the generating function for each of the following sequences:

  (a) $(1, 2, 4, \ldots, 2^n, \ldots)$ (b) $(1, 5, 11, \ldots, n^2 + 3n + 1, \ldots)$

  (c) $(1, 5, 15, \ldots, C(n+4, 4), \ldots)$ (d) $(1, 3, 5, \ldots, 2n + 1, \ldots)$

28. Consider the function $b : \mathbb{N} \to \mathbb{Z}$ defined recursively by $b(0) = 1$, $b(n) = 2b(n-1) + n$, $n \geq 1$.

  (a) Show that

$$g(x) = \frac{1 - x + x^2}{(1 - 2x)(1 - x)^2}$$

  is the generating function for the sequence $(b(0), b(1), \ldots, b(n), \ldots)$.
  (b) Find an explicit formula for $b(n)$.

29. Consider the function $c : \mathbb{N} \to \mathbb{Z}$ defined recursively by $c(0) = 9$, $c(n) = 27(n^2 + n) - 2c(n-1)$, $n \geq 1$.

  (a) Show that the generating function $g(x)$ for the sequence $(c(0), c(1), \ldots, c(n), \ldots)$ satisfies $(1 + 2x)g(x) = 9 + 54x/(1 - x^3)$.
  (b) Find an explicit formula for $c(n)$.

30. Consider the function $a : \mathbb{N} \to \mathbb{Z}$ defined recursively by $a(0) = 1$, $a(n) = 3a(n-1) + 8C(n, 2)$, $n \geq 1$.

  (a) Show that

$$g(x) = \frac{-x^3 + 11x^2 - 3x + 1}{(1 - 3x)(1 - x)^3}$$

  is the generating function for the sequence $(a(0), a(1), \ldots, a(n), \ldots)$.
  (b) Find an explicit formula for $a(n)$.

31. Let n be a nonnegative integer and let U be the set of solutions $(x_1, x_2, \ldots, x_6)$ in nonnegative integers to

$$x_1 + x_2 + x_3 + x_4 + x_5 + x_6 = n$$

  (a) Let $c_1(n)$ be the number of solutions with x_k a multiple of k, $1 \leq k \leq 6$. Give the generating function for the sequence $(c_1(0), c_1(1), \ldots, c_1(n), \ldots)$.
  (b) Let $c_2(n)$ be the number of solutions such that each $x_k \leq 3$. Give the generating function for the sequence $(c_2(0), c_2(1), \ldots, c_2(n), \ldots)$.
  (c) Find $c_2(17)$.

32. A particle is initially at the origin of a number line. At each step, the particle can move 1 unit right, remain where it is, or move 1 unit left. The probability that the particle moves right at any step is r, the probability that it remains at the same place is s, and the probability that it moves left is $g = 1 - r - s$. Let $P_n(x)$ be the probability that the partitcle is at position x after n steps.

(a) Find the initial values $P_0(x)$.

(b) Find a recurrence equation for $P_n(x)$, $n \geq 1$ (in terms of $P_{n-1}(x-1)$, $P_{n-1}(x)$, and $P_{n-1}(x+1)$).

(c) Find $P_3(1)$.

33. Show that the number of partitions of m into three parts is equal to the number of partitions of $2m$ into three parts each less than m.

34. Show that the number of partitions of m into even parts is the same as the number of partitions of m in which each part occurs with even multiplicity (i.e., each part is repeated an even number of times).

35. Find:

(a) $p(15, 8)$ (b) $p(18, 9)$

36. Show that if a partition π of m has distinct parts, then its transpose has consecutive parts. What about the converse?

37. Let $q(m, n)$ denote the number of partitions of m into n distinct parts. Show that:

(a) $q(m, 1) = 1$

(b) $q(m, n) = 0$ if $m < C(n + 1, 2)$

(c) $q(m, n) = q(m - n, n) + q(m - n, n - 1)$, $1 < n < m$

(d) $n! \, q(m, n) \leq C(m - 1, n - 1)$

38. Show that the number of partitions of m into at most n parts is $q(m + C(n + 1, 2), n)$. (See the preceding problem.)

39. Let $q(n)$ denote the number of partitions of n into distinct parts and let $q_1(n)$ denote the number of partitions of n into odd parts.

(a) Find the generating function for the sequence $(0, 1, \ldots, q(n), \ldots)$.

(b) Find the generating function for the sequence $(0, 1, \ldots, q_1(n), \ldots)$.

(c) Show that $q(n) = q_1(n)$.

★40. Show that the number of partitions of m into an even number of distinct parts differs from the number of partitions of m into an odd number of distinct parts by at most 1, and determine those values of m for which the difference is 0. This result was first discovered by Euler. (Hint: Given a partition $\pi = x_1 + x_2 + \cdots + x_n$ of m with $x_1 > x_2 > \cdots > x_n$, let $\sigma(\pi)$ denote the largest j for which $x_j = x_1 - j + 1$; if $x_n \leq \sigma(\pi)$, map π to the partition π' of m obtained by adding 1 to each of the x_n largest parts of π and deleting x_n; if $x_n > \sigma(\pi)$, map π to the partition π' of m obtained by subtracting 1 from each of the $\sigma(\pi)$ largest parts of π and inserting a new smallest part of size $\sigma(\pi)$. Note that if π has an even number of distinct parts, then π' has an odd number of distinct parts.)

41. Suppose that the function g has the following representation as a power series on $(-r, r)$, where r is a positive real number:

$$\sum_{n=0}^{\infty} c(n) \frac{x^n}{n!}$$

Then g is called an **exponential generating function** (egf) for the sequence $(c_0, c_1, \ldots, c_n, \ldots)$. Show the following:

(a) $g_1(x) = 1/(1 - x)$ is the egf for $(0, 1, \ldots, n!, \ldots)$
(b) $g_2(x) = (1 + x)^m$ is the egf for $(1, m, \ldots, P(m, n), \ldots)$
(c) $g_3(x) = e^{mx}$ is the egf for $(1, m, \ldots, m^n, \ldots)$
(d) $g_4(x) = e^{-x}/(1 - x)$ is the egf for $(1, 0, \ldots, d(n), \ldots)$, where $d(0) = 1$ and, for $n \geq 1$, $d(n)$ is the number of derangements of $\{1, 2, \ldots, n\}$

42. Let m and n be integers with $1 \leq n \leq m$. Consider the number of partitions of $\{1, 2, \ldots, m\}$ into n permutations; for example, $\{(1, 4, 2), (3, 5)\}$ is a partition of $\{1, 2, 3, 4, 5\}$ into 2 permutations. Equivalently, consider the number of ways to distribute m distinct balls to n identical boxes such that the order in which balls are distributed to a box matters and no box is empty. Denote this number by $pp(m, n)$.

(a) Show that $pp(m, n) = m! C(m - 1, n - 1)/n!$.
(b) Show that $pp(m, n) = C(m, n) P(m - 1, m - n)$.

43. With $pp(m, n)$ as in the preceding problem, note that $pp(m, 1) = m!$ and $pp(m, m) = 1$. Find a recurrence equation for $pp(m, n)$ in terms of $pp(m-1, n-1)$ and $pp(m - 1, n)$ that holds for $1 < n < m$.

44. The number

$$P(n + k - 1, k) = n(n + 1) \cdots (n + k - 1)$$

is also denoted by $(n)^k$ and is called a **rising factorial power of n of degree k** (see Exercise 6 in Exercise Set 2.4). With $pp(m, k)$ as in the preceding problem, show that

$$(n)^m = \sum_{k=0}^{m} pp(m, k)(n)_k$$

45. As noted in Exercise 6 of Exercise Set 2.4, the falling factorial power $(n)_k$ is a polynomial of degree k in n. Hence, there exist integers $s(m, k)$, $0 \leq k \leq m$, such that

$$(n)_m = \sum_{k=0}^{m} s(m, k) n^k$$

The numbers $s(m, k)$ are called **Stirling numbers of the first kind**.

(a) Find $s(4, 0)$, $s(4, 1)$, $s(4, 2)$, $s(4, 3)$, and $s(4, 4)$.
(b) Show that $s(m, 0) = 0$ and $s(m, m) = 1$.
(c) Show that $s(m, k) = s(m - 1, k - 1) - (m - 1)s(m - 1, k)$, $\quad 0 < k < m$.

46. Let $X = \{1, 2, \ldots, m\}$ and let $f : X \to X$ be a permutation of X. For $x \in X$, look at the sequence

$$x, f(x), f(f(x)), \ldots$$

of iterates of f at x, and suppose $f^i(x)$, denoting the ith iterate, is the first repeated element of the sequence. Then the circular permutation $[x, f(x), \ldots, f^{i-1}(x)]$ is called a **cycle** of f.

   (a) Show that, if $[x, f(x), \ldots, f^{i-1}(x)]$ is a cycle of f, then $f^i(x) = x$.
   (b) Show that each $x \in X$ belongs to exactly one cycle of f.
   (c) Let $c(m, n)$ be the number of permutations of $\{1, 2, \ldots, m\}$ having exactly n cycles. Show that

$$c(m, 1) = (m - 1)!, \quad c(m, m) = 1,$$

$$c(m, n) = c(m - 1, n - 1) + (m - 1)c(m - 1, n) \quad 1 < n < m$$

   (d) Define $c(m, 0) = 0$ so that the above recurrence equation holds for $0 < n < m$. Show that $c(m, n) = |s(m, n)|$, where $s(m, n)$ is a Stirling number of the first kind, as defined in the preceding problem.

47. Let $S_0(x) = 1$ and, for $n \geq 1$, define

$$S_n(x) = \sum_{m=0}^{\infty} S(m, n)x^m$$

   (a) Show that $S_n(x) = xS_{n-1}(x) + nxS_n(x)$ for $n \geq 1$.
   (b) Show that

$$S_n(x) = \frac{x^n}{(1 - x)(1 - 2x) \cdots (1 - nx)}$$

48. Use generating functions to find the number of solutions (x_1, x_2, x_3, x_4) of $x_1 + x_2 + x_3 + x_4 = n$ with each x_i a nonnegative integer, x_1 even, x_2 a multiple of 5, $x_3 \leq 4$, and $x_4 \in \{0, 1\}$.

49. For k a fixed positive integer, find the generating function for the number of solutions $(x_1, x_2, \ldots, x_k)$ of $x_1 + x_2 + \cdots + x_k = n$ with each $x_i \in \{1, 3, 5, \ldots\}$.

50. Consider the equation $x_1 + x_2 + \cdots + x_{2n} = 0$.

   (a) Find the number of solutions $(x_1, x_2, \ldots, x_{2n})$ with each $x_i \in \{-1, 1\}$.
   (b) Show that the number of solutions with each $x_i \in \{-1, 1\}$, satisfying the condition that

$$x_1 + x_2 + \cdots + x_k \geq 0$$

for each k, $1 \leq k \leq 2n$, is the Catalan number $C(2n, n)/(n + 1)$.

51. In the directed graph G_n (see Exercise 20 in Exercise Set 1.6), find the number of paths from $(0, 0)$ to (n, n) that stay on or below the line $y = x$. (Hint: Apply part (b) of the preceding exercise.)

52. Use generating functions to show that, for positive integers m and n:

(a) $\displaystyle\sum_{k=0}^{n}(-1)^k C(m,k)C(m+2n-2k-1, m-1) = C(m, 2n)$

(b) $\displaystyle\sum_{n=1}^{\infty} n2^{-n} = 2$

(c) $\displaystyle\sum_{n=0}^{\infty} 2^{-n}C(n+m-1, n) = 2^m$

53. Show (for $m > 2$) that

$$S(m, m-2) = \sum_{k=1}^{m-2} kC(k+1, 2)$$

54. Develop a formula for $S(m, 3)$.

55. Show that the number of partitions of $2m + n$ into $m + n$ parts is independent of n.

56. Let $f(n)$ denote the nth Fibonacci number.

(a) Prove that $f(n-1)$ is even if and only if n is a multiple of 3.

(b) Prove that $f(n-1)$ is a multiple of 3 if and only if n is a multiple of 4.

⋆(c) Find similar results to and/or generalize the results of parts (a) and (b).

(d) Prove: If m divides n, then $f(m-1)$ divides $f(n-1)$. What about the converse?

(e) Prove: If $d = \gcd(m, n)$, then $f(d-1) = \gcd(f(m-1), f(n-1))$. What about the converse?

57. For a positive integer n, let $2n$ distinct points be given on a circle. Let $c(n)$ be the number of ways to choose n nonintersecting chords so that each chord connects two of the given points. Then $c(1) = 1$, $c(2) = 2$, $c(3) = 5$, and so on. Find a recurrence equation for $c(n)$.

58. Using calculus, we can show that

$$f_1(n) = \sqrt{2\pi n}\left(\frac{n}{e}\right)^n < n! < \sqrt{2\pi n}\left(\frac{n}{e}\right)^n \sqrt{1 + \frac{1}{2n}} = f_2(n)$$

for every positive integer n.

(a) Make a table of $f_1(n)$, $n!$, and $f_2(n)$ for several values of n, including $n = 2, 4, 8, 16$.

(b) Show that

$$\lim_{n\to\infty} \frac{n!}{f_1(n)} = 1$$

Thus, $f_1(n)$ is approximately equal to $n!$ for n sufficiently large. This approximation is called Stirling's approximation; it was developed about 1730 by James Stirling.

(c) Use Stirling's approximation to approximate 32! and 64!.

(d) Use Stirling's approximation to develop an approximation to $C(2n, n)$.

⋆**59.** Design an algorithm that inputs integers n and k with $0 \le k \le n$ and outputs $C(n, k)$. Note that such an algorithm could be based on (1) the formula for $C(n, k)$ given by Corollary 1.3.7, or (2) Pascal's identity (Theorem 1.6.3), or (3) the relation $C(n, k) = nC(n - 1, k)$ div $(n - k)$, or (4) the relation $C(n, k) = (n - k + 1)C(n, k - 1)$ div k. Discuss the advantages and/or disadvantages of each approach.

⋆**60.** Design algorithms, such as those in Sections 2.5 and 2.6, for generating, ranking, and unranking:

   (a) Ordered partitions of n.
   (b) Unordered partitions of n.
   (c) Ordered partitions of $\{1, 2, \ldots, n\}$.
   (d) Unordered partitions of $\{1, 2, \ldots, n\}$.

Chapter *3*

Graph Theory

3.1 INTRODUCTION TO GRAPHS

This chapter introduces some of the basic concepts and results of graph theory and indicates several of the more important applications of the subject.

Many people date the beginning of graph theory to the year 1736, when the great Swiss mathematician Leonhard Euler published an analysis of the **Königsberg bridges problem**. Although Euler did not use graph theory as we know it, it is clear that he was thinking in graphical terms. Another famous problem that came to involve graph theory is the **four-color problem**, which was first posed in 1852 in a letter from Augustus De Morgan to Sir William Rowan Hamilton. We discuss the Königsberg bridges problem in the next section and return to the four-color problem later in the chapter. Obviously, it was a long time before graph theory developed into the popular and fertile area of mathematical research that it is today; for a long time its main use was as an aid in the solution of mathematical puzzles and games.

Perhaps a good place to start our discussion is with the definition of "graph." We can find several meanings for this term in the mathematical literature; for now, we consider two kinds of "graph" that are particularly natural and useful.

Figure 3.1.1 shows a **simple graph** G_1 and a **directed graph** G_2. Note that both G_1 and G_2 have four distinguished points labeled u, v, x, and y. These points are called **vertices** (the plural of **vertex**). Certain pairs of vertices of G_1 are joined by line segments or simple curves called **edges**, whereas in G_2 certain pairs of vertices are joined by directed line segments or simple curves called **directed edges** or **arcs**. There is also a special type of arc called a **loop** from the vertex u to itself. Basically, the presence of an arc in a directed graph is meant to indicate that one vertex is related to another, whereas an edge in a simple graph indicates that two vertices are symmetrically related.

We now formally define the terms *simple graph* and *directed graph*.

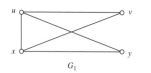
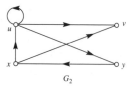

Figure 3.1.1   A simple graph G_1 and a directed graph G_2.

DEFINITION 3.1.1

1. A (**simple**) **graph** G consists of a finite nonempty set V and a set E of 2-element subsets of V. The set V is called the **vertex set** of G, the set E is called the **edge set** of G, and we write $G = (V, E)$ to denote the graph G with vertex set V and edge set E.
2. A **directed graph** or **digraph** G consists of a finite nonempty set V together with a subset A of the product set $V \times V$. We call V the **vertex set** of G and A the **arc set** of G, and we write $G = (V, A)$ to denote the digraph G with vertex set V and arc set A.   ∎

For the remainder of this introduction we will be concerned only with graphs and will return to digraphs in a later section. In general, it can be said that most of the terminology for graphs introduced here and in Chapter 0 carries over to digraphs in a natural way.

Given a graph $G = (V, E)$, each element v of V is called a **vertex** of G (the elements of V are called **vertices**), whereas each element e of E is called an **edge** of G. If $e = \{u, v\}$ is an edge, we denote it simply by $e = uv$ (or $e = vu$). In this case, u and v are referred to as **adjacent vertices**, u and v are said to be **incident** with e, and e is **incident** with u and v. Similarly, if distinct edges e_1 and e_2 of G have a vertex in common, then e_1 and e_2 are called **adjacent edges**. Also, if several graphs are under discussion at the same time, then to avoid confusion we use $V(G)$ and $E(G)$ to denote the vertex set and edge set, respectively, of a particular graph G.

An alternate term for vertex is **node**. Also, some authors use the terms *point* and *line* in place of vertex and edge, respectively, but it is a good idea to reserve the terms point and line for geometry.

As illustrated by Figure 3.1.1, a graph can be represented geometrically, or "drawn" (in a plane), if each vertex is made to correspond to a point and an edge $e = uv$ is represented by joining the points corresponding to u and v by a line segment or simple curve. We make no distinction between a graph $G = (V, E)$ and a drawing of it; sometimes a graph is presented by giving its vertex set and edge set and sometimes by giving a drawing of the graph.

For a graph $G = (V, E)$, the relation of "adjacency" is an irreflexive and symmetric relation on V. It is for this reason that an irreflexive and symmetric relation has a "graphical representation," as discussed in Chapter 0.

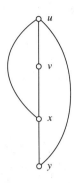

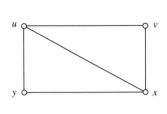

Figure 3.1.2 Alternate ways of drawing the graph G_1.

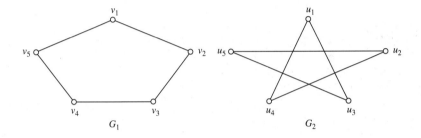

Figure 3.1.3 Two graphs, G_1 and G_2.

■ *Example 3.1.1* For the graph G_1 of Figure 3.1.1, $G_1 = (V, E)$, where $V = \{u, v, x, y\}$ and $E = \{uv, ux, uy, vx, xy\}$. Thus, the only nonadjacent vertices of G_1 are v and y. The edges uv and vx are adjacent since both are incident with the vertex v. The edges uv and xy are nonadjacent. Two alternate ways of drawing the graph G_1 are shown in Figure 3.1.2. There is no unique way of drawing a graph; the relative positions of the points and curves have no special significance. ■

Two graphs G_1 and G_2 are called **equal**, written $G_1 = G_2$, provided $V(G_1) = V(G_2)$ and $E(G_1) = E(G_2)$. Keeping this in mind, consider the graphs G_1 and G_2 shown in Figure 3.1.3. It is clear that $G_1 \neq G_2$ since the vertex sets are different. However, suppose G_2 is drawn as in Figure 3.1.4. Then, except for the labeling of the vertices, the two graphs appear alike.

In fact, we can establish a one-to-one function $\phi : V(G_1) \to V(G_2)$ satisfying the condition

$$xy \in E(G_1) \leftrightarrow \phi(x)\phi(y) \in E(G_2)$$

for all $x, y \in V(G_1)$. In words, two vertices x and y are adjacent in G_1 if and only if their images $\phi(x)$ and $\phi(y)$ are adjacent in G_2. One such function is defined by

$$\phi(v_1) = u_1, \quad \phi(v_2) = u_3, \quad \phi(v_3) = u_5, \quad \phi(v_4) = u_2, \quad \phi(v_5) = u_4$$

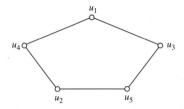

Figure 3.1.4   An alternate drawing of the graph G_2.

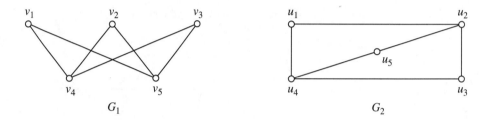

Figure 3.1.5   Two isomorphic graphs.

A function such as ϕ establishes a very strong relationship between the graphs G_1 and G_2. Indeed, even though G_1 and G_2 are not equal, the existence of ϕ implies that the graphs are "structurally" the same.

> **DEFINITION 3.1.2**   Two graphs G_1 and G_2 are called **isomorphic**, written $G_1 \simeq G_2$, provided there is a bijection $\phi : V(G_1) \to V(G_2)$ satisfying the condition that, for all $x, y \in V(G_1)$,
>
> $$xy \in E(G_1) \leftrightarrow \phi(x)\phi(y) \in E(G_2)$$
>
> Such a function ϕ is called an **isomorphism**.   ∎

In words, the condition of the definition states that the function ϕ **preserves adjacency**.

■ *Example 3.1.2*   Show that the two graphs G_1 and G_2 of Figure 3.1.5 are isomorphic.

Solution
Let $\phi : V(G_1) \to V(G_2)$ be defined by

$$\phi(v_1) = u_1, \quad \phi(v_2) = u_3, \quad \phi(v_3) = u_5, \quad \phi(v_4) = u_2, \quad \phi(v_5) = u_4$$

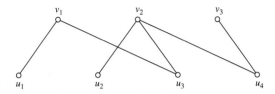

Figure 3.1.6   A bipartite graph.

We can readily verify that ϕ is one-to-one and onto. We can also check that ϕ preserves adjacency. For example, $v_2 v_5 \in E(G_1)$ and $\phi(v_2)\phi(v_5) = u_3 u_4 \in E(G_2)$.

You may have found a different isomorphism; if $V_1 = \{v_1, v_2, v_3\}$, $V_2 = \{v_4, v_5\}$, $U_1 = \{u_1, u_3, u_5\}$, and $U_2 = \{u_2, u_4\}$, then any bijection $f : V(G_1) \rightarrow V(G_2)$ with $f(V_1) = U_1$ and $f(V_2) = U_2$ is an isomorphism. ∎

Definiton 3.1.1 states explicitly that the vertex set of a graph is a nonempty finite set. If a graph G has n vertices and m edges, then G is said to have **order** n and **size** m.

Throughout this chapter several special graphs are frequently encountered; these are now introduced.

A graph in which every two vertices are adjacent is called a **complete graph**. Clearly, any two complete graphs of the same order are isomorphic, so we allow ourselves to be somewhat imprecise and refer to *the* complete graph of order n; this graph is denoted K_n.

DEFINITION 3.1.3   A graph $G = (V, E)$ is called a **bipartite graph** if there exists a partition $\{V_1, V_2\}$ of V such that each edge $e \in E$ is incident with a vertex of V_1 and a vertex of V_2. The sets V_1 and V_2 are called **partite sets** of G. ∎

The graph shown in Figure 3.1.6 is a bipartite graph with partite sets $V_1 = \{u_1, u_2, u_3, u_4\}$ and $V_2 = \{v_1, v_2, v_3\}$, and edge set $E = \{u_1 v_1, u_2 v_2, u_3 v_1, u_3 v_2, u_4 v_2, u_4 v_3\}$.

Let G be a bipartite graph with partite sets V_1 and V_2, where $|V_1| = r$ and $|V_2| = s$. If each $u \in V_1$ is adjacent to each $v \in V_2$, then G is called a **complete bipartite graph**. Let H be another complete bipartite graph with partite sets X_1 and X_2 such that $|X_2| = r$ and $|X_2| = s$. Then it is not difficult to see that $G \simeq H$, and, thus, up to isomorphism, there is a unique complete bipartite graph with partite sets of cardinalities r and s. This graph is denoted $K_{r,s}$. As an example, both of the graphs in Figure 3.1.5 are isomorphic to $K_{2,3}$.

■ *Example 3.1.3*   Find all nonisomorphic complete bipartite graphs of order 6.

Solution
The graphs are $K_{1,5}$, $K_{2,4}$, and $K_{3,3}$. ∎

■ ***Example 3.1.4***   Show that the graph G_1 of Figure 3.1.3 is not bipartite.

Solution
We proceed by contradiction. Suppose that G_1 is bipartite with partite sets V_1 and V_2. Without loss of generality, assume that $v_1 \in V_1$. Since v_2 is adjacent to v_1, it must be that $v_2 \in V_2$. Similarly, we can argue that $v_3 \in V_1$ and $v_4 \in V_2$. But now consider v_5. Since v_5 is adjacent to v_1, we have $v_5 \notin V_1$, and, since v_5 is adjacent to v_4, we have $v_5 \notin V_2$. This contradicts the requirement that $\{V_1, V_2\}$ be a partition of $V(G_1)$.   ■

In considering a given graph G, we are often interested in determining its structural properties. Of particular importance is the number of edges incident with a given vertex v. This number is called the **degree** of v and is denoted by $\deg(v)$. If several graphs are under consideration at the same time, then we use $\deg(v, G)$ to denote the degree of the vertex v in the graph G. The maximum and minimum vertex degrees in G are denoted by $\Delta(G)$ and $\delta(G)$, respectively. The graph G of Figure 3.1.6, for example, has $\deg(u_1) = 1 = \delta(G)$, $\deg(v_1) = 2$, and $\deg(v_2) = 3 = \Delta(G)$.

If we sum the degrees of the vertices in a graph, then each edge is counted twice, once for each of its incident vertices. This observation yields the following basic and very useful result:

THEOREM 3.1.1   Let G be a graph of order n and size m with vertex set $V = \{v_1, v_2, \ldots, v_n\}$. Then

$$\deg(v_1) + \deg(v_2) + \cdots + \deg(v_n) = 2m$$ □

COROLLARY 3.1.2   In any graph G, the number of vertices of odd degree is even.

Proof   Let $U = \{u_1, u_2, \ldots, u_r\}$ be the set of vertices of odd degree and let $W = \{w_1, w_2, \ldots, w_s\}$ be the set of vertices of even degree. Also, let

$$d_1 = \sum_{i=1}^{r} \deg(u_i) \qquad \text{and} \qquad d_2 = \sum_{j=1}^{s} \deg(w_j)$$

If G has size m, then $d_1 + d_2 = 2m$ by Theorem 3.1.1. Since $\deg(w_j)$ is even for $1 \le j \le s$, we see that d_2 is even. It follows that d_1 must also be even. Then, since $\deg(u_i)$ is odd for $1 \le i \le r$, we have r even.   □

A graph G is called **k-regular** (or **regular of degree k**) provided $\deg(v) = k$ for all $v \in V(G)$. For example, the complete graph K_n is $(n-1)$-regular and the complete bipartite graph $K_{r,r}$ is r-regular. Also, note that G is k-regular if and only if $\Delta(G) = \delta(G) = k$.

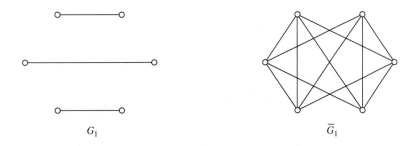

G_1 $\overline{G}_1$

Figure 3.1.7 A 1-regular graph and its 4-regular complement.

DEFINITION 3.1.4 Given a graph $G = (V, E)$, the **complement** of G is the graph $\overline{G} = (V, \overline{E})$, where

$$\overline{E} = \{uv \mid uv \notin E\}$$ ■

Note that if G is a k-regular graph of order n, then its complement $\overline{G}$ is $(n-1-k)$-regular. This observation is useful for the next example.

■ ***Example 3.1.5*** Give an example of a k-regular graph of order 6 for each k, $0 \le k \le 5$.

Solution
Since K_6 is 5-regular, its complement $\overline{K}_6$ is 0-regular. Figure 3.1.7 shows a 1-regular graph G_1 of order 6 and its 4-regular complement $\overline{G}_1$. Also, since $K_{3,3}$ is a 3-regular graph of order 6, its complement $\overline{K}_{3,3}$ is 2-regular. ■

Let $G = (V, E)$ be a graph, with $u, v \in V$. A u-v **walk** in G is a finite, alternating sequence of vertices and edges,

$$u, uu_1, u_1, u_1u_2, \ldots, u_{n-1}, u_{n-1}v, v$$

beginning with u and ending with v, such that each edge in the sequence is, as indicated, incident with the vertices that immediately precede and follow it. The **length** of a walk is the number of edges it contains, with repeated edges counted. A u-v walk is **closed** if $u = v$ and **open** if $u \neq v$. A u-v **trail** is a u-v walk in which no edge is repeated. A **path** is a trail with no repeated vertices; a **circuit** is a closed trail, and a **cycle** is a circuit in which the only repeated vertex is the first vertex, this being the same as the last vertex. A walk is called **trivial** if it contains no edges, that is, if it has length zero.

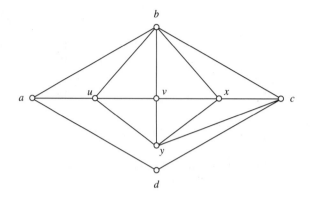

Figure 3.1.8 The graph of Freedonia.

Note that it is superfluous to include the edges when listing the vertices and edges of a walk, for it is assumed that two consecutive vertices in a walk are adjacent. Thus, we agree to list only the vertices in a walk.

■ *Example 3.1.6*   In the country of Freedonia (a small, tropical island paradise) there are eight cities, a, b, c, d, u, v, x, y, with the usual property that highways exist between certain pairs of cities. We can depict a roadmap of Freedonia as a graph by representing each city as a vertex and joining two vertices with an edge if the corresponding cities are (directly) linked by a highway. Assume that this process yields the graph H shown in Figure 3.1.8. Find examples of the following walks in the graph H:

(a) A u-v walk that is not a trail

(b) A u-v trail that is not a path

(c) A u-v path of length 5

(d) A u-u circuit that is not a cycle

(e) A u-u cycle of length 8

Solution

(a) The walk u, v, x, y, v, b, u, v is not a trail since the edge uv is repeated. This walk has length 7.

(b) The trail u, b, x, y, c, x, v is not a path since the vertex x is repeated. This trail has length 6.

(c) The path u, a, d, c, x, v has length 5.

(d) The circuit u, b, v, y, x, v, u is not a cycle since the vertex v is repeated. This circuit has length 6.

(e) The cycle $u, v, x, y, c, d, a, b, u$ has length 8.   ■

Let n be a positive integer and let $V = \{v_1, v_2, \ldots, v_n\}$. Define the graph $G_1 = (V, E_1)$, where

$$E_1 = \{v_i v_{i+1} \mid 1 \leq i < n\}$$

and the graph $G_2 = (V, E_2)$, for $n \geq 3$, by

$$E_2 = E_1 \cup \{v_1 v_n\}$$

Notice that the edges of G_1 form a path of length $n - 1$ from v_1 to v_n; for this reason, a graph isomorphic to G_1 is called a **path of order n** and is denoted by P_n. The edges of G_2 form a cycle of length n; any graph isomorphic to G_2 is called a **cycle of order n** or **n-cycle** and is denoted by C_n. For instance, the graphs in Figure 3.1.3 are 5-cycles. Note that removing an edge from C_n results in the graph P_n.

Graphs are often used to model communication or transportation networks. In the graph of a communication network, it is important that there be a way for any vertex to communicate with any other vertex. Similarly, in the graph of a transportation network, it is important that there be a way to travel from any vertex to any other. This leads to the next definition.

DEFINITION 3.1.5 A graph $G = (V, E)$ is called **connected** provided, for any $u, v \in V$, there is a u-v walk in G. If G is not connected, then it is called **disconnected**. ∎

■ *Example 3.1.7* We can observe that the graph H in Figure 3.1.8 is connected. Suppose that a hurricane hits Freedonia (one of the drawbacks of living in the tropics), and a resulting flood destroys several bridges, making highways ab, bc, bu, cx, uv, vy, and xy impassable. The graph of the resulting highway network H_1 is shown in Figure 3.1.9. We note that H_1 is disconnected; for example, there is no walk between a and b. ∎

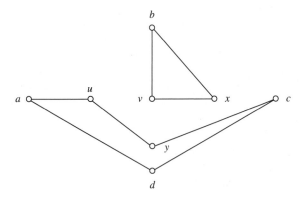

Figure 3.1.9 The graph of Freedonia after a flood.

EXERCISE SET 3.1

1. Give an example of a graph G_1 of order 6 that has one vertex of degree 1, two vertices of degree 2, and three vertices of degree 3. What is the size of such a graph?

2. If a graph G has no adjacent edges, what can be said about the degrees of its vertices?

3. Show the following:

   (a) A graph G of order 5 cannot have two vertices of degree 4 and one vertex of degree 1.

   (b) A graph G of order n cannot have two vertices of degree $n-1$ and one vertex of degree 1.

4. Show that, in any set of two or more people, there are always two people with the same number of acquaintances in the set.

5. For an arbitrary positive integer n, give an example of a graph having:

   (a) Order n and size $n(n-1)/2$.   (b) Order $2n$ and size n.

   (c) Order n and size n ($n \geq 3$).   (d) Order $2n$ and size $3n$ ($n \geq 3$).

6. Let G_1 and G_2 be two isomorphic graphs and let $\phi : V(G_1) \rightarrow V(G_2)$ be an isomorphism. Prove each of these facts:

   (a) G_1 and G_2 have the same order.

   (b) For each $v \in V(G_1)$, $\deg(\phi(v)) = \deg(v)$.

   (c) G_1 and G_2 have the same size.

7. Each of the following parts lists three graphs. You are to (i) draw the graphs, (ii) find the two graphs that are isomorphic, (iii) find an isomorphism, and (iv) give a reason why the third graph is not isomorphic to the other two.

   (a) $G_1 = (\{u_1, u_2, u_3, u_4, u_5, u_6\}, \{u_1u_2, u_1u_5, u_1u_6, u_2u_3, u_2u_5, u_3u_4, u_4u_5\})$
   $G_2 = (\{v_1, v_2, v_3, v_4, v_5, v_6\}, \{v_1v_5, v_1v_6, v_2v_4, v_2v_5, v_2v_6, v_3v_4, v_4v_6\})$
   $G_3 = (\{x_1, x_2, x_3, x_4, x_5, x_6\}, \{x_1x_2, x_1x_5, x_2x_3, x_3x_4, x_3x_6, x_4x_5\})$

   (b) $G_4 = (\{u_1, u_2, u_3, u_4, u_5\}, \{u_1u_2, u_1u_3, u_1u_4, u_1u_5, u_2u_4, u_2u_5, u_3u_4, u_3u_5\})$
   $G_5 = (\{v_1, v_2, v_3, v_4, v_5\}, \{v_1v_2, v_1v_5, v_2v_3, v_2v_4, v_2v_5, v_3v_4, v_3v_5, v_4v_5\})$
   $G_6 = (\{x_1, x_2, x_3, x_4, x_5\}, \{x_1x_2, x_1x_3, x_1x_4, x_1x_5, x_2x_3, x_2x_5, x_3x_4, x_4x_5\})$

   (c) $G_7 = (\{u_1, u_2, u_3, u_4, u_5, u_6\}, \{u_1u_2, u_1u_4, u_1u_5, u_2u_3, u_3u_4, u_3u_6, u_5u_6\})$
   $G_8 = (\{v_1, v_2, v_3, v_4, v_5, v_6\}, \{v_1v_2, v_1v_6, v_2v_3, v_2v_6, v_3v_4, v_3v_5, v_4v_5\})$
   $G_9 = (\{x_1, x_2, x_3, x_4, x_5, x_6\}, \{x_1x_2, x_1x_5, x_2x_3, x_2x_6, x_3x_4, x_4x_5, x_4x_6\})$

8. Let G be a graph with vertex set $V = \{v_1, v_2, \ldots, v_n\}$. The list of numbers

   $$(\deg(v_1), \deg(v_2), \ldots, \deg(v_n))$$

   is called a **degree sequence** of G. Each of the following parts gives a list of nonnegative integers; give an example of a graph having that degree sequence or argue that no such graph exists.

   (a) $(3, 2, 2, 2, 1)$ (b) $(3, 2, 2, 2, 1, 1)$ (c) $(4, 3, 2, 1, 0)$

   (d) $(4, 4, 3, 3, 2, 2)$ (e) $(5, 4, 3, 3, 2, 1)$ (f) $(5, 5, 5, 4, 4, 3, 2)$

9. Give an example (preferably of smallest possible order) of nonisomorphic graphs G_1 and G_2 satisfying the following conditions:

(a) G_1 and G_2 have the same order.

(b) G_1 and G_2 have the same order and the same size.

(c) G_1 and G_2 have the same order n, $V(G_1) = \{u_1, u_2, \ldots, u_n\}$, $V(G_2) = \{v_1, v_2, \ldots, v_n\}$, and $\deg(u_i) = \deg(v_i)$ for each i, $1 \leq i \leq n$.

10. Show that the relation "is isomorphic to" is an equivalence relation on the set of graphs.

11. Let $G_1 = (\{v_1, v_2, v_3, v_4, v_5, v_6, v_7\}, \{v_1v_2, v_1v_4, v_2v_3, v_2v_5, v_3v_4, v_3v_6, v_4v_7, v_5v_6, v_6v_7\})$ and $G_2 = (\{v_1, v_2, v_3, v_4, v_5, v_6\}, \{v_1v_2, v_1v_4, v_1v_6, v_2v_3, v_2v_6, v_3v_4, v_4v_5, v_5v_6\})$.

(a) Draw the graph G_1.   (b) Show that G_1 is bipartite.

(c) Draw the graph G_2.   (d) Show that G_2 is not bipartite.

12. Given a graph G of order n, size m, minimum degree δ, and maximum degree Δ, prove that $n\delta \leq 2m \leq n\Delta$.

13. Let G be a k-regular graph of order n and size m.

(a) Give an identity involving k, n, and m.

(b) Show that k or n is even.

14. Let G be a bipartite graph of order n and size m.

(a) Prove that $4m \leq n^2$.

(b) If $4m = n^2$, to what graph is G isomorphic?

15. Let G be a k-regular bipartite graph ($k > 0$) with partite sets V_1 and V_2. Prove that $|V_1| = |V_2| \geq k$.

16. Give an example of each of the following:

(a) A 1-regular bipartite graph of order 4.

(b) A 2-regular bipartite graph of order 6.

(c) A 3-regular bipartite graph of order 8.

(d) An r-regular bipartite graph of order $2r + 2$, for an arbitrary $r \in \mathbb{N}$.

17. Given that G is a k-regular graph of order n and size m, what can be said about G (up to isomorphism) under the following conditions?

(a) $k = 3$ and $m = n + 3$ (b) $k = 4$ and $m = 3n - 5$

18. Prove: If G_1 and G_2 are isomorphic graphs and G_1 is bipartite, then G_2 is bipartite.

19. Determine, up to isomorphism, all graphs of order 5.

20. The intramural hockey league at a small college has seven teams: t, u, v, w, x, y, z. Each team is scheduled to play four games over a 5-week season, against four different opponents. During each of the first 4 weeks, six of the teams play a game while the remaining team has a bye. Then, for the last week, the three teams that have already played four games have a bye, while the remaining four teams play.

(a) Design a schedule for the hockey league, using a graph to indicate which teams each team plays.

    (b) Suppose the season is extended to 6 weeks. Is it possible for each of the seven teams to play five games against five different opponents?

21. Find $\overline{G}_1$, $\overline{G}_4$, and $\overline{G}_7$ for the graphs G_1, G_4, and G_7 of Exercise 7.

22. Let G be a graph of order n, size m, minimum degree δ, and maximum degree Δ.

    (a) Determine the size of $\overline{G}$.

    (b) If d is the degree of vertex v in G, determine the degree of v in $\overline{G}$.

    (c) Determine the minimum degree of $\overline{G}$.

    (d) Determine the maximum degree of $\overline{G}$.

23. Let $G = (\{t, u, v, w, x, y, z\}, \{tu, tv, tw, ux, vw, vy, uz, wx, wz, xy, yz\})$. Give examples of the following walks in G:

    (a) A u-v walk that is not a trail.

    (b) A u-v trail that is not a path.

    (c) A u-v path of minimum length.

    (d) A u-v path of maximum length.

    (e) A u-u circuit that is not a cycle.

    (f) A u-u cycle of minimum length.

    (g) A u-u cycle of maximum length.

24. Draw each of the following graphs: (a) K_6, (b) $K_{3,5}$, (c) P_7, (d) C_7.

25. Show that a graph G and its complement $\overline{G}$ cannot both be disconnected.

26. Prove: If G_1 and G_2 are isomorphic graphs and G_1 is connected, then G_2 is connected.

27. Prove: If $G = (V, E)$ is a connected bipartite graph, then there is a unique partition $\{V_1, V_2\}$ of V into partite sets.

3.2 EULERIAN GRAPHS

This section introduces the Königsberg bridges problem, mentioned at the outset of the chapter.

    Figure 3.2.1 shows a map of the town of Königsberg as it appeared in 1736. The river Pregel ran through the town and was spanned by seven bridges, which connected two islands in the river with each other and with the opposite banks.

    The townsfolk amused themselves with the following problem: Is it possible to start at some point in the town and take a walk that crosses each bridge exactly once and returns to the starting point? We can represent the situation as a kind of graph by representing each land mass as a vertex and joining two vertices with a number of edges equal to the number of bridges that join the corresponding land masses. The resulting "graph" is shown in Figure 3.2.2(a).

    Of course, what we have in Figure 3.2.2(a) is not really a simple graph, because some pairs of vertices are joined by more than one edge; such a graph is referred to as a "multigraph." However, we can obtain a graph from a multigraph by inserting a new vertex (of degree 2) into every edge of the multigraph. For example, the graph obtained from the multigraph of Königsberg is shown in Figure 3.2.2(b).

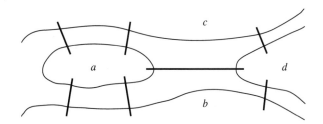

Figure 3.2.1   The bridges of Königsberg.

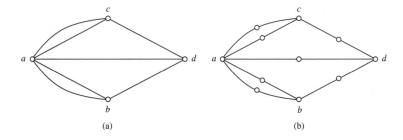

(a)   (b)

Figure 3.2.2   The multigraph (a) and graph (b) of Königsberg.

Note that, as far as the problem is concerned, the insertion of these extra vertices of degree 2 has no effect. This allows us to state the problem in terms of simple graphs.

Let $G = (V, E)$ be a graph, with $u, v \in V$. Recall that a u-v trail is a u-v walk in which no edge is repeated, and that a circuit is a closed trail. With this terminology, we can state the Königsberg bridges problem in graphical terms:

Is there a circuit in the graph of Königsberg that includes every edge?

In general, a circuit in a graph G that includes every edge is called an **eulerian circuit**, and an open trail with this property is called an **eulerian trail**. Thus, we ask the more general question:

When does a graph possess an eulerian circuit or trail?

■   ***Example 3.2.1***   Recall the graph H of Freedonia from Example 3.1.6; this is repeated for convenience in Figure 3.2.3. The Freedonia Highway Patrol might be interested in finding an eulerian circuit or trail for this graph, because a patrol car following such a trail would be able to cover each highway exactly once. As we will see shortly, the graph H does not contain an eulerian circuit; however, it does contain an eulerian trail, for instance

$$a, b, c, d, a, u, b, x, c, y, u, v, x, y, v, b$$ ■

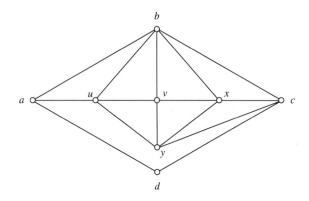

Figure 3.2.3 The graph of Freedonia.

Now let G be a graph, let v be a vertex of G, and suppose that G contains an eulerian circuit. Then each occurrence of v on the circuit contributes 2 to the degree of v, because the edges of the circuit that immediately precede and follow v are distinct and both are incident with v. Thus, v has even degree in G, and we have the following lemma:

LEMMA 3.2.1 Let G be a graph. If G contains an eulerian circuit, then every vertex of G has even degree.   □

In other words, the condition that every vertex has even degree is a necessary condition for a graph to have an eulerian circuit. It is also sufficient, provided the graph is connected. In fact, if G contains an eulerian circuit C and G has no vertices of degree 0, then G is certainly connected. This is because, given any two distinct vertices x and y, they can be located on C, and then there is a part of C that is an x-y trail.

A connected graph that contains an eulerian circuit is called an **eulerian graph**. Thus, we wish to prove that a connected graph G is eulerian if and only if every vertex of G has even degree. To do this, it is useful to have the following lemma.

LEMMA 3.2.2 Let G be a graph such that every vertex of G has even degree. If u and v are adjacent vertices of G, then there is a circuit of G that contains the edge uv.

Proof   Consider the set S of all trails of G that begin

$$u, v, \ldots$$

Clearly, S is nonempty. Moreover, since $E(G)$ is finite, so is S. Thus, we may consider the set of lengths of trails in S, and since this is a nonempty finite set of positive integers, it has a maximum element. Thus, let W be a trail in S of maximum length, say,

$$u, v, \ldots, x$$

We claim that W is a circuit, namely, that $x = u$. If not, then note that the number of edges of W incident with x is odd, because each occurrence of x on W, except for the last occurrence, accounts for two edges of W incident with x, whereas the last occurrence of x accounts for one edge of W incident with it. However, x has even degree in G. Hence, there is an edge xy of G that is not an edge of W. But then W can be extended (by one edge and one vertex) to the trail

$$u, v, \ldots, x, y$$

which has length 1 greater than that of W. This contradicts the choice of W as a trail in S of maximum length, thereby proving the result. $\square$

THEOREM 3.2.3 Let G be a connected graph. Then G is eulerian if and only if every vertex of G has even degree.

Proof The result is trivial if G is isomorphic to K_1 or K_2, so we may assume that G has at least three vertices. Moreover, in view of Lemma 3.2.1, it suffices to prove sufficiency. So let $G = (V, E)$ be a connected graph such that every vertex of G has (positive) even degree. We wish to prove that G contains an eulerian circuit.

Consider the set S of all nontrivial circuits of G. As in the proof of Lemma 3.2.2, S is a nonempty finite set, and so it makes sense to let W be an element of S of maximum length. We claim that W is an eulerian circuit of G. If not, then there is an edge uv of G that is not in W. Since G is connected, we may assume, without loss of generality, that u is on W. (If u is not on W, let w be a vertex on W. Since G is connected, there is a u-w walk in G; let v_1u_1 be the first edge of this walk such that v_1 is not on W and u_1 is on W. Then replace u by u_1 and v by v_1 in the argument.) Now let $G_1 = (V, E_1)$ be the graph obtained from G by removing the edges of W, that is, $E_1 = E - \{$edges of $W\}$. Then every vertex of G_1 has even degree, so by Lemma 3.2.2 there is a circuit W_1 of G_1 that includes the edge uv. It is then possible to describe a u-u circuit W' of G that extends W by "inserting" W_1 into W at u; namely, proceed from u to u by following (all) the edges of W, and then proceed from u to u by following the edges of W_1. But W' has length greater than that of W, contradicting our choice of W. This proves the result. $\square$

The preceding proof suggests an algorithm for finding an eulerian circuit in an eulerian graph. A formal presentation of such an algorithm is left to the next section, where it is given for the case of directed graphs. For now, let's consider an example that illustrates the main idea of the algorithm.

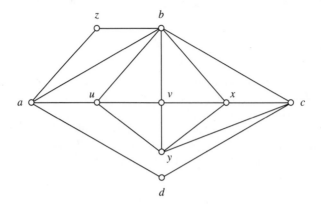

Figure 3.2.4 An eulerian graph.

■ *Example 3.2.2* Consider the graph H of Figure 3.2.3. Since vertices a and b have odd degree, this graph is not eulerian. Let us add to this graph a new vertex z, together with the edges az and bz. The resulting graph, call it G, is shown in Figure 3.2.4.

Since G is connected and each vertex has even degree, G is eulerian, and we wish to construct an eulerian circuit W. We choose a starting vertex, say, a, and set W equal to the trivial circuit a. We now construct a circuit W_1 that includes every edge incident with vertex a. Doing this, suppose we obtain the following circuit:

$$a, b, c, d, a, u, b, z, a$$

The circuit W_1 is inserted into W at a, so that W is now equal to W_1. Next, we look for a vertex of W that is incident with an edge of G not in W. If no such vertex exists, then W is an eulerian circuit. However, in our case we find that vertex b is incident with edges of G that are not edges of W, namely, bv and bx. We may then, as in the proof of Theorem 3.2.3, find a b-b circuit W_1 of G that has no edges in common with W. (In fact, we may construct W_1 so that, together, W and W_1 include every edge of G incident with b.) For example, suppose W_1 is the circuit

$$b, v, u, y, v, x, b$$

We then revise W to be the larger circuit obtained by inserting W_1 into W at (the last occurrence of) b; this yields the circuit

$$a, b, c, d, a, u, b, v, u, y, v, x, b, z, a$$

We next repeat the step of searching W for a vertex that is incident with an edge of G not in W. Note that x is such a vertex and that W_1:

$$x, c, y, x$$

is an x-x circuit of G having no edges in common with W. Again, we replace W by the larger circuit obtained by inserting W_1 into W at x, yielding the circuit

$$a, b, c, d, a, u, b, v, u, y, v, x, c, y, x, b, z, a$$

This time we find that no vertex of W is incident to an edge of G not in W, so that W is an eulerian circuit.

Note that, if we delete from the eulerian circuit W the edges az and bz and the vertex z, we obtain the a-b eulerian trail

$$a, b, c, d, a, u, b, v, u, y, v, x, c, y, x, b$$

of the graph H representing the Freedonia highway network. ∎

EXERCISE SET 3.2

1. For each of the following graphs: (i) draw it, (ii) determine whether it is eulerian, and (iii) if it is, find an eulerian circuit, preferably using the algorithm suggested by Example 3.2.2.

   (a) $G_1 = (\{u_1, u_2, u_3, u_4, u_5\}, \{u_1u_2, u_1u_3, u_1u_4, u_1u_5, u_2u_4, u_2u_5, u_3u_4, u_3u_5\})$
   (b) $G_2 = (\{v_1, v_2, v_3, v_4, v_5\}, \{v_1v_2, v_1v_5, v_2v_3, v_2v_4, v_2v_5, v_3v_4, v_3v_5, v_4v_5\})$
   (c) $G_3 = (\{u_1, u_2, u_3, u_4, u_5, u_6\}, \{u_1u_2, u_1u_3, u_1u_4, u_1u_5, u_2u_3, u_3u_4, u_3u_6, u_5u_6\})$
   (d) $G_4 = (\{v_1, v_2, v_3, v_4, v_5, v_6\}, \{v_1v_2, v_1v_6, v_2v_3, v_2v_6, v_3v_4, v_3v_5, v_4v_5\})$
   (e) $G_5 = (\{v_1, v_2, v_3, v_4, v_5, v_6, v_7\}, \{v_1v_2, v_1v_4, v_2v_3, v_3v_4, v_5v_6, v_5v_7, v_6v_7\})$
   (f) $G_6 = (\{v_1, v_2, v_3, v_4, v_5, v_6\}, \{v_1v_2, v_1v_3, v_1v_4, v_1v_6, v_2v_3, v_2v_4, v_2v_6, v_3v_4,$
   $v_3v_6, v_4v_5, v_5v_6\})$

2. A connected graph that contains an eulerian trail is called **edge traceable**. Let G be a connected graph of order at least 2. Prove: G is edge traceable if and only if G has exactly two vertices of odd degree.

3. Determine which of the graphs of Exercise 1 are edge traceable.

4. (a) Show that the graph of Königsberg (Figure 3.2.2(b)) does not contain an eulerian trail nor an eulerian circuit.
   (b) In Königsberg (Kaliningrad), there are two additional bridges, joining regions b and c and regions b and d. Is it now possible to take a walk through the town that crosses each bridge exactly once? If so, devise such a walk.

5. Prove: If G_1 and G_2 are isomorphic graphs and G_1 is eulerian, then G_2 is eulerian.

6. Give an example of a multigraph G' of order 5 that has two vertices of degree 4 and one vertex of degree 1. Compare with Exercise 3 in Exercise Set 3.1. Is the result of Corollary 3.1.2 true for multigraphs?

7. Consider a standard set of dominoes with the doubles removed. Show how to arrange these 21 dominoes in a rectangular pattern so that adjacent squares match. (Hint: Look at the sequence of edges of an eulerian circuit of the complete graph with vertex set $\{0, 1, 2, 3, 4, 5, 6\}$.)

8. Look up Fleury's algorithm—see, for example, page 126 of *Graphs, An Intro-ductory Approach*, by Robin J. Wilson and John J. Watkins—and compare it with the algorithm suggested by Example 3.2.2.

3.3 DIRECTED GRAPHS

Consider the problem of traffic flow in a large city, where some streets are one way and others are two way. As part of modeling such a situation, a directed graph may be constructed as follows: Associate a vertex with each intersection of streets, and then place an arc from vertex a to vertex b if it is possible to travel directly from a to b without passing through a third intersection. Note that it is possible for both the arc from a to b and the arc from b to a to be included. (When does this happen?)

   Related to the traffic flow problem, we may wish to design a travel route from some intersection u to some other intersection v. Is it possible to get from u to v? If so, how much time does the trip take and what is the taxicab fare? Note that the answer to this last question might depend on the number of intersections someone must pass through on the way from u to v. These issues are related to the notions of "reachability" and "distance" in directed graphs, which are ideas we discuss in this section and the next.

   Let's begin by reviewing and introducing some additional terminology for directed graphs.

   As we know, given a digraph $G = (V, A)$, the elements of V are called **vertices** (the plural of **vertex**) and the elements of A are called **arcs** or **directed edges**. An arc of the form (u, u), where u is a vertex, is called a **loop**. For distinct vertices u and v, the arcs (u, v) and (v, u) are called **symmetric arcs**. Given $(u, v) \in A$, we say that u is **adjacent to** v, that v is **adjacent from** u, and that u and v are **incident** with the arc (u, v). For distinct vertices u and v, if neither u is adjacent to v nor v adjacent to u, then u and v are said to be **nonadjacent**.

■ *Example 3.3.1*   A representation, or "drawing," of the digraph $G = (V, A)$, with $V = \{u, v, x, y, z\}$ and $A = \{(v, u), (v, x), (x, y), (y, x), (y, u), (z, u), (z, z)\}$ is shown in Figure 3.3.1. This digraph has one loop, (z, z), and one pair of symmetric arcs, (x, y) and (y, x). The vertex v is adjacent to the vertex x, x is adjacent from v, and both v and x are incident with the arc (v, x). The vertices u and x are nonadjacent.   ■

   Notice that a digraph $G = (V, A)$ determines a relation on V, namely, A. It is for this reason that a relation on a finite set has a "digraph representation," as discussed in Chapter 0. If the relation A is irreflexive, then G has no loops and is called **loopless**. Moreover, if A is symmetric, then $(v, u) \in A$ whenever $(u, v) \in A$, and we may as well replace the symmetric pair of arcs (u, v) and (v, u) by the edge uv. Doing this for each symmetric pair of arcs results in a simple graph.

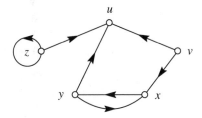

Figure 3.3.1   The digraph G of Example 3.3.1.

For vertices x and y in a directed graph G, an **x-y walk** in G is a finite alternating sequence W of vertices and arcs of the form

$$x = x_0, (x_0, x_1), x_1, (x_1, x_2), x_2, \ldots, (x_{n-1}, x_n), x_n = y$$

beginning with x and ending with y. We call n, the number of arcs in W, the **length** of W. Note that each arc in W joins the vertex immediately preceding it to the vertex immediately following it. For this reason, we may conveniently exhibit a walk by listing only its vertices. An x-y walk is **open** or **closed** according as $x \neq y$ or $x = y$, respectively. A walk with no repeated arcs is called a **trail**, and a trail with no repeated vertices is called a **path**. A **circuit** is a closed trail, and a **cycle** is a circuit in which the only repeated vertex is the first vertex, which is the same as the last vertex.

■   *Example 3.3.2*   For the digraph G_1 shown in Figure 3.3.2, find:

(a)   A u-v walk that is not a trail.
(b)   A u-v trail that is not a path.
(c)   A u-v path of maximum length.
(d)   A v-v circuit that is not a cycle.
(e)   A v-v cycle of maximum length.

Solution

(a)   The walk u, x, w, z, y, x, w, v is not a trail since the arc (x, w) is repeated. The length of this walk is 7.

(b)   The trail u, t, w, z, y, x, w, v is not a path since the vertex w is repeated. The length of this trail is 7.

(c)   The u-v path u, t, w, z, y, v has length 5; there are several other u-v paths having length 5, but no u-v paths of length 6, so 5 is the maximum length of a u-v path.

(d)   The circuit v, t, w, z, y, x, w, v is not a cycle since the vertex w is repeated.

(e)   The v-v cycle v, t, w, z, y, v, with length 5, is a v-v cycle of maximum length.   ■

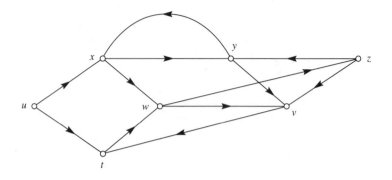

Figure 3.3.2 The digraph G_1 of Example 3.3.2.

DEFINITION 3.3.1 Let $G = (V, A)$ be a digraph and let $x, y \in V$. If G contains an x-y walk, then we say that y is **reachable** from x. In this case, the **distance** from x to y in G is defined as the minimum length among all x-y walks in G, and is denoted by $d(x, y, G)$, or simply by $d(x, y)$ if the graph under consideration is understood. If y is not reachable from x, then we write $d(x, y) = \infty$. If G contains an x-y walk for every ordered pair of vertices x and y, then G is called **strongly connected** or **strong**. ∎

■ **Example 3.3.3** Consider the digraph $G_1 = (V_1, A_1)$ of Figure 3.3.2.

(a) Find $d(u, a)$ for all $a \in V_1$.

(b) Find $d(a, z)$ for all $a \in V_1$.

(c) Show that G_1 is not strong.

(d) Let $G_2 = (V_1, A_2)$, where $A_2 = A_1 \cup \{(z, u)\}$. Show that G_2 is strong.

Solution

(a) We find that $d(u, u) = 0$, $d(u, t) = d(u, x) = 1$, $d(u, w) = d(u, y) = 2$, and $d(u, v) = d(u, z) = 3$.

(b) We find that $d(u, z) = d(v, z) = d(y, z) = 3$, $d(x, z) = d(t, z) = 2$, $d(w, z) = 1$, and $d(z, z) = 0$.

(c) Note that u is not reachable from any other vertex of G_1, showing that G_1 is not strong.

(d) In G_1, every vertex is reachable from u and every vertex can reach z, so this same property holds in G_2. Thus, to show that G_2 is strong, it suffices to observe that u is reachable from z. (Note that the relation "is reachable from" on the vertex set of a digraph is reflexive and transitive but not necessarily symmetric.) ∎

Next we state two basic but important results concerning reachability, paths, and distance in directed graphs.

THEOREM 3.3.1 Let $G = (V, A)$ be a digraph and let $x, y \in V$. If y is reachable from x, then G contains an x-y path.

Proof Let $x, y \in V$, and assume y is reachable from x. The result of the theorem is trivial if $x = y$, so we assume $x \neq y$. Let S be the set of x-y walks in G; note that S is nonempty. By considering the lengths of the walks in S as a nonempty set of positive integers, we may apply the principle of well ordering to obtain a walk W in S of minimum length. We claim that W is an x-y path. If not, then some vertex of W is repeated. If y is repeated, then W has the form

$$x, \ldots, y, z, \ldots, y$$

and that part of W from x to the first occurrence of y is an x-y walk whose length is less than that of W. This contradicts our choice of W. So we may assume that W has the form

$$x = u_0, u_1, \ldots, u_i, u_{i+1}, \ldots, u_j, u_{j+1}, \ldots, u_n = y$$

where $u_i = u_j$, $0 \leq i < j < n$. Then the walk

$$x = u_0, u_1, \ldots, u_i, u_{j+1}, \ldots, u_n = y$$

is an x-y walk whose length is less than that of W, contradicting our choice of W. This completes the proof. □

Now let x and y be vertices of a digraph G such that y is reachable from x. As a consequence of Theorem 3.3.1, $d(x, y)$ may be computed as the minimum length among all x-y paths in G.

THEOREM 3.3.2 Let $G = (V, A)$ be a digraph and let $x, y, z \in V$. If y is reachable from x and z is reachable from y, then z is reachable from x and

$$d(x, z) \leq d(x, y) + d(y, z)$$

Proof Let $x, y, z \in V$ and assume y is reachable from x and z is reachable from y. Let P_1 be an x-y path of length $d(x, y)$ and P_2 be a y-z path of length $d(y, z)$. Define the x-z walk W by appending P_2 to P_1, that is, proceed from x to y along P_1 and then from y to z along P_2. Then the length of W is $d(x, y) + d(y, z)$. Hence, since $d(x, z)$ is defined as the minimum length among all x-z walks, we have

$$d(x, z) \leq d(x, y) + d(y, z)$$ □

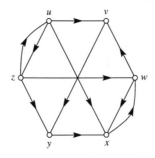

Figure 3.3.3   The digraph G_2 of Example 3.3.4.

What about other properties of the distance function? If $G = (V, A)$ is a digraph with $x, y \in V$ and y reachable from x, then it is clear that $d(x, y) \geq 0$. However, as illustrated by the digraph G_1 of Figure 3.3.2, the distance function is not symmetric, in general. Notice in this digraph that $d(x, w) = 1$, whereas $d(w, x) = 3$.

■ *Example 3.3.4*   Show that the digraph $G_2 = (V, A)$ of Figure 3.3.3 is not strong, but that it is possible to obtain a strong digraph G_2' by reversing the direction of exactly one arc of G_2.

Solution
Let $U = \{u, z\}$. Note that each arc incident with a vertex of U and a vertex of $V - U$ is directed from U to $V - U$. Hence, neither of the two vertices in U is reachable from any of the four vertices in $V - U$, so that G_2 is not strong. Observe that, to obtain a strong digraph G_2', it suffices to reverse the direction of any one of the arcs (u, v), (u, x), (z, w), or (z, y). For example, let $G_2' = (V, A')$, where

$$A' = (A - \{(u, v)\}) \cup \{(v, u)\}$$

We claim that G_2' is strong. An easy way to see this is to observe that

$$W : u, x, w, v, y, x, w, v, u, z, u$$

is a closed walk of G_2' that includes every vertex. Thus, given any two vertices a and b of G_2', we may locate a and b on W, and then that portion of W from a to b is an a-b walk of G_2'.   ■

Taking our lead from the preceding example, note the following useful characterization of strongly connected directed graphs.

THEOREM 3.3.3 A digraph G is strongly connected if and only if G has a closed walk that includes every vertex.

Proof We assume that G is a strong digraph and let W be a closed walk of G that includes a maximum number of distinct vertices. We claim that W includes every vertex of G. If not, we suppose that y is a vertex of G that is not on W. We then let x be a vertex of G on W; then we may consider W as an x-x walk. Since G is strong, it contains an x-y walk W_1 and a y-x walk W_2. But then we may define an x-x walk W' by following W from x to x, then W_1 from x to y, and then W_2 from y to x. Thus, W' includes more distinct vertices of G than does W. This contradicts our choice of W and thus proves necessity.

For sufficiency, let W be a closed walk of G that includes every vertex. Given arbitrary vertices u and v of G, we may locate u and v on W, and then that part of W from u to v is a u-v walk of G. This shows that G is strong and thus proves sufficiency. $\square$

Recall that, given a simple graph G and a vertex v of G, the degree of v is the number of vertices of G with which v is adjacent. Given a digraph $G = (V, A)$ and $v \in V$, then possibly there are vertices of G adjacent to v and vertices adjacent from v. Thus, it makes sense to assign to v two different degree values. The **indegree** of v, written id(v), is the number of vertices of G adjacent to v, while the **outdegree** of v, written od(v), is the number of vertices of G adjacent from v. For example, the vertices of the digraph G_1 of Figure 3.3.2 have the following indegrees and outdegrees:

$$\text{id}(t) = 2, \text{id}(u) = 0, \text{id}(v) = 3, \text{id}(w) = \text{id}(x) = \text{id}(y) = 2, \text{id}(z) = 1$$
$$\text{od}(t) = 1, \text{od}(u) = 2, \text{od}(v) = 1, \text{od}(w) = \text{od}(x) = \text{od}(y) = \text{od}(z) = 2$$

Just as with graphs, the number of vertices and the number of arcs in a digraph are referred to as its **order** and **size**, respectively. Our next result gives the digraph analogue of Theorem 3.1.1; its proof is left for you to develop in Exercise 10.

THEOREM 3.3.4 Let $G = (V, A)$ be a directed graph of order n and size m and let $V = \{v_1, v_2, \ldots, v_n\}$. Then

$$\sum_{i=1}^{n} \text{id}(v_i) = m = \sum_{i=1}^{n} \text{od}(v_i) \qquad \square$$

To complete this section, let's consider the eulerian problem for digraphs. Just as for graphs, a circuit in a digraph G that includes every arc is called an **eulerian circuit**. Note that, if G contains an eulerian circuit and has no vertex v such that id(v) = od(v) = 0, then G is strong. Thus, we ask:

When does a strong digraph possess an eulerian circuit?

A strong digraph that has an eulerian circuit is called **eulerian**. We wish to know when a strong digraph is eulerian. Our analysis of this problem parallels the graph version.

Let G be a strong digraph, let v be a vertex of G, and suppose that G contains an eulerian circuit. Then each occurrence of v on the circuit contributes 1 to the indegree of v and 1 to the outdegree of v, because the arc of the circuit preceding a given occurrence of v leads into v and the arc following v leads out of v. It follows that $\text{id}(v) = \text{od}(v)$. Conversely, it can be shown that if $\text{id}(v) = \text{od}(v)$ for every vertex v of G, then G is eulerian. The following lemma and theorem are the digraph versions of Lemma 3.2.2 and Theorem 3.2.3. Their proofs are completely analogous to the graph versions, and are left for you to work out in Exercises 18 and 19.

LEMMA 3.3.5   Let $G = (V, A)$ be a digraph such that $\text{id}(v) = \text{od}(v)$ for all $v \in V$. If $(u, v) \in A$, then there is a circuit of G that contains the arc (u, v). $\square$

THEOREM 3.3.6   Let $G = (V, A)$ be a strong digraph. Then G is eulerian if and only if $\text{id}(v) = \text{od}(v)$ for all $v \in V$. $\square$

Just as for the graph version, the proofs of this lemma and theorem suggest an algorithm for finding an eulerian circuit in an eulerian digraph. To formally discuss this and other algorithms for directed graphs, assume that we have at our disposal a package for the abstract data type DIGRAPH.

The abstract data type DIGRAPH is based on a more primitive notion, that of a "list of vertices." Given a digraph $G = (V, A)$, we can represent G by presenting a list of the elements of V and, for each $v \in V$, giving a list of those vertices adjacent from v; such a list is called an **adjacency list for** v. For example, consider the digraph $G = (V, A)$ of Figure 3.3.4, where $V = \{1, 2, 3, 4, 5, 6\}$ and $A = \{(1, 2), (1, 3), (2, 3), (2, 6), (3, 1), (3, 4), (3, 5), (4, 1), (5, 2), (6, 3)\}$. We can represent G by giving a list of its vertices:

$$(1, 2, 3, 4, 5, 6)$$

and then, for each vertex, giving an adjacency list for that vertex:

1: (2,3)   2: (3,6)   3: (1,4,5)   4: (1)   5: (2)   6: (3)

(where the notation

$$v : (u_1, u_2, \ldots, u_k)$$

is used to indicate that $(u_1, u_2, \ldots, u_k)$ is an adjacency list for the vertex v.) Moreover, a walk in a digraph can be represented by giving the vertices of the walk in a list. For instance, note that

$$(1, 2, 3, 1, 3, 5, 2, 6, 3, 4, 1)$$

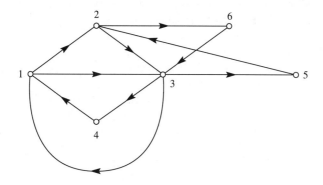

Figure 3.3.4   An eulerian digraph.

represents an eulerian circuit of *G*. The idea of a list of vertices is quite general and useful.

Thus, we assume the existence of a package for the abstract data type VER-TEX_LIST, where each vertex is of the type VERTEX_TYPE and is a natural number as declared, in Ada for instance, by

    subtype VERTEX_TYPE is NATURAL;

This package provides, among others, the following operations:

1. The function EMPTY_LIST returns the empty list (of vertices).
2. The function EMPTY(*L*) determines whether the list *L* is empty.
3. The procedure PUSH(*v*, *L*) modifies the list *L* by inserting the vertex *v* at the first position.
4. The procedure POP(*L*) modifies the list *L* by deleting the first element (assuming *L* is not empty).
5. The function FIRST(*L*) returns the first element of the list *L* (assuming *L* is not empty).
6. The function LENGTH(*L*) returns the length of the list *L*.
7. The procedure SEARCH(*v*, *L*, FOUND) searches the list *L* for (the first occurrence of) the vertex *v*; if *v* is FOUND, then *v* is moved to the first position in *L*.

Relative to these operations, except for the SEARCH procedure, an object *L* of type VERTEX_LIST is viewed as a "stack" or "LIFO (last-in, first-out) list." An element may be "pushed" onto *L* via the PUSH operation; the first or "top" element of *L* may be "popped" from *L* via the POP operation; and the top element may be examined via the FIRST operation.

The type VERTEX_LIST may be implemented as a "linked list." In Ada, for example, the type VERTEX_LIST can be declared as follows:

> type NODE_TYPE;
> type NODE_POINTER is access NODE_TYPE;
> type NODE_TYPE is record
> > VERTEX: VERTEX_TYPE;
> > NEXT: NODE_POINTER;
> > end record;
> type VERTEX_LIST is record
> > LENGTH: NATURAL;
> > START: NODE_POINTER;
> > end record;

With this approach, each of the operations EMPTY_LIST, EMPTY, PUSH, POP, FIRST, and LENGTH is $\Theta(1)$, and the SEARCH operation is $O(n)$, where n is the length of the list being searched.

Before describing the DIGRAPH abstract data type, let's consider an example to see how vertex lists are used to construct an eulerian circuit of an eulerian digraph.

■   *Example 3.3.5*   We construct an eulerian circuit of the digraph G of Figure 3.3.4. It is easily verified that G is strong and $id(v) = od(v)$ for each vertex v. Thus, G is eulerian by Theorem 3.3.6.

Our algorithm uses two vertex lists L_1 and L_2 and a copy G_1 of G. The list L_1 is to contain (eventually) an eulerian circuit of G and is initially empty. The list L_2 is to contain the vertices of a closed circuit of G, in reverse order. We may begin at any vertex u, say, $u = 1$. In phase 1 we use the idea of Lemma 3.3.5 to find a u-u circuit of G_1; as this circuit is traversed, its vertices are pushed onto L_2 and the arcs of the circuit are deleted from G_1. Thus, L_2 holds the vertices of the circuit in reverse order. For instance, in our example it might be that

$$L_2 = (1, 4, 3, 1, 3, 2, 1)$$

so that the arc set of G_1 has been reduced to $\{(2, 6), (3, 5), (5, 2), (6, 3)\}$. In phase 2 we repeat the following step until L_2 is empty or the first vertex u of L_2 is such that $od(u) > 0$ in G_1: pop u from L_2 and push it onto L_1. In our example this phase yields

$$L_1 = (4, 1) \quad \text{and} \quad L_2 = (3, 1, 3, 2, 1)$$

Since L_2 is not empty, we begin again at the first vertex u of L_2, in this case $u = 3$, and return to phase 1 to find a u-u circuit of G_1. In our example this results in

$$L_2 = (3, 6, 2, 5, 3, 1, 3, 2, 1)$$

and G_1 is now empty. We then return to phase 2, where we repeat the step of popping the first vertex u of L_2 and pushing it onto L_1 until od$(u) > 0$ in G_1 or L_2 is empty. This phase yields

$$L_1 = (1, 2, 3, 1, 3, 5, 2, 6, 3, 4, 1) \qquad \text{and} \qquad L_2 = (\ )$$

Since L_2 is empty, L_1 represents an eulerian circuit of G and the algorithm terminates.
∎

The preceding example indicates some of the operations that we would like to have for the abstract data type DIGRAPH. Further thought leads to the following specification of a package of operations for this abstract data type.

1. The procedure COPY_DIGRAPH(G_1, G_2) places a copy of the digraph G_2 in G_1.

2. The procedure INITIALIZE(G, v) initializes G to the digraph with a single vertex v (and no arcs).

3. The function IS_VERTEX(v, G) determines whether v is a vertex of the digraph G.

4. The procedure ADD_VERTEX(v, G) adds the vertex v to the digraph G. (If v is already a vertex of G, then G is unchanged.)

5. The procedure DELETE_VERTEX(v, G) deletes the vertex v and any arcs incident with v from the digraph G. (If v is not a vertex of G, then G is unchanged.)

6. The procedure RELABEL_VERTEX(u, v, G) relabels the vertex u as v in the digraph G. (If u is not a vertex or v is already a vertex of G, then G is unchanged.)

7. The function ADJACENT_TO(u, v, G) determines whether vertex u is adjacent to vertex v in the digraph G.

8. The procedure ADD_ARC(u, v, G) adds the arc (u, v) to the digraph G, if the arc is not already present. (If u or v is not a vertex of G, then G is unchanged.)

9. The procedure DELETE_ARC(u, v, G) deletes the arc (u, v) from the digraph G, if the arc is present. (If u or v is not a vertex of G, then G is unchanged.)

10. The function ORDER(G) returns the order of the digraph G.

11. The function INDEGREE(v, G) returns the indegree of the vertex v in the digraph G.

12. The function OUTDEGREE(v, G) returns the outdegree of the vertex v in the digraph G.

13. The function SIZE(G) returns the size of the digraph G.

14. The function LIST_OF_VERTICES(G) returns a list of the vertices of the digraph G.

15. The function FIRST_VERTEX(G) returns

    FIRST(LIST_OF_VERTICES(G))

    (without having to compute LIST_OF_VERTICES(G)).

16. The function ADJACENCY_LIST(v, G) returns an adjacency list for the vertex v in the digraph G.

17. The function FIRST_NEIGHBOR(v, G) returns

   FIRST(ADJACENCY_LIST(v, G))

(without having to compute ADJACENCY_LIST(v, G)).

One way to implement the type DIGRAPH is the "adjacency list method," in which we keep track of an adjacency list for each vertex. Since the vertices themselves may be linked in a list, this gives us a view of a digraph as a "list of lists." For example, we may view the digraph G of Figure 3.3.4 as follows:

   $((1, (2, 3)), (2, (3, 6)), (3, (1, 4, 5)), (4, (1)), (5, (2)), (6, (3)))$

where each element of the list is a sublist of the form

   $(v, (u_1, u_2, \ldots, u_k))$

indicating that v is a vertex of G and that $(u_1, u_2, \ldots, u_k)$ is an adjacency list for v.
   In Ada, the type DIGRAPH can be declared as follows:

```
type NODE_TYPE;
type NODE_POINTER is access NODE_TYPE;
type NODE_TYPE is record
                VERTEX: VERTEX_TYPE;
                ADJACENCY_LIST: VERTEX_LIST;
                NEXT: NODE_POINTER;
          end record;
type DIGRAPH is record
                ORDER: POSITIVE;
                START: NODE_POINTER;
          end record;
```

Further details of this method of implementation and of the resulting execution times for the operations are discussed later or are left to the exercises. For now, it is best for you to focus on understanding the specification of the operations.
   We can now look at an algorithm for finding an eulerian circuit of an eulerian digraph.

ALGORITHM 3.3.1   Given an eulerian digraph $G = (V, A)$, the function

   EULER_CIRCUIT(G)

returns an eulerian circuit of G, represented as a list of vertices.

EULER_CIRCUIT uses the following local variables:

G_1 is a digraph and is initialized to a copy of G;
L_1 and L_2 are lists of vertices, both initialized to EMPTY_LIST;
The vertex u is initialized to FIRST_VERTEX(G);
v is a vertex;

```
begin
    PUSH(u, L₂);
    MAIN: loop
        while OUTDEGREE(u, G₁) > 0 loop--phase 1
            v := FIRST_NEIGHBOR(u, G₁);
            PUSH(v, L₂);
            DELETE_ARC(u, v, G₁);
            u := v;
        end loop;
        loop--phase 2
            PUSH(u, L₁);
            POP(L₂);
            exit MAIN when EMPTY(L₂);
            u := FIRST(L₂);
            exit when OUTDEGREE(u, G₁) > 0;
        end loop;
    end loop MAIN;
    return L₁;
end EULER_CIRCUIT;
```

■ *Example 3.3.6* Let's trace Algorithm 3.3.1 for the digraph G of Figure 3.3.4.

Initially, G is copied into G_1, L_1 is empty, and, assuming FIRST_VERTEX(G) returns 1, then $u := 1$ and $L_2 = (1)$. Phase 1 is now repeated until OUTDEGREE(u, G_1) = 0. At the first iteration, assume $v :=$ FIRST_NEIGHBOR($1, G_1$) = 2. Then $L_2 = (2, 1)$, the arc $(1, 2)$ is deleted from G_1, and $u := 2$. At the second iteration, assume $v :=$ FIRST_NEIGHBOR($2, G_1$) = 3. Then $L_2 = (3, 2, 1)$, the arc $(2, 3)$ is deleted (from G_1), and $u := 3$. Next, suppose $v :=$ FIRST _NEIGHBOR($3, G_1$) = 1. Then $L_2 = (1, 3, 2, 1)$, the arc $(3, 1)$ is deleted, and $u := 1$. Next, $v :=$ FIRST_NEIGHBOR($1, G_1$) = 3. This results in $L_2 = (3, 1, 3, 2, 1)$, the arc $(1, 3)$ being deleted, and $u := 3$. Next, suppose $v :=$ FIRST_NEIGHBOR($3, G_1$) = 4; then $L_2 = (4, 3, 1, 3, 2, 1)$, the arc $(3, 4)$ is deleted, and $u := 4$. Next, $v :=$ FIRST_NEIGHBOR($4, G_1$) = 1, $L_2 = (1, 4, 3, 1, 3, 2, 1)$, the arc $(4, 1)$ is deleted, and $u := 1$. At this stage the arc set of G_1 is $\{(2, 6), (3, 5), (5, 2), (6, 3)\}$, so that OUTDEGREE($1, G_1$) = 0. Thus, we move to phase 2.

In phase 2, $u = 1$ is pushed onto L_1 and popped from L_2, so that $L_1 = (1)$ and $L_2 = (4, 3, 1, 3, 2, 1)$. Since L_2 is not empty, $u :=$ FIRST(L_2) = 4. Since OUTDEGREE($4, G_1$) = 0, 4 is pushed onto L_1 and popped from L_2, so that $L_1 = (4, 1)$ and $L_2 = (3, 1, 3, 2, 1)$. Again, since L_2 is not empty, $u :=$ FIRST(L_2) = 3. Since OUTDEGREE($3, G_1$) > 0, we return to phase 1.

At the first iteration, $v := \text{FIRST\_NEIGHBOR}(3, G_1) = 5$, $L_2 = (5, 3, 1, 3, 2, 1)$, the arc $(3, 5)$ is deleted, and $u := 5$. At the second iteration, $v := \text{FIRST\_NEIGHBOR}(5, G_1) = 2$, $L_2 = (2, 5, 3, 1, 3, 2, 1)$, the arc $(5, 2)$ is deleted, and $u := 2$. Next, $v := \text{FIRST\_NEIGHBOR}(2, G_1) = 6$, $L_2 = (6, 2, 5, 3, 1, 3, 2, 1)$, the arc $(2, 6)$ is deleted, and $u := 6$. Finally, $v := \text{FIRST\_NEIGHBOR}(6, G_1) = 3$, $L_2 = (3, 6, 2, 5, 3, 1, 3, 2, 1)$, the arc $(6, 3)$ is deleted, and $u := 3$. Notice that the arc set of G_1 is now empty, so that $\text{OUTDEGREE}(3, G_1) = 0$. Thus, phase 2 is entered.

In phase 2, each vertex of L_2 now has outdegree 0 in G_1. Thus, the vertices of L_2 are popped one by one and pushed onto L_1 until L_2 is empty. At this point the algorithm terminates, with

$$L_1 = (1, 2, 3, 1, 3, 5, 2, 6, 3, 4, 1)$$

returned. This list represents an eulerian circuit of G.   ∎

EXERCISE SET 3.3

1. For the digraph G_2' of Example 3.3.4, exhibit each of the following walks:
   (a) A cycle of length 4.
   (b) A path of length 5 whose first vertex is adjacent to its last vertex.
   (c) A trail of maximum length.
   (d) A walk that includes every arc.

2. Find all distances in the digraph $G_3 = (V, A)$, where $V = \{u, v, x, y, z\}$ and
   $$A = \{(u, v), (v, z), (x, v), (y, x), (y, y), (z, u), (z, y)\}$$

3. Find all distances in the digraph G_2' of Example 3.3.4.

4. Give an example of a strong digraph G that does not have a circuit that includes every vertex.

5. Find the indegree and outdegree of each vertex in the digraph G_2 of Example 3.3.3.

6. Prove or disprove: In any digraph, the number of vertices with odd outdegree is even. (Compare with Corollary 3.1.2.)

7. Give an example of a digraph $G = (V, A)$ (of order n), with $V = \{v_1, v_2, \ldots, v_n\}$, such that $\text{od}(v_i) = i - 1$, $1 \le i \le n$. (Compare your example with the result of Exercise 4 of Exercise Set 3.1.)

8. Prove or disprove: If G is a strong digraph, then there is a closed walk of G that includes every arc of G.

9. A digraph $G = (V, A)$ is called **r-regular** provided $\text{id}(v) = \text{od}(v) = r$ for all $v \in V$.
   (a) Give an example of a 2-regular digraph of order 5.
   (b) For arbitrary integers r and n with $0 \le r < n$, construct an r-regular digraph of order n.

10. Prove Theorem 3.3.4.

11. Given a digraph $G = (V, A)$, the **complement** of G is the digraph $\overline{G} = (V, \overline{A})$, where $\overline{A} = (V \times V) - A$. Find the complement of each of the following digraphs:

    (a) The digraph G of Figure 3.3.1.

    (b) The digraph G_2 of Figure 3.3.3.

12. Let G be a digraph of order n. Prove: If $od(u) + id(v) \geq n - 1$ whenever u and v are distinct vertices and u is not adjacent to v, then G is strong. Also, show that this result is the best possible result in the sense that the bound $n - 1$ cannot be lowered to $n - 2$.

13. Let $G_1 = (V_1, A_1)$ and $G_2 = (V_2, A_2)$ be two digraphs. We say that G_1 and G_2 are **isomorphic** provided there is a bijection $\phi : V_1 \rightarrow V_2$ such that, for all $x, y \in V_1$, $(x, y) \in A_1 \leftrightarrow (\phi(x), \phi(y)) \in A_2$. Such a function ϕ is called an **isomorphism**.

    (a) Show that G_1 and G_2 are isomorphic digraphs, where

    $$G_1 = (\{a, b, c, d, e\}, \{(a, b), (a, d), (a, e), (b, c), (c, d), (c, e), (d, e), (e, b)\})$$

    and

    $$G_2 = (\{u, v, x, y, z\}, \{(u, v), (v, x), (x, u), (x, z), (y, u), (y, v), (y, z), (z, u)\})$$

    (b) Determine whether the digraph

    $$G_3 = (\{a, b, c, d, e\}, \{(a, a), (a, e), (b, c), (c, b), (d, c), (d, e), (e, b)\})$$

    is isomorphic to the digraph G of Figure 3.3.1.

14. Let $G_1 = (V_1, A_1)$ and $G_2 = (V_2, A_2)$ be isomorphic digraphs and let $\phi : V_1 \rightarrow V_2$ be an isomorphism. Prove:

    (a) G_1 and G_2 have the same order.

    (b) For each $v \in V_1$, $id(\phi(v)) = id(v)$.

    (c) For each $v \in V_1$, $od(\phi(v)) = od(v)$.

    (d) G_1 and G_2 have the same size.

    (e) If $v_0, v_1, \ldots, v_k$ is a walk (trail, path, circuit, cycle) in G_1, then $\phi(v_0), \phi(v_1), \ldots, \phi(v_k)$ is a walk (trail, path, circuit, cycle) in G_2.

15. Give an example (preferably of smallest possible order) of nonisomorphic digraphs $G_1 = (V_1, A_1)$ and $G_2 = (V_2, A_2)$ such that:

    (a) G_1 and G_2 have the same order.

    (b) G_1 and G_2 have the same order and the same size.

    (c) G_1 and G_2 have the same order n, $V_1 = \{u_1, u_2, \ldots, u_n\}$, $V_2 = \{v_1, v_2, \ldots, v_n\}$, and $id(u_i) = id(v_i)$ and $od(u_i) = od(v_i)$ for each i, $1 \leq i \leq n$.

16. Show that the relation "is isomorphic to" is an equivalence relation on the set of directed graphs.

17. Determine, up to isomorphism, all loopless directed graphs of order 3.

18. Prove Lemma 3.3.5.

19. Prove Theorem 3.3.6.

20. Apply Algorithm 3.3.1 to each of the following digraphs:

    (a) $G_1 = (\{1, 2, 3, 4, 5, 6, 7\}, \{(1, 2), (1, 4), (2, 1), (2, 5), (3, 1), (3, 2), (3, 7), (4, 3),$
    $(4, 7), (5, 4), (5, 6), (6, 3), (6, 5), (7, 3), (7, 6)\})$

(b) $G_2 = (\{1, 2, 3, 4, 5, 6\}, \{(1, 2), (1, 6), (2, 4), (2, 5), (3, 2), (3, 4), (4, 3), (4, 5),$
$(5, 1), (5, 6), (6, 1), (6, 3)\})$

(c) $G_3 = (\{1, 2, 3, 4, 5, 6, 7, 8, 9\}, \{(1, 2), (2, 3), (2, 6), (2, 8), (3, 4), (3, 9), (4, 5),$
$(5, 1), (5, 2), (6, 5), (6, 7), (7, 2), (7, 8), (8, 3), (8, 9), (9, 6), (9, 7)\})$

21. Let $G_1 = (V_1, A_1)$ be the digraph of Figure 3.3.2 and let $G_1' = (V_1, A_1')$, where
$A_1' = A_1 \cup \{(t, z), (v, u), (v, z), (z, u)\}$. Find an eulerian circuit of G_1'.

22. Let $G = (V, A)$ be a loopless digraph such that, for some $k \in \mathbb{N}$, od$(v) \geq k$ for all
$v \in V$. Prove that G contains:

(a) A path whose length is at least k.
(b) A cycle whose length is at least $k + 1$.

23. Given a digraph $G = (V, A)$, the **converse** of G is the digraph $G^c = (V, A^c)$, where
$A^c = \{(v, u) \mid (u, v) \in A\}$. Find the converse of each of the following digraphs:

(a) The digraph G of Figure 3.3.1.
(b) The digraph G_2 of Figure 3.3.3.

24. Let $G = (V, A)$ be a loopless digraph such that, for some $k \in \mathbb{N}$, id$(v) \geq k$ for all
$v \in V$. Use the result of Exercise 22 and the idea of the converse (Exercise 23)
to show that G contains:

(a) A path whose length is at least k.
(b) A cycle whose length is at least $k + 1$.

25. Consider applying Algorithm 3.3.1 to a digraph $G = (V, A)$ that is not eulerian.

(a) What happens if G is strong but id$(v) \neq$ od(v) for some $v \in V$?
(b) What happens if id$(v) =$ od(v) for all $v \in V$ but G is not strong?

26. With respect to the DIGRAPH operations used in Algorithm 3.3.1, justify the
following observations, where n is the order of G:

(a) COPY_DIGRAPH(G_1, G) is $O(n^2)$.
(b) FIRST_VERTEX(G) is $\Theta(1)$.
(c) OUTDEGREE(v, G) is $O(n)$ in general, but is $\Theta(1)$ if
$v =$ FIRST_VERTEX(G).
(d) FIRST_NEIGHBOR(v, G) is $O(n)$ in general, but is $\Theta(1)$ if
$v =$ FIRST_VERTEX(G).
(e) DELETE_ARC(u, v, G) is $O(n)$ in general and is $\Theta(1)$ if
$u =$ FIRST_VERTEX(G) and $v =$ FIRST_NEIGHBOR(u).

27. Let G be an eulerian digraph of order n and size m.

(a) Justify the observation that Algorithm 3.3.1 is $O(mn)$.
(b) Suppose that the package of operations for DIGRAPH includes a procedure VERTEX_SEARCH$(v, G,$ FOUND$)$ that searches the digraph G for the
vertex v; if FOUND, then v becomes the first vertex of G (so that $v =$
FIRST_VERTEX(G)). Use this operation to improve (slightly) the efficiency
of Algorithm 3.3.1.

3.4 *STRONG COMPONENTS, REACHABILITY, AND DISTANCE*

Given a digraph $G = (V, A)$ and $u \in V$, we may be interested in knowing which vertices are reachable from u, or in finding $d(u, v)$ for all $v \in V$. More generally, we may be interested in knowing for which pairs of vertices (u, v) v is reachable from u, or in finding $d(u, v)$ for all $u, v \in V$. So, we now consider several algorithms for answering questions of this sort.

As is usually the case when a particular kind of mathematical structure is studied, we encounter the important notion of "substructure." In this regard, graphs and digraphs are no exception.

> **DEFINITION 3.4.1**   Let $G_1 = (V_1, A_1)$ and $G_2 = (V_2, A_2)$ be two digraphs. We call G_1 a **subdigraph** of G_2 provided $V_1 \subseteq V_2$ and $A_1 \subseteq A_2$. In this case we also call G_2 a **superdigraph** of G_1. The fact that G_1 is a subdigraph of G_2 is denoted by writing $G_1 \subseteq G_2$.   ∎

For example, let $G = (V, A)$ be a digraph. For $v \in V$, the notation $G - v$ is used to denote the subdigraph G_1 of G obtained by deleting v and all arcs of G incident with v; that is, $G_1 = (V_1, A_1)$, where $V_1 = V - \{v\}$ and $A_1 = A - \{(x, y) \mid x = v$ or $y = v\}$. Similarly, if $(u, v) \in A$, then $G - (u, v)$ denotes the subdigraph of G with vertex set V and arc set $A - \{(u, v)\}$. On the other hand, if $x, y \in V$ and $(x, y) \notin A$, then $G + (x, y)$ denotes the superdigraph of G with vertex set V and arc set $A \cup \{(x, y)\}$. In general, if $H = (V', A')$ is a subdigraph of G with $V' = V$, then H is called a **spanning subdigraph** of G. Thus, for example, if $u, v, x, y \in V$, $(u, v) \in A$, and $(x, y) \notin A$, then $G - (u, v)$ is a spanning subdigraph of G, which in turn is a spanning subdigraph of $G + (x, y)$.

■ *Example 3.4.1*   Recall the digraph

$$G_1 = (\{t, u, v, w, x, y, z\}, \{(t, w), (u, t), (u, x), (v, t), (w, v), (w, z), (x, w),$$
$$(x, y), (y, v), (y, x), (z, v), (z, y)\})$$

shown in Figure 3.3.2. The subdigraphs $H_1 = G_1 - w$ and $H_2 = G_1 - (y, x)$, and the superdigraph $H_3 = G_1 + (w, u)$ are shown in Figure 3.4.1(a), (b), and (c), respectively.   ■

Given a digraph $G = (V, A)$, recall that the relation A on V is transitive provided it satisfies the condition

$$(u, v) \in A \text{ and } (v, w) \in A \text{ implies } (u, w) \in A$$

for all $u, v, w \in V$. If A is not transitive, then there are vertices u_1, v_1, and w_1 (not necessarily distinct) such that $(u_1, v_1) \in A$ and $(v_1, w_1) \in A$ but $(u_1, w_1) \notin A$. Suppose we let $A_1 = A \cup \{(u_1, w_1)\}$. We may then check whether A_1 is a transitive relation on V.

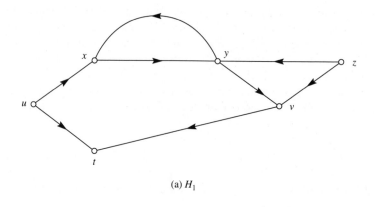

(a) H_1

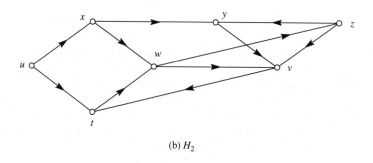

(b) H_2

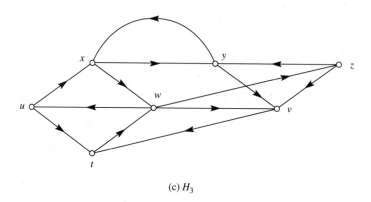

(c) H_3

Figure 3.4.1

If A_1 is not transitive, then there are vertices u_2, v_2, and w_2 such that $(u_2, v_2) \in A$, and $(v_2, w_2) \in A$, but $(u_2, w_2) \notin A$. By letting $A_2 = A_1 \cup \{(u_2, w_2)\}$, we may check whether A_2 is transitive, and so on. Since the complete relation $V \times V$ is transitive, this process must eventually come to a relation A^t such that A^t is transitive and $A \subseteq A^t$. What we are seeking is, in some sense, the "smallest" transitive relation on V that contains A.

DEFINITION 3.4.2 Let V be a set and let A be a relation on V. The **transitive closure** of A is the relation A^t on V defined by the following three conditions:

1. $A \subseteq A^t$
2. A^t is transitive.
3. For any relation A' on V, if $A \subseteq A'$ and A' is transitive, then $A^t \subseteq A'$.

If V is finite, then the digraph $G^t = (V, A^t)$ is called the **transitive closure** of the digraph $G = (V, A)$. ∎

Given a digraph $G = (V, A)$, note that G is a spanning subdigraph of its transitive closure $G^t = (V, A^t)$. Condition 3 of the definition makes precise the sense in which A^t is the smallest transitive relation that contains A; it requires that A^t be a subset of any transitive relation A' such that A is a subset of A'.

The conditions of Definition 3.4.2 make the transitive closure well defined (see Exercise 2) but give no hint how to find it. In the case that V is finite, we can give an alternate, recursive, definition based on the discussion preceding Definition 3.4.2.

ALTERNATE DEFINITION 3.4.2 Let V be a finite set and let A be a relation on V. The **transitive closure** of A is the relation A^t on V defined recursively as follows:

1. If A is transitive, then $A^t = A$.
2. If A is not transitive, let $u, v, w \in V$ such that $(u, v) \in A$ and $(v, w) \in A$, but $(u, w) \notin A$; then

$$A^t = (A \cup \{(u, w)\})^t$$ ∎

This alternate definition indicates how to find the transitive closure but does not indicate that the transitive closure is well defined. If A is not transitive, then condition 2 of the alternate definition is applied, and there may be several different choices for the arc (u, w) that is added to the arc set A^t being constructed. We must verify that the final result is independent of the order in which such arcs are chosen. One way to do this is as follows: let A^t be the transitive closure of A as defined by Definition 3.4.2 and let A_1^t be a relation that is constructed from A according to Alternate Definition 3.4.2; show that $A_1^t = A^t$. Clearly, $A \subseteq A_1^t$, and it is also not difficult to see that A_1^t is transitive. Therefore, to complete the verification, we must show that A_1^t satisfies condition 3 of Definition 3.4.2. The details are left for you to work out in Exercise 2. Theorem 3.4.1 provides the following helpful result, which relates the idea of transitivity to the notion of reachability.

THEOREM 3.4.1   For any digraph $G = (V, A')$, the relation A' is transitive if and only if it satisfies the following condition for any $u, w \in V$:

$(u, w) \in A'$ if and only if there is a nontrivial u-w walk in G

Proof   First, observe that if $(u, v) \in A'$, then clearly there is a nontrivial u-v walk in G (of length 1).

To see that the condition is necessary, assume A' is transitive. We must show that if there is a nontrivial u-w walk W in G, then $(u, w) \in A'$. We proceed by induction of the length n of W. The result is obvious if $n = 1$. So let k be an arbitrary positive integer; the induction hypothesis is that if there is a nontrivial u-w walk of length k, then $(u, w) \in A'$. Suppose that there is in G a u-w walk W of length $k + 1$, say,

$$u = u_0, u_1, u_2, \ldots, u_{k+1} = w$$

Since $(u_0, u_1) \in A'$ and $(u_1, u_2) \in A'$ and A' is transitive, we have $(u_0, u_2) \in A'$. Hence,

$$u_0, u_2, \ldots, u_{k+1}$$

is a u-w walk of length k. Thus, by the induction hypothesis, $(u, w) \in A'$, and this completes the proof of necessity.

To prove sufficiency, suppose that A' satisfies the condition of the theorem; we must show that A' is transitive. Let $u, v, w \in V$ and suppose that $(u, v) \in A'$ and $(v, w) \in A'$. We must show that $(u, w) \in A'$. This is obvious if $u = v$ or $v = w$, so we may assume that $u \neq v$ and $v \neq w$. Then u, v, w is a u-w walk of length 2; hence, by the condition, $(u, w) \in A$.   □

We next consider Warshall's algorithm, which is an efficient method for finding the transitive closure of a given digraph $G = (V, A)$ [See S. Warshall, A theorem on Boolean matrices, *Journal of the ACM*, 9 (1962): 11–12]. The following definition is useful in this context and in other parts of graph theory.

DEFINITION 3.4.3   Let $G = (V, A)$ be a digraph of order n with $V = \{v_1, v_2, \ldots, v_n\}$. The **adjacency matrix** of G is the n by n, $(0, 1)$-matrix $M = [m_{ij}]$ defined by

$$m_{ij} = \begin{cases} 0 & \text{if } (v_i, v_j) \notin A \\ 1 & \text{if } (v_i, v_j) \in A \end{cases}$$

(Note: a matrix, each of whose entries is 0 or 1, is called a **(0, 1)-matrix**.)   ∎

■ ***Example 3.4.2*** Let $G = (V, A)$, where $V = \{v_1, v_2, v_3, v_4, v_5\}$ and $A = \{(v_1, v_4),$ $(v_2, v_1), (v_2, v_3), (v_3, v_4), (v_4, v_3), (v_5, v_1), (v_5, v_5)\}$. The adjacency matrix of G is

$$M = \begin{bmatrix} 0 & 0 & 0 & 1 & 0 \\ 1 & 0 & 1 & 0 & 0 \\ 0 & 0 & 0 & 1 & 0 \\ 0 & 0 & 1 & 0 & 0 \\ 1 & 0 & 0 & 0 & 1 \end{bmatrix}$$

■

Let $G = (V, A)$ be a digraph with $V = \{v_1, v_2, \ldots, v_n\}$ and let M be the adjacency matrix of G. Let $G^t = (V, A^t)$ be the transitive closure of G, and let M^t denote the adjacency matrix of G^t. We can effectively compute A^t by computing M^t from M; thus, let m_{ij} denote the (i, j)-entry of M^t. Initially, $M^t = M$. This reflects the fact that $(v_i, v_j) \in A^t$ if $(v_i, v_j) \in A$. We then search for subscripts i, j, k such that $m_{ij} = 1 = m_{jk}$ and $m_{ik} = 0$. If no such subscripts are found, then M^t is the adjacency matrix of a transitive relation and we are done. Otherwise, we set $m_{ik} = 1$ and repeat this last step. Note that we are simply applying the recursive definition of the transitive closure. This is the essential idea of Warshall's algorithm, except that Warshall's algorithm does the "subscript search" in a rather efficient manner. Before describing the algorithm, let's look at an example.

■ ***Example 3.4.3*** Let $G = (\{1, 2, 3, 4\}, \{(1, 2), (2, 2), (2, 3), (3, 1), (3, 4)\})$. Use the method described in the preceding paragraph to compute the adjacency matrix M^t of G^t.

Solution
We begin by initializing M^t to the adjacency matrix for G, so

$$M^t = \begin{bmatrix} 0 & 1 & 0 & 0 \\ 0 & 1 & 1 & 0 \\ 1 & 0 & 0 & 1 \\ 0 & 0 & 0 & 0 \end{bmatrix}$$

We then search for subscripts i, j, k such that $m_{ij} = 1 = m_{jk}$ and $m_{ik} = 0$. Note that this happens when $i = 3$, $j = 1$, and $k = 2$. We thus set $m_{32} = 1$, yielding

$$M^t = \begin{bmatrix} 0 & 1 & 0 & 0 \\ 0 & 1 & 1 & 0 \\ 1 & 1 & 0 & 1 \\ 0 & 0 & 0 & 0 \end{bmatrix}$$

Again we search for subscripts i, j, k such that $m_{ij} = 1 = m_{jk}$ and $m_{ik} = 0$. This time we find $i = 1$, $j = 2$, and $k = 3$. Hence, we set $m_{13} = 1$, and so

$$M^t = \begin{bmatrix} 0 & 1 & 1 & 0 \\ 0 & 1 & 1 & 0 \\ 1 & 1 & 0 & 1 \\ 0 & 0 & 0 & 0 \end{bmatrix}$$

Next we find $i = 3$, $j = 2$, and $k = 3$. It is important to note that here we are using m_{32}, which was changed to 1 in a previous step. Hence, we set $m_{33} = 1$. Continuing in this way, we find:

$$i = 1, j = 3, k = 1, \qquad \text{so that } m_{11} = 1$$
$$i = 1, j = 3, k = 4, \qquad \text{so that } m_{14} = 1$$
$$i = 2, j = 3, k = 1, \qquad \text{so that } m_{21} = 1$$
$$i = 2, j = 3, k = 4, \qquad \text{so that } m_{24} = 1$$

At this point,

$$M^t = \begin{bmatrix} 1 & 1 & 1 & 1 \\ 1 & 1 & 1 & 1 \\ 1 & 1 & 1 & 1 \\ 0 & 0 & 0 & 0 \end{bmatrix}$$

We can check that there are no more subscripts i, j, k such that $m_{ij} = m_{jk} = 1$ and $m_{ik} = 0$, so that this last matrix is the adjacency matrix for G^t. Thus, $G^t = (\{1, 2, 3, 4\}, \{1, 2, 3\} \times \{1, 2, 3, 4\})$. ∎

Our presentation of Warshall's algorithm makes use of the abstract data type DIGRAPH and the package of operations provided for it, as described at the end of the last section. Moreover, we assume that $G = (V, A)$, where $V = \{1, 2, \ldots, n\}$ for some positive integer n. (This is not really a limitation, because a digraph with some other vertex set could have its vertices (temporarily) relabeled, using the RELABEL operation.) Given such a digraph, it is a simple matter to use the operations provided by the package to implement a function

ADJACENCY_MATRIX(G)

that returns the adjacency matrix of a given digraph G. The adjacency matrix of G is a useful data structure for Warshall's algorithm because the algorithm repeatedly asks the question

Is vertex u adjacent to vertex v?

The adjacency matrix allows us to answer this question in time $\Theta(1)$ by simply consulting the (u, v)-entry of the adjacency matrix. Recall that the package for DIGRAPH provides the operation

ADJACENT_TO(u, v, G)

for answering the same question, but this operation is $O(n)$ in the worst case. This is because the list structure of G must be searched for the vertex u and then the adjacency list for u must be searched for v.

One final point before we present Warshall's algorithm. The algorithm contains the assignment statement

$m_{uw} := m_{uw}$ or $(m_{uv}$ and $m_{vw})$;

Here we are thinking of the adjacency matrix as a "Boolean matrix," with the values 0 and 1 representing the Boolean values FALSE and TRUE, respectively. Thus, this statement can change the value of m_{uw} only from 0 (FALSE) to 1 (TRUE), and this will happen only when m_{uv} and m_{vw} are both 1 (TRUE).

ALGORITHM 3.4.1 (Warshall's algorithm)   Given a digraph $G = (V, A)$, with $V = \{1, 2, \ldots, n\}$, the function

TRANSITIVE_CLOSURE(G)

returns the transitive closure $G^t = (V, A^t)$ of G.
TRANSITIVE_CLOSURE uses the following local variables:

n is a positive integer and is initialized to ORDER(G);
M is an $n \times n$ $(0, 1)$-matrix and is initialized to ADJACENCY_MATRIX(G);

```
begin
    --Compute the adjacency matrix of G'.
    for v in 1 .. n loop
        for u in 1 .. n loop
            for w in 1 .. n loop
                m_uw := m_uw or (m_uv and m_vw);
            end loop;
        end loop;
    end loop;
    --Build G'.
    INITIALIZE(G', 1);
    for v in 2 .. n loop
        ADD_VERTEX(v, G');
    end loop;
```

```
    for u in reverse 1 .. n loop
        for v in reverse 1 .. n loop
            if m_uv = 1 then
                ADD_ARC(u, v, G^t);
            end if;
        end loop;
    end loop;
    return G^t;
end TRANSITIVE_CLOSURE;
```

Example 3.4.3 essentially employs Warshall's algorithm to compute the transitive closure, so you may wish to review that example at this point.

It is not difficult to see that, if Algorithm 3.4.1 sets $m_{uw} = 1$ at some point, then G contains a nontrivial u-w walk; hence, $(u, w) \in A^t$. What is not clear is the converse assertion that, if $(x, z) \in A^t$, then Algorithm 3.4.1 sets $m_{xz} = 1$ at some point. This is certainly true if $(x, z) \in A$, but what if G contains an x-z walk of length 2 or more? In this case we let P be an x-z path (or cycle for the case $x = z$) of length 2 or more, and we let y be the largest intermediate vertex on P. (If $u_0, u_1, \ldots, u_{k-1}, u_k$ is a path or cycle of length $k \geq 2$, then $u_1, \ldots, u_{k-1}$ are its intermediate vertices.) We claim that $m_{xz} = 1$ after the statement

$$m_{uw} := m_{uw} \text{ or } (m_{uv} \text{ and } m_{vw}); \tag{*}$$

is executed with $v = y$, $u = x$, and $w = z$. We now prove this claim by induction (the strong form) on y.

If $y = 1$, then necessarily P has length 2. Thus, $(x, y), (y, z) \in A$, so that $m_{xy} = m_{yz} = 1$, initially. Hence, when statement (*) is executed with $v = y = 1$, $u = x$, and $w = z$, m_{xz} is set equal to 1.

Now let y be a positive integer, $y > 1$. The induction hypothesis is that, for all $a, c \in V$, if G contains an a-c path (or cycle) of length 2 or more and b is the largest intermediate vertex on this path, where $1 \leq b < y$, then $m_{ac} = 1$ after statement (*) of the algorithm is executed with $v = b$, $u = a$, and $w = c$.

Now let P' and P'' be those parts of P from x to y and from y to z, respectively, so that P' is an x-y path and P'' is a y-z path. If P' has length 1, then $(x, y) \in A$, and so $m_{xy} = 1$, initially. On the other hand, if the length of P' is greater than 1, we let b_1 be the largest intermediate vertex on P'. Since b_1 is also an intermediate vertex of P, $b_1 < y$, and so, by the induction hypothesis, $m_{xy} = 1$ after statement (*) is executed with $v = b_1$, $u = x$, and $w = y$. In a similar manner, working with P'', we can argue that $m_{yz} = 1$ either initially or after statement (*) is executed with $v = b_2$, $u = y$, and $w = z$, where $b_2 < y$. Hence, when statement (*) is executed with $v = y$, $u = x$, and $w = z$, we have $m_{xy} = m_{yz} = 1$, and so m_{xz} is set to 1. This completes the proof of the correctness of Warshall's algorithm.

Again, we let $G = (V, A)$ be a digraph and let U be a nonempty subset of V. The **subdigraph of G induced by** U, denoted by $\langle U \rangle$, has vertex set U and arc set consisting of those arcs of G joining vertices of U, that is, $\langle U \rangle = (U, A')$, where

$$A' = \{(x, y) \in A \mid x \in U \text{ and } y \in U\}$$

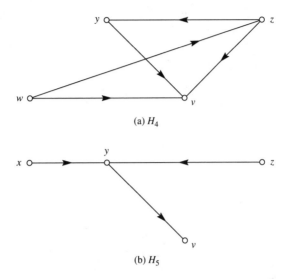

(a) H_4

(b) H_5

Figure 3.4.2

A subdigraph H of G is called an **induced subdigraph** provided $H = \langle U \rangle$ for some subset U of V. Similarly, for a nonempty subset B of A, the **subdigraph of G induced by B**, denoted by $\langle B \rangle$, has arc set B and vertex set consisting of those vertices of V incident with some arc of B. A subdigraph H of G is called **arc induced** provided $H = \langle B \rangle$ for some subset B of A.

■ *Example 3.4.4*   Again, we let G_1 be the digraph of Figure 3.3.2. The subdigraphs $H_4 = \langle \{v, w, y, z\} \rangle$ and $H_5 = \langle \{(x, y), (y, v), (z, y)\} \rangle$ are shown in Figure 3.4.2(a) and (b), respectively.   ■

Now we let $G = (V, A)$ be a digraph and let $v \in V$. Then $\langle \{v\} \rangle$, the subdigraph of G consisting of v alone, is certainly strong. Thus, it makes sense to look for a maximal subset U of V such that $\langle U \rangle$, the subdigraph of G induced by U, is strong.

DEFINITION 3.4.4   Let $G = (V, A)$ be a digraph and let U be a nonempty subset of V. If $\langle U \rangle$ is strong, but $\langle X \rangle$ is not strong for any subset X of V that properly contains U, then U is called a **strong set** and $\langle U \rangle$ is called a **strong component** of G.   ■

■ *Example 3.4.5*   Let $G = (V, A)$ be the digraph shown in Figure 3.4.3 and let $U = \{1, 3, 6, 8\}$. Note that $1, 3, 6, 1, 3, 8, 1$ is a closed walk of G, so $\langle U \rangle$ is strong. Furthermore, it can be checked that for any $v \in V - U$, either v cannot reach some vertex of U or some vertex of U cannot reach v. It follows that U is a strong set of G and that $\langle U \rangle = (\{1, 3, 6, 8\}, \{(1, 3), (3, 6), (3, 8), (6, 1), (8, 1)\})$ is a strong component of G.   ■

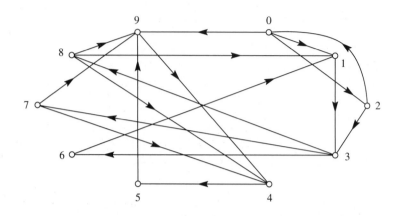

Figure 3.4.3

THEOREM 3.4.2   The strong sets of a digraph $G = (V, A)$ form a partition of V.

Proof   From the remark made just before Definition 3.4.4 and the definition, it follows that the strong sets of G are nonempty and that each vertex of G belongs to some strong set. It remains to show that the strong sets of G are pairwise disjoint. So let U_1 and U_2 be strong sets of G and suppose that $v \in U_1 \cap U_2$ for some $v \in V$. We claim that $\langle U_1 \cup U_2 \rangle$ is strong. To see this, let $u \in U_1$ and $w \in U_2$. Since $u, v \in U_1$ and $\langle U_1 \rangle$ is strong, v is reachable from u. Similarly, since $v, w \in U_2$ and $\langle U_2 \rangle$ is strong, w is reachable from v. It follows that w is reachable from u. By a similar argument, u is reachable from w, and so the claim is verified. It now follows from the maximality of strong sets that $U_1 = U_1 \cup U_2$, namely, $U_1 = U_2$. This shows that the strong sets of G are pairwise disjoint.   □

As a consequence of Theorem 3.4.2, the vertex set V of a given digraph G may be partitioned as $\{U_1, U_2, \ldots, U_k\}$, where each U_i is a strong set of G. Of course, if G itself is strong, then $k = 1$ and $\{V\}$ is such a partition, whereas, if G is not strong, then $k \geq 2$.

We need an algorithm for finding the strong sets of a digraph $G = (V, A)$. The algorithm will be based on two observations, presented as lemmas. Before proceeding to these, we need the following definition. For $Y \subseteq V$, the **outdegree** of Y, written od(Y), is defined as the number of arcs from vertices of Y to vertices of $V - Y$. The **indegree** of Y, written id(Y), is defined analogously.

LEMMA 3.4.3   Let $G = (V, A)$ be a digraph and let $Y_1, Y_2, \ldots, Y_k$ be nonempty subsets of V such that each $\langle Y_i \rangle$ is strong and, for $1 \leq j < k$, there is an arc of G from some vertex $u \in Y_j$ to some vertex $v \in Y_{j+1}$. Then:

1.   If od(Y_k) = 0, then Y_k is a strong set of G.

2. If $od(Y_k) > 0$ and a vertex of Y_k is adjacent to a vertex of Y_j for some j, $1 \le j < k$, then $\langle Y_j \cup Y_{j+1} \cup \cdots \cup Y_k \rangle$ is strong. □

The proof of Lemma 3.4.3 is not difficult and is left for you to work out in Exercise 4. The following example illustrates an algorithm, based on Lemma 3.4.3, which finds one strong set of a given directed graph.

■ **Example 3.4.6** Consider the digraph $G = (V, A)$ of Figure 3.4.3. Initially, start at any vertex u, say, $u = 0$; let $Y_0 = V - \{u\}$, $Y_1 = \{u\}$, and $k = 1$. We now repeat the following steps until $od(Y_k) = 0$:

1. Choose $u \in Y_k$ and $v \in Y_j$, $j < k$, such that $(u, v) \in A$.
2. If $j = 0$, then $k := k + 1$, $Y_k := \{v\}$, and $Y_0 := Y_0 - \{v\}$; otherwise [by applying Lemma 3.4.3(2)], $Y_j := Y_j \cup Y_{j+1} \cup \cdots \cup Y_k$ and $k := j$.

We trace these steps as follows:

Iteration 1: Choose $u = 0 \in Y_1$ and $v = 1 \in Y_0$. Since $j = 0$, we have $k := 2$, $Y_2 := \{1\}$, and $Y_0 := \{2, 3, 4, 5, 6, 7, 8, 9\}$.

Iteration 2: Our only choice is $u = 1 \in Y_2$ and $v = 3 \in Y_0$. Since $j = 0$, we have $k := 3$, $Y_3 := \{3\}$, and $Y_0 := \{2, 4, 5, 6, 7, 8, 9\}$.

Iteration 3: Suppose $u = 3 \in Y_3$ and $v = 6 \in Y_0$ are chosen; then $k := 4$, $Y_4 := \{6\}$, and $Y_0 := \{2, 4, 5, 7, 8, 9\}$.

Iteration 4: Next, $u = 6 \in Y_4$ and $v = 1 \in Y_2$. Since $j = 2 \ne 0$, $Y_2 := Y_2 \cup Y_3 \cup Y_4 = \{1, 3, 6\}$, and $k := 2$.

Iteration 5: Suppose $u = 3 \in Y_2$ and $v = 7 \in Y_0$; then $k := 3$, $Y_3 := \{7\}$, and $Y_0 := \{2, 4, 5, 8, 9\}$.

Iteration 6: Next, suppose $u = 7 \in Y_3$ and $v = 4 \in Y_0$; then $k := 4$, $Y_4 := \{4\}$, and $Y_0 := \{2, 5, 8, 9\}$.

Iteration 7: Next, $u = 4 \in Y_4$ and $v = 5 \in Y_0$; thus, $k := 5$, $Y_5 := \{5\}$, and $Y_0 := \{2, 8, 9\}$.

Iteration 8: Next, $u = 5 \in Y_5$ and $v = 9 \in Y_0$, so $k := 6$, $Y_6 := \{9\}$, and $Y_0 := \{2, 8\}$.

Iteration 9: Next, $u = 9 \in Y_6$ and $v = 4 \in Y_4$. Since $j = 4$, we have $Y_4 := Y_4 \cup Y_5 \cup Y_6 = \{4, 5, 9\}$ and $k := 4$.

At this point, $od(Y_4) = 0$, and so it follows from Lemma 3.4.3(1) that $Y_4 = \{4, 5, 9\}$ is a strong set of G. ■

Algorithm 3.4.2 finds a strong set of a given directed graph G. The algorithm uses sets $Y_1, Y_2, \ldots, Y_k$ with the properties described in Lemma 3.4.3, with $Y_0 = V - (Y_1 \cup Y_2 \cup \cdots \cup Y_k)$. Since each Y_j is a set of vertices of G and since such a set can be represented as a list of vertices, our algorithm uses the type VERTEX_LIST and the package of operations for this type as described in Section 3.3. In addition to these operations, we need the additional function

COPY_LIST(L)

that returns a copy of the list L.

ALGORITHM 3.4.2   Given a directed graph $G = (V, A)$, the function

STRONG_SET(G)

returns, as a list of vertices, a strong set of G.
   STRONG_SET uses the following local variables:

> $(Y(0), Y(1), \ldots, Y(\mathrm{ORDER}(G)))$ is an array of vertex lists;
> k is a positive integer and is initially 1;
> j is a natural number;

STRONG_SET also uses the following local function and procedure:
   The function TEST determines whether $\mathrm{od}(Y(k)) = 0$; if not, it finds a vertex v in $Y(j)$, for some $j < k$, such that v is adjacent from some vertex u in $Y(k)$ and makes $v = \mathrm{FIRST}(Y(j))$.
   TEST uses the following local variables:

> S is a vertex list and is intialized to COPY_LIST($Y(k)$);
> T is a vertex list;
> v is a vertex;
> V_FOUND is of type BOOLEAN;

```
begin
   MAIN: while not EMPTY(S) loop
      T := ADJACENCY_LIST(FIRST(S), G);
      while not EMPTY(T) loop
         v := FIRST(T);
         j := 0;
         while j < k loop
            SEARCH(v, Y(j), V_FOUND);
            exit MAIN when V_FOUND;
            j := j + 1;
         end loop;
         POP(T);
      end loop;
      POP(S);
   end loop MAIN;
      return not V_FOUND;
end TEST;
```

The procedure UNION makes $Y(j)$ the union of the sets

$$Y(j), Y(j + 1), \ldots, Y(k).$$

```
            begin
                for i in j + 1 .. k loop
                    while not EMPTY(Y(i)) loop
                        PUSH(FIRST(Y(i)), Y(j));
                        POP(Y(i));
                    end loop;
                end loop;
            end UNION;
```

STRONG_SET is then defined as follows:

```
        begin
            Y(0) := LIST_OF_VERTICES(G);
            Y(1) := EMPTY_LIST;
            PUSH(FIRST(Y(0)), Y(1));
            POP(Y(0));
            while not TEST loop
                if j = 0 then
                    k := k + 1;
                    Y(k) := EMPTY_LIST;
                    PUSH(FIRST(Y(0)), Y(k));
                    POP(Y(0));
                else
                    UNION;
                    k := j;
                end if;
            end loop;
            return Y(k);
        end STRONG_SET;
```

■ *Example 3.4.7*   Trace Algorithm 3.4.2 for the digraph G of Figure 3.4.3.

Solution

Assume LIST_OF_VERTICES(G) = $(0, 1, 2, 3, 4, 5, 6, 7, 8, 9)$, so that $Y(0) = (1, 2, 3, 4, 5, 6, 7, 8, 9)$, $Y(1) = (0)$, and $k = 1$, initially. The main while loop is now repeated as long as TEST returns false.

Iteration 1: Assume TEST makes $j = 0$ and FIRST($Y(0)$) = 1. Then $k := 2$, $Y(2) = (1)$, and $Y(0) = (2, 3, 4, 5, 6, 7, 8, 9)$.

Iteration 2: As a result of TEST, $j = 0$ and FIRST($Y(0)$) = 3. Thus, $k := 3$, $Y(3) = (3)$, and $Y(0) = (2, 4, 5, 6, 7, 8, 9)$.

Iteration 3: Assume TEST makes $j = 0$ and FIRST($Y(0)$) = 6. Then $k := 4$, $Y(4) = (6)$, and $Y(0) = (2, 4, 5, 7, 8, 9)$.

Iteration 4: As a result of TEST, $j = 2$. Thus, UNION is called, resulting in $Y(2) = (6, 3, 1)$. Also, $k := 2$.

Iteration 5: Assume TEST makes $j = 0$ and FIRST($Y(0)$) $= 7$. Then $k := 3$, $Y(3) = (7)$, and $Y(0) = (2, 4, 5, 8, 9)$.

Iteration 6: Assume TEST makes $j = 0$ and FIRST($Y(0)$) $= 4$. Then $k := 4$, $Y(4) = (4)$, and $Y(0) = (2, 5, 8, 9)$.

Iteration 7: As a result of TEST, $j = 0$ and FIRST($Y(0)$) $= 5$. Then $k := 5$, $Y(5) = (5)$, and $Y(0) = (2, 8, 9)$.

Iteration 8: As a result of TEST, $j = 0$ and FIRST($Y(0)$) $= 9$. Then $k := 6$, $Y(6) = (9)$, and $Y(0) = (2, 8)$.

Iteration 9: As a result of TEST, $j = 4$. Thus, UNION is called, resulting in $Y(4) = (9, 5, 4)$. Also, $k := 4$.

At this point, $Y(4)$ has outdegree 0, and so TEST returns TRUE. Hence, $Y(4)$ is returned as a strong set of G. $\blacksquare$

LEMMA 3.4.4   Let $G = (V, A)$ be a digraph. If U_1 is a strong set of G and U_2 is a strong set of $\langle V - U_1 \rangle$, then U_2 is a strong set of G. $\square$

The proof of Lemma 3.4.4 is left for you to provide in Exercise 6. Our result here suggests a method to find the strong sets of $G = (V, A)$ by applying Algorithm 3.4.2 recursively. First, we apply Algorithm 3.4.2 to find a strong set U_1 of G. Next, if $U_1 \neq V$, we apply Algorithm 3.4.2 to the digraph $\langle V - U_1 \rangle$ to find a strong set U_2. By Lemma 3.4.4, U_2 is also a strong set of G. Next, if $U_1 \cup U_2 \neq V$, we apply Algorithm 3.4.2 to the digraph $\langle V - (U_1 \cup U_2) \rangle$, and so on.

However, if we desire to find all the strong sets of G, then recursively calling Algorithm 3.4.2 is somewhat inefficient. The reason for this is that, when Algorithm 3.4.2 terminates, not only is Y_k a strong set, but each of $Y_1, \ldots, Y_{k-1}$ is a subset of some strong set of G. Thus, Algorithm 3.4.2 not only finds one strong set of G, it may make significant progress toward finding other strong sets as well. Realizing this, we can modify Algorithm 3.4.2 so that it finds all the strong sets of G in a rather efficient manner. This is left to Exercise 11.

$\blacksquare$ ***Example 3.4.8***   We illustrate an algorithm to find the strong sets of a given digraph such as the digraph $G = (V, A)$ of Figure 3.4.3.

Our algorithm follows the steps of Algorithm 3.4.2, as illustrated in Example 3.4.6, finding the strong set $U_1 = \{4, 5, 9\}$. At this point, $k = 4$, $Y_0 = \{2, 8\}$, $Y_1 = \{0\}$, $Y_2 = \{1, 3, 6\}$, and $Y_3 = \{7\}$. We now decrement k to 3 and repeat the main loop of Algorithm 3.4.2. Since there is no arc $(7, v)$ with $v \in Y_0 \cup Y_1 \cup Y_2$, the set Y_3 has outdegree 0 (in $\langle V - U_1 \rangle$). Hence, $U_2 = Y_3 = \{7\}$ is a strong set of G. The value of k is now decremented to 2 and the main loop of Algorithm 3.4.2 is repeated. We may choose $u = 3 \in Y_2$ and $v = 8 \in Y_0$, so that $k = 3$, and $Y_3 = \{8\}$. Next, $u = 8 \in Y_3$ and $v = 1 \in Y_2$; since $j = 2$, Y_2 is replaced by $Y_2 \cup Y_3 = \{1, 3, 6, 8\}$ and k is assigned the value 2. At this point, there is no arc (u, v) with $u \in Y_2$ and $v \in Y_0 \cup Y_1$, and so $U_3 = Y_2 = \{1, 3, 6, 8\}$ is a strong set. The value of k is decremented to 1 and we continue. By finding $u = 0 \in Y_1$ and $v = 2 \in Y_0$, we have $k = 2$ and $Y_2 = \{2\}$. Next,

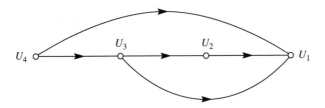

Figure 3.4.4   The condensation of G.

$u = 2 \in Y_2$ and $v = 0 \in Y_1$; since $j = 1$, Y_1 is replaced by $Y_1 \cup Y_2 = \{0, 2\}$ and $k = 1$. Now there is no arc (u, v) with $u \in Y_1$ and $v \in Y_0$ (in fact, Y_0 is empty at this point). Thus, $U_4 = Y_1 = \{0, 2\}$ is a strong set. The value of k is decremented to 0; since $Y_0 = \{ \ \}$, the algorithm terminates. (If Y_0 were not empty at this point, we would choose $u \in Y_0$, let $Y_1 = \{u\}$, remove u from Y_0, let $k = 1$, and continue.)

In summary, the algorithm has found the following strong sets of G:

$$U_1 = \{4, 5, 9\}, \quad U_2 = \{7\}, \quad U_3 = \{1, 3, 6, 8\}, \quad U_4 = \{0, 2\} \qquad \blacksquare$$

DEFINITION 3.4.5   Let $G = (V, A)$ be a digraph. The **condensation** of G is the digraph $G^* = (V^*, A^*)$, where $V^* = \{U_1, U_2, \ldots, U_k\}$ is the set of strong sets of G and

$$A^* = \{(U_i, U_j) \mid i \neq j \text{ and there is } (u_i, u_j) \in A \text{ with } u_i \in U_i \text{ and } u_j \in U_j\} \qquad \blacksquare$$

■ *Example 3.4.9*   For the digraph G of Figure 3.4.3, the condensation $G^* = (V^*, A^*)$, where $V^* = \{U_1, U_2, U_3, U_4\}$, and $A^* = \{(U_4, U_3), (U_4, U_1), (U_3, U_2), (U_3, U_1), (U_2, U_1)\}$ (U_1, U_2, U_3, and U_4 are as found in Example 3.4.8). The digraph G^* is shown in Figure 3.4.4. The elements of A^* may be obtained as follows:

$0 \in U_4$ and $1 \in U_3$ and $(0, 1) \in A \rightarrow (U_4, U_3) \in A^*$

$0 \in U_4$ and $9 \in U_1$ and $(0, 9) \in A \rightarrow (U_4, U_1) \in A^*$

$3 \in U_3$ and $7 \in U_2$ and $(3, 7) \in A \rightarrow (U_3, U_2) \in A^*$

$8 \in U_3$ and $4 \in U_1$ and $(8, 4) \in A \rightarrow (U_3, U_1) \in A^*$

$7 \in U_2$ and $4 \in U_1$ and $(7, 4) \in A \rightarrow (U_2, U_1) \in A^* \qquad \blacksquare$

Note that the condensation G^* of a digraph G has no loops. It consists of a single vertex if and only if G is strong. Furthermore (as you are asked to show in Exercise 12), the digraph G^* contains no cycles. Digraphs that contain no cycles are called **acyclic**; these form an important class of digraphs and are explored further in the next section.

We now consider another question mentioned at the beginning of this section, that of finding the distances from a fixed vertex of a digraph G to the other vertices of G. However, we wish to consider this question in the more general context of finding the "shortest paths" from a fixed vertex to the other vertices of a "weighted directed graph." A **weighted directed graph** is a directed graph $G = (V, A)$ together with a "weight function" w from A to the set of nonnegative real numbers. In many applications the weight of an arc represents some "cost" associated with its incident vertices. For instance, in the traffic flow problem mentioned at the beginning of Section 3.3, the weight of an arc (u, v) might be the time it takes, on average, to travel from intersection u to intersection v during a weekday morning rush hour. A weighted directed graph G with vertex set V, arc set A, and weight function w is denoted by (V, A, w).

Given a weighted directed graph $G = (V, A, w)$ and a walk W of G, the **weight** of W is defined to be the sum of the weights of the arcs of W. For $u, v \in V$, the **(weighted) distance** from u to v is written $d(u, v)$ and is defined as the minumum weight among all u-v paths in G; a u-v path P whose weight is $d(u, v)$ is called a **shortest u-v path**. (If v is not reachable from u, then $d(u, v) = \infty$.)

Suppose $V = \{u_0, u_1, \ldots, u_{n-1}\}$. For $1 \leq i \leq n - 1$, we wish to find a shortest u_0-u_i path in G. The following algorithm that does this is due to E. W. Dijkstra [Two problems in connexion with graphs, *Numerische Mathematik*, 1 (1959): 269–271].

As an example of this type of problem, suppose that a company has manufacturing plants in cities $c_1, c_2, c_3, c_4, c_5, c_6$ and its headquarters in city c_0. Frequently, the company must send executives from c_0 to one of the cities c_i, $1 \leq i \leq 6$. For this problem, we could form a weighted directed graph G with vertex set $\{c_0, c_1, \ldots, c_6\}$, where an arc from c_i to c_j indicates the existence of a direct flight from c_i to c_j and the weight of such an arc is the cost of this flight. For the specific example we are going to consider, the arc set is symmetric, and so G is represented as the weighted graph shown in Figure 3.4.5 (each edge represents a pair of symmetric arcs of equal weight), where $w(c_0, c_1) = 185$, $w(c_0, c_2) = 75$, $w(c_0, c_3) = 68$, $w(c_0, c_5) = 190$, $w(c_1, c_2) = 101$, $w(c_1, c_4) = 90$, $w(c_1, c_5) = 84$, $w(c_1, c_6) = 51$, $w(c_2, c_3) = 93$, $w(c_2, c_6) = 77$, $w(c_3, c_4) = 72$, $w(c_3, c_6) = 45$, and $w(c_4, c_5) = 48$. (Note that we have simplified the notation by denoting the weight of the arc (u, v) by $w(u, v)$ rather than by $w((u, v))$.)

The company would very much like to determine the cheapest routes from c_0 to each of the other cities. Notice that such routes correspond to the shortest c_0-c_i paths in the weighted directed graph G, where $1 \leq i \leq 6$.

We now discuss Dijkstra's algorithm for finding shortest u_0-u_i paths in a weighted directed graph $G = (V, A, w)$, where $V = \{u_0, u_1, \ldots, u_{n-1}\}$. In order to comprehend this algorithm, it is necessary to understand an important property of shortest paths. Suppose that U is a proper subset of V with $u_0 \in U$. Define a shortest path from u_0 to $V - U$ to be a path of minimum weight among all u_0-v paths with $v \in V - U$, and let $d(u_0, V - U)$ be the weight of such a path. Suppose that

$$P : u_0 = v_0, v_1, \ldots, v_{k-1}, v_k = y$$

is such a path in G. It then is not difficult to argue the following (see Exercise 22):

1. $v_0, v_1, \ldots, v_{k-1} \in U$ and $y \in V - U$

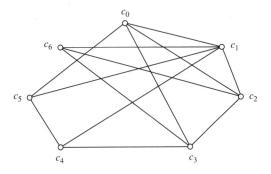

Figure 3.4.5

2.   The path

$$P' : u_0 = v_0, v_1, \ldots, v_{k-1}$$

representing that portion of P from u_0 to v_{k-1}, is a shortest u_0-v_{k-1} path in G. It follows that P is a shortest u_0-y path and that

$$d(u_0, V - U) = d(u_0, y) = \min\{d(u_0, v) + w(v, y)\}$$

where the minimum is taken over all vertices v and y with $v \in U$, $y \in V - U$, and $(v, y) \in A$.

The ideas of the preceding paragraph form the gist of Dijkstra's algorithm. During its application, U denotes the set of vertices for which shortest paths from u_0 have been found; initially, $U = \{u_0\}$. At each iteration, another vertex x is added to U; in fact, x is that vertex such that

$$d(u_0, x) = d(u_0, V - U)$$

The algorithm continues until $U = V$. The algorithm is to compute $d(u_0, v)$ for each vertex v. We denote this weighted distance simply by $d(v)$; initially, $d(u_0) := 0$ and $d(v) := \infty$ for $v \neq u_0$. (In practice, ∞ may be replaced by some appropriately large positive number.) In addition, we want the algorithm to compute shortest u_0-v paths. A slick way to handle this problem is to have the algorithm compute, for each vertex v, the predecessor $p(v)$ of v on a shortest u_0-v path. Then, given a vertex v, we may find a shortest u_0-v path by working backward from v to $p(v)$, to $p(p(v))$, and so on, until u_0 is reached. For convenience, the algorithm initially sets $p(u_0) = u_0$.

Before we consider the formal presentation of Dijkstra's algorithm, let's illustrate its application with an example.

■ *Example 3.4.10*   We now apply Dijkstra's algorithm to find $d(c_0, c_i)$ and short-est c_0-c_i paths for the weighted digraph G of Figure 3.4.5.

Initially, $U = \{c_0\}$, $d(c_0) = 0$, $d(c_i) = \infty$, $1 \leq i \leq 6$, $p(c_0) = c_0$, and $x = c_0$. We then repeat the following steps until $U = V$ (that is, $n - 1$ times, where n is the order of G):

1.   For each $y \in V - U$, if $(x, y) \in A$ and $d(x) + w(x, y) < d(y)$, then $d(y) := d(x) + w(x, y)$ and $p(y) := x$.

2.   Among the vertices in $V - U$, let x be such that $d(x)$ is a minimum; $U := U \cup \{x\}$.

We trace these steps as follows.

Iteration 1: Each $y \in V - U = \{c_1, c_2, c_3, c_4, c_5, c_6\}$ that is adjacent from c_0 has $d(y)$ and $p(y)$ updated. Thus, $d(c_1) := 185$, $p(c_1) := c_0$, $d(c_2) := 75$, $p(c_2) := c_0$, $d(c_3) := 68$, $p(c_3) := c_0$, $d(c_5) := 190$, and $p(c_5) := c_0$. The vertex in $V - U$ with the minimum distance from c_0 is c_3, so $x := c_3$ and $U := U \cup \{x\} = \{c_0, c_3\}$.

Iteration 2: Each $y \in V - U = \{c_1, c_2, c_4, c_5, c_6\}$ that is adjacent from $x = c_3$ may have $d(y)$ and $p(y)$ updated. For $y = c_2$, $d(y) = 75$ whereas $d(x) + w(x, y) = 68 + 93 = 161$. So $d(c_2)$ and $p(c_2)$ are not changed. On the other hand, for $y = c_4$, $d(y) = \infty$ whereas $d(x) + w(x, y) = 68 + 72 = 140$. So $d(c_4) := 140$ and $p(c_4) := c_3$. Similarly, $d(c_6) := 68 + 45 = 113$ and $p(c_6) := c_3$. The vertex in $V - U$ with minimum distance from c_0 is c_2, so $x := c_2$ and $U := U \cup \{x\} = \{c_0, c_2, c_3\}$.

Iteration 3: The vertices in $V - U = \{c_1, c_4, c_5, c_6\}$ that are adjacent from $x = c_2$ are c_1 and c_6. For $y = c_1$, $d(x) + w(x, y) = 75 + 101 = 176$ whereas $d(y) = 185$. Thus, $d(c_1) := 176$ and $p(c_1) := c_2$. For $y = c_6$, $d(x) + w(x, y) = 75 + 77 = 152 > 113 = d(y)$, so $d(c_6)$ is not changed. The vertex in $V - U$ with minimum distance from c_0 is c_6, so $x := c_6$ and $U := U \cup \{x\} = \{c_0, c_2, c_3, c_6\}$.

Iteration 4: The only vertex $y \in V - U = \{c_1, c_4, c_5\}$ that is adjacent from $x = c_6$ is c_1. Since $d(c_6) + w(c_6, c_1) = 113 + 51 = 164 < 176 = d(c_1)$, we have $d(c_1) := 164$ and $p(c_1) := c_6$. The vertex in $V - U$ with minimum distance from c_0 is c_4, so $x := c_4$ and $U := U \cup \{x\} = \{c_0, c_2, c_3, c_4, c_6\}$.

Iteration 5: We must check $d(y)$ and $p(y)$ for each $y \in \{c_1, c_5\}$. The values updated are $d(c_5) := 140 + 48 = 188$ and $p(c_5) := c_4$. The vertex in $V - U$ with minimum distance from c_0 is c_1, so $x := c_1$ and $U := U \cup \{x\} = \{c_0, c_1, c_2, c_3, c_4, c_6\}$.

Iteration 6: For $y = c_5$, $d(x) + w(x, y) = 164 + 84 > 188 = d(y)$, so no values are updated. Now, $x := c_5$ and $U := U \cup \{x\} = V$.

Since $U = V$, the algorithm terminates. The final computed values are:

$$d(c_1) = 164, \quad d(c_2) = 75, \quad d(c_3) = 68, \quad d(c_4) = 140, \quad d(c_5) = 188, \quad d(c_6) = 113$$

$$p(c_1) = c_6, \quad p(c_2) = c_0, \quad p(c_3) = c_0, \quad p(c_4) = c_3, \quad p(c_5) = c_4, \quad p(c_6) = c_3$$

As pointed out before, the predecessor values may be used to find shortest c_0-c_i paths, $1 \leq i \leq 6$. To illustrate, suppose we desire a shortest c_0-c_1 path. Since

$p(c_1) = c_6$, $p(c_6) = c_3$, and $p(c_3) = c_0$, one such path is c_0, c_3, c_6, c_1. Exercise 13 asks you to find shortest paths for the other vertices of G.   ∎

We are now ready for a formal presentation of Dijkstra's algorithm (as a procedure). Input to the procedure are a directed graph $G = (V, A)$, a weight function w, and a vertex u of G. It is assumed that $V = \{0, 1, \ldots, n-1\}$ for some positive integer n. The procedure is to compute and return $d(v) = d(u, v)$ and $p(v)$ for each $v \in V$. (It is assumed that each $v \in V$ is reachable from u.)

We also assume, for the sake of concreteness, that weights are nonnegative integers. With these assumptions, notice that a weight function w can be represented as a matrix (two-dimensional array) of weight values; for $0 \leq u$, $v \leq n - 1$, if $(u, v) \in A$, then $w(u, v)$ gives the weight of (u, v). Moreover, $(d(0), d(1), \ldots, d(n-1))$ is an array of weights and $(p(0), p(1), \ldots, p(n-1))$ is an array of vertices. Thus, let us assume that appropriate types WEIGHT_TYPE, WEIGHT_FUNCTION, WEIGHT_ARRAY, and VERTEX_ARRAY are declared in the obvious manner. In Ada, for example, such types can be declared as follows:

subtype WEIGHT_TYPE is NATURAL;
type WEIGHT_FUNCTION is array (VERTEX_TYPE range <>,
                        VERTEX_TYPE range <>) of WEIGHT_TYPE;
type WEIGHT_ARRAY is array (VERTEX_TYPE range <>) of WEIGHT_TYPE;
type VERTEX_ARRAY is array (VERTEX_TYPE range <>) of VERTEX_TYPE;

Our procedure follows the description given in Example 3.4.10 with the minor modification that a list L is used to hold the vertices in $V - U$.

ALGORITHM 3.4.3 (Dijkstra's algorithm)   Given a weighted directed graph $G = (V, A, w)$ with $V = \{0, 1, \ldots, n-1\}$ and $u \in V$, the procedure

DIJKSTRA(G, w, u, d, p)

computes, for each $v \in V$, $d(v) = d(u, v)$ and $p(v) =$ the predecessor of v on a shortest u-v path.

The procedure DIJKSTRA uses the following local variables:

L is a list of vertices initialized to LIST_OF_VERTICES(G);
L' is a list of vertices;
FOUND is of type BOOLEAN;
x is a vertex and is initially u;
y and x' are vertices;
n is a positive integer and is initially ORDER(G);
MIN is a positive integer;

```
begin
    for v in 0 .. n − 1 loop
        d(v) := ∞;
    end loop;
    d(u) := 0;
    p(u) := u;
    SEARCH(u, L, FOUND);
    POP(L);
    for i in 1 .. n − 1 loop
        MIN := ∞;
        L' := COPY_LIST(L);
        while not EMPTY(L') loop
            y := FIRST(L');
            POP(L');
            if ADJACENT_TO(x, y, G) then
                if d(x) + w(x, y) < d(y) then
                    d(y) := d(x) + w(x, y);
                    p(y) := x;
                end if;
            end if;
            if d(y) < MIN then
                MIN := d(y);
                x' := y;
            end if;
        end loop;
        x := x';
        SEARCH(x, L, FOUND);
        POP(L);
    end loop;
end DIJKSTRA;
```

■   *Example 3.4.11*   Apply Algorithm 3.4.3 to find $d(0, v)$ and shortest 0-v paths for the digraph G of Figure 3.4.6, given the weight function defined by $w(0, 1) = 12$, $w(0, 3) = 6$, $w(0, 4) = 1$, $w(1, 2) = 6$, $w(2, 3) = 10$, $w(3, 1) = 5$, $w(3, 2) = 12$, $w(3, 4) = w(4, 3) = w(4, 5) = 9$, and $w(5, 0) = 5$.

Solution

Initially, $L = (1, 2, 3, 4, 5)$, $x = 0$, $d(0) = 0$, $d(v) = ∞$, $1 \leq v \leq 5$, and $p(0) = 0$. The main loop is now repeated five times.

Iteration 1: The vertices y in L adjacent from x are 1, 3, and 4. Hence, $d(1) := 12$, $d(3) := 6$, and $d(4) := 1$; $p(1)$, $p(3)$, and $p(4)$ are each assigned 0. Also, MIN = 1 and $x' = 4$, so $x := 4$ and 4 is popped from L, yielding $L = (1, 2, 3, 5)$.

Iteration 2: The vertices y in L adjacent from x are 3 and 5. For $y = 3$, $d(y) = 6$, whereas $d(x) + w(x, y) = 1 + 9 = 10$. So $d(3)$ is not changed. On the other hand, for

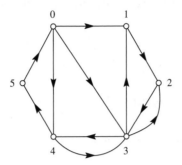

Figure 3.4.6 The digraph G for Example 3.4.11.

$y = 5$, $d(y) = \infty$ whereas $d(x) + w(x, y) = 1 + 9 = 10$. Thus, $d(5) := 10$ and $p(5) := 4$. Also, MIN $= d(3) = 6$ and $x' = 3$, so $x := 3$ and $L = (1, 2, 5)$.

Iteration 3: The vertices y in L adjacent from x are 1 and 2. For $y = 1$, $d(y) = 12$ whereas $d(x) + w(x, y) = 6 + 5 = 11$. So $d(1) := 11$ and $p(1) := 3$. For $y = 2$, $d(y) = \infty$ whereas $d(x) + w(x, y) = 6 + 12 = 18$. So $d(2) := 18$ and $p(2) := 3$. Also, MIN $= d(5) = 10$, so $x := x' = 5$ and $L = (1, 2)$.

Iteration 4: Neither vertex in L is adjacent from x, so their d and p values are not changed. Since MIN $= d(1) = 11$, we have $x := x' = 1$ and $L = (2)$.

Iteration 5: Note that $y = 2$ is adjacent from $x = 1$ and that $d(x) + w(x, y) = 11 + 6 = 17 < 18 = d(y)$, so $d(2) := 17$ and $p(2) := 1$.

The algorithm now terminates, returning the following values:

$$d(0) = 0, \quad d(1) = 11, \quad d(2) = 17, \quad d(3) = 6, \quad d(4) = 1, \quad d(5) = 10$$
$$p(0) = 0, \quad p(1) = 3, \quad p(2) = 1, \quad p(3) = 0, \quad p(4) = 0, \quad p(5) = 4$$

We may use the predecessor values to find shortest 0-v paths, $1 \leq v \leq 5$. For example, since $p(1) = 3$ and $p(3) = 0$, a shortest 0-1 path is $0, 3, 1$. You are asked in Exercise 15 to find shortest paths for the other vertices of G. ∎

EXERCISE SET 3.4

1. Apply Algorithm 3.4.1 to the directed graph G_2 of Figure 3.3.3.

2. Let $G = (V, A)$ be a digraph, let A^t be the transitive closure of A as defined by Definition 3.4.2 and let A_1^t be a relation that is constructed from A according to Alternate Definition 3.4.2.

   (a) Show that A^t is uniquely determined and, hence, that Definition 3.4.2 makes the transitive closure well defined.

   (b) Show that $A_1^t = A^t$, thus showing that the two definitions of transitive closure are equivalent in the case when V is finite.

3. Apply Algorithm 3.4.1 to each of the directed graphs in Figure 3.4.7.

4. Prove Lemma 3.4.3.

5. Draw each of the following subdigraphs of the digraph G_2 in Figure 3.3.3.

   (a) $H_1 = G_2 - z$
   (b) $H_2 = (G_2 - z) - w$
   (c) $H_3 = \langle\{u, v, x, y\}\rangle$
   (d) $H_4 = G_2 - (w, x)$
   (e) $H_5 = \langle\{(u, v), (u, z), (v, y), (y, x)\}\rangle$

6. Prove Lemma 3.4.4.

7. Apply Algorithm 3.4.2 to each of the following directed graphs and find its condensation:

   (a) $G_1 = (\{1, 2, 3, 4, 5, 6, 7, 8\}, \{(1, 2), (2, 5), (3, 4), (4, 5), (4, 7), (5, 1), (5, 6), (6, 1),$
   $(7, 3), (7, 6), (8, 1), (8, 3), (8, 7)\})$
   (b) $G_2 = (\{1, 2, 3, 4, 5, 6, 7, 8, 9\}, \{(1, 2), (2, 3), (2, 5), (3, 1), (3, 4), (4, 6), (5, 4),$
   $(6, 5), (7, 8), (8, 9), (9, 1), (9, 3), (9, 4), (9, 8)\})$

8. Let $G = (V, A)$ be a digraph. Identify each of the following statements as true or false:

   (a) Any induced subdigraph of G can be obtained by successively deleting vertices.
   (b) Any arc-induced subdigraph of G can be obtained by successively deleting arcs.

9. Apply Algorithm 3.4.2 to the digraph in Figure 3.4.3, only assume that $Y(0) = (0, 1, 2, 3, 4, 5, 6, 7, 8, 9)$ and $Y(1) = (7)$, initially.

10. Let $G = (V, A)$ be a digraph. Define a relation $\sim$ on V by $x \sim y$ if and only if there is a closed walk of G containing both x and y.

    (a) Show that $\sim$ is an equivalence relation on V.
    (b) Let U be an equivalence class for the relation $\sim$. Show that U is a strong set of G.

11. Concerning Example 3.4.8 and the remarks preceding it:

    (a) Develop an algorithm, based on Algorithm 3.4.2, which finds the strong sets of a given digraph $G = (V, A)$.
    (b) Test this algorithm using the digraphs G_1 and G_2 in Exercise 7.

12. Let $G = (V, A)$ be a digraph and let G^* be the condensation of G.

    (a) Prove that G^* is acyclic.
    (b) State and prove a necessary and sufficient condition for G and G^* to be isomorphic.

13. Complete Example 3.4.10 by finding shortest c_0-c_i paths for $2 \le i \le 6$.

14. Let $G = (V, A)$, where $V = \{v_1, v_2, \ldots, v_n\}$, and let M be the adjacency matrix for G. Consider the sum of the elements in the ith row of M. What information about G does this give us? What about the sum of the elements in the jth column?

15. Complete Example 3.4.11 by finding shortest 0-v paths for $2 \le v \le 5$.

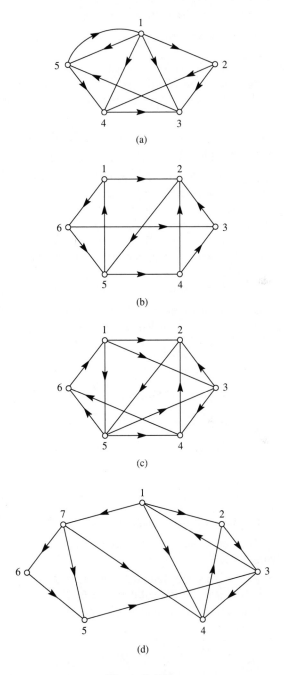

(a)

(b)

(c)

(d)

Figure 3.4.7

16. Let $G = (V, A, w)$ be a weighted directed graph with $V = \{0, 1, \ldots, n-1\}$ such that no two arcs of G have the same weight.

    (a) Show that it is not always the case that the arc of smallest weight is used in some shortest 0-v path.
    (b) Give a sufficient condition under which the arc of smallest weight must be used in some shortest 0-v path.
    (c) For each $v \in \{1, 2, \ldots, n-1\}$, is it true that G contains a unique shortest 0-v path?

17. Apply Algorithm 3.4.3 to each of the directed graphs in Figure 3.4.7 to find shortest 1-v paths, given the following weight functions:

    (a) $w(1, 2) = 2$, $w(1, 3) = 3$, $w(1, 4) = 5$, $w(1, 5) = 6$, $w(2, 3) = 2$, $w(2, 4) = 4$, $w(3, 5) = 1$, $w(4, 3) = 1$, $w(5, 1) = 7$, $w(5, 4) = 3$
    (b) $w(1, 2) = 15$, $w(1, 6) = 1$, $w(2, 5) = 3$, $w(3, 2) = 9$, $w(4, 2) = 5$, $w(4, 3) = 4$, $w(5, 1) = 9$, $w(5, 4) = 2$, $w(6, 3) = 4$, $w(6, 5) = 7$
    (c) $w(1, 2) = 13$, $w(1, 3) = 2$, $w(1, 5) = 7$, $w(2, 5) = 6$, $w(3, 2) = 3$, $w(3, 4) = 5$, $w(4, 2) = 8$, $w(4, 6) = 10$, $w(5, 3) = 4$, $w(5, 4) = 15$, $w(5, 6) = 25$, $w(6, 1) = 40$
    (d) $w(1, 2) = 17$, $w(1, 4) = 11$, $w(1, 7) = 9$, $w(2, 3) = 1$, $w(3, 1) = 18$, $w(3, 4) = 3$, $w(4, 2) = 5$, $w(5, 3) = 3$, $w(6, 5) = 1$, $w(7, 4) = 3$, $w(7, 5) = 7$, $w(7, 6) = 9$

18. Let $G = (V, A)$ be a digraph of order n with $V = \{v_1, v_2, \ldots, v_n\}$. The **reachability matrix** of G is the $n \times n$, $(0, 1)$-matrix $R = [r_{ij}]$ defined by

    $$r_{ij} = \begin{cases} 0 & \text{if } v_j \text{ is not reachable from } v_i \\ 1 & \text{if } v_j \text{ is reachable from } v_i \end{cases}$$

    For example, the digraph G of Example 3.3.1 has the following reachability matrix R:

    $$R = \begin{bmatrix} 1 & 0 & 1 & 1 & 0 \\ 1 & 1 & 1 & 1 & 0 \\ 0 & 0 & 1 & 1 & 0 \\ 0 & 0 & 1 & 1 & 0 \\ 1 & 0 & 1 & 1 & 1 \end{bmatrix}$$

    Modify Algorithm 3.4.1 so that it computes the reachability matrix of G.

19. Apply the algorithm developed in the preceding exercise to each of the directed graphs in Figure 3.4.7.

20. Given G as in Exercise 18, the **distance matrix** of G is the $n \times n$, $(0, 1)$-matrix $D = [d_{ij}]$, where $d_{ij} = d(v_i, v_j)$. (Recall that $d_{ij} = \infty$ if v_j is not reachable from v_i.) For example, the digraph G of Example 3.3.1 has the following distance matrix:

    $$D = \begin{bmatrix} 0 & \infty & 2 & 1 & \infty \\ 1 & 0 & 1 & 2 & \infty \\ \infty & \infty & 0 & 1 & \infty \\ \infty & \infty & 1 & 0 & \infty \\ 1 & \infty & 3 & 2 & 0 \end{bmatrix}$$

Modify Algorithm 3.4.1 so that it computes the distance matrix of G. (Optional: Allow G to be a weighted directed graph.)

21. Apply the algorithm of Exercise 20 to each of the directed graphs in Figure 3.4.7. (Ignore the weights, unless you did the optional part of Exercise 20.)

22. Let $G = (V, A, w)$ be a weighted directed graph and let U be a proper subset of V with $u_0 \in U$. Suppose that

$$P : u_0 = v_0, v_1, \ldots, v_{n-1}, v_n = y$$

is a shortest path from u_0 to $V - U$ in G. Show the following:

   (a) $v_0, v_1, \ldots, v_{n-1} \in U$ and $y \in V - U$
   (b) The path $P' : u_0 = v_0, v_1, \ldots, v_{n-1}$, representing that portion of P from u_0 to v_{n-1}, is a shortest u_0-v_{n-1} path in G.
   (c) $d(u_0, V - U) = \min\{d(u_0, v) + w(v, y)\}$, where the minimum is taken over all vertices v and y with $v \in U$ and $y \in V - U$ and $(v, y) \in A$.

23. Let M be the adjacency matrix of the digraph

$$G = (\{v_1, v_2, v_3, v_4\}, \{(v_1, v_2), (v_1, v_3), (v_2, v_1), (v_2, v_2), (v_2, v_3), (v_3, v_4), (v_4, v_1)\})$$

   (a) Verify that the $(2, 3)$-entry of M^2 is the number of distinct walks of length 2 from v_2 to v_3.
   (b) Verify that the $(2, 3)$-entry of M^3 is the number of distinct walks of length 3 from v_2 to v_3.
   (c) Verify that the $(2, 3)$-entry of M^4 is the number of distinct walks of length 4 from v_2 to v_3.

24. Let $G = (V, A)$ be a directed graph with $V = \{v_1, v_2, \ldots, v_n\}$, and let M be the adjacency matrix of G.

   (a) Show that the (i, j)-entry of M^2 is the number of distinct v_i-v_j walks of length 2 in G.
   (b) Generalizing the result of part (a), show that, for $p \in \mathbb{Z}^+$, the (i, j)-entry of M^p is the number of v_i-v_j walks of length p in G. (Hint: Employ induction on p.)
   (c) Let M^0 denote the $n \times n$ identity matrix. Prove: G is strong if and only if each entry of $M^0 + M^1 + \cdots + M^{n-1}$ is positive.

3.5 *ACYCLIC DIGRAPHS AND TREES*

An **acyclic digraph** is one having no cycles. In this section we discuss acyclic digraphs, including an important special class of such digraphs, the rooted trees. Simple graphs that are connected and acyclic are called **trees** and are also discussed.

Let $G = (V, A)$ be a loopless digraph. To begin, we know that the relation "is reachable from" on V is both reflexive and transitive. If G is acyclic, then this relation is also antisymmetric. That is, suppose that u and v are distinct vertices of G and that v is reachable from u and u is reachable from v. Then G contains a u-v walk W_1 and a v-u walk W_2. If W denotes the u-u walk obtained by appending W_2 to W_1, then W is a nontrivial closed walk. Now, in a manner similar to the proof of Theorem 3.3.1, we may show that G contains a cycle, contradicting the fact that G is acyclic. Thus, if $G = (V, A)$ is an acyclic digraph, then the relation "is reachable from" on V is a partial ordering of V. It is easy to see that the converse is also true.

> **THEOREM 3.5.1** Let $G = (V, A)$ be a loopless digraph. Then G is acyclic if and only if the relation "is reachable from" on V is antisymmetric.   □

An application where acyclic directed graphs arise concerns the decomposition of a large, complicated problem into smaller, more manageable tasks. For example, we often decompose a complicated algorithm into smaller steps, some of which might conceivably be executed "in parallel." To introduce some consistent terminology into our discussion, we shall call such a smaller step, which is always executed as a unit, a "task." We form a directed graph G whose vertex set is the set of tasks and which contains an arc from task u to task v provided u must be executed before v. This might be required, for instance, because u calculates some results that are needed by v's computation, or because u and v require independent access to the same external device. Notice that such a digraph must be acyclic, for otherwise there would exist tasks u and v such that u must be executed prior to v and v must be executed prior to u.

Suppose now that a schedule is desired for executing the tasks; that is, we desire a list

$$(u_1, u_2, \ldots, u_n)$$

of the tasks, indicating an order in which they are to be executed, such that, whenever u_i is required to be executed prior to u_j, then $i < j$. More generally, we let $G = (V, A)$ be an acyclic digraph of order n. We desire a list $(u_1, u_2, \ldots, u_n)$ of the vertices such that, whenever u_j is reachable from u_i, then $i \leq j$. One method of finding such a list is called the **topological** (or **selection**) **sort**. This method is based on the following lemma. (See Exercise 22 of Exercise Set 3.3.)

> **LEMMA 3.5.2** Let $G = (V, A)$ be a digraph. If $\text{od}(v) > 0$ for every $v \in V$, then G contains a cycle. Hence, if G is acyclic, then G contains a vertex v such that $\text{od}(v) = 0$.

Proof Assume $\text{od}(v) > 0$ for every $v \in V$. Let P be a path in G of maximum length, say,

$$v_0, v_1, \ldots, v_k$$

Since $\text{od}(v_k) > 0$, there is a vertex w such that $(v_k, w) \in A$. We claim that w is on P, namely, that $w = v_i$ for some i, $0 \leq i \leq k$; for otherwise, the path

$$v_0, v_1, \ldots, v_k, w$$

has length 1 greater than that of P, contradicting the choice of P. Thus,

$$v_i, v_{i+1}, \ldots, v_k, v_i$$

is a cycle. □

Given an acyclic digraph $G = (V, A)$, let $u_n \in V$ such that $od(u_n) = 0$. Clearly, no other vertex of G is reachable from u_n. Next, consider the subdigraph of G induced by $V - \{u_n\}$. Note that this digraph is also acyclic. Thus, we may choose $u_{n-1} \in V - \{u_n\}$ such that $od(u_{n-1}) = 0$ in this digraph, and then no vertex of $V - \{u_{n-1}, u_n\}$ is reachable from u_{n-1}. Continuing in this fashion gives the desired list $(u_1, u_2, \ldots, u_n)$ of the vertices of G such that u_j is reachable from u_i only if $i \leq j$.

ALGORITHM 3.5.1 Let $G = (V, A)$ be an acyclic digraph of order n. The function

SORT(G)

returns a list $(u_1, u_2, \ldots, u_n)$ of the vertices of G such that u_j is reachable from u_i only if $i \leq j$.

The function SORT uses the following local variables:

RESULT is a list of vertices and is initialized to EMPTY_LIST;
G' is a digraph and is initialized to a copy of G;
u is a vertex;

```
begin
    for i in 1 .. ORDER(G') loop
        --Find a vertex u such that od(u) = 0 in G'.
        u := FIRST_VERTEX(G');
        while OUTDEGREE(u, G') > 0 loop
            u := FIRST_NEIGHBOR(u, G');
        end loop;
        --Push u onto RESULT and delete u from G'.
        PUSH(u, RESULT);
        DELETE_VERTEX(u, G');
    end loop;
    return RESULT;
end SORT;
```

■ *Example 3.5.1*   Illustrate Algorithm 3.5.1 for the digraph $G = (\{1, 2, 3, 4, 5, 6\}, \{(1, 2), (1, 4), (2, 3), (2, 5), (3, 6), (4, 5), (5, 6)\})$.

Solution

Initially, RESULT = ( ) and $G' = G$. The main loop is executed ORDER(G) = six times.

Iteration 1: Assume FIRST_VERTEX(G') = 1, so $u := 1$. Since OUTDEGREE $(1, G') = 2 > 0$, the body of the while loop is executed; assume FIRST_NEIGHBOR $(1, G') = 2$, so that $u := 2$. Since OUTDEGREE$(2, G') > 0$, the body of the while loop is executed again; assume FIRST_NEIGHBOR$(2, G') = 3$, so that $u := 3$. Again, OUTDEGREE$(3, G') > 0$, so $u :=$ FIRST_NEIGHBOR$(3, G') = 6$. Now, OUTDEGREE$(6, G') = 0$, so the while loop is exited. Thus, 6 is pushed onto RESULT and deleted from G', giving RESULT= (6) and $G' = (\{1, 2, 3, 4, 5\}, \{(1, 2), (1, 4), (2, 3), (2, 5), (4, 5)\})$.

Iteration 2: Assume FIRST_VERTEX(G') = 1, so $u := 1$. Since OUTDEGREE $(1, G') > 0$, the body of the while loop is executed. Assume FIRST_NEIGHBOR $(1, G') = 2$, so that $u := 2$. Next, OUTDEGREE$(2, G') > 0$; assume FIRST_NEIGHBOR$(2, G') = 3$, so that $u := 3$. Then OUTDEGREE$(3, G') = 0$, so 3 is pushed onto RESULT and deleted from G'. This yields RESULT= (3, 6) and $G' = (\{1, 2, 4, 5\}, \{(1, 2), (1, 4), (2, 5), (4, 5)\})$.

Iteration 3: Assume FIRST_VERTEX(G') = 1, so $u := 1$. Since OUTDEGREE $(1, G') > 0$, the body of the while loop is executed. Assume FIRST_NEIGHBOR $(1, G') = 2$, so that $u := 2$. Next, OUTDEGREE$(2, G') > 0$, so $u :=$ FIRST_NEIGHBOR$(2, G') = 5$. Then OUTDEGREE$(5, G') = 0$, so 5 is pushed onto RESULT and deleted from G'. This gives RESULT= (5, 3, 6) and $G' = (\{1, 2, 4\}, \{(1, 2), (1, 4)\})$.

Iteration 4: Assume FIRST_VERTEX(G') = 1, so $u := 1$. Since OUTDEGREE $(1, G') > 0$, the body of the while loop is executed. Assume FIRST_NEIGHBOR $(1, G') = 2$, so that $u := 2$. Then OUTDEGREE$(2, G') = 0$, so 2 is pushed onto RESULT and deleted from G', giving RESULT= (2, 5, 3, 6) and $G' = (\{1, 4\}, \{(1, 4)\})$.

Iteration 5: Assume FIRST_VERTEX(G') = 1, so $u := 1$. Since OUTDEGREE $(1, G') > 0$, we have $u :=$ FIRST_NEIGHBOR$(1, G') = 4$. Then OUTDEGREE$(4, G') = 0$, so 4 is pushed onto RESULT and deleted from G', giving RESULT= (4, 2, 5, 3, 6) and $G' = (\{1\}, \{ \})$.

Iteration 6: We have $u :=$ FIRST_VERTEX(G') = 1. Since OUTDEGREE$(1, G') = 0$, 1 is pushed onto RESULT, giving RESULT= (1, 4, 2, 5, 3, 6).

Thus, the list (1, 4, 2, 5, 3, 6) is returned. Note that the other correct results are (1, 2, 3, 4, 5, 6), (1, 2, 4, 3, 5, 6), (1, 2, 4, 5, 3, 6), and (1, 4, 2, 3, 5, 6). ∎

One class of acyclic digraphs that is particularly important in the study of computer science is the class of rooted trees.

DEFINITION 3.5.1 An acyclic digraph $T = (V, A)$ is called a **rooted tree** provided the following properties are satisfied:

1. There is a unique vertex $r \in V$ such that id(r) = 0; the vertex r is called the **root** of T.
2. For every vertex v, $v \neq r$, id(v) = 1; that vertex u for which $(u, v) \in T$ is called the **parent** of v. ∎

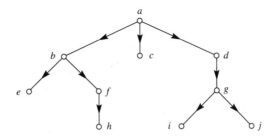

Figure 3.5.1   A rooted tree.

An example of a rooted tree is the digraph $T = (V, A)$, where $V = \{a, b, c, d, e, f, g,$ $h, i, j\}$ and $A = \{(a, b), (a, c), (a, d), (b, e), (b,f), (d, g), (f, h), (g, i), (g,j)\}$, shown in Figure 3.5.1. The vertex a is the root. Note, for instance, that f is the parent of h, b is the parent of f, and a is the parent of b.

Now let $T = (V, A)$ be a rooted tree with root r. Note that, for every $v \in V$, there is a unique path in T from r to v. That is, suppose that

$$r = v_0, v_1, \ldots, v_n = v$$

is such a path. Then necessarily v_{n-1} is the parent of v, v_{n-2} is the parent of v_{n-1}, and so on, and r is the parent of v_1. Conversely, if $T = (V, A)$ is an acyclic digraph having a unique vertex r with indegree 0, and for every $v \in V$ there is a unique r-v path in T, then T is a rooted tree. The proof of this last statement is left for you to develop in Exercise 2.

There is quite a bit of standard terminology associated with rooted trees. Let $T = (V, A)$ be a rooted tree and let $v \in V$. Those vertices adjacent from v are called **children** of v; if v has no children, then v is called a **leaf** of T. Vertices of T that have the same parent are called **siblings**. The notions of "child" and "parent" extend naturally to the notions of "**descendant**" and "**ancestor**."

As mentioned, there is a unique path in T from the root to v. The length of this path is the **level** of v in T. Thus, the root of T has level 0, the children of the root have level 1, and so on. The **height** of T is the maximum level among its vertices.

■   *Example 3.5.2*   For the rooted tree T of Figure 3.5.1, find (a) the children of b, (b) the ancestors of i, (c) the vertices having level 2, and (d) the height of T.

Solution

(a) The children of b are e and f.

(b) The ancestors of i are g, d, and a.

(c) The vertices at level 2 are e, f, and g.

(d) The height of T is 3.   ■

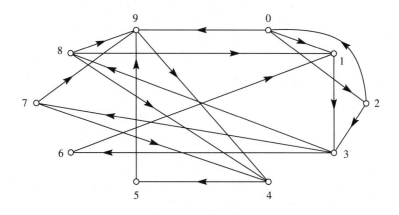

Figure 3.5.2

Now let m be a positive integer. A rooted tree T is said to be an ***m*-ary rooted tree** (or simply an ***m*-ary tree**) provided each vertex of T has at most m children. If $m = 1$, then a 1-ary tree is simply a path (from the root to the leaf). A 2-ary tree is also called a **binary tree**. Note that the tree of Figure 3.5.1 is a 3-ary tree.

Let $G = (V, A)$ be a digraph and let $u \in V$. Then there is a subdigraph T_u of G that is a rooted tree with root u such that the vertex set of T_u is the set of vertices of G that are reachable from u. One method for finding such a rooted tree is called a **depth-first search**, and a rooted tree so found is called a **depth-first search tree** of G. If every vertex of G is reachable from u, in particular, if G is strong, then T_u is a spanning subdigraph of G, called a **depth-first search spanning tree** of G.

It is easier to illustrate how to find a depth-first search tree of a given directed graph than to define what one is. Thus, before giving a definition, let's consider an example.

■ *Example 3.5.3*   The directed graph $G = (V, A)$ of Figure 3.4.3 is repeated for convenience in Figure 3.5.2.

We wish to find a depth-first search tree $T_0 = (V_0, A_0)$ of G having root 0. Initially, $V_0 = \{0\}$ and $A_0 = \{\ \}$. Since $(0, 1) \in A$, we add 1 to V_0 and $(0, 1)$ to A_0. Next, since $(1, 3) \in A$, we add 3 to V_0 and $(1, 3)$ to A_0. Here we can see one property of a depth-first search: it always prefers to move to the next level of the rooted tree being constructed, rather than staying at the same level. That is, suppose the depth-first search has just added the vertex v and the arc (u, v) to the tree; it next prefers to add a vertex x that is adjacent from v rather than adding another vertex y adjacent from u. Continuing with G, since $(3, 6) \in A$, we add 6 to V_0 and $(3, 6)$ to A_0. At this point, $V_0 = \{0, 1, 3, 6\}$ and $A_0 = \{(0, 1), (1, 3), (3, 6)\}$. Now 6 is adjacent to no vertex of $V - V_0$, so the depth-first search "backs up" the tree to the parent of 6, namely, 3. From 3 we may proceed to 7, then from 7 to 4, from 4 to 5, and from 5 to 9, adding the vertices 7, 4, 5, and 9 to V_0 and the arcs $(3, 7)$, $(7, 4)$, $(4, 5)$, and $(5, 9)$ to A_0. At this point, 9 is adjacent to no vertex of $V - V_0$, so we back up to 5, the parent of 9. However, 5 also is not adjacent to any vertex of $V - V_0$, so we back up to its parent, 4. Similarly, we must back up from 4 to 7 and then from 7 to 3. From 3 we may again proceed, this time to 8, so that 8 is added to V_0 and $(3, 8)$ to A_0. Then we must back up from 8 to 3, then from 3 to 1, and then from 1 to 0. Finally, the vertex 2 is added to V_0 and the arc $(0, 2)$ is added to A_0, so that $V_0 = V$ and $A_0 = \{(0, 1), (0, 2), (1, 3), (3, 6), (3, 7), (3, 8), (4, 5), (5, 9), (7, 4)\}$. Note that T_0 is a spanning subdigraph of G. ∎

DEFINITION 3.5.2 Let $G = (V, A)$ and let $u \in V$. A **depth-first search tree of G with root u** is a rooted tree $T = (V_0, A_0)$ defined recursively as follows:

0. Let V_0 be the subset of V consisting of those vertices reachable from u. If $V_0 = \{u\}$, then $A_0 = \{ \ \}$.

1. Otherwise, let $u_1 \in V_0 - \{u\}$ such that $(u, u_1) \in A$. Let V_1 be the subset of $V_0 - \{u\}$ consisting of those vertices reachable from u_1, and let $T_1 = (V_1, A_1)$ be a depth-first search tree of $\langle V_1 \rangle$ with root u_1. If $V_1 \cup \{u\} = V_0$, then let $A_0 = A_1 \cup \{(u, u_1)\}$.

2. Otherwise, let $u_2 \in V_0 - (V_1 \cup \{u\})$ such that $(u, u_2) \in A$. Let V_2 be the subset of $V_0 - (V_1 \cup \{u\})$ consisting of those vertices reachable from u_2, and let $T_2 = (V_2, A_2)$ be a depth-first search tree of $\langle V_2 \rangle$ with root u_2. If $V_1 \cup V_2 \cup \{u\} = V_0$, then $A_0 = A_1 \cup A_2 \cup \{(u, u_1), (u, u_2)\}$.

3. Otherwise, continue this process, finding depth-first search trees $T_1, T_2, \ldots, T_m$, where $T_i = (V_i, A_i)$ and has root u_i, until

$$V_1 \cup V_2 \cup \cdots \cup V_m \cup \{u\} = V_0$$

Then $A_0 = A_1 \cup A_2 \cup \cdots \cup A_m \cup \{(u, u_i) \mid 1 \leq i \leq m\}$. ∎

■ *Example 3.5.4* Let us apply Definition 3.5.2 to the digraph $G = (V, A)$ of Figure 3.5.2 to find a depth-first search tree with root $u = 0$. In step 0, we find that $V_0 \neq \{0\}$, because, for instance, $(0, 1) \in A$. Thus, in step 1, we let $u_1 = 1$. Without going into the details of the recursion, suppose that we find that $T_1 = (V_1, A_1)$ is a depth-first search tree of $\langle V_1 \rangle$, where $V_1 = \{1, 3, 4, 5, 6, 7, 8, 9\}$ and $A_1 = \{(1, 3), (3, 6), (3, 7), (3, 8), (4, 5), (5, 9), (7, 4)\}$. Note that $V_1 \cup \{0\} \neq V_0$, because $(0, 2) \in A$ and $2 \notin V_1 \cup \{0\}$. Thus, in step 2, we let $u_2 = 2$. By applying the definition recursively, we find that $T_2 = (V_2, A_2)$ is a depth-first search tree of $\langle V_2 \rangle$, where $V_2 = \{2\}$ and $A_2 = \{ \ \}$. Since $V_1 \cup V_2 \cup \{0\} = V = V_0$, we have $A_0 = A_1 \cup A_2 \cup \{(0, 1), (0, 2)\}$. This gives the tree $T = (V_0, A_0)$ found in Example 3.5.3. ∎

The next algorithm finds a depth-first search tree with given root u in a given digraph G. We can regard such an algorithm as "searching" for all the vertices of G that are reachable from u, and this is the reason for the word *search* in the name "depth-first search."

Again, we make the assumption that $V(G) = \{0, 1, \ldots, n-1\}$ for some positive integer n. The algorithm uses an array PARENT. If v is a vertex of T, then PARENT(v) is the parent of v in T, except that PARENT(u) = u for the root u. If v is not (yet) a vertex of T, then PARENT(v) = -1; this allows us to keep track of the vertex set of T as the algorithm progresses.

ALGORITHM 3.5.2   Given a digraph $G = (V, A)$ with $V = \{0, 1, \ldots, n-1\}$ and $u \in V$, the function

DEPTH_FIRST_SEARCH(G, u)

returns a depth-first search tree $T = (V_0, A_0)$ of G with root u.
   The function DEPTH_FIRST_SEARCH uses the following local variables:

T is a directed graph;
G' is a directed graph and is initialized to a copy of G;
(PARENT(0), PARENT(1), ..., PARENT(ORDER(G) $-$ 1)) is an array of
   integers and is initialized to the constant array $(-1, -1, \ldots, -1)$;
v is a vertex and is initialized to u;
w is a vertex;

```
begin
   INITIALIZE(T, u);
   PARENT(u) := u;
   loop
      while OUTDEGREE(v, G') > 0 loop
         w := FIRST_NEIGHBOR(v, G');
         DELETE_ARC(v, w, G');
         if PARENT(w) = −1 then
            ADD_VERTEX(w, T);
            ADD_ARC(v, w, T);
            PARENT(w) := v;
            v := w;
         end if;
      end loop;
      exit when v = u;
      v := PARENT(v);
   end loop;
   return T;
end DEPTH_FIRST_SEARCH;
```

■   *Example 3.5.5*   Illustrate Algorithm 3.5.2 for the digraph $G = (V, A)$ of Figure 3.5.2, using $u = 0$.

Solution

Initially, a copy of G is placed in G', $T = (\{0\}, \{ \})$, and $v := u = 0$. Also, PARENT $= (0, -1, -1, \ldots, -1)$. The main loop is entered where the while loop is iterated as follows:

Iteration 1: Assume FIRST_NEIGHBOR$(0, G') = 1$, so that $w := 1$ and the arc $(0, 1)$ is deleted from G'. Since PARENT$(1) = -1$, the vertex 1 and the arc $(0, 1)$ are added to T, PARENT$(1) := v = 0$, and $v := w = 1$.

Iteration 2: Here $w := $ FIRST_NEIGHBOR$(1, G') = 3$ and the arc $(1, 3)$ is deleted from G'. Since PARENT$(3) = -1$, the vertex 3 and the arc $(1, 3)$ are added to T, PARENT$(3) := v = 1$, and $v := w = 3$.

Iteration 3: Assume FIRST_NEIGHBOR$(3, G') = 6$, so that $w := 6$ and $(3, 6)$ is deleted from G'. Since PARENT$(6) = -1$, the vertex 6 and the arc $(3, 6)$ are added to T, PARENT$(6) := 3$, and $v := 6$.

Iteration 4: Here $w := $ FIRST_NEIGHBOR$(6, G') = 1$ and the arc $(6, 1)$ is deleted from G'. However, since PARENT$(1) = 0 \neq -1$, no changes are made to T. Now since OUTDEGREE$(6, G') = 0$, the while loop is exited. At this point $v := $ PARENT$(v) = $ PARENT$(6) = 3$ and the while loop is repeated. Note that $T = (\{0, 1, 3, 6\}, \{(0, 1), (1, 3), (3, 6)\})$ and that the arc set of G' is $\{(0, 2), (0, 9), (2, 0), (2, 3), (3, 7), (3, 8), (4, 5), (5, 9), (7, 4), (7, 9), (8, 1), (8, 4), (8, 9), (9, 4)\}$ at this stage.

Iteration 5: Assume FIRST_NEIGHBOR$(3, G') = 7$, so that $w := 7$ and $(3, 7)$ is deleted from G'. Since PARENT$(7) = -1$, the vertex 7 and the arc $(3, 7)$ are added to T, PARENT$(7) := 3$, and $v := 7$.

Iteration 6: Assume FIRST_NEIGHBOR$(7, G') = 4$, so that $w := 4$ and $(7, 4)$ is deleted from G'. Since PARENT$(4) = -1$, the vertex 4 and the arc $(7, 4)$ are added to T, PARENT$(4) := 7$, and $v := 4$.

Iteration 7: Here $w := $ FIRST_NEIGHBOR$(4, G') = 5$, so $(4, 5)$ is deleted from G'. Then the vertex 5 and the arc $(4, 5)$ are added to T, PARENT$(5) := 4$, and $v := 5$.

Iteration 8: At this step $w := $ FIRST_NEIGHBOR$(5, G') = 9$, so $(5, 9)$ is deleted from G'. The vertex 9 and the arc $(5, 9)$ are added to T; PARENT$(9) := 5$, and $v := 9$.

Iteration 9: Here $w := $ FIRST_NEIGHBOR$(9, G') = 4$, so $(9, 4)$ is deleted from G'. Since PARENT$(4) = 7 \neq -1$, no changes are made to T. Now OUTDEGREE $(9, G') = 0$, so the while loop is exited. At this point $v := $ PARENT$(v) = $ PARENT$(9) = 5$, and the while loop is reentered. However, OUTDEGREE$(5, G') = 0$ also. Thus, $v := $ PARENT$(5) = 4$, and the while loop is repeated. Again OUTDEGREE$(4, G') = 0$, so $v := $ PARENT$(4) = 7$. At this point $T = (\{0, 1, 3, 4, 5, 6, 7, 9\}, \{(0, 1), (1, 3), (3, 6), (3, 7), (4, 5), (5, 9), (7, 4)\})$, and the arc set of G' is $\{(0, 2), (0, 9), (2, 0), (2, 3), (3, 8), (7, 9), (8, 1), (8, 4), (8, 9)\}$.

Iteration 10: We have $w := $ FIRST_NEIGHBOR$(7, G') = 9$, so $(7, 9)$ is deleted from G'; T is not changed since PARENT$(9) = 5$. Now OUTDEGREE$(7, G') = 0$, so the while loop is exited. Then $v := $ PARENT$(7) = 3$ and the while loop is repeated.

Iteration 11: We have $w := $ FIRST_NEIGHBOR$(3, G') = 8$, so $(3, 8)$ is deleted from G'. Since PARENT$(8) = -1$, the vertex 8 and the arc $(3, 8)$ are added to T, PARENT$(8) := 3$, and $v := 8$. Then, since OUTDEGREE$(8, G') = 0$, the while loop is exited, leading to $v := $ PARENT$(8) = 3$. Next, OUTDEGREE$(3, G') = 0$, so $v := $ PARENT$(3) = 1$. Then OUTDEGREE$(1, G') = 0$, so $v := $ PARENT$(1) = 0$.

Iteration 12: Assume FIRST_NEIGHBOR$(0, G') = 2$, so that $w := 2$ and $(0, 2)$ is deleted from G'. Since PARENT$(2) = -1$, the vertex 2 and the arc $(0, 2)$ are added to T, PARENT$(2) := 0$, and $v := 2$.

Iteration 13: Assume FIRST_NEIGHBOR$(2, G') = 0$, so $w := 0$ and $(2, 0)$ is deleted from G'. The tree T is not changed.

Iteration 14: Next, $w :=$ FIRST_NEIGHBOR$(2, G') = 3$ and $(2, 3)$ is deleted from G'. Now OUTDEGREE$(2, G') = 0$, so $v := $ PARENT$(2) = 0$.

Iteration 15: Next, $w :=$ FIRST_NEIGHBOR$(0, G') = 9$ and $(0, 9)$ is deleted from G'. Now OUTDEGREE$(0, G') = 0$, so the while loop is exited. Note that $v = 0 = u$, so the main loop is exited and the algorithm terminates.

The tree

$$T = (\{0, 1, \ldots, 9\}, \{(0, 1), (0, 2), (1, 3), (3, 6), (3, 7), (3, 8), (4, 5), (5, 9), (7, 4)\})$$

is returned. ∎

Rooted trees are seen in this section to be acyclic digraphs that are "weakly connected" in the sense that every vertex of a rooted tree is reachable from the root by a unique path. We now turn our attention to the analogous concept for simple graphs.

DEFINITION 3.5.3 An **acyclic graph** is one with no cycles. Such a graph is termed a **forest**, and a connected forest is called a **tree**. ∎

Several examples of trees are shown in Figure 3.5.3. It is not surprising that the term *tree* is used to describe such graphs. Note that T_1 is the complete bipartite graph $K_{1,6}$ and that T_4 is a path of order 5. In general, a graph isomorphic to $K_{1,n-1}$ is called a **star** of order n. Also recall that there is a unique path of order n, up to isomorphism; it is denoted by P_n.

In many ways, the graphs P_n and $K_{1,n-1}$ are extreme examples of trees of order n. For example, consider the maximum degree among trees of order $n \geq 4$. The path P_n is the unique tree with maximum degree 2, whereas $K_{1,n-1}$ is the only such tree having maximum degree $n - 1$ (see Exercise 20).

For $n = 1, 2, 3$, there is a unique tree of order n (up to isomorphism), namely, P_n. The nonisomorphic trees of order 4 are P_4 and $K_{1,3}$.

■ *Example 3.5.6* Find (a) three nonisomorphic trees of order 5 and (b) three nonisomorphic trees of order 6 having maximum degree 3.

Solution

(a) The nonisomorphic trees of order 5 are P_5, $K_{1,4}$, and the tree T_3 of Figure 3.5.3.

(b) The trees shown in Figure 3.5.4 are three nonisomorphic trees of order 6 having maximum degree 3. ∎

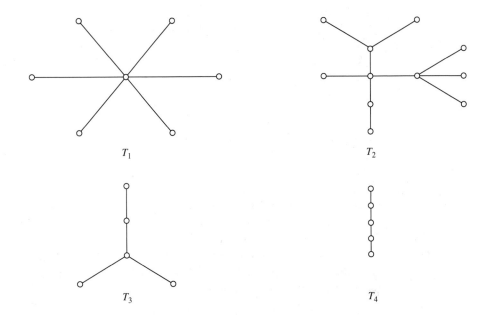

T_1 T_2

T_3 T_4

Figure 3.5.3   Several trees.

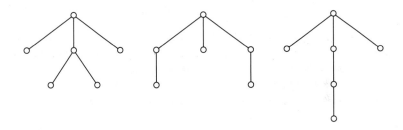

Figure 3.5.4   Three nonisomorphic trees of order 6 with maximum degree 3.

An important characterization of a tree is that it is a graph in which every two vertices are joined by a unique path. (Notice how this fact conforms with the analogous property for rooted trees.) In order to prove this result, we use the following lemma.

LEMMA 3.5.3   A graph $G = (V, E)$ contains a cycle if and only if it contains two different u-v paths for some distinct vertices $u, v \in V$.

Proof   Let

$$P : u = x_0, x_1, \ldots, x_s = v \qquad \text{and} \qquad Q : u = y_0, y_1, \ldots, y_t = v$$

be two distinct u-v paths in G. Without loss of generality, we may assume that

$x_1 \neq y_1$. Then there is a least subscript $i > 0$ such that $x_i = y_j$ for some j, $1 \leq j \leq t$. Let $w = x_i$. Then that portion of P from u to w, followed by that part of Q from w back to u, that is,

$$u = x_0, x_1, \ldots, x_i = y_j, y_{j-1}, \ldots, y_0 = u$$

is a cycle in G.

Conversely, suppose that

$$u = u_0, u_1, u_2, \ldots, u_t = u$$

is a cycle (of length $t \geq 3$). Then

$$P : u, u_1 \qquad \text{and} \qquad Q : u, u_{t-1}, \ldots, u_2, u_1$$

are two distinct u-u_1 paths. $\qquad\qquad\qquad\qquad\qquad\qquad\qquad\qquad\qquad\square$

THEOREM 3.5.4   A graph $T = (V, E)$ is a tree if and only if there is a unique u-v path in T for all $u, v \in V$.

Proof   Necessity follows from Lemma 3.5.3 and the fact that a tree is acyclic.

Conversely, assume that T contains a unique u-v path for all $u, v \in V$. Then T is clearly connected and it follows, again from Lemma 3.5.3, that T is acyclic. Therefore, T is a tree. $\qquad\qquad\qquad\qquad\qquad\qquad\qquad\qquad\qquad\qquad\qquad\qquad\square$

As a consequence of Theorem 3.5.4, if T is a tree and e is any edge of T, then the graph $T - e$ is disconnected. (Why?) In fact, if $e = uv$, then $T - e$ contains exactly two connected parts, one containing u and the other containing v. To make this last statement more precise and to introduce several ideas needed in subsequent sections, we need the concepts of "subgraph" and "component."

DEFINITION 3.5.4   Let $G_1 = (V_1, E_1)$ and $G_2 = (V_2, E_2)$ be two graphs. We call G_1 a **subgraph** of G_2 provided $V_1 \subseteq V_2$ and $E_1 \subseteq E_2$. The fact that G_1 is a subgraph of G_2 is denoted by writing $G_1 \subseteq G_2$. $\qquad\qquad\qquad\qquad\blacksquare$

Much of the terminology and notation for subgraphs is analogous to that for subdigraphs (see Section 3.4). In particular, for a graph $G = (V, E)$, $v \in V$, and $e \in E$, we use $G - v$ to denote the subgraph of G obtained by deleting the vertex v (and all edges incident with v) and $G - e$ to denote the subgraph obtained by deleting the edge e. Also, for $U \subseteq V$, the notation $\langle U \rangle$ denotes the **subgraph of G induced by** U, that is, the subgraph with vertex set U and edge set consisting of those edges of G joining vertices of U.

Now, given a graph $G = (V, E)$ and $v \in V$, it is clear that the subgraph $\langle \{v\} \rangle$, consisting of only the vertex v, is connected. It thus makes sense to consider those subgraphs of G that are maximal with respect to the property of being connected. It turns out that such subgraphs are always induced.

DEFINITION 3.5.5 Let $G = (V, E)$ be a graph and let $U \subseteq V$. The induced subgraph $\langle U \rangle$ is called a **component** of G provided it is maximal with respect to the property of being connected; that is, $\langle U \rangle$ is connected and $\langle U_1 \rangle$ is disconnected for any subset U_1 of V such that $U \subset U_1$. ∎

■ **Example 3.5.7** Let $G = (V, E)$, where $V = \{p, q, r, s, t, u, v, w, x, y, z\}$ and $E = \{pq, pu, rs, st, su, uv, uw, vx, vy, vz\}$. Then G is isomorphic to the tree T_2 in Figure 3.5.3. Since G is connected, G itself is its only component. Let $G_1 = G - uv$. As mentioned above, this subgraph of G is disconnected and consists of two components, each of which is a tree. The component containing u is $\langle \{p, q, r, s, t, u, w\} \rangle$, which has edge set $\{pq, pu, rs, st, su, uw\}$, and the component containing v is $\langle \{v, x, y, z\} \rangle$, which has edge set $\{vx, vy, vz\}$. Let $G_2 = G - u$; this subgraph is a forest with four components: $\langle \{p, q\} \rangle$, $\langle \{r, s, t\} \rangle$, $\langle \{w\} \rangle$, and $\langle \{v, x, y, z\} \rangle$. In general, we can see that if u is a vertex of a tree T, then $T - u$ is a forest with $\deg(u)$ components. ∎

In general, an edge e of a connected graph G is called a **bridge** provided $G - e$ is disconnected. Thus, every edge of a tree is a bridge; it turns out that trees are partly characterized by this property.

THEOREM 3.5.5 A graph $T = (V, E)$ is a tree if and only if T is connected and every edge of T is a bridge.

Proof Necessity follows from the comments made preceding the theorem.

Conversely, let $T = (V, E)$ be a connected graph with the property that every edge is a bridge. We wish to show that T is acyclic. If not, then T contains a cycle, say,

$$u = u_0, v = u_1, u_2, \ldots, u_n = u$$

Now consider the graph $T - e$, where $e = uv$. Note that this graph contains the u-v path

$$u = u_n, u_{n-1}, \ldots, u_2, u_1 = v$$

Therefore, $T - e$ is connected, showing that e is not a bridge. This gives a contradiction, thus proving the result. □

As an important corollary to Theorem 3.5.5 we can show that every connected graph G contains a **spanning tree**. This is a subgraph T of G such that T is a tree and $V(T) = V(G)$.

COROLLARY 3.5.6 If $G = (V, E)$ is a connected graph, then G contains a spanning tree.

Proof   Since G is connected, G is a connected spanning subgraph of itself. Thus, it makes sense to let $T = (V, E_0)$ be a connected spanning subgraph of G of minimum size (that is, $|E_0|$ is a minimum). We claim that T is a tree. If not, then T contains an edge e that is not a bridge by Theorem 3.5.5. This means that the graph $T_1 = T - e$ is connected, contradicting the choice of T as a connected spanning subgraph of G of minimum size. This proves the result.   □

■ **Example 3.5.8**   Consider the graph $G = (V, E)$, where $V = \{u, v, x, y, z\}$ and $E = \{uv, ux, vy, vz, xy, xz\}$.

(a) Find a spanning tree T_1 isomorphic to P_5.

(b) Find a spanning tree T_2 isomorphic to the tree T_3 of Figure 3.5.3.

(c) Give a reason why G cannot have a spanning tree isomorphic to $K_{1,4}$.

Solution

(a) There are six possibilities; for example, let $T_1 = (V, E_1)$, where $E_1 = \{uv, vz, xy, xz\}$.

(b) Again there are six possibilities, one being $T_2 = (V, E_2)$, where $E_2 = \{uv, ux, vy, vz\}$.

(c) Since $\Delta(K_{1,4}) = 4$ and $\Delta(G) = 3$, G cannot have a spanning tree isomorphic to $K_{1,4}$.   ■

Let us summarize the characterizations of trees that we have so far; namely, a graph G is a tree if and only if G satisfies one of the following conditions:

1. G is connected and acyclic. (Definition 3.5.3)

2. There is a unique path joining any two vertices of G. (Theorem 3.5.4)

3. G is connected and every edge is a bridge. (Theorem 3.5.5)

Two additional characterizations are given in the next theorem, with the proof left for you to develop in Exercise 23.

THEOREM 3.5.7   A graph G of order n and size m is a tree if and only if G satisfies one of the following conditions:

4. G is connected and $m = n - 1$.

5. G is acyclic and $m = n - 1$.   □

At this point in our discussion of trees, let's consider an application to transportation and communication networks. For this we need the concept of a weighted graph, which is analogous to the concept of a weighted directed graph introduced in Section 3.4.

Given a graph $G = (V, E)$, suppose there is associated with each $e \in E$ a positive number $w(e)$, called the **weight** of the edge e. The graph G, together with the function $w : E \to (0, \infty)$, is called a **weighted graph**. Given a subgraph $H = (V_0, E_0)$ of G, we define its **weight** to be the sum of the weights of the edges in E_0 and denote this weight by $w(H)$.

Suppose we are given the problem of connecting n computers with a network of communications links, so that, perhaps indirectly, any computer may transmit data to any other computer. By associating a vertex with each computer, if it is feasible to install a direct communications link between vertices c_1 and c_2 at a cost $w(c_1 c_2)$, then this number becomes the weight of the edge $c_1 c_2$ in our weighted graph. On the other hand, if it is determined not to be feasible to directly link c_1 and c_2, then c_1 and c_2 are nonadjacent in our weighted graph. Our goal may then be to build a network in which these conditions are met:

1. It is possible (perhaps indirectly) to send data from any computer to any other computer in the network.

2. The cost of installing the communications links is a minimum.

This means that we desire a connected spanning subgraph of our weighted graph of minimum weight. Clearly, such a subgraph is a spanning tree, and it is called a **minimum spanning tree** of the weighted graph.

■ *Example 3.5.9*   Find a minimum spanning tree of the graph G of Example 3.5.8 if the weights of the edges are $w(uv) = 2$, $w(ux) = 4$, $w(vy) = 4$, $w(vz) = 6$, $w(xy) = 3$, and $w(xz) = 5$.

Solution
The graph G has order 5 and size 6, so, to obtain a spanning tree, two edges of G must be excluded. We would like to exclude the two edges of largest weight, namely, vz and xz; however, the resulting subgraph is not connected. The best we can do is exclude vz and one of the edges with weight 4, say, ux. The resulting spanning tree T is a minimum spanning tree of G having $w(T) = w(uv) + w(vy) + w(xy) + w(xz)$ $= 14$.   ■

Now we let $G = (V, E)$ be a graph and $u, v \in V$ be nonadjacent vertices. Then $G + uv$ denotes the graph obtained by adding the edge uv to G. Note that G is a proper spanning subgraph of $G + uv$. Now let $T = (V, E_0)$ be a spanning tree of G and let $e \in E - E_0$. It can be shown (see Exercise 24) that $T + e$ contains a unique cycle. If f is an edge of this cycle, $f \neq e$, then $T_1 = (T + e) - f$ is another spanning tree of G. By repeating this operation of "edge replacement" a finite number of times, it is possible to transform a given spanning tree T_1 into any other spanning tree T_2 of G (see Exercise 26).

■ *Example 3.5.10*   Find all minimum spanning trees of the weighted graph $G = (V, E)$ of Figure 3.5.5, where $w(rs) = 8$, $w(rx) = 7$, $w(rz) = 4$, $w(st) = 10$, $w(tx) = 2$, $w(xy) = 1$, and $w(yz) = 7$.

Solution
Note that G has seven edges, whereas a spanning tree of G has five edges. Thus, to obtain a spanning tree, two edges of G must be excluded. If we exclude the

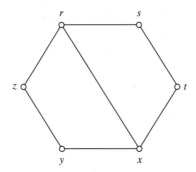

Figure 3.5.5　The graph G of Example 3.5.10.

two edges of largest weight, namely, rs and st, the result is not a tree. So let's try $T_0 = (G - st) - yz$. This is a spanning tree of G, and it is not hard to see that $T_0 = (V, E_0)$ is a minimum spanning tree, with weight 22. Now we ask, is it possible to replace an edge e_0 of T_0 with an edge e_1 of $E - E_0$ such that $w(e_0) = w(e_1)$ and the resulting graph $T_1 = (T_0 + e_1) - e_0$ is acyclic? If so, then T_1 is another minimum spanning tree of G. Note that $T_0 + yz$ contains the cycle r, x, y, z, r. On this cycle, the edge rx has the same weight as the edge yz. Thus, $T_1 = (T_0 + yz) - rx$ is also a minimum spanning tree of G. Finally, it can be argued that T_0 and T_1 are the only minimum spanning trees of G. ■

　　Given a connected weighted graph G, we desire an algorithm that produces a minimum spanning tree of G. One well-known algorithm for this problem was discovered by J. B. Kruskal [On the shortest spanning subtree of a graph and the traveling salesman problem, *Proceedings of the AMS* 1 (1956): 48–50]. Another algorithm is due to R. C. Prim [Shortest connection networks and some generalizations, *Bell System Technical Journal* 36 (1957): 1389–1401]. We illustrate both algorithms with an example, and then present and analyze Prim's algorithm. For a discussion of Kruskal's algorithm, see Chapter 7 of *Foundations of Discrete Mathematics*, 2nd ed., by A. D. Polimeni and H. J. Straight.

■ *Example 3.5.11*　We'll describe Kruskal's algorithm informally by applying it to the weighted graph $G = (V, E)$ of Example 3.5.10. Our goal is to find a minimum spanning tree $T = (V, E_0)$ of G. We begin by sorting the edges of G into a list that is in increasing order by weight; this gives the list $L = (xy, tx, rz, rx, yz, rs, st)$. Initially, $E_0 = \{\ \}$; the main idea of the algorithm is to add to E_0, at each iteration, an edge of L of smallest weight such that (V, E_0) remains acyclic. In this sense, Kruskal's algorithm is an example of an important class of algorithms known as "greedy" algorithms. In our example, the first four edges of L may be added to E_0; this yields the forest $(V, \{rx, rz, tx, xy\})$. At this point, the next edge of L is yz;

however, the graph $(V, \{rx, rz, tx, xy, yz\})$ contains the cycle r, x, y, z, r. Thus, the edge yz is excluded from consideration. The next edge of L, namely, rs, causes no such problem, so this edge may be added to E_0. This yields the minimum spanning tree $T = T_0 = (V, E_0)$, where $E_0 = \{rs, rx, rz, tx, xy\}$.

Prim's algorithm constructs a minimum spanning tree $T = (V, E_0)$ of a graph $G = (V, E)$ as follows. Initially, $E_0 = \{ \ \}$, and we begin at an arbitrary vertex v of G, letting $U = V - \{v\}$. At each iteration, one vertex is removed from U and one edge is added to E_0. The edge chosen is an edge uv of smallest weight such that $u \in U$ and $v \in V - U$; the vertex u is removed from U. This process continues until $U = \{ \ \}$. In our example, we let $U = V - \{r\}$ initially. At the first iteration, the edge rz is selected, so that $U = \{s, t, x, y\}$ and $E_0 = \{rz\}$. At the second iteration, either rx or zy could be selected; suppose rx is selected so that $U = \{s, t, y\}$ and $E_0 = \{rx, rz\}$. Then xy is chosen at the third iteration and xt at the fourth, so that $U = \{s\}$ and $E_0 = \{rx, rz, tx, xy\}$. At the fifth iteration, the edge of smallest weight is yz; however, $y \notin U$ and $z \notin U$, so that this edge can be excluded from further consideration. The edge chosen at the fifth stage is rs, so that $U = \{ \ \}$ and $E_0 = \{rs, rx, rz, tx, xy\}$. Now U is empty, so the algorithm terminates. ■

Prim's algorithm is based on the following observation. Suppose that $T = (V, E_0)$ is a tree and that u and v are vertices with $u \notin V$ and $v \in V$. Then $T' = (V \cup \{u\}, E_0 \cup \{uv\})$ is also a tree (see Exercise 36). Thus, at each iteration during an application of Prim's algorithm, the subgraph $(V - U, E_0)$ of G is, in fact, a subtree of G. Like Kruskal's algorithm, Prim's algorithm is also greedy in the sense that it extends the subtree $(V - U, E_0)$ at each iteration by using an edge of smallest possible weight.

We can now give a description (using pseudocode) of Prim's algorithm; some of the details of implementation are discussed in Exercise 40.

ALGORITHM 3.5.3 (Prim's algorithm)   Given a connected weighted graph $G = (V, E, w)$, this algorithm finds a minimum spanning tree $T = (V, E_0)$ of G.

> begin
>   Choose any vertex $v \in V$;
> $U := V - \{v\}$;
> $E_0 := \{ \ \}$;
>   while $U \neq \{ \ \}$ loop
>   Let $e = uv$ be an edge of smallest weight such that $u \in U$ and $v \in V - U$;
> $U := U - \{u\}$;
> $E_0 := E_0 \cup \{uv\}$;
>   end loop;
> end;

■ ***Example 3.5.12***   Illustrate Prim's algorithm using the weighted graph $G = (V, E, w)$ of Figure 3.5.6, where $w(xy) = 1$, $w(qx) = w(qy) = 2$, $w(qs) = 3$, $w(st) = 4$, $w(qt) = w(tx) = 5$, $w(qz) = 6$, $w(yz) = 7$, $w(rs) = w(rz) = 8$, and $w(qr) = 9$.

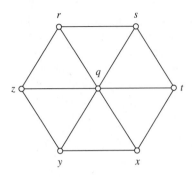

Figure 3.5.6   The graph *G* of Example 3.5.12.

Solution

Initially, we choose $v = q$, so that $U := \{r, s, t, x, y, z\}$ and $E_0 := \{\ \}$. The main loop is now executed six times.

Iteration 1: We have $e = uv = xq$, so that $U := U - \{x\} = \{r, s, t, y, z\}$ and $E_0 := E_0 \cup \{xq\} = \{qx\}$.

Iteration 2: This time $e = uv = yx$, so that $U := U - \{y\} = \{r, s, t, z\}$ and $E_0 := E_0 \cup \{yx\} = \{qx, xy\}$.

Iteration 3: Now $e = uv = sq$, so that $U := U - \{s\} = \{r, t, z\}$ and $E_0 := E_0 \cup \{sq\} = \{qs, qx, xy\}$.

Iteration 4: Here $e = uv = ts$, so that $U := U - \{t\} = \{r, z\}$ and $E_0 := E_0 \cup \{ts\} = \{qs, qx, st, xy\}$.

Iteration 5: This time $e = uv = zq$, so that $U := U - \{z\} = \{r\}$ and $E_0 := E_0 \cup \{zq\} = \{qs, qx, qz, st, xy\}$.

Iteration 6: Finally, $e = uv = rs$ or $e = uv = rz$. In either case, $U := U - \{r\} = \{\ \}$. Suppose rs is the edge chosen; then $E_0 := \{qs, qx, qz, rs, st, xy\}$.

Thus, $(V, \{qs, qx, qz, rs, st, xy\})$ is found to be a minimum spanning tree of *G*. If the edge rz is chosen at iteration 6, then the minimum spanning tree found is $(V, \{qs, qx, qz, rz, st, xy\})$. Since it is only at iteration 6 that there is a choice as to which edge to include in E_0, we can say that these two trees are the only minimum spanning trees of *G*.   ∎

That Algorithm 3.5.3 produces a spanning tree of *G* is clear, because $(V - U, E_0)$ is a subtree of *G* at each iteration, with *U* empty at the end. The next result states that a spanning tree (V, E_0) produced by Algorithm 3.5.3 is, in fact, a minimum spanning tree.

THEOREM 3.5.8   Let $G = (V, E, w)$ be a connected weighted graph and let $T = (V, E_0)$ be a spanning tree of *G* that is produced by an application of Algorithm 3.5.3. Then *T* is a minimum spanning tree of *G*.

Proof Let $E_0 = \{e_1, e_2, \ldots, e_{n-1}\}$, where the edges are listed in the order chosen by Algorithm 3.5.3. We proceed by contradiction, supposing that T is not a minimum spanning tree. Let $k + 1$ be the smallest integer such that there does not exist a minimum spanning tree of G containing each edge e_i, $1 \leq i \leq k + 1$; so $0 \leq k < n - 1$. Let $T_1 = (V, E_1)$ be a minimum spanning tree of G such that $e_i \in E_1$, $1 \leq i \leq k$. Then $e_{k+1} \notin E_1$, so that the graph $T_1 + e_{k+1}$ contains a unique cycle, say, C. Since the subgraph of G induced by $\{e_1, e_2, \ldots, e_k\}$ is acyclic, there must be an edge f of C that is not in $\{e_1, e_2, \ldots, e_k\}$. We claim that f can be chosen so that it is adjacent with one of the edges $e_1, e_2, \ldots, e_k$. To see this, let $e_{k+1} = vu$ and let vz be the other edge of C incident with v. If $k = 0$, then let $f = vz$. If $k > 0$, then v is incident with one of the edges $e_i = wv$, $1 \leq i \leq k$. If $w \neq z$, again let $f = vz$. Otherwise, we may follow C in the direction u, v, w, and so on, to find f. Now consider the graph $T_2 = (T_1 + e_{k+1}) - f$. By our earlier remarks concerning the operation of edge replacement, T_2 is also a spanning tree of G and its weight is

$$w(T_2) = w(T_1) + w(e_{k+1}) - w(f)$$

Since T_1 is a minimum spanning tree, it must be that $w(e_{k+1}) \geq w(f)$. In addition, the subgraph of G induced by $\{e_1, \ldots, e_k, f\}$ is a subtree of G by the way f was chosen. If $w(f) < w(e_{k+1})$, then f would have been chosen instead of e_{k+1} as an edge of T, a contradiction to our assumption that T is produced by an application of Algorithm 3.5.3. Hence, we must conclude that $w(e_{k+1}) = w(f)$, so that $w(T_2) = w(T_1)$. Therefore, T_2 is a minimum spanning tree of G. But T_2 contains each of the edges $e_1, \ldots, e_k, e_{k+1}$, contradicting the definition of k. This completes the proof. $\square$

EXERCISE SET 3.5

1. For the digraph $G = (V, A)$ of Figure 3.5.2, apply Algorithm 3.5.2 to find a depth-first search tree with root:

   (a) 1   (b) 4   (c) 7   (d) 8

2. Prove: If $T = (V, A)$ is an acyclic digraph with a unique vertex r with indegree 0, and for every $v \in V$ there is a unique r-v path in T, then T is a rooted tree.

3. Apply Algorithm 3.5.2 to each of the following directed graphs to find a depth-first search tree with root 0.

   (a) $G_1 = (\{0, 1, 2, 3, 4, 5, 6, 7\}, \{(0, 1), (0, 2), (0, 3), (0, 5), (1, 4), (1, 7), (2, 4), (2, 5)$
   $(3, 1), (3, 6), (4, 0), (4, 7), (5, 4), (6, 2), (6, 7), (7, 0), (7, 1)\})$

   (b) $G_2 = (\{0, 1, 2, 3, 4, 5, 6, 7, 8\}, \{(0, 1), (0, 2), (0, 6), (1, 3), (1, 4), (1, 8), (2, 0),$
   $(2, 3), (2, 5), (2, 6), (2, 7), (4, 1), (4, 3), (4, 7), (5, 4), (6, 7), (7, 8), (8, 0),$
   $(8, 2), (8, 3)\})$

4. Let G be a digraph. Prove: If G contains a nontrivial closed walk, then G contains a cycle.

5. Let G be a digraph and let H be a subdigraph of G. Prove: If G is acyclic, then H is acyclic.

6. Let $T = (V_0, A_0)$ be a rooted tree of order n (that is, $|V_0| = n$).

   (a) Show that the size of T is $n - 1$ (that is, $|A_0| = n - 1$).
   (b) Prove or disprove: If H is an acyclic digraph of order n and size $n - 1$, then H is a rooted tree.

7. Let T be an m-ary tree with height h, and let v be a vertex of T at level k, $k \geq 1$.

   (a) How many ancestors does v have?
   (b) Give an upper bound on the number of descendants of v.

8. An m-ary tree is called **full** provided every vertex that is not a leaf has m children.

   (a) Give an example of a full binary tree with height 3 that has as many vertices as possible.
   (b) Give an example of a full 3-ary tree with height 3 that has as few vertices as possible.
   (c) Give a lower bound on the order of a full m-ary tree with height h.

9. A **complete m-ary tree with height h** is a full m-ary tree such that every leaf is at level h; thus, each vertex at level k, $k < h$, has m children. What is the order of such a rooted tree?

10. A rooted tree with height h is called **balanced** provided every leaf is at level h or level $h - 1$. Let T be a full balanced m-ary tree with height h such that T has exactly k leaves. Show that $m^{h-1} + m - 1 \leq k \leq m^h$.

11. Apply Algorithm 3.5.1 to the digraph $G = (\{1, 2, 3, 4, 5, 6, 7, 8, 9\}, \{(9, 8), (9, 7), (9, 6), (8, 5), (8, 4), (7, 5), (7, 4), (7, 3), (6, 3), (5, 2), (4, 1), (3, 1), (2, 1)\})$.

12. What happens if a digraph G input to Algorithm 3.5.1 contains a cycle? For instance, let G be the digraph of Exercise 11 and try applying Algorithm 3.5.1 to the digraph $G + (1, 8)$.

13. Find all nonisomorphic trees of order 7 with maximum degree 4.

14. Find all nonisomorphic trees of order 7.

15. Find all nonisomorphic spanning trees of the graph $G = (V, E)$, where $V = \{u, v, w, x, y, z\}$ and $E = \{uv, ux, uy, uz, vy, wx, wz, xy, yz\}$.

16. For each weighted graph of Figure 3.5.7, apply Algorithm 3.5.3 to find a minimum spanning tree.

   (a) $w(st) = 2$, $w(sx) = 3$, $w(sy) = 5$, $w(sz) = 7$, $w(tx) = 2$, $w(ty) = 4$, $w(xy) = 9$, $w(xz) = 1$, $w(yz) = 3$
   (b) $w(rs) = 15$, $w(ry) = 9$, $w(rz) = 1$, $w(st) = 9$, $w(sx) = 5$, $w(sy) = 3$, $w(tx) = 4$, $w(tz) = 4$, $w(xy) = 2$, $w(yz) = 7$
   (c) $w(rs) = 13$, $w(rt) = 2$, $w(ry) = 7$, $w(rz) = 40$, $w(st) = 3$, $w(sx) = 8$, $w(sy) = 6$, $w(tx) = 5$, $w(ty) = 4$, $w(xy) = 15$, $w(xz) = 10$, $w(yz) = 25$
   (d) $w(qr) = 17$, $w(qt) = 11$, $w(qz) = 9$, $w(rs) = 1$, $w(rt) = 5$, $w(st) = w(sx) = w(tz) = 3$, $w(xy) = 1$, $w(xz) = 7$, $w(yz) = 9$

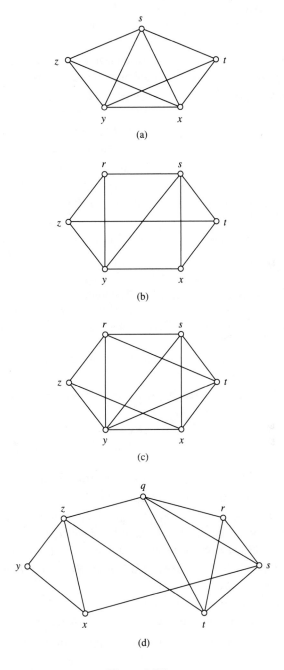

(a)

(b)

(c)

(d)

Figure 3.5.7

17. Apply Algorithm 3.5.3 to find a minimum spanning tree of the weighted graph $G = (V, E, w)$, where $V = \{r, s, t, x, y, z\}$, $E = \{rs, rx, ry, rz, sy, sx, tx, xy, yz\}$, and $w(rs) = 12$, $w(rx) = 6$, $w(ry) = 1$, $w(rz) = 5$, $w(sy) = 6$, $w(sx) = 5$, $w(tx) = 10$, $w(xy) = 9$, and $w(yz) = 3$.

18. Prove or disprove: If the edges of the connected weighted graph $G = (V, E, w)$ have distinct weights, then G has a unique minimum spanning tree.

19. Find a minimum spanning tree of the graph of Figure 3.2.4, assuming that $w(bv) = w(uv) = w(vx) = w(vy) < w(au) = w(az) = w(bz) = w(cx) = w(uy) = w(xy) < w(bu) = w(bx) < w(cy) < w(ab) = w(bc) < w(ad) = w(cd)$.

20. Let T be a tree of order $n > 3$ and maximum degree Δ.
    (a) Show that T is isomorphic to P_n if and only if $\Delta = 2$.
    (b) Show that T is isomorphic to $K_{1,n-1}$ if and only if $\Delta = n - 1$.

21. For $n > 4$, find the unique tree of order n, up to isomorphism, with maximum degree $n - 2$.

22. For $n > 5$, find the three nonisomorphic trees of order n with maximum degree $n - 3$.

23. Let G be a graph of order n and size m.
    (a) Prove characterization 4 given in Theorem 3.5.7: G is a tree if and only if G is connected and $m = n - 1$. (Hint: To prove necessity, proceed by induction on n; to prove sufficiency, let T be a spanning tree of G and show that $G = T$.)
    (b) Prove characterization 5 given in Theorem 3.5.7: G is a tree if and only if G is acyclic and $m = n - 1$.

24. Let $G = (V, E)$ be a connected graph and let $T = (V, E_0)$ be a spanning tree of G. Let $e \in E - E_0$.
    (a) Show that $T + e$ contains a unique cycle; call it C.
    (b) Let f be an edge of $C, f \neq e$. Show that $T_1 = (T + e) - f$ is a spanning tree of G.
    (c) If G has order n, how many edges do the trees T and T_1 have in common?

25. Prove that every tree is a bipartite graph.

26. Let $G = (V, E)$ be a connected graph and let $T_1 = (V, E_1)$ and $T_2 = (V, E_2)$ be spanning trees of G. Show that T_2 may be obtained from T_1 by performing a finite sequence of edge replacement operations.

27. Vertices of degree 1 in a tree are called **leaves**. Every tree T of order $n > 1$ has at least two leaves. Prove this fact:
    (a) Using the fact that T has $n - 1$ edges.
    (b) By considering the first and last vertices of a longest path.

28. With regard to the result of Exercise 27, suppose further that T has maximum degree Δ.
    (a) Show that T has at least Δ leaves.
    (b) What can be said about T if it has exactly Δ leaves?

29. Let $T_1 = (V_1, E_1)$ and $T_2 = (V_2, E_2)$ be trees, where $V_1 = \{u_1, u_2, \ldots, u_n\}$, $V_2 = \{v_1, v_2, \ldots, v_n\}$, and $\deg(u_i) = \deg(v_i)$, $1 \leq i \leq n$.

    (a) If $n \leq 5$, show that T_1 is isomorphic to T_2.

    (b) For $n \geq 6$, give an example of two nonisomorphic such trees.

30. Let F be a forest and let u and v be distinct nonadjacent vertices of F. Show that $F + uv$ is a forest if and only if u and v belong to different components of F.

31. Given that a forest F has order n and k components, what is its size m?

32. Let G be a graph. Prove that G is a forest if and only if every connected subgraph is an induced subgraph.

33. Use Kruskal's algorithm to do Exercise 16.

34. Use Kruskal's algorithm to do Exercise 17.

35. Recall that a leaf in a rooted tree is a vertex having no children. A **penultimate vertex** is a vertex with the property that it is not a leaf and each of its children is a leaf. (For example, in the tree of Figure 3.5.1, vertices f and g are the penultimate vertices.) Prove that every nontrivial rooted tree contains a penultimate vertex.

36. Suppose that $T = (V, E_0)$ is a tree and that u and v are vertices with $u \notin V$ and $v \in V$. Show that $T' = (V \cup \{u\}, E_0 \cup \{uv\})$ is also a tree.

37. Let $G = (V, E, w)$ be a connected weighted graph of order $n > 1$ and let $T = (V, E_0)$ be a minimum spanning tree of G. Is it true, for some choice of the edges $e_1, e_2, \ldots, e_{n-1}$ in Prim's algorithm, where e_i is the edge added at the ith iteration, that $E_0 = \{e_1, e_2, \ldots, e_{n-1}\}$? In other words, can every minimum spanning tree of G be produced by Algorithm 3.5.3?

38. Discuss what happens if Algorithm 3.5.3 is applied to a disconnected weighted graph $G = (V, E, w)$. Does the algorithm produce, or can it be modified to produce, a minimum spanning forest of G of maximum size?

39. Let $G = (V, E, w)$ be a connected weighted graph of order $n > 1$ and let $T = (V, E_0)$ be a minimum spanning tree of G produced by Algorithm 3.5.3. If G has a unique edge e of smallest weight, show that $e \in E_0$. In other words, the edge of smallest weight must be used in any minimum spanning tree produced by Algorithm 3.5.3.

40. Concerning Algorithm 3.5.3:

    (a) How might we keep track of which vertices are in U and which are in $V - U$?

    (b) How might we search efficiently for an edge $e = uv$ of smallest weight such that $u \in U$ and $v \in V - U$? In performing this search, suppose we encounter an edge $v_1 v_2$ with $v_1, v_2 \in V - U$. Can this edge be deleted from further consideration? What about an edge $u_1 u_2$ with $u_1, u_2 \in U$?

3.6 MATCHINGS

Consider a computer dating service that has a set $G = \{g_1, g_2, \ldots, g_m\}$ of m gentlemen clients and a set $L = \{w_1, w_2, \ldots, w_n\}$ of n lady clients. Suppose each of the ladies wishes to have a compatible date for an upcoming social occasion. For each i, $1 \leq i \leq n$, let G_i be the subset of G consisting of those gentlemen deemed to be acceptable dates for lady w_i. Then the dating service must come up with an n-permutation $(x_1, x_2, \ldots, x_n)$ of G such that $x_i \in G_i$ for each i, $1 \leq i \leq n$. Such a permutation is called a **system of distinct representatives** (SDR) for the ordered collection of sets $(G_1, G_2, \ldots, G_n)$.

> **DEFINITION 3.6.1** Let Y be a set and let $(X_1, X_2, \ldots, X_n)$ be an ordered collection of n subsets of Y. An n-tuple $(x_1, x_2, \ldots, x_n)$ with $x_i \in X_i$ for each i, $1 \leq i \leq n$, is called a **system of representatives** for $(X_1, X_2, \ldots, X_n)$. If, in addition, the x_i's are distinct, that is, if $(x_1, x_2, \ldots, x_n)$ is an n-permutation of Y, then $(x_1, x_2, \ldots, x_n)$ is called a **system of distinct representatives** (or **transversal**) for $(X_1, X_2, \ldots, X_n)$. ■

■ **Example 3.6.1** Let $X_1 = \{1, 2\}$, $X_2 = \{1, 3, 4\}$, $X_3 = \{3, 4\}$, $X_4 = \{1, 3, 5\}$, and $X_5 = \{4, 5\}$. Then $(1, 3, 4, 5, 4)$ is a system of representatives for $(X_1, X_2, X_3, X_4, X_5)$ (but is not a system of distinct representatives since 4 is used to represent both X_3 and X_5), and $(2, 1, 4, 3, 5)$ is a system of distinct representatives for $(X_1, X_2, X_3, X_4, X_5)$. ■

Let $(X_1, X_2, \ldots, X_n)$ be an ordered collection of finite sets and let $Y = X_1 \cup X_2 \cup \cdots \cup X_n$. The problem of finding an SDR for this ordered collection can be represented using a bipartite graph G as follows. The partite sets for G are Y and $\{X_1, X_2, \ldots, X_n\}$. The edges of G indicate which elements of Y belong to which subsets—there is an edge joining x and X_i if and only if $x \in X_i$. Now suppose that $(x_1, x_2, \ldots, x_n)$ is an SDR for $(X_1, X_2, \ldots, X_n)$. Then $x_1 X_1, x_2 X_2, \ldots, x_n X_n$ are edges of G and no two of these edges are adjacent. In general, a set of pairwise nonadjacent edges in a graph is called a **matching**.

■ **Example 3.6.2** The ordered collection $(X_1, X_2, X_3, X_4, X_5)$ of Example 3.6.1 is represented by the bipartite graph G shown in Figure 3.6.1. The SDR $(2, 1, 4, 3, 5)$ corresponds to the matching $\{2X_1, 1X_2, 4X_3, 3X_4, 5X_5\}$ of G; the edges of this matching are marked in the figure. ■

> **DEFINITION 3.6.2** Let $G = (V, E)$ be a graph. A subset M of E is called a **matching** of G provided no two edges of M are adjacent. A matching M is said to be a **maximal matching** provided it is not a proper subset of some other matching of G. A matching M is called a **maximum matching** provided it has maximum cardinality among the matchings of G; if each vertex of G is incident with an edge of M, then M is called a **perfect matching**. ■

Any perfect matching of G is a maximum matching, whereas any maximum matching is also maximal.

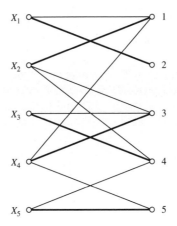

Figure 3.6.1

■ ***Example 3.6.3***   The matching $\{2X_1, 1X_2, 4X_3, 3X_4, 5X_5\}$ of the graph G of Figure 3.6.1 is a perfect matching and hence is a maximum matching; it has cardinality 5 and corresponds to the SDR $(2, 1, 4, 3, 5)$ for $(X_1, X_2, X_3, X_4, X_5)$. In fact, any maximum matching of G corresponds to an SDR for $(X_1, X_2, X_3, X_4, X_5)$; for instance, the maximum matching $\{2X_1, 4X_2, 3X_3, 1X_4, 5X_5\}$ corresponds to the SDR $(2, 4, 3, 1, 5)$. ■

Thus, we can see that the problem of finding an SDR for an ordered collection of finite sets $(X_1, X_2, \ldots, X_n)$ reduces to the problem of finding a suitable maximum matching in the corresponding bipartite graph—one that has an edge incident with each vertex of $X = \{X_1, X_2, \ldots, X_n\}$. We now focus our attention on the problem of finding maximum matchings in graphs.

Let $G = (V, E)$ be a graph. If M is a matching of G, then $2|M| \leq |V|$, because each vertex of G is incident with at most one edge of M. By a similar reason, if G is a bipartite graph with partite sets X and Y, then $|M| \leq \min(|X|, |Y|)$. Suppose that $|X| \leq |Y|$ and $|M| = |X|$. In this case we say that M **matches** X to a subset of Y and that X is **matched** to a subset of Y **under** M.

For a graph $G = (V, E)$ and S a nonempty subset of V, the set of vertices adjacent to at least one vertex of S is called the **neighborhood** of S and is denoted by $N(S, G)$ or simply by $N(S)$ if the graph under consideration is understood. Let G be bipartite with partite sets $X = \{x_1, x_2, \ldots, x_m\}$ and $Y = \{y_1, y_2, \ldots, y_n\}$, where $m \leq n$, and suppose that $M = \{x_1 y_{i_1}, x_2 y_{i_2}, \ldots, x_m y_{i_m}\}$ matches X to a subset of Y. For $S = \{x_1, x_2, \ldots, x_k\}$, note that $\{y_{i_1}, y_{i_2}, \ldots, y_{i_k}\} \subseteq N(S)$. In fact, we see that $|S| \leq |N(S)|$ for any nonempty subset S of X and that this is a necessary condition for X to be matched to a subset of Y. It turns out that it is also a sufficient condition, a result shown by P. Hall [On representatives of subsets, *Journal of the London Mathematical Society* 10 (1935): 26–30].

THEOREM 3.6.1 (Hall's Theorem) Let G be a bipartite graph with partite sets X and Y. Then X can be matched to a subset of Y if and only if $|S| \leq |N(S)|$ for every nonempty subset S of X.

Proof Necessity follows by the remarks preceding the theorem.

For sufficiency, we proceed by induction on the cardinality m of X. The result is obvious for $m = 1$. Let $m > 1$; the induction hypothesis is that for any bipartite graph G_1 having partite sets X_1 and Y_1 with $|X_1| < m$, if $|S| \leq |N(S)|$ for every nonempty subset S of X_1, then X_1 can be matched to a subset of Y_1. We consider two cases.

Case 1: Suppose, for any nonempty proper subset S of X, that

$$|S| < |N(S)|$$

Let $x_1 y_1$ be an edge of G with $x_1 \in X$, $y_1 \in Y$, and let $G_1 = (G - x_1) - y_1$. Then G_1 is bipartite with partite sets $X_1 = X - \{x_1\}$ and $Y_1 = Y - \{y_1\}$. Furthermore, for any nonempty subset S of X_1,

$$|N(S, G_1)| \geq |N(S, G)| - 1 \geq |S|$$

so that G_1 satisfies the induction hypothesis. Thus, X_1 can be matched to a subset of Y_1, say, under the matching M_1. Then $M = M_1 \cup \{x_1 y_1\}$ matches X to a subset of Y.

Case 2: The negation of the first case is that for some nonempty proper subset S_1 of X,

$$|S_1| = |N(S_1)|$$

In this case the bipartite subgraph of G with partite sets S_1 and $N(S_1)$ satisfies the induction hypothesis, so that S_1 can be matched to $N(S_1)$, say, under the matching $M_1 = \{x_1 y_1, x_2 y_2, \ldots, x_k y_k\}$, where $1 \leq k < m$, $S_1 = \{x_1, x_2, \ldots, x_k\}$, and $N(S_1) = \{y_1, y_2, \ldots, y_k\}$. Now consider the bipartite subgraph G_1 of G with partite sets $X_1 = X - \{x_1, x_2, \ldots, x_k\}$ and $Y_1 = Y - \{y_1, y_2, \ldots, y_k\}$. We claim that G_1 satisfies the induction hypothesis. To see this, let S be a nonempty subset of X_1. Then

$$
\begin{aligned}
|N(S, G_1)| &= |N(S, G) - N(S_1, G)| \\
&= |(N(S, G) \cup N(S_1, G)) - N(S_1, G)| \\
&= |N(S, G) \cup N(S_1, G)| - |N(S_1, G)| \\
&\geq |S \cup S_1| - |S_1| \\
&= |S|
\end{aligned}
$$

Hence, X_1 can be matched to a subset of Y_1, say under the matching M_2. Then $M_1 \cup M_2$ matches X to a subset of Y. This completes the proof. $\square$

Given an ordered collection $(X_1, X_2, \ldots, X_n)$ of finite sets, we have seen how to construct a bipartite graph G with partite sets $X = \{X_1, X_2, \ldots, X_n\}$ and $Y = X_1 \cup X_2 \cup \cdots \cup X_n$ whose edges indicate which elements of Y belong to which sets in X. Applying Theorem 3.6.1 to this bipartite graph yields the following corollary, whose proof is left for you to develop in Exercise 2.

COROLLARY 3.6.2   Let n be a positive integer and let $(X_1, X_2, \ldots, X_n)$ be an ordered collection of n finite sets. Then $(X_1, X_2, \ldots, X_n)$ has a system of distinct representatives if and only if for any choice of k with $1 \leq k \leq n$ and $i_1, i_2, \ldots, i_k$ with $1 \leq i_1 < i_2 < \cdots < i_k \leq n$,

$$|X_{i_1} \cup X_{i_2} \cup \cdots \cup X_{i_k}| \geq k \qquad\qquad \square$$

Stated another way, Corollary 3.6.2 says that $(X_1, X_2, \ldots, X_n)$ has an SDR if and only if the union of any k of these sets has at least k elements. We call this condition the **matching condition**. The matching condition is also known widely as the **marriage condition**, and Hall's theorem is often called the **marriage theorem**. This terminology comes from another popular version of the problem of the computer dating service stated at the beginning of this section. Suppose the "matchmaker" in a small, isolated village has a set $G = \{g_1, g_2, \ldots, g_m\}$ of m single gentlemen and a set $L = \{w_1, w_2, \ldots, w_n\}$ of n ladies who wish to have husbands. For each i, $1 \leq i \leq n$, let G_i be the subset of G consisting of those gentlemen deemed to be acceptable mates for lady w_i. Then the matchmaker must come up with an SDR for the ordered collection $(G_1, G_2, \ldots, G_n)$.

Theorem 3.6.1 is an "existence theorem"—it gives us a condition for checking whether, in a bipartite graph G with partite sets X and Y, it is possible for X to be matched to a subset of Y, but it does not provide us with a method for constructing such a matching should one exist. Furthermore, checking whether $|N(S)| \geq |S|$ for each subset S of X involves checking 2^n conditions, where $n = |X|$.

What is desired is an efficient algorithm for finding a maximum matching in a graph. Such an algorithm is based on a result of C. Berge [Two theorems in graph theory, *Proceedings of the National Academy of Sciences* 43 (1957): 842–844] and is a special case of the "network flow" algorithm of L. R. Ford and D. R. Fulkerson [A simple algorithm for finding maximal network flows and an application to the Hitchcock problem, *Canadian Journal of Mathematics* 9 (1957): 210–218; see also J. Edmonds and R. M. Karp, Theoretic improvements in algorithmic efficiency for network flow problems, *Journal of the ACM* 19 (1972): 248–264].

Let $G = (V, E)$ be a graph and let M be a matching of G. An **alternating path** for M is a path whose edges are alternately in M and not in M. Let us say that a vertex v is **covered** by M if v is incident with an edge of M. An **augmenting path** for M is an alternating path P for M such that neither the first nor last vertex of P is covered by M. This means that the first and last edges of such a path are not in M, and thus an augmenting path for M necessarily has odd length.

Suppose that P is an augmenting path for M. Let M_1 be the set of edges of P that are not in M and let M_2 be the set of edges of P that are in M. Then $M' = (M - M_2) \cup$

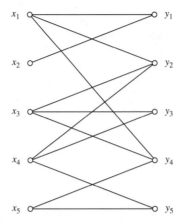

Figure 3.6.2

M_1 is a matching of G and

$$|M'| = |M| - |M_2| + |M_1| = |M| + 1$$

Hence, M is not a maximum matching.

■ ***Example 3.6.4*** The bipartite graph $G = (\{x_1, x_2, x_3, x_4, x_5\} \cup \{y_1, y_2, y_3, y_4, y_5\}, \{x_1y_1, x_1y_2, x_1y_4, x_2y_1, x_3y_2, x_3y_3, x_3y_4, x_4y_2, x_4y_3, x_4y_5, x_5y_4, x_5y_5\})$ is shown in Figure 3.6.2. The matching $M'' = \{x_1y_1, x_3y_3, x_5y_5\}$ is not maximal, because it is properly contained in the matching $M = \{x_1y_1, x_3y_3, x_4y_2, x_5y_5\}$. The matching M is maximal; show that M is not a maximum matching by finding an augmenting path P for M. Use P to construct a matching M' such that $|M'| = |M| + 1 = 5$. It follows that M' is a perfect matching of G.

Solution
Because the initial vertex of P is not covered by M and P begins in one partite set and ends in the other, we may assume that x_2y_1 is the initial edge of P. Since y_1 is incident with an edge of M, the length of P must be at least 3. The second edge of P is an edge of M; thus, the second edge of P is y_1x_1. The third edge of P is not an edge of M; either x_1y_2 or x_1y_4 could be this edge. If we take x_1y_4, then, since y_4 is not covered by M, we obtain the augmenting path

$$x_2, y_1, x_1, y_4$$

However, for the purpose of illustration, let us take x_1y_2 as the third edge of P. Then the fourth edge is y_2x_4 and the fifth edge is x_4y_3 or x_4y_5. If the fifth edge is x_4y_3, then the sixth edge is y_3x_3, the seventh edge is x_3y_4, and we obtain the augmenting path

$$P: x_2, y_1, x_1, y_2, x_4, y_3, x_3, y_4$$

(If the fifth edge is x_4y_5, then the sixth edge is y_5x_5, the seventh edge is x_5y_4, and we obtain the augmenting path $x_2, y_1, x_1, y_2, x_4, y_5, x_5, y_4$.)

The set of edges of P not in M is $M_1 = \{x_2y_1, x_1y_2, x_4y_3, x_3y_4\}$. The set of edges of P in M is $M_2 = \{x_1y_1, x_3y_3, x_4y_2\}$. Thus,

$$M' = (M - M_2) \cup M_1 = \{x_1y_2, x_2y_1, x_3y_4, x_4y_3, x_5y_5\}$$

is a matching of G of cardinality greater than the cardinality of M.   ∎

So we see that if there is an augmenting path for a matching in a graph, then the matching is not maximum. Conversely, we wish to prove that if M is a matching in a graph G and there is no augmenting path for M in G, then M is a maximum matching.

THEOREM 3.6.3 (Berge's Theorem)   A matching M in a graph G is a maximum matching if and only if there does not exist an augmenting path for M in G.

Proof   Necessity has already been demonstrated. For sufficiency we prove the contrapositive: If the matching M is not maximum, then G contains an augmenting path for M.

Let M be a matching of G that is not maximum; then there is a matching M' of G such that $|M| < |M'|$. Consider the subgraph G' of G induced by $M \cup M'$. Because any vertex is incident with at most one edge of M and at most one edge of M', we have that each component of G' either is a path or a cycle whose edges alternate between M and M' or consists of a single edge that is in $M \cap M'$. Since $|M| < |M'|$, there is a component of G' that contains more edges of M' than M. It is then easy to see that this component is an augmenting path for M.   □

As mentioned, Berge's theorem is the basis for an algorithm to find a maximum matching of a given graph $G = (V, E)$. Let M be any matching of G. The algorithm searches for an augmenting path P for M. If no such path exists, then M is a maximum matching by the theorem. Otherwise, let A be the set of edges of P and replace M by $(M - M_2) \cup M_1$, where $M_1 = A - M$ and $M_2 = A \cap M$. Then repeat these steps.

The crux of the algorithm is the search for an augmenting path. For this we use an auxiliary algorithm that returns the set A of edges of an augmenting path for the matching M; if no such path exists, the empty set is returned. This auxiliary algorithm uses the following local variables:

- F is the set of potential odd edges of an augmenting path and is initialized to $E - M$;
- S is the set of possible initial vertices of an augmenting path, initialized to the set of vertices not covered by M;

- u is the initial vertex, and x, y, and z are successive vertices of a potential augmenting path;

- $p(x)$ is the predecessor of x on a potential augmenting path.

Initially, $p(v) = v$ for each vertex v; also, $p(u) = u$ whenever u is the initial vertex of a potential augmenting path. We are now ready to present these algorithms (in pseudocode).

ALGORITHM 3.6.1 Given a graph $G = (V, E)$ and a matching M of G, this algorithm returns in A the set of edges of an augmenting path for M; if no such path exists, the empty set is returned. The algorithm uses local variables F, S, u, x, y, z, and p as described above.

```
begin
A := { };
MAIN: while S ≠ { } loop
    u := an element of S chosen arbitrarily;
    S := S − {u};
    x := u;
        loop
            if possible, choose a vertex y such that y ≠ u, p(y) = y, and xy ∈ F; then
                F := F − {xy};
                p(y) := x;
                if y ∈ S then--augmenting path found
                    loop
                        A := A ∪ {xy};
                        y := x;
                        x := p(x);
                        exit loop MAIN when x = y;
                    end loop;
                else--y is covered by M
                    z := that vertex such that yz ∈ M;
                    p(z) := y;
                    x := z;
                end if;
            else
                exit when x = u;--no augmenting path starting at u
                y := p(x);
                p(x) := x;
                x := p(y);
                p(y) := y;
            end if;
        end loop;
    end loop;
    return A;
end;
```

ALGORITHM 3.6.2 Given a graph $G = (V, E)$, this algorithm returns a maximum matching M of G. The algorithm uses local variables A, M_1, and M_2 as described above.

```
begin
    M := { };
    loop
        A := the result of applying Algorithm 3.6.1 to the matching M in G;
        exit when A = { };
        M₁ := A − M;
        M₂ := A ∩ M;
        M := (M − M₂) ∪ M₁;
    end loop;
    return M;
end;
```

Algorithm 3.6.2 begins by initializing M to the empty set; this is merely a convenience. In fact, M can be initialized to any known matching of G. From the standpoint of efficiency, it may be desirable to begin by initializing M to a maximal matching of G. Such a matching is easy to construct (the development of an algorithm to find a maximal matching of a given graph G is left for you to work out in Exercise 4).

In a sense, Algorithm 3.6.1 uses a technique reminiscent of depth-first search to search for an augmenting path. The most subtle point in verifying its correctness concerns the use of the variable F to keep track of which edges of $E - M$ have been tried as odd edges of a potential augmenting path. (Note that once an edge $xy \in F$ is tried as an edge of a potential augmenting path, say, with initial vertex u, it is removed from further consideration.) We must prove that if G contains an augmenting path for M, then such a path is found by the algorithm. (The proof of this fact is left for you to develop in Exercise 14.)

■ *Example 3.6.5* We trace Algorithm 3.6.2 for the bipartite graph

$$G = (\{x_1, x_2, x_3, x_4, x_5\} \cup \{y_1, y_2, y_3, y_4, y_5\},$$

$$\{x_1y_1, x_1y_2, x_1y_4, x_2y_1, x_3y_2, x_3y_3, x_3y_4, x_4y_2, x_4y_3, x_4y_5, x_5y_4, x_5y_5\})$$

of Example 3.6.4. In particular, suppose at some stage that $M = \{x_1y_1, x_3y_3, x_4y_2, x_5y_5\}$; we focus on the use of Algorithm 3.6.1 to search for an augmenting path. So F is initially $E - M = \{x_1y_2, x_1y_4, x_2y_1, x_3y_2, x_3y_4, x_4y_3, x_4y_5, x_5y_4\}$ and S is initially $\{x_2, y_4\}$. Assume that we begin by choosing $u = x_2 \in S$; so $S := S - \{u\} = \{y_4\}$ and $x := x_2$. The inner loop is then repeated as follows:

Iteration 1: Note that $p(y_1) = y_1$ and $x_2y_1 \in F$. Thus, let $y = y_1$, so $F := F - \{xy\} = \{x_1y_2, x_1y_4, x_3y_2, x_3y_4, x_4y_3, x_4y_5, x_5y_4\}$ and $p(y_1) := x_2$. Since $y \notin S$, we have $z := x_1, p(x_1) := y_1$, and $x := x_1$.

Iteration 2: Here we have two choices for y: y_2 or y_4. Let $y = y_2$; then $F := F - \{xy\} = \{x_1y_4, x_3y_2, x_3y_4, x_4y_3, x_4y_5, x_5y_4\}$ and $p(y_2) := x_1$. Since $y \notin S$, we have $z := x_4$, $p(x_4) := y_2$, and $x := x_4$.

Iteration 3: Again there are two choices, y_3 or y_5, for y. By letting $y = y_3$, we have $F := F - \{xy\} = \{x_1y_4, x_3y_2, x_3y_4, x_4y_5, x_5y_4\}$ and $p(y_3) := x_4$. Since $y \notin S$, we obtain $z := x_3$, $p(x_3) := y_3$, and $x := x_3$.

Iteration 4: Here $y = y_4$, so $F := \{x_1y_4, x_3y_2, x_4y_5, x_5y_4\}$ and $p(y_4) := x_3$. Now $y \in S$, so an augmenting path for M has been found. We use the p values to trace our way back along the path, adding the edges of the path to A as we go. For example, we first have $A := A \cup \{xy\} = \{x_3y_4\}$, $y := x = x_3$, and $x := p(x) = y_3$. Next, $A := A \cup \{xy\} = \{y_3x_3, x_3y_4\}$, $y := x = y_3$, and $x := p(x) = x_4$, and so on. This continues until $x = x_2$ and $y = y_1$, whence $A := A \cup \{xy\} = \{x_2y_1, y_1x_1, x_1y_2, y_2x_4, x_4y_3, y_3x_3, x_3y_4\}$, $y := x = x_2$, and $x := p(x) = x_2$. At this point, $x = p(x)$, so the initial vertex of the path has been reached. The main loop is then exited and A is returned.

Back in Algorithm 3.6.2, we have $M_1 := A - M = \{x_2y_1, x_1y_2, x_4y_3, x_3y_4\}$, $M_2 := A \cap M = \{x_1y_1, x_4y_2, x_3y_3\}$, and then

$$M := (M - M_2) \cup M_1 = \{x_1y_2, x_2y_1, x_3y_4, x_4y_3, x_5y_5\}$$

Next, when Algorithm 3.6.1 is called with this matching M in G, note that $S = \{\ \}$, so $A = \{\ \}$ is returned. Thus, M is a maximum matching of G; in fact, M is a perfect matching. ∎

■ **Example 3.6.6**   We trace Algorithm 3.6.2 for the graph $G = (\{x_1, x_2, \ldots, x_6\} \cup \{y_1, y_2, y_3, y_4, y_5\}, \{x_1y_1, x_1y_2, x_2y_2, x_3y_2, x_3y_3, x_4y_2, x_5y_1, x_5y_3, x_5y_4, x_5y_5, x_6y_3\} \cup \{y_2y_3\})$ shown in Figure 3.6.3. (Note that $G - y_2y_3$ is bipartite.) In particular, suppose at some stage that $M = \{x_1y_1, x_4y_2, x_5y_3\}$; we focus on the use of Algorithm 3.6.1 to search for an augmenting path. So F is initialized to $E - M = \{x_1y_2, x_2y_2, x_3y_2, x_3y_3, x_5y_1, x_5y_4, x_5y_5, x_6y_3, y_2y_3\}$ and S is initialized to $\{x_2, x_3, x_6, y_4, y_5\}$. Assume that $u := x_2$, so $S := S - \{u\} = \{x_3, x_6, y_4, y_5\}$ and $x := x_2$. The inner loop is then repeated as follows:

Iteration 1.1: Note that $p(y_2) = y_2$ and $xy_2 \in F$. Thus, let $y = y_2$, so $F := F - \{xy\} = \{x_1y_2, x_3y_2, x_3y_3, x_5y_1, x_5y_4, x_5y_5, x_6y_3, y_2y_3\}$ and $p(y_2) := x_2$. Since $y \notin S$, we have $z := x_4$, $p(x_4) := y_2$, and $x := x_4$.

Iteration 1.2: There are no choices for y and $x \neq u$, so we must "back up"— $y := p(x_4) = y_2$, $p(x_4) := x_4$, $x := p(y_2) = x_2$, and $p(y_2) := y_2$.

Iteration 1.3: There are no choices for y and $x = u$, so this means that there is no augmenting path with initial vertex $u = x_2$. Thus the inner loop is exited.

Back at the top of the main loop, let $u = x_3$, so that $S := S - \{u\} = \{x_6, y_4, y_5\}$ and $x := u = x_3$. The inner loop is repeated as follows:

Iteration 2.1: There are two choices for y: y_2 or y_3. Let $y = y_2$; then $F := \{x_1y_2, x_3y_3, x_5y_1, x_5y_4, x_5y_5, x_6y_3, y_2y_3\}$ and $p(y_2) := x_3$. Since $y \notin S$, we have $z := x_4$, $p(x_4) := y_2$, and $x := x_4$.

Iteration 2.2: Just as in iteration 1.2, we are forced to back up—$p(x_4) := x_4$, $p(y_2) := y_2$, and $x := x_3$.

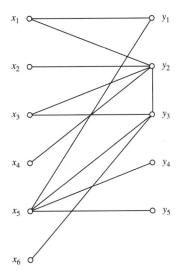

Figure 3.6.3

Iteration 2.3: Now let $y = y_3$, so $F := \{x_1y_2, x_5y_1, x_5y_4, x_5y_5, x_6y_3, y_2y_3\}$ and $p(y_3) := x_3$. Since $y \notin S$, we have $z := x_5$, $p(x_5) := y_3$, and $x := x_5$.

Iteration 2.4: The choices for y are y_1, y_4, and y_5. You should see that if y_1 is chosen, then the path proceeds to x_1, then to y_2, then to x_4 and gets stuck. So let $y = y_4$, so that x_5y_4 is removed from F and $p(y_4) := x_5$. Since $y = y_4 \in S$, an augmenting path for M has been found. By using the p values to trace back along the path, you should verify that

$$A = \{x_3y_3, y_3x_5, x_5y_4\}$$

is returned.

Back in Algorithm 3.6.2, A is used to update M so that

$$M := \{x_1y_1, x_3y_3, x_4y_2, x_5y_4\}$$

Next, when Algorithm 3.6.1 is called with this matching M, note that S is initialized to $\{x_2, x_6, y_5\}$ and F is initialized to $\{x_1y_2, x_2y_2, x_3y_2, x_5y_1, x_5y_3, x_5y_5, x_6y_3, y_2y_3\}$. You are asked to trace Algorithm 3.6.1 and verify that there is no augmenting path for M, so that M is a maximum matching of G. ∎

Now let $G = (V, E)$ be a graph. The cardinality of a maximum matching in G is written $\beta_1(G)$. As a related notion, a subset C of V is said to **cover** E if each edge is incident with some vertex in C. Note that V covers E, so it makes sense to consider the minimum cardinality among those subsets of V that cover E; this parameter is denoted by $\alpha(G)$. If M is a maximum matching and C is a cover of E, then for each

$e \in M$ there is $v \in C$ such that v is incident with e and no $v \in C$ is incident with two edges of M. It follows that

$$\beta_1(G) \leq \alpha(G)$$

For example, let G be the graph of Figure 3.6.3. As shown in Example 3.6.6, $\beta_1(G) = 4$, so that $\alpha(G) \geq 4$. In fact, observe that $C = \{x_1, x_5, y_2, y_3\}$ covers $E(G)$, so that $\alpha(G) = 4$.

It turns out that $\beta_1(G) = \alpha(G)$ for any bipartite graph G. This result was discovered independently by J. Egerváry and D. König in 1931.

THEOREM 3.6.4 If G is a bipartite graph, then $\beta_1(G) = \alpha(G)$.

Proof Let M be a maximum matching of $G = (V, E)$, where G is bipartite with partite sets X and Y. To prove the theorem, it suffices to find a subset C of V such that C covers E and $|C| = |M|$.

Let S_1 be the subset of X consisting of those vertices not covered by M. Let T be the set of vertices of G that are joined to some vertex of S_1 by an alternating path for M. (T includes the vertices of S_1.) Let $T_1 = X \cap T$ and $T_2 = Y \cap T$. Consider any $y \in T_2$. Then y is covered by M, for otherwise there is an augmenting path for M, contradicting the fact that M is a maximum matching. Hence, there is a unique $x \in T_1$ with $xy \in M$. Thus, we see that $T_1 - S_1$ is matched by some subset of M to T_2, so that $|T_1| - |S_1| = |T_2|$. Moreover, $N(T_1) = T_2$.

Let $C = (X - T_1) \cup T_2$. Then C covers E, for otherwise there is an edge xy such that $x \in T_1$ and $y \notin T_2$. Furthermore,

$$|C| = |X| - |T_1| + |T_2| = |X| - |S_1| = |M|$$

which completes the proof. $\square$

Just as Algorithm 3.6.2 is a special case of a more general algorithm for finding the maximum flow through a network, the König–Egerváry theorem is a special case of the "max flow–min cut" theorem of Ford and Fulkerson equating the maximum flow through a network with the capacity of a minimum cut [Maximal flow through a network, *Canadian Journal of Mathematics* 8 (1956): 399–404]. For a discussion of networks, see *Graphs and Digraphs*, 2nd ed., by G. Chartrand and L. Lesniak, and *Graph Theory*, by R. J. Gould.

We end this section with a classic theorem of W. T. Tutte that gives a necessary and sufficient condition for a graph to have a perfect matching; this theorem is used in the subsequent section on factorizations. The proof is due to I. Anderson [Perfect matchings of a graph, *Journal of Combinatorial Theory*, 10B (1971): 183–186] and uses Hall's theorem.

THEOREM 3.6.5 (Tutte's Theorem) A nontrivial graph $G = (V, E)$ has a perfect matching if and only if the number of components of odd order of $\langle V - S \rangle$ does not exceed $|S|$ for every proper subset S of V.

Proof   To prove necessity, we proceed by contradiction. Let M be a perfect matching of G and suppose that, for some proper subset S of V, that the number of odd components (components of odd order) of $\langle V - S \rangle$ is greater than $|S|$. For any such odd component H, there must be an edge of M that joins a vertex of H with a vertex in S. This means, however, that some vertex of S is incident with at least two edges of M, a contradiction.

For sufficiency, we assume that the number of odd components of $\langle V - S \rangle$ is less than or equal to $|S|$ for every $S \subset V$, and we wish to show that G has a perfect matching. In particular, if $S = \{ \ \}$, then this condition implies that every component of G has even order, and so G has even order, say, $2n$. We now proceed by induction on n.

For $n = 1$, our condition implies that G is isomorphic to K_2, and so clearly G has a perfect matching. We let n be a positive integer, $n > 1$, and assume that the result holds for every graph $G' = (V', E')$ having even order less than $2n$; that is, if the number of odd components of $\langle V' - S' \rangle$ does not exceed $|S'|$ for every proper subset S' of V', then G' has a perfect matching. We consider two cases.

Case 1: Suppose that the number of odd components of $\langle V - S \rangle$ is strictly less than $|S|$ for $2 \leq |S| < 2n$. Then, since the number of odd components of $\langle V - S \rangle$ and $|S|$ must have the same parity, we have that the number of odd components of $\langle V - S \rangle$ is at most $|S| - 2$ for $2 \leq |S| < 2n$. Let $uv \in E$ and $G' = (G - u) - v$. For any proper subset S' of $V' = V - \{u, v\}$, we claim that the number c' of odd components of $\langle V' - S' \rangle$ (as an induced subgraph of G') does not exceed $|S'|$; hence, by the induction hypothesis, G' has a perfect matching and, thus, so does G. To verify our claim, suppose to the contrary that $c' > |S'|$. Let $S = S' \cup \{u, v\}$ and let c be the number of odd components of $\langle V - S \rangle$. Then

$$c = c' > |S'| = |S| - 2$$

contradicting our supposition.

Case 2: Suppose there is some subset T of V, with $2 \leq |T| < 2n$, such that the number of odd components of $\langle V - T \rangle$ is equal to $|T|$. Among all such T, choose one having the maximum cardinality c; call it S. Let $G_1, G_2, \ldots, G_c$ be the odd components of $\langle V - S \rangle$. Then these are the only components of $\langle V - S \rangle$, for if G_0 is an even component and u_0 is a vertex of G_0, then the number of odd components of $\langle V - (S \cup \{u_0\}) \rangle = c + 1 = |S \cup \{u_0\}|$, contradicting the choice of S. For $1 \leq i \leq c$, let S_i denote the set of vertices in S adjacent to one or more vertices in G_i. Note that each S_i is nonempty; moreover, we claim that the union of any k of these sets contains at least k vertices. To see this, choose any k, $1 < k \leq c$, and some $1 \leq i_1 < \cdots < i_k \leq c$. Then $|S_{i_1} \cup \cdots \cup S_{i_k}|$ is at least the number of odd components of $H = \langle V - (S_{i_1} \cup \cdots \cup S_{i_k}) \rangle$, and this number in turn is at least k, because $G_{i_1}, \ldots, G_{i_k}$ are odd components of H. Thus, by Corollary 3.6.2, $(S_1, S_2, \ldots, S_c)$ has a system of distinct representatives, say, $(v_1, v_2, \ldots, v_c)$. For $1 \leq i \leq c$, let u_i be a vertex of G_i such that $u_i v_i \in E$. We will now be done if we can show that each graph $G_i - u_i$ has a perfect matching; for these perfect matchings, along with the edges $u_i v_i$, $1 \leq i \leq c$, together form a perfect matching of G. Fix i, $1 \leq i \leq c$, and let $V_i = V(G_i - u_i)$; to show that $G_i - u_i$ has a perfect matching, it suffices by the induction hypothesis to show that the number c' of odd components of $\langle V_i - S' \rangle$ (as an induced subgraph of $G_i - u_i$) is at most $|S'|$ for any proper subset S' of V_i. Suppose to the contrary that this fails for some proper subset S' of V_i. Since c' and $|S'|$ must have the same parity, we can then say that $c' \geq |S'| + 2$. Hence, the number of odd components of $\langle V - (S \cup S' \cup \{u_i\}) \rangle$ is $c - 1 + c'$, and this is at least

$$c - 1 + |S'| + 2 = c + |S'| + 1 = |S| + |S'| + 1 = |S \cup S' \cup \{u_i\}|$$

But this contradicts the choice of S, thus completing the proof. $\qquad\qquad\square$

EXERCISE SET 3.6

1. Find an SDR for $(X_1, X_2, X_3, X_4, X_5)$ or give a reason why there is no SDR.
   (a) $X_1 = \{2, 5\}$, $X_2 = \{1, 3, 4\}$, $X_3 = \{3, 5, 6\}$, $X_4 = \{5\}$, $X_5 = \{2, 4\}$
   (b) $X_1 = \{1, 2\}$, $X_2 = \{2, 3\}$, $X_3 = \{1, 3\}$, $X_4 = \{1, 2, 3\}$, $X_5 = \{4, 5\}$
   (c) $X_1 = \{1, 2, 3\}$, $X_2 = \{1, 2, 4\}$, $X_3 = \{1, 3, 4\}$, $X_4 = \{2, 3, 4\}$, $X_5 = \{1, 4\}$
   (d) $X_1 = \{1\}$, $X_2 = \{1, 2\}$, $X_3 = \{1, 2, 3\}$, $X_4 = \{1, 2, 3, 4\}$, $X_5 = \{1, 2, 3, 4, 5\}$

2. Prove Corollary 3.6.2.

3. Assume that the ordered collection of sets $(X_1, X_2, \ldots, X_n)$ satisfies the matching condition, and let $x \in X_1 \cup X_2 \cup \cdots \cup X_n$.
   (a) Show that $(X_1, X_2, \ldots, X_n)$ has an SDR $(x_1, x_2, \ldots, x_n)$ such that $x_i = x$ for some i, $1 \leq i \leq n$.
   (b) Prove or disprove: If $x \in X_1$, then $(X_1, X_2, \ldots, X_n)$ has an SDR of the form $(x, x_2, \ldots, x_n)$.

4. Develop an algorithm to find a maximal matching of a given graph $G = (V, E)$.

5. Let $(X_1, X_2, \ldots, X_n)$ be an ordered collection of sets such that for any k, $1 \leq k \leq n$, the union of any k of these sets has cardinality at least $k + 1$. Let $x \in X_1$. Show that $(X_1, X_2, \ldots, X_n)$ has an SDR of the form $(x, x_2, \ldots, x_n)$.

6. Let m and n be positive integers and let $X_1 = X_2 = \cdots = X_n = \{1, 2, \ldots, m\}$. Find the number of systems of distinct representatives for $(X_1, X_2, \ldots, X_n)$.

7. Let n be a positive integer. For $1 \leq i \leq n$, let $X_i = \{1, 2, \ldots, n\} - \{i\}$. Show that the number of systems of distinct representatives for $(X_1, X_2, \ldots, X_n)$ is the nth derangement number $d(n)$.

8. Consider the bipartite graph
   $$G = (\{x_1, x_2, \ldots, x_6\} \cup \{y_1, y_2, \ldots, y_6\},$$
   $$\{x_1 y_1, x_1 y_2, x_1 y_3, x_2 y_2, x_2 y_4, x_3 y_2, x_3 y_4, x_3 y_5, x_4 y_4, x_5 y_1, x_5 y_3, x_5 y_5,$$
   $$x_5 y_6, x_6 y_4\})$$
   (a) Find a maximal matching of G (preferably using the algorithm developed in Exercise 4).

(b) Apply Algorithm 3.6.2 to find a maximum matching of G.

9. Give an example of a graph G such that $\beta_1(G) < \alpha(G)$.

10. Let $G = (V, E)$ be a bipartite graph with partite sets X and Y. Assume that there is a positive integer k such that $\deg(x) \geq k$ for each $x \in X$ whereas $\deg(y) \leq k$ for each $y \in Y$.

   (a) Show that $|X| \leq |Y|$.
   (b) Show that X can be matched to a subset of Y.
   (c) Suppose G is k-regular; show that G has a perfect matching.

11. Let $G = (V, E)$ be a bipartite graph with partite sets X and Y with $|X| \leq |Y|$. Prove or disprove: If $\delta(G) \geq k \in \mathbb{Z}^+$, then X can be matched to a subset of Y.

12. Let $X_0 = \{0, 7, 9\}$, $X_1 = \{4, 7\}$, $X_2 = \{1, 2, 3\}$, $X_3 = \{0, 8\}$, $X_4 = \{0, 4, 7\}$, $X_5 = \{0, 3, 4, 6, 9\}$, $X_6 = \{7, 9\}$, $X_7 = \{4, 5, 9\}$, $X_8 = \{5, 7, 9\}$, and $X_9 = \{0, 4, 7, 9\}$. Find the maximum n such that, for some $0 \leq i_1 \leq i_2 \leq \cdots \leq i_n \leq 9$, the ordered collection $(X_{i_1}, X_{i_2}, \ldots, X_{i_n})$ has an SDR.

13. A small surveying company has five positions to be filled and six applicants for these positions; number the positions $1, 2, 3, 4, 5$ and the applicants $1, 2, 3, 4, 5, 6$. For $1 \leq i \leq 5$, let X_i be the set of applicants who are deemed qualified for position i; suppose $X_1 = \{2, 5\}$, $X_2 = \{1, 3, 4\}$, $X_3 = \{1, 3, 5, 6\}$, $X_4 = \{5\}$, and $X_5 = \{4, 5\}$. Find an assignment of applicants to positions so that each position is filled with a qualified applicant.

14. Let $G = (V, E)$ be a graph and let M be a matching of G. Prove: If G contains an augmenting path for M, then Algorithm 3.6.1 finds such a path.

★15. Let T be a tree of order n. Find the number of maximum matchings of T.

16. Use Berge's Theorem (Theorem 3.6.3) to prove Hall's Theorem (Theorem 3.6.1). (Hint: For sufficiency, prove the contrapositive. Suppose that X cannot be matched to a subset of Y and let M be a maximum matching of G; then there is a vertex $x \in X$ that is not covered by M. Let T be the set of all vertices of G that are joined to x by an alternating path for M; let $T_1 = T \cap X$ and $T_2 = T \cap Y$. Argue that $|N(T_1)| = |T_2| < |T_1|$.)

17. Determine the number of maximum matchings of K_n.

18. Let G be a connected bipartite graph with partite sets X and Y. Show that

$$\beta_1(G) = |X| - \max\{0, |S| - |N(S)|\}$$

   where the maximum is over all $S \subseteq X$.

19. Determine the number of maximum matchings of $K_{m,n}$ $(m \leq n)$.

20. Here is another problem involving systems of distinct representatives or, equivalently, matchings in bipartite graphs. Consider a chessboard with mn squares arranged into m rows and n columns. Remove some d squares from consideration, $0 \leq d \leq mn$; the resulting board is called a "pruned chessboard." Suppose we have a supply of dominoes, each of which can cover two adjacent squares of the chessboard. Is it possible to cover the $mn - d$ squares of the pruned chessboard with $(mn - d)/2$ dominoes (so that each square is covered by exactly

*	b_1	w_1	b_2
b_3	w_2	*	w_3
w_4	b_4	*	*
b_5	w_5	b_6	w_6

Figure 3.6.4

one domino)? In general, let $B = \{b_1, b_2, \ldots, b_r\}$ be the set of black squares and $W = \{w_1, w_2, \ldots, w_s\}$ the set of white squares of the pruned chessboard, and let an asterisk denote a pruned square. Note that a necessary condition for a covering with dominoes to exist is that $r = s$. Indicate how to solve this problem by formulating it as a problem that involves asking whether a certain bipartite graph has a perfect matching. In particular, solve the problem for the pruned chessboard represented in Figure 3.6.4.

21. For the bipartite graph G of Exercise 8, find (preferably using the method contained in the proof of Theorem 3.6.4) a subset C of $V(G)$ such that C covers $E(G)$ and $|C| = \alpha(G)$.

⋆22. Refer to the terminology introduced in Exercise 20.

   (a) Suppose mn is odd and the number of white squares is one more than the number of black squares. Prune $d = 1$ white square (any one) from the board. Show that the resulting pruned chessboard can be covered with dominoes.

   (b) Suppose mn is even and $mn \geq 4$. Prune $d = 2$ squares from the board, (any) one white square and (any) one black square. Show that the resulting pruned chessboard can be covered with dominoes.

23. Apply the directions of Exercise 21 to the bipartite graph $G - y_2y_3$ of Example 3.6.6.

24. Let $G = (V, E)$, where $V = \{a, b, c, d, u, v, w, x, y, z\}$ and $E = \{ac, ad, bc, bd, cd, cu, cx, cz, dw, dx, dz, uv, uw, vw, xy, yz\}$. Use (Tutte's) Theorem 3.6.5 to show that G does not have a perfect matching.

3.7 HAMILTONIAN GRAPHS

In a previous section we considered the problem of determining whether a given graph (or digraph) contains a circuit that includes each of its edges, that is, an eulerian circuit. Another famous problem in graph theory is to determine whether a given graph (or digraph) contains a cycle through each of its vertices. This question, considered for graphs only, is the focus of this section.

DEFINITION 3.7.1   Let G be a graph or digraph. A path of G that includes every vertex is called a **hamiltonian path**, and a cycle that includes every vertex is called a **hamiltonian cycle**. A graph or digraph that has a hamiltonian path is called **traceable**, whereas one that has a hamiltonian cycle is called **hamiltonian**. ∎

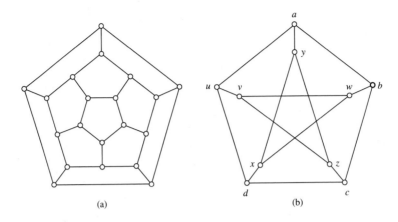

Figure 3.7.1 (a) The graph of the dodecahedron; (b) Petersen's graph.

Hamiltonian graphs are named after the famous Irish mathematician, Sir William Rowan Hamilton. Hamilton had an idea for a puzzle in 1857. The puzzle was a solid wooden dodecahedron, with its 20 vertices labeled with the names of important cities of the day. The object was to find a route along the edges of the dodecahedron that passed through each city exactly once, starting and ending at a given city. To keep track of which cities had been visited, the player was given 20 pegs and a long piece of string; the pegs were inserted at the vertices of the dodecahedron and the string used to connect the pegs. Hamilton's puzzle was not commercially successful, but it was a forerunner of such modern-day puzzles as Instant Insanity and Rubik's Cube. (Perhaps Hamilton's puzzle was too easy to solve; attempt to find a hamiltonian cycle in the graph of the dodecahedron shown in Figure 3.7.1(a).)

Of course, not all graphs are hamiltonian. The graph shown in Figure 3.7.1(b) is known as **Petersen's graph**. This graph is traceable, with one hamiltonian path being $a, b, c, d, u, v, w, x, y, z$. However, the Petersen graph is not hamiltonian, although this fact is surprisingly difficult to verify. You are encouraged to experiment with it for a while to be convinced that no hamiltonian cycle exists.

Unlike the situation for eulerian graphs, there is no useful necessary and sufficient condition known for a graph to be hamiltonian. We can develop some necessary conditions and some sufficient conditions but, in general, the problem of determining whether a given graph is hamiltonian is hard. In fact, it is a special case of a more general problem known as the "traveling salesman problem," that of finding a "shortest" hamiltonian cycle in a weighted graph. The hamiltonian problem is a member of an important class of problems in computer science known as "NP-complete" problems. These are very interesting for several reasons, one being the following: no (deterministic) polynomial-time algorithms are known for any of them; yet, if a polynomial-time algorithm is found for any one of the problems in the class, then polynomial-time algorithms can be constructed for all of them, or, if it is proven that no polynomial-time algorithm exists for some problem in the class, then no polynomial-time algorithms exist for any of them.

As far as necessary conditions for hamiltonian graphs are concerned, it is an immediate consequence of the definition that a hamiltonian graph is connected and has order at least 3. In fact, the removal of any vertex from a hamiltonian graph results in a graph that still contains a hamiltonian path and thus is still connected. In general, a graph $G = (V, E)$ of order at least 3 is called **2-connected** provided $G - v$ is connected for each $v \in V$. Thus, we may remark that if a graph G is hamiltonian, then G is 2-connected.

So, let $G = (V, E)$ be a hamiltonian graph and let $H = (V, E_1)$ be a spanning subgraph of G induced by the edges of some hamiltonian cycle of G. Suppose $|V| = n$ and $v_1, v_2, \ldots, v_m$ are distinct vertices, where $m < n$. Note that the graph $H - v_1$ is connected and that $(H - v_1) - v_2$ has at most two components. Similarly, $((H - v_1) - v_2) - v_3$ has at most three components, and so on. In general, if m vertices are removed from H, the resulting graph has at most m components. This observation leads to the following necessary condition for a graph G to be hamiltonian.

THEOREM 3.7.1 Let $G = (V, E)$ be a graph of order at least 3. If G is hamiltonian, then, for every proper subset U of V, the subgraph of G induced by $V - U$ has at most $|U|$ components.

Proof   Let $G = (V, E)$ be a hamiltonian graph, and let $H = (V, E_1)$ be a spanning subgraph of G induced by the edges of some hamiltonian cycle of G. Let U be a proper subset of V, let k_1 denote the number of components in the subgraph of G induced by $V - U$, and let k_2 denote the number of components in the subgraph of H induced by $V - U$. Since H is a spanning subgraph of G, it is clear that $k_1 \leq k_2$. Also, by the remarks preceding the theorem, we have $k_2 \leq |U|$. Hence, $k_1 \leq |U|$, as was to be shown. $\square$

■ *Example 3.7.1*   Show that the graph $G = (V, E)$ of Figure 3.7.2 is not hamiltonian.

Solution
Let $U = \{t, w, y\}$ and apply Theorem 3.7.1. Note that $|U| = 3$, whereas the subgraph of G induced by $V - U$ has four components: $\langle\{s, z\}\rangle$, $\langle\{u\}\rangle$, $\langle\{v\}\rangle$, and $\langle\{x\}\rangle$. Thus, G is not hamiltonian.   ■

We now turn to sufficient conditions for hamiltonian cycles. Here one basic approach is to attempt to force the graph to have a relatively large number of edges, the idea being that a sufficiently "dense" graph is certain to possess a hamiltonian cycle. Of such results, one of the simplest to apply goes back to O. Ore [Note on Hamilton circuits, *American Mathematical Monthly* 67 (1960): 55].

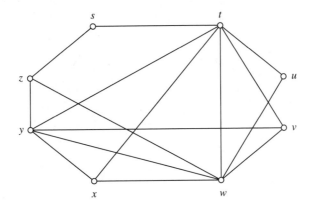

Figure 3.7.2 A nonhamiltonian graph.

THEOREM 3.7.2 (Ore's Theorem) Let G be a graph of order $n \geq 3$ such that, for every pair of distinct nonadjacent vertices u and v,

$$\deg(u) + \deg(v) \geq n$$

Then G is hamiltonian. □

In looking at the proof of Ore's theorem, J. A. Bondy and V. Chvátal observed that, in fact, a much stronger statement could be proved [A method in graph theory, *Discrete Mathematics* 15 (1976): 111–135]. (Recall that, if u and v are nonadjacent vertices of a graph G, then $G + uv$ denotes the graph obtained from G by adding the edge uv.)

THEOREM 3.7.3 (Bondy and Chvátal's Theorem) Let G be a graph of order $n \geq 3$ and suppose that u and v are distinct nonadjacent vertices of G such that

$$\deg(u) + \deg(v) \geq n$$

Then G is hamiltonian if and only if $G + uv$ is hamiltonian.

Proof Let $G = (V, E)$ and let u and v be as in the statement of the theorem. It is clear that, if G is hamiltonian, then $G + uv$ is hamiltonian.

To show the converse, we proceed by contradiction; we assume that $G + uv$ is hamiltonian and that G is not hamiltonian. Then G is traceable and contains a spanning u-v path, say,

$$u = u_1, u_2, \ldots, u_n = v$$

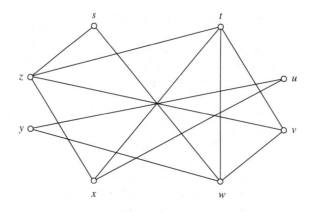

Figure 3.7.3

We claim that there is some vertex u_k, $1 < k < n - 1$, such that $u_1 u_{k+1} \in E$ and $u_k u_n \in E$. Suppose this is not the case. Then for each of the $\deg(u_1)$ values of k for which $u_1 u_{k+1} \in E$, we have $u_k u_n \notin E$. Thus,

$$
\begin{aligned}
\deg(u) + \deg(v) &= \deg(u_1) + \deg(u_n) \\
&\leq \deg(u_1) + (n - 1 - \deg(u_1)) \\
&= n - 1
\end{aligned}
$$

contradicting the hypothesis of the theorem. Now, notice that the cycle

$$u_1, u_{k+1}, u_{k+2}, \ldots, u_n, u_k, u_{k-1}, \ldots, u_2, u_1$$

is a hamiltonian cycle of G. □

■ **Example 3.7.2**   Consider the graph $G = G_0 = (V, E_0)$ shown in Figure 3.7.3. Note that G_0 has order 8 and that w and z are nonadjacent vertices such that

$$\deg(w) + \deg(z) = 8$$

Hence, by Theorem 3.7.3, G_0 is hamiltonian if $G_1 = G_0 + wz$ is hamiltonian. Now consider G_1; in this graph, vertices w and x are nonadjacent with $\deg(w) + \deg(x) = 8$. Again by using Theorem 3.7.3, we see that G_1 is hamiltonian if $G_2 = G_1 + wx$ is hamiltonian. We may continue to use this procedure, obtaining a sequence of graphs $G_0, G_1, G_2, \ldots, G_{16}$, such that G_{16} is isomorphic to K_8. Each of $G_1, G_2, \ldots, G_{16}$ is obtained from its predecessor by the addition of a single edge, and each of $G_0, G_1, \ldots, G_{15}$ is hamiltonian provided its successor in the sequence is hamiltonian. Thus, since any complete graph is obviously hamiltonian, we may conclude that G_0 is hamiltonian. ■

Let us generalize the process of the preceding example. Given a graph G of order n, we let $G_0 = G$. If G_0 contains two nonadjacent vertices u_0 and v_0 such that $\deg(u_0, G_0) + \deg(v_0, G_0) \geq n$, we let $G_1 = G_0 + u_0 v_0$. Next, if G_1 contains two nonadjacent vertices u_1 and v_1 such that $\deg(u_1, G_1) + \deg(v_1, G_1) \geq n$, then we let $G_2 = G_1 + u_1 v_1$. We continue this process, obtaining a sequence of graphs $G = G_0, G_1, \ldots, G_k$, $k \geq 0$, such that G_k does not contain two nonadjacent vertices whose degree sum in G_k is at least n. Then G is hamiltonian if G_k is hamiltonian. In particular, if G_k is a complete graph, then G is hamiltonian.

Bondy and Chvátal called G_k the "closure" of G. To be a bit more precise, let's give a recursive definition of the closure operation.

DEFINITION 3.7.2   Given a graph G of order n, the **closure** of G is the graph $Cl(G)$ defined recursively as follows: If G contains two distinct nonadjacent vertices u and v such that $\deg(u) + \deg(v) \geq n$, then $Cl(G) = Cl(G + uv)$; otherwise, $Cl(G) = G$.   ■

It can be shown that the closure operation is well defined (see Exercise 17). Also, note that G is always a spanning subgraph of $Cl(G)$.

The next result is an immediate corollary of Theorem 3.7.3.

COROLLARY 3.7.4   Let G be a graph of order $n \geq 3$. Then G is hamiltonian if and only if $Cl(G)$ is hamiltonian.   □

The next result presents a series of sufficient conditions for a graph to be hamiltonian. These are listed so that the corresponding theorems are in order of decreasing strength; that is, each condition from the second on implies the condition preceding it. Moreover, notice that the theorems are, with one exception, in reverse chronological order as to date of discovery.

COROLLARY 3.7.5   Let $G = (V, E)$ be a graph of order $n \geq 3$ and let $d_1, d_2,$ $\ldots, d_n$ be the degrees of the vertices of G, listed so that $d_1 \leq d_2 \leq \cdots \leq d_n$. If G satisfies any of the following conditions, then G is hamiltonian.

1.   (Bondy and Chvátal, 1976) $Cl(G)$ is isomorphic to K_n.
2.   (Las Vergnas, 1971)   There exists a bijection $f : V \to \{1, 2, \ldots, n\}$ such that, if u and v are distinct nonadjacent vertices with $f(u) < f(v)$, $f(u) + f(v) \geq n$, $\deg(u) \leq f(u)$, and $\deg(v) < f(v)$, then $\deg(u) + \deg(v) \geq n$.
3.   (Chvátal, 1972)   For each i with $2 \leq 2i < n$, if $d_i \leq i$, then $d_{n-i} \geq n - i$.
4.   (Bondy, 1969)   For all i, j with $1 \leq i < j \leq n$, if $d_i \leq i$ and $d_j \leq j$, then $d_i + d_j \geq n$.
5.   (Pósa, 1962)   For each i with $2 \leq 2i < n$, $|\{v \mid \deg(v) \leq i\}| < i$.
6.   (Ore, 1960)   For any two distinct nonadjacent vertices u and v, $\deg(u) + \deg(v) \geq n$.
7.   (Dirac, 1952)   For any vertex v, $2 \deg(v) \geq n$.   □

(The references, besides the two already given, are as follows: M. Las Vergnas, Sur une propriété des arbres maximaux dans un graphe, *Comptes Rendus de l'Académie des Sciences Paris* 272 (1971), 1297–1300; V. Chvátal, On Hamilton's ideals, *Journal of Combinatorial Theory* 12B (1972): 163–168; J. A. Bondy, Properties of graphs with constraints on degrees, *Studia Scientarum Mathematicarum Hungarica* 4 (1969): 473–475; L. Pósa, A theorem concerning Hamilton lines, *Magyar Tudományos Akademia Matematikai Kutató Intézetének Közleményei* 7 (1962): 225–226; G. A. Dirac, Some theorems on abstract graphs, *Proceedings of the London Mathematical Society* 2 (1952): 69–81.)

As mentioned, each of conditions 2 through 7 in Corollary 3.7.5 implies the conditions preceding it, that is,

$$7 \to 6 \to 5 \to 4 \to 3 \to 2 \to 1$$

It is a direct consequence of Corollary 3.7.4 that condition 1 is sufficient for a graph G to be hamiltonian, and it then follows that each of conditions 2 through 7 is sufficient as well.

It can also be verified that no two of the conditions 1 through 7 are equivalent. To say this another way, let S_1 denote the set of graphs (of order at least 3) that satisfy condition 1, let S_2 denote the set of graphs that satisfy condition 2, and define sets $S_3, S_4, \ldots, S_7$ in a similar manner. Then what we are saying is that

$$S_7 \subset S_6 \subset \cdots \subset S_2 \subset S_1$$

(recall that $\subset$ denotes "is a proper subset of"). Verification of most of the above inclusions is left to the exercises, but we prove a few of them as examples to illustrate the general approach. (Also recall that, to show that $S \subset T$, we need to show that $S \subseteq T$ and to exhibit an element $x \in T - S$.)

■   *Example 3.7.3*   We show that $S_4 \subseteq S_3$. To do this, we let G be a graph of order n and let $d_1 \le d_2 \le \cdots \le d_n$ be as in the statement of Corollary 3.7.5. We proceed by contrapositive, showing that if G does not satisfy Chvátal's condition 3, then G does not satisfy Bondy's condition 4. If G does not satisfy condition 3, then for some i, $2 \le 2i < n$, we have $d_i \le i$ and $d_{n-i} < n - i$. We let $j = n - i$; then $1 \le i < j \le n$, $d_i \le i$, $d_j \le j$, but $d_i + d_j < i + (n - i) = n$. Thus, G does not satisfy condition 4.   ■

■   *Example 3.7.4*   We give an example of a graph $G_1 \in S_1 - S_2$ and an example of a graph $G_2 \in S_2 - S_3$.

The graph $G_1 = (V_1, E_1)$ is shown in Figure 3.7.4(a). It is easy to verify that $Cl(G_1) \simeq K_7$, so that $G_1 \in S_1$. To show that $G_1 \notin S_2$, let f be any bijection from V_1 to $\{1, 2, 3, 4, 5, 6, 7\}$. Of the two vertices of degree 2, let x denote the one with the larger f value. Among the four vertices not adjacent to x, let y denote the one with the largest f value. If $f(x) < f(y)$, then we have $f(x) \ge 2 = \deg(x)$, $f(y) - 1 \ge 3 \ge \deg(y)$, and $f(x) + f(y) \ge 7$. On the other hand, if $f(y) < f(x)$, then we have $f(y) \ge 4 > \deg(y)$,

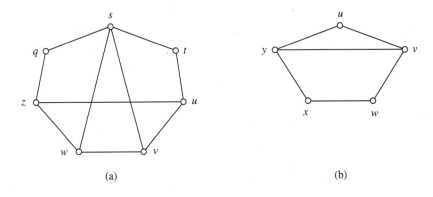

Figure 3.7.4 (a) The graph G_1; (b) The graph G_2.

$f(x) - 1 \geq 4 > \deg(x)$, and $f(y) + f(x) \geq 7$. However, in either case, $\deg(x) + \deg(y) \leq 5$, so that G_1 does not satisfy condition 2.

The graph $G_2 = (V_2, E_2)$ is shown in Figure 3.7.4(b). Define $f : V_2 \rightarrow \{1, 2, 3, 4, 5\}$ by $f(u) = 1$, $f(v) = 5$, $f(w) = 3$, $f(x) = 2$, $f(y) = 4$. Then it can be checked that f satisfies condition 2. For condition 3, we have $d_1 = d_2 = d_3 = 2$, $d_4 = d_5 = 3$. Let $i = 2$; note that $d_2 \leq 2$ and that $d_3 < 3$. Thus, G_2 does not satisfy condition 3, and so $G_2 \in S_2 - S_3$. ∎

It should be mentioned that none of conditions 1 through 7 in Corollary 3.7.5 is a necessary condition for a graph to be hamiltonian. In fact, a result of C. St. J. A. Nash-Williams (1971) says that if G is an r-regular graph of order $2r + 1$, then G is hamiltonian. Note that such a graph G does not satisfy Dirac's condition 7.

Note that the Bondy–Chvátal condition, besides being easy to apply, can be developed into an algorithm for finding a hamiltonian cycle in any graph G of "Bondy–Chvátal type," that is, one for which $Cl(G)$ is isomorphic to K_n. First of all, it is a simple matter to develop an algorithm for finding the closure $Cl(G)$ of a graph G of order n. Moreover, knowing that $G = (V, E_0)$ has size q_0 and that $Cl(G) = (V, E_1)$ has size q_1, we can have the closure-finding algorithm construct a "weight function" $w : E_1 \rightarrow \{1, 2, \ldots, q_0, \ldots, q_1\}$ so that w is a bijection and $w(E_0) = \{1, 2, \ldots, q_0\}$. In this way, we can tell from the weight of an edge whether it is an edge of G or not. (Development of such an algorithm is left for you to work out in Exercise 18.)

We next find, by some manner, a hamiltonian cycle C_1 in $Cl(G)$. Of course, if $Cl(G)$ is complete, then finding such a cycle is trivial. What is needed at this point is a method for modifying the cycle C_1 into a hamiltonian cycle C_0 of the original graph G. For this, we use the idea of the proof of Theorem 3.7.3.

ALGORITHM 3.7.1 Let $G = (V, E_0)$ be a graph of order n and size q_0, let $Cl(G) = (V, E_1)$ be the closure of G, of size q_1, let $w : E_1 \rightarrow \{1, 2, \ldots, q_1\}$ be a bijection with $w(E_0) = \{1, 2, \ldots, q_0\}$, and let C_1 be a hamiltonian cycle of $Cl(G)$.

This algorithm modifies C_1 to a hamiltonian cycle of G.

> begin
>> loop
>>> $uv :=$ an edge of C_1 for which $w(uv)$ is a maximum;
>>> exit when $w(uv) \leq q_0$;
>>> Let
>>>
>>> $$u = u_1, u_2, \ldots, u_n = v$$
>>>
>>> be the hamiltonian path obtained by deleting uv from C_1;
>>> As in the proof of Theorem 3.7.3, find a subscript k, $1 < k < n - 1$,
>>> such that $uu_{k+1} \in E_1$, $u_k v \in E_1$, $w(uu_{k+1}) < w(uv)$, and $w(u_k v) < w(uv)$;
>>> As in the proof of Theorem 3.7.3, replace C_1 by the cycle
>>> $$u, u_{k+1}, u_{k+2}, \ldots, v, u_k, u_{k-1}, \ldots, u_2, u;$$
>> end loop;
> end;

It can be argued that Algorithm 3.7.1 is $O(n^3)$; this and a verification of correctness are left for you to develop in Exercise 19.

■ *Example 3.7.5*   We apply Algorithm 3.7.1 to the graph $G = (V, E_0)$, where $V = \{r, s, t, x, y, z\}$ and $E_0 = \{rs, rt, ry, sx, sz, ty, xy, xz\}$. Here $q_0 = 8$ and $Cl(G)$ is complete; suppose that $w : E_0 \rightarrow \{1, 2, \ldots, 15\}$ is given by $w(rs) = 1$, $w(rt) = 2$, $w(ry) = 3$, $w(sx) = 4$, $w(sz) = 5$, $w(ty) = 6$, $w(xy) = 7$, $w(xz) = 8$, $w(rx) = 9$, $w(rz) = 10$, $w(st) = 11$, $w(sy) = 12$, $w(tx) = 13$, $w(tz) = 14$, $w(yz) = 15$. Also, suppose that C_1 is the cycle

$$r, s, t, x, y, z, r$$

The loop of Algorithm 3.7.1 is iterated as follows.

Iteration 1: We find that yz is the edge of C_1 for which w is a maximum. By considering the path

$$y = u_1, x = u_2, t = u_3, s = u_4, r = u_5, z = u_6$$

we note that $w(ty) < w(yz)$ and $w(xz) < w(yz)$, so that $k = 2$. Thus, C_1 becomes the cycle

$$y, t, s, r, z, x, y$$

Iteration 2: We find that $w(st) = 11$ is the maximum weight of an edge of C_1. By considering the path

$$s = u_1, r = u_2, z = u_3, x = u_4, y = u_5, t = u_6$$

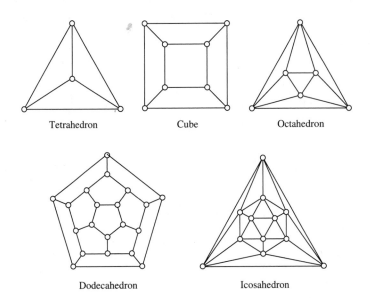

Figure 3.7.5

we find that $w(sz) < w(st)$ and $w(rt) < w(st)$, so again $k = 2$. Thus, C_1 becomes the cycle

$$s, z, x, y, t, r, s$$

Iteration 3: Now we find that $w(xz) = 8$ is the maximum weight of an edge of C_1. Since $q_0 = 8$, this means that each edge of C_1 is an edge of G. Hence, the loop is exited, and the algorithm terminates, having found the hamiltonian cycle s, z, x, y, t, r, s of G. ∎

EXERCISE SET 3.7

1. Show that the graphs of the five "regular polyhedra" are hamiltonian. These are the graphs of the dodecahedron, icosahedron, octahedron, cube, and tetrahedron shown in Figure 3.7.5.

2. Let m and n be integers with $1 \leq m \leq n$, and consider the complete bipartite graph $K_{m,n}$.
   (a) Show that $K_{3,3}$ is hamiltonian.
   (b) Show that $K_{3,4}$ is not hamiltonian.
   (c) Give a necessary and sufficient condition for $K_{m,n}$ to be hamiltonian.

3. Let G be a graph of order $n \geq 3$ and size q, where $2q \geq n^2 - 3n + 6$.
   (a) Show that G satisfies Ore's condition; thus, G is hamiltonian.

   (b) Show that G need not necessarily satisfy Dirac's condition.

4. Find the closure of the graph of:

   (a) Figure 3.7.2.   (b) Figure 3.5.6.   (c) Figure 3.1.8.

5. Find the closure of each graph in Figure 3.5.7.

6. Each of the following parts gives a graph having vertex set $V = \{t, u, v, w, x, y, z\}$. You are to (i) find the closure of the graph and (ii) determine whether the graph is hamiltonian, giving a reason for your answer.

   (a) $G_1 = (V, E_1)$, where $E_1 = \{tu, tz, uy, uz, vw, vx, vy, wx, wy, xy, yz\}$
   (b) $G_2 = (V, E_2)$, where $E_2 = \{tu, tw, tx, tz, ux, uy, vy, vz, wz\}$
   (c) $G_3 = (V, E_3)$, where $E_3 = \{tv, ty, uv, uw, uz, vw, vx, wy, xy, xz, yz\}$

7. Prove or give a counterexample:

   (a) If $G = (V, E)$, where $V = \{v_1, v_2, \ldots, v_9\}$, $\deg(v_1) = 2$, $\deg(v_2) = \deg(v_3) = \deg(v_4) = 3$, $\deg(v_5) = 5$, $\deg(v_6) = \deg(v_7) = 6$, and $\deg(v_8) = \deg(v_9) = 7$, then G is hamiltonian.
   (b) If $G = (V, E)$, where $V = \{v_1, v_2, \ldots, v_7\}$, $\deg(v_1) = 2$, $\deg(v_2) = \deg(v_3) = \deg(v_4) = 3$, $\deg(v_5) = \deg(v_6) = 4$, and $\deg(v_7) = 5$, then G is hamiltonian.
   (c) If $G = (V, E)$, where $V = \{v_1, v_2, \ldots, v_8\}$, $\deg(v_1) = \deg(v_2) = \deg(v_3) = 3$, $\deg(v_4) = \deg(v_5) = \deg(v_6) = 4$, $\deg(v_7) = 5$, and $\deg(v_8) = 6$, then G is hamiltonian.

8. Let G be a graph of order $n \geq 3$. With reference to Corollary 3.7.5, prove:

   (a) If G satisfies the Las Vergnas condition 2, then G satisfies the Bondy–Chvátal condition 1.
   (b) If G satisfies Chvátal's condition 3, then G satisfies the Las Vergnas' condition 2.
   (c) If G satisfies Pósa's condition 5, then G satisfies Bondy's condition 4.
   (d) If G satisfies Ore's condition 6, then G satisfies Pósa's condition 5.
   (e) If G satisfies Dirac's condition 7, then G satisfies Ore's condition 6.

9. With reference to Corollary 3.7.5, give examples of the following:

   (a) A graph G_3 that satisfies Chvátal's condition 3 but not Bondy's condition 4.
   (b) A graph G_4 that satisfies Bondy's condition 4 but not Pósa's condition 5.
   (c) A graph G_5 that satisfies Pósa's condition 5 but not Ore's condition 6.
   (d) A graph G_6 that satisfies Ore's condition 6 but not Dirac's condition 7.

10. Determine whether the digraph G of Figure 3.5.2 is hamiltonian or traceable.

11. For each digraph shown in Figure 3.4.7, determine whether it is hamiltonian or traceable.

12. Let G be a loopless digraph of order $n \geq 2$. Let the "closure" of G be the digraph $Cl(G)$ defined recursively as follows: If G contains distinct vertices u and v such that u is not adjacent to v and $\mathrm{od}(u) + \mathrm{id}(v) \geq n$, then $Cl(G) = Cl(G + (u, v))$; otherwise, $Cl(G) = G$. Prove or disprove: If $Cl(G)$ is a complete symmetric loopless digraph, then G is hamiltonian.

13. State an analogue of Theorem 3.7.1 for digraphs. (Hint: For a digraph G, define a "traceable component" of G to be a subdigraph that is maximal with respect to the property of being traceable.)

14. Let $G = (V, E)$ be a graph. Prove: If G is traceable, then, for every proper subset U of V, the number of components of the subgraph of G induced by $V - U$ is at most $|U| + 1$.

15. For a graph G of order $n \geq 2$, define the graph $Cl_1(G)$ recursively as follows: If G contains distinct nonadjacent vertices u and v for which $\deg(u) + \deg(v) \geq n - 1$, then $Cl_1(G) = Cl_1(G + uv)$; otherwise, $Cl_1(G) = G$.

    (a) Prove: If $Cl_1(G) \simeq K_n$, then G is traceable.
    (b) For each $n \geq 3$, give an example of a graph G of order n for which $Cl_1(G) \simeq K_n$ and G is not hamiltonian.

16. Let $G = (V, E)$ be a graph of order $n \geq 4$ and let $v \in V$.

    (a) Prove: If $\deg(v) > (n - 1)/2$ and $G - v$ is hamiltonian, then G is hamiltonian.
    (b) Does the result of part (a) help in showing that the graph of Figure 3.7.3 is hamiltonian?
    (c) Does the result of part (a) help in showing that the graph of Figure 3.1.8 is hamiltonian?

17. Show that the operation of forming the closure of a graph is well defined. (Hint: Let $G = (V, E_0)$ be a graph of order n and suppose that $G_1 = (V, E_1)$ and $G_2 = (V, E_2)$ are two graphs formed by applying the closure process to G; we must show that $E_1 = E_2$. This is clearly the case if $E_0 = E_1 = E_2$, so suppose that $E_1 - E_0 = \{e_1, e_2, \ldots, e_k\}$ and $E_2 - E_0 = \{f_1, f_2, \ldots, f_m\}$, where the edges in the sets $E_1 - E_0$ and $E_2 - E_0$ are listed in the order in which they were added to G. We wish to show that $E_1 - E_0 = E_2 - E_0$; if this is not the case, then, without loss of generality, there is a smallest subscript t such that $e_t \notin E_2$. Consider the subgraph $H = (V, E_0 \cup \{e_i \mid 1 \leq i < t\})$ and obtain a contradiction.)

⋆18. Develop an algorithm for finding the closure $Cl(G)$ of a graph G of order n. Moreover, knowing that $G = (V, E_0)$ has size q_0 and that $Cl(G) = (V, E_1)$ has size q_1, have the closure-finding algorithm construct a "weight function" $w :$ $E_1 \rightarrow \{1, 2, \ldots, q_0, \ldots, q_1\}$ so that w is a bijection and $w(E_0) = \{1, 2, \ldots, q_0\}$.

⋆19. With respect to Algorithm 3.7.1:

    (a) Argue that Algorithm 3.7.1 is $O(n^3)$.
    (b) Give a justification for its correctness.

20. Apply Algorithm 3.7.1 to the graph $G = (V, E_0)$ of Figure 3.7.3. (Let $Cl(G) = (V, E_1)$ be the closure of G, let $w : E_1 \rightarrow \{1, 2, \ldots, 28\}$ be a bijection with $w(E_0) = \{1, 2, \ldots, 12\}$, and let C_1 be a hamiltonian cycle of $Cl(G)$. Use the algorithm to modify C_1 to a hamiltonian cycle of G.)

3.8 VERTEX COLORING AND PLANAR GRAPHS

In this section we introduce a part of graph theory that arose from a very famous mathematics problem, the **four-color problem**. This problem seems to have first surfaced in a letter from Augustus De Morgan to Sir William Rowan Hamilton, two mathematicians whose names have come up before in this text. The letter is dated October 23, 1852, and part of it is reproduced in the book *Graph Theory 1736–1936* by Norman L. Biggs, Keith Lloyd, and Robin J. Wilson. It is sufficiently interesting to bear repeating here.

> A student of mine asked me today to give him a reason for a fact which I did not know was a fact—and do not yet. He says that if a figure be anyhow divided and the compartments differently coloured so that figures with any portion of common boundary line are differently coloured—four colors may be wanted, but not more—the following is the case in which four are wanted [see Figure 3.8.1(a)]. Query cannot a necessity for five or more be invented. As far as I see at this moment if four ultimate compartments have each boundary line in common with one of the others, three of them inclose the fourth and prevent any fifth from connexion with it. If this be true, four colours will colour any possible map without any necessity for colour meeting colour except at a point.
>
> Now it does seem that drawing three compartments with common boundary ABC two and two you cannot make a fourth take boundary from all, except by enclosing one [see Figure 3.8.1(b) and (c)]. But it is tricky work, and I am not sure of the convolutions—what do you say? And has it, if true, been noticed? My pupil says he guessed it in colouring a map of England. The more I think of it, the more evident it seems. If you retort with some very simple case which makes me out a stupid animal, I think I must do as the Sphynx did....

In ordinary language, the four-color problem can be stated as follows:

> A map consisting of *n* countries with well-defined boundaries is given in a plane. Two countries are said to be adjacent if their common boundary contains a nontrivial segment (not the empty set or a set of isolated points).

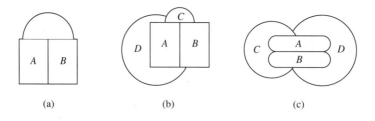

(a) (b) (c)

Figure 3.8.1

Can this map be colored with four (or fewer) different colors so that no two adjacent countries are of the same color?

For more than one hundred years, many attempts were made to solve the problem, and several fallacious solutions were submitted to professional journals. In some cases such alleged solutions were first judged to be valid and were actually published. As is often the case in mathematics, this work, though unsuccessful in solving the four-color problem, introduced a great deal of interesting and useful mathematics. Finally, in 1977, the four-color problem became the **four-color theorem** (in other words, the answer to the question posed above is yes) with the publication of the celebrated paper, "Every Planar Map Is Four-Colorable," by Kenneth Appel, Wolfgang Haken, and John Koch. The article appeared in the *Illinois Journal of Mathematics* [21 (1977): 429–567] and, besides generating much excitement, was quite controversial. The reason for this was that the solution of the four-color problem was an important instance in which a computer was used to help do a mathematical proof. Moreover, this use was truly significant, as more than 1200 hours of computing time were required to analyze a large number of special cases! Those of you interested in learning more about the solution of the four-color problem may wish to consult an article that Appel and Haken wrote for *Scientific American* [237 (1977): 108–121] entitled, "The Solution of the Four-Color Map Problem."

As will be seen shortly, it is possible to associate a graph with a plane map in such a way that each vertex of the graph corresponds to a country of the map, and so that two vertices are adjacent if and only if the two corresponding countries are adjacent. Such a graph is a "planar graph," which is a term defined later in this section, and we may then consider the problem of whether it is possible to assign one of four given colors to each vertex in such a way that any two adjacent vertices receive different colors.

More generally, given a graph $G = (V, E)$ and a set $P = \{1, 2, \ldots, k\}$ of k "colors," does there exist a "coloring function" $c: V \to P$ such that, whenever u and v are distinct vertices and $uv \in E$, $c(u) \neq c(v)$? If so, then we speak more informally of "coloring the vertices" of G with k (or fewer) colors, and the assignment c of colors to vertices is called a **k-coloring** of G. We also say that G is **k-colorable**.

The idea of coloring the vertices of a graph arises naturally in certain kinds of scheduling problems. For example, suppose that a department of mathematics at some university is concerned with scheduling its courses for the next semester. There are several (sections of) courses to be offered and, no doubt, a smaller number of time periods during which courses may be given. Two different courses may be intended for roughly the same audience of students and thus should not be given during the same time period. This suggests modeling the situation with a graph G whose vertices correspond to the courses to be offered, such that two vertices are adjacent provided the two corresponding courses should not be offered during the same time period. Suppose now that a k-coloring of G exists. Since all the vertices of a given color are mutually nonadjacent, the corresponding courses could all be offered during the same time period; hence, all the courses offered by the department can be accommodated using k time periods.

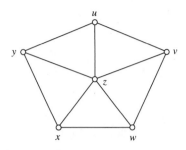

Figure 3.8.2   A wheel of order 6.

DEFINITION 3.8.1   The minimum positive integer k for which a graph G is k-colorable is called the **chromatic number** of G and is denoted by $\chi(G)$.   ∎

Note that if G has order n, then $1 \leq \chi(G) \leq n$. Moreover, $\chi(G) = 1$ if and only if G is empty $(G \simeq \overline{K_n})$, and $\chi(G) = n$ if and only if G is complete $(G \simeq K_n)$.

The chromatic number of a cycle is also easily determined. If C_p denotes the cycle of order p, then for $n \in \mathbb{N}$, we call C_{2n} an **even cycle** and C_{2n+1} an **odd cycle**. It is readily verfied that $\chi(C_{2n}) = 2$ and $\chi(C_{2n+1}) = 3$.

■ *Example 3.8.1*   Consider the graph G shown in Figure 3.8.2. Such a graph is called a **wheel of order 6**. Find $\chi(G)$.

Solution
Notice that $G - z$ is isomorphic to an odd cycle C_5. Also, z is adjacent to each of the five vertices on this cycle. Thus, since $\chi(C_5) = 3$ and z requires its own fourth color, $\chi(G) = 4$.

In general, let W_p denote the wheel of order p. Then it can be shown, for $n \geq 2$, that $\chi(W_{2n}) = 4$ and $\chi(W_{2n+1}) = 3$. (See Exercise 11.)   ∎

A standard technique for showing that $\chi(G) = k$ is to exhibit a k-coloring of G and to argue that G is not $(k - 1)$-colorable. This technique is illustrated in the following example.

■ *Example 3.8.2*   Find $\chi(G)$ for the graph G shown in Figure 3.8.3.

Solution
We first note that $\langle \{x_1, x_7, x_8, x_9\} \rangle$ is an induced subgraph of G isomorphic to K_4. This shows that $\chi(G) \geq 4$. To show that $\chi(G) = 4$, we must exhibit a 4-coloring of G. One possibility is to color x_1 and x_{10} with one color, say, red; x_2, x_4, x_6, and x_8 with blue; x_3 and x_9 with green; and yellow for x_5 and x_7.   ∎

Suppose that a graph $G = (V, E)$ is k-colorable and a particular k-coloring of G is given. Define a relation on V as follows: for $x, y \in V$, say that x is related to y

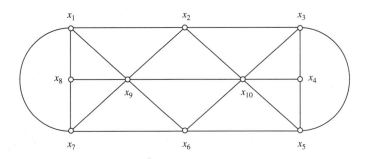

Figure 3.8.3

provided x and y are the same color. This relation is an equivalence relation on V and, as such, partitions V into equivalence classes, with each class consisting of all vertices of a particular color. These classes are called the **color classes** of the k-coloring of G. If the color classes are $V_1, V_2, \ldots, V_k$, then each V_i, $1 \le i \le k$, has the property that no two of its vertices are adjacent.

DEFINITION 3.8.2 Let $G = (V, E)$ be a graph. A nonempty subset U of V is called an **independent set** (of vertices) provided no two of its vertices are adjacent in G. An independent set U is called a **maximal independent set** provided U is not a proper subset of some other independent set of G. ∎

Thus, the color classes for a given k-coloring of G are independent sets in G. Note that the chromatic number of G is the minimum number of independent sets into which the vertex set of G may be partitioned.

∎ **Example 3.8.3** The sets $V_1 = \{x_1, x_{10}\}$, $V_2 = \{x_2, x_4, x_6, x_8\}$, $V_3 = \{x_3, x_9\}$, and $V_4 = \{x_5, x_7\}$, which are the color classes for the 4-coloring of Example 3.8.2, are independent sets of the graph $G = (V, E)$ of Figure 3.8.3. In fact, V_1, V_2, and V_3 are maximal independent sets, but V_4 is not maximal since it is properly contained in the independent set $V_4 \cup \{x_2\}$. Notice that if we let $U_1 = V_1$, $U_2 = V_2 - \{x_2\}$, $U_3 = V_3$, and $U_4 = V_4 \cup \{x_2\}$, then U_1, U_2, U_3, and U_4 are independent sets in G whose union is V. ∎

As was mentioned earlier, a graph has chromatic number 1 if and only if the graph is empty. The following theorem characterizes those graphs having chromatic number 2.

THEOREM 3.8.1 A graph $G = (V, E)$ has chromatic number 2 if and only if G is a nonempty bipartite graph.

Proof If G is a nonempty bipartite graph, then clearly $\chi(G) \ge 2$. Moreover, if V_1 and V_2 are partite sets of G, then V_1 and V_2 are independent sets in G whose union is V. Thus, $\chi(G) = 2$.

Conversely, if $\chi(G) = 2$, then E is nonempty. Let G be 2-colored, with U_1 and U_2 the resulting color classes. Then each of U_1 and U_2 is an independent set in G, and so every edge of G must join a vertex of U_1 with a vertex of U_2. It follows that G is bipartite.   □

In Exercise 6 you are asked to show that a graph G is bipartite if and only if G has no odd cycles. With this result, we can restate Theorem 3.8.1 as follows:

A graph G has chromatic number 2 if and only if G is nonempty and has no odd cycles.

Thus, if G has an odd cycle, then $\chi(G) \geq 3$; conversely, if $\chi(G) \geq 3$, then G must contain an odd cycle. Unfortunately, for $k \geq 3$, no nice characterization is known of those graphs with chromatic number k.

There is a simple and efficient algorithm for deciding whether an arbitrary given graph is bipartite (see Exercise 6). However, it is curious and interesting that the problem of deciding whether an arbitrary given graph is even 3-colorable is NP-complete, as described in the last section. Thus, at present we have no algorithm for finding the chromatic number of a given input graph that runs in polynomial time for all graphs.

We next survey a few of the known results concerning chromatic number. For these, the following definition is helpful.

DEFINITION 3.8.3   A graph $G = (V, E)$ is called **critical** (with respect to χ) provided $\chi(G - v) < \chi(G)$ for every $v \in V$. Specifically, G is called **k-critical**, where $k \geq 2$, provided G is critical and $\chi(G) = k$.   ∎

As a consequence of this definition, we can readily see that if G is k-critical, then $\chi(G - v) = k - 1$ for all $v \in V$.

It is not difficult to verify that K_2 is the only 2-critical graph, whereas the only 3-critical graphs are the odd cycles (see Exercise 12). The wheels W_{2n}, where $n \geq 2$, are 4-critical graphs; we illustrate the verification of this fact using the wheel W_6, an instance of which is the graph G of Figure 3.8.2. In Example 3.8.1, we showed that $\chi(G) = 4$. Since $G - z$ is the odd cycle C_5, we have $\chi(G - z) = 3$. Now consider any vertex on the "rim" of the wheel, such as u. Note that $(G - u) - z$ is a path and hence is bipartite, so that $\chi((G - u) - z) = 2$. Since z is adjacent to each of v, w, x, and y, a third color is required for z; hence, $\chi(G - u) = 3$. It follows that G is 4-critical.

For $k \geq 4$, the k-critical graphs have not as yet been determined. However, we can obtain the following useful result. (Recall that $\delta(G)$ denotes the minimum degree among the vertices of G.)

THEOREM 3.8.2   If G is a k-critical graph, then $\delta(G) \geq k - 1$.

Proof   We proceed by contradiction; let G be a k-critical graph and suppose that u is a vertex of G such that $\deg(u) < k - 1$. We know that $\chi(G - u) = k - 1$, so let a $(k - 1)$-coloring of $G - u$ be given. Since $\deg(u) < k - 1$, one of the $k - 1$ colors used, say green, is such that none of the vertices adjacent to u in G is colored green. We may then color u with green to obtain a $(k - 1)$-coloring of G. This contradicts the fact that $\chi(G) = k$ and thus establishes the result. $\quad\square$

If $\chi(G) = k$, then G may or may not be k-critical, but G does contain a k-critical subgraph. For if G is not k-critical, then there is some vertex u of G such that $\chi(G - u) = k$. If $G - u$ is k-critical, then we have the sought-after subgraph; if not, then the foregoing process may be repeated and, after a finite number of steps, we arrive at a k-critical subgraph of G. (In fact, note that we actually obtain a k-critical induced subgraph.)

By applying Theorem 3.8.2 to a $\chi(G)$-critical subgraph of G, we obtain the following upper bound on $\chi(G)$. (Recall that $\Delta(G)$ denotes the maximum degree among the vertices of G.)

COROLLARY 3.8.3 Let G be a graph. Then $\chi(G) \leq \Delta(G) + 1$.

Proof Let $\chi(G) = k$ and let H be a k-critical subgraph of G. By Theorem 3.8.2, $\delta(H) \geq k - 1$, and so $\Delta(H) \geq k - 1$. Certainly $\Delta(H) \leq \Delta(G)$, and so we have $k \leq \Delta(H) + 1 \leq \Delta(G) + 1$. $\quad\square$

There is a stronger result than Corollary 3.8.3, proved by R. L. Brooks [On coloring the nodes of a network, *Proceedings of the Cambridge Philosophical Society* 37 (1941): 194–197].

THEOREM 3.8.4 (Brooks's Theorem) Let G be a connected graph that is not isomorphic to either a complete graph or an odd cycle. Then $\chi(G) \leq \Delta(G)$. $\quad\square$

It should be mentioned that the bound provided by Theorem 3.8.4 is not particularly satisfying for certain classes of graphs. For example, $\chi(K_{1,n}) = 2$, whereas $\Delta(K_{1,n}) = n$.

■ **Example 3.8.4** By applying Brooks's theorem to the graph G of Figure 3.8.3, we find that $\chi(G) \leq \Delta(G) = 6$. For a lower bound, we observe that G contains a subgraph isomorphic to K_4; thus, $4 \leq \chi(G)$. Another useful observation is the following (see Exercise 20): if v is a vertex of degree less than k in G, then G is k-colorable if and only if $G - v$ is k-colorable. Let us apply this observation to the graph G, using $k = 4$. Note that x_4 has degree 3 in G, so G is 4-colorable if and only if $G_1 = G - x_4$ is 4-colorable. Next, x_8 has degree 3 in G_1, so G_1 is 4-colorable if and only if $G_2 = G_1 - x_8$ is 4-colorable. Applying Brooks's theorem to G_2, we find that $\chi(G_2) \leq \Delta(G_2) = 5$, so let's continue removing vertices of "small" degree. Note

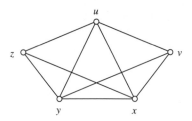

Figure 3.8.4   A planar graph *H*.

that both x_1 and x_3 have degree 3 in G_2, so let $G_3 = G_2 - x_1$ and $G_4 = G_3 - x_3$. Then G_2 is 4-colorable if and only if G_4 is 4-colorable. Now $\chi(G_4) \leq \Delta(G_4)$, which shows that G_4 is 4-colorable. Therefore, $\chi(G) = 4$.   ■

At this point it is appropriate to return to a discussion of the four-color problem with which we began this section. In particular, we wish to explain how the map-coloring problem translates into a vertex-coloring problem for a particular class of graphs. A rigorous and formal treatment would take considerable space and would, perhaps, simply cloud the issue. Because it is our intent to provide only an intuitive description, we are necessarily imprecise, but we hope it is clear.

DEFINITION 3.8.4   A graph *G* is called a **planar graph** provided it can be represented in the plane so that its edges meet only at incident vertices. Such a representation of a planar graph *G* is called a **plane embedding** of *G*. A graph that is not planar is called **nonplanar**.   ■

■ *Example 3.8.5*   Figure 3.8.3 shows a plane embedding of a planar graph *G*. Likewise, Figure 3.8.2 shows a plane embedding of the wheel W_6; thus, W_6 is planar. The graph *H*, shown in Figure 3.8.4, is isomorphic to K_5 minus an edge. Show that *H* is planar by giving a plane embedding of *H*.

Solution
A plane embedding of *H* is shown in Figure 3.8.5.   ■

Two noteworthy and interesting examples of nonplanar graphs are K_5 and $K_{3,3}$; you are asked to show that these graphs are nonplanar in Exercise 24. Indeed, by a famous result of K. Kuratowski [Sur le problème des courbes gauches en topologie, *Fundamenta Mathematicae* 15 (1930): 271–283], nonplanar graphs can be characterized in terms of K_5 and $K_{3,3}$. To state his result, we require some additional terminology. An **elementary subdivision** of a nonempty graph $G = (V, E)$ is a graph $(V \cup \{w\}, (E - uv) \cup \{uw, vw\})$, obtained from *G* by removing an edge $e = uv$ of *G* and then adding a new vertex $w \notin V$, along with the edges uw and vw. For example, Figure 3.8.6(b) shows an elementary subdivision of the graph in Figure 3.8.6(a). A

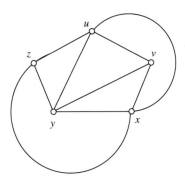

Figure 3.8.5   A plane embedding of *H*.

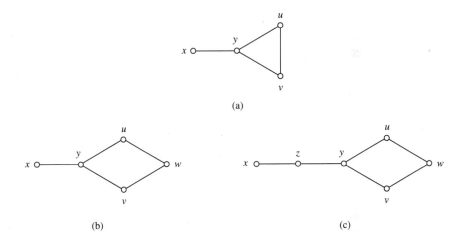

(a)

(b)   (c)

Figure 3.8.6   Graph (a) with an elementary subdivision (b) and subdivision (c).

subdivision of *G* is a graph obtained from *G* by applying the following recursive definition:

1. *G* is a subdivision of *G*.

2. If H_1 is a subdivision of *G* and H_2 is an elementary subdivision of H_1, then H_2 is a subdivision of *G*.

Figure 3.8.6(c) shows a subdivision of the graph in Figure 3.8.6(a).

  We now state Kuratowski's result.

> **THEOREM 3.8.5 (Kuratowski's Theorem)**   A graph *G* is planar if and only if *G* does not contain a subgraph *H* that is isomorphic to a subdivision of K_5 or $K_{3,3}$.

□

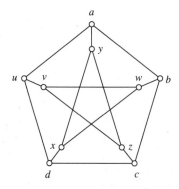

Figure 3.8.7 Petersen's graph.

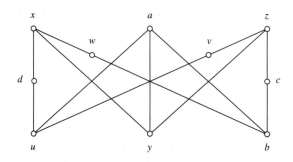

Figure 3.8.8 A subdivision of $K_{3,3}$ in Petersen's graph.

■ *Example 3.8.6* The graph G shown in Figure 3.8.7 is known as Petersen's graph and was introduced in the last section. Show that G is nonplanar by finding a subgraph H of G that is isomorphic to a subdivision of $K_{3,3}$.

Solution
One possibility is the subgraph H shown in Figure 3.8.8, which is easily seen to be a subdivision of $K_{3,3}$. ■

Given a planar graph G, a plane embedding of G partitions the plane into a finite number of regions, where each region is bounded by a closed walk of G. This is shown, for example, by the plane embedding of the graph G in Figure 3.8.9. Here there are four regions, numbered 1 through 4 as shown. Note that region 2 is bounded by the cycle t, y, z, t, whereas region 4 is bounded by the closed walk $s, t, y, x, u, v, w, x, y, z, s$. In general, we refer to the regions of a (plane embedding of a) planar graph as its **faces**. The closed walk that bounds a given face f is called the **boundary** of f.

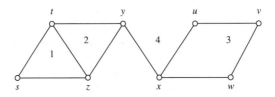

Figure 3.8.9

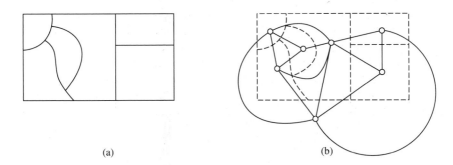

(a)   (b)

Figure 3.8.10   A plane map M (a) and its corresponding planar graph G (b).

We now describe how the four-color problem for plane maps is equivalent to the problem of finding the maximum chromatic number among all planar graphs. Let a plane map M, such as the one in Figure 3.8.10(a), be given. We construct a planar graph G from M as follows. In each region of M we place one vertex of G and make two vertices adjacent provided their associated regions have a nontrivial segment of their boundaries in common. Figure 3.8.10(b) shows the planar graph G corresponding to the map M of Figure 3.8.10(a).

Our problem is to color the "countries" of a given plane map M so that any two countries sharing a common boundary receive different colors. Note that this problem is equivalent to that of coloring the vertices of the associated planar graph so that adjacent vertices are colored differently. Thus, the four-color problem, which asks whether four colors suffice to "properly color" the countries of any plane map, is equivalent to asking whether any planar graph has chromatic number at most 4. We can now restate the celebrated result of Appel, Haken, and Koch.

THEOREM 3.8.6 (The Four-Color Theorem)   For any planar graph G, we have $\chi(G) \leq 4$.   □

EXERCISE SET 3.8

1. Draw a plane map with six countries such that each country shares a boundary with exactly four others.

2. Consider plane maps, such as the one shown in Figure 3.8.11(a), which are formed by some finite number of line segments joining points on different sides (or opposite corners) of a given rectangle.

   (a) Color the map of Figure 3.8.11(a) using just two colors.
   (b) Use induction (on the number of line segments) to show that any such map may be colored using two colors.

3. Consider the map of Figure 3.8.11(b); note that each country is adjacent to exactly five others.

   (a) Color the map using four colors.
   (b) Give the planar graph G corresponding to this map.
   (c) Show that it is not possible to color the map (or to color the vertices of G) with only three colors.

4. Consider maps such as the one in Figure 3.8.11(c); such maps are allowed to contain "disconnected" countries, such as A, whose territory consists of two (or more) disconnected regions.

   (a) Show that such maps must be excluded in order that the four-color theorem holds.
   (b) Give an example of a map with six countries, two of which are disconnected, which requires six colors.

5. Find the chromatic number of each graph shown in Figure 3.8.12.

6. Let $G = (V, E)$ be a graph.

   (a) Show that G is bipartite if and only if G has no odd cycles.
   (b) Develop a "fast" algorithm to determine whether a given input graph is bipartite.

7. Find $\chi(G)$ for the graph G of Figure 3.8.7 (Petersen's graph).

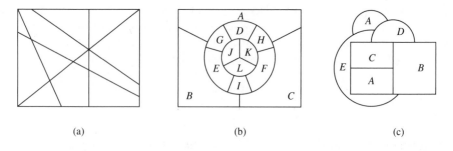

(a)   (b)   (c)

Figure 3.8.11

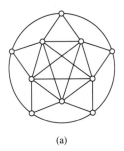 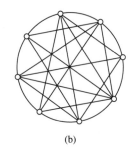 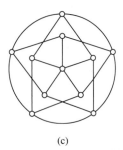

(a)   (b)   (c)

Figure 3.8.12

8. Suppose that each of n students $s_1, s_2, \ldots, s_n$ is taking some subset of the courses $c_1, c_2, \ldots, c_p$. We wish to schedule a final exam for each course in such a way that no student has the conflict of two finals scheduled for the same time period. What is the minimum number of time periods required?

   (a) Formulate this problem as one determining the chromatic number of a graph.
   (b) Suppose the set of students is {Al, Dana, Luis, Joe, Nancy}; Al is taking MA 332, MA 337, and MA 351; Dana is taking MA 332 and MA 457; Luis is taking MA 332, MA 351, and MA 423; Joe is taking MA 423 and MA 457; and Nancy is taking MA 332, MA 351, and MA 423. Construct the graph and solve the problem.

9. Let T be a nonempty tree. Find $\chi(T)$.

10. Given two graphs $G_1 = (V_1, E_1)$ and $G_2 = (V_2, E_2)$ with $V_1 \cap V_2 = \{\ \}$, their **join** is the graph $G = (V, E)$ defined by $V = V_1 \cup V_2$ and $E = E_1 \cup E_2 \cup \{v_1 v_2 \mid v_1 \in V_1$ and $v_2 \in V_2\}$. We write $G = G_1 + G_2$ to denote that G is the join of G_1 and G_2.

    (a) Given $G_1 = (\{u\}, \{\ \})$ and $G_2 = (\{w, x, y, z\}, \{wx, wz, xy, yz\})$, find $G = G_1 + G_2$. (If $G_1 \simeq K_1$ and $G_2 \simeq C_{p-1}$, then $G_1 + G_2 \simeq W_p$, the wheel of order p. In part (a), for example, $G \simeq W_5$.)
    (b) Given $G_1 = (\{u, v\}, \{uv\})$ and $G_2 = (\{w, x, y, z\}, \{wx, xy, yz\})$, find $G = G_1 + G_2$.
    (c) Prove: If $G = G_1 + G_2$, then $\chi(G) = \chi(G_1) + \chi(G_2)$.

11. Let W_p denote the wheel of order p. Show, for $n \geq 2$, that $\chi(W_{2n}) = 4$ and $\chi(W_{2n+1}) = 3$.

12. Concerning critical graphs:

    (a) Show that the 3-critical graphs are precisely the odd cycles.
    (b) Give an example of a 4-critical graph that is not isomorphic to any wheel.
    (c) Give an example of a 5-critical graph (other than K_5).

13. Let $U = \{0, 1, 2, \ldots, 9\}$ and consider the following subsets of U:

    $$A = \{0, 1, 8\}, \quad B = \{2, 5\}, \quad C = \{0, 3, 7\}, \quad D = \{1, 4, 9\}$$
    $$E = \{2, 6, 8\}, \quad F = \{5, 7, 9\}, \quad G = \{3, 4, 6\}$$

Form a graph H whose vertex set is $\{A, B, C, D, E, F, G\}$ such that two vertices of H are adjacent provided they are not disjoint.

(a) Find $\chi(H)$.

(b) Given a $\chi(H)$-coloring of H, what property do the color classes have?

(c) Is there a subset of $\{A, B, C, D, E, F, G\}$ that is a partition of U?

14. Show that if G is k-critical, then $K_1 + G$ is $(k + 1)$-critical. (See Exercise 10.)

15. Show that if H is a subgraph of G, then $\chi(H) \leq \chi(G)$.

16. Prove or disprove the following statements about a graph G:

   (a) If G contains a unique odd cycle, then $\chi(G) = 3$.

   (b) If any two odd cycles of G have a vertex in common, then $\chi(G) \leq 5$.

17. Show that $\chi(G) \leq 1 + \max \delta(H)$, where the maximum is taken over all induced subgraphs H of G. (Hint: Modify the proof of Corollary 3.8.3.)

18. Let G be a graph of order n and let $\beta(G)$ denote the maximum cardinality of an independent set of vertices in G. Show that

$$n/\beta(G) \leq \chi(G) \leq n + 1 - \beta(G)$$

19. Recall that a graph $G = (V, E)$ is 2-connected provided G is connected and, for every $v \in V$, $G - v$ is connected. Prove: If G is k-critical, where $k \geq 2$, then G is 2-connected.

20. Let G be a graph and let v be a vertex of G. Prove: If $\deg(v) < k$, then G is k-colorable if and only if $G - v$ is k-colorable.

21. Prove **Euler's formula**: Given a plane embedding of a connected planar graph G with n vertices, m edges, and r faces, then $n - m + r = 2$. (Hint: Proceed by induction on m. If $m = n - 1$, then G is a tree, so that $r = 1$ and the formula holds. If $m \geq n$, let e be a cycle edge of G and apply the induction hypothesis to $G - e$.)

22. Let G be a planar graph of order $n \geq 3$ and size m. Prove that $m \leq 3n - 6$. (Hint: Let r be the number of faces in some plane embedding of G; count the number of edges on the boundary of each face and sum over all faces. Since each edge is on the boundary of exactly two faces and each face boundary contains at least three edges, we have $3r \leq 2m$. Now apply Euler's formula from Exercise 21.)

23. Modify the proof of the result in Exercise 22 to show that, if G is a planar graph of order $n \geq 3$ and size m, and G does not contain a subgraph isomorphic to K_3, then $m \leq 2n - 4$.

24. Concerning the remarks preceding Theorem 3.8.5:

   (a) Use the result of Exercise 22 to show that K_5 is nonplanar.

   (b) Use the result of Exercise 23 to show that $K_{3,3}$ is nonplanar.

25. Prove: If G is a planar graph, then $\delta(G) \leq 5$.

26. Give an example of a planar graph G with $\delta(G) = 5$.

27. Use the result of Exercise 25 to prove that if G is a planar graph, then $\chi(G) \leq 6$.

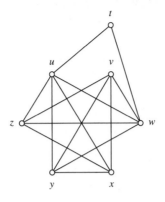

Figure 3.8.13

28. Determine whether each of the following graphs is planar or nonplanar.

(a) $K_{2,4}$

(b) $G - ay$, where G is the graph of Figure 3.8.7

(c) A graph isomorphic to the "Grötzsch graph" of Figure 3.8.12(c)

(d) A graph isomorphic to the 4-regular subgraph of K_6

29. Find the maximal independent sets in the graph G of Figure 3.8.13. Try to do this in a systematic way.

CHAPTER 3 PROBLEMS

1. Let $G = (V, E)$ be a graph. Prove that G is connected if and only if, for every partition of V as $\{V_1, V_2\}$, there is $v_1 v_2 \in E$ with $v_1 \in V_1$ and $v_2 \in V_2$.

2. Each of the following parts gives two graphs G_1 and G_2. Draw the two graphs and then either exhibit an isomorphism between them or give a reason why they are not isomorphic.

(a) $G_1 = (\{u_1, u_2, u_3, u_4, u_5, u_6\}, \{u_1 u_4, u_1 u_5, u_1 u_6, u_2 u_4, u_2 u_5, u_2 u_6, u_3 u_4, u_3 u_5, u_3 u_6\})$

   $G_2 = (\{v_1, v_2, v_3, v_4, v_5, v_6\}, \{v_1 v_2, v_1 v_3, v_1 v_6, v_2 v_3, v_2 v_5, v_3 v_4, v_4 v_5, v_4 v_6, v_5 v_6\})$

(b) $G_1 = (\{u_1, u_2, u_3, u_4, u_5, u_6, u_7\}, \{u_1 u_2, u_1 u_3, u_1 u_4, u_1 u_5, u_1 u_6, u_1 u_7, u_2 u_5, u_2 u_7, u_3 u_4, u_3 u_5, u_4 u_5, u_4 u_6, u_4 u_7, u_5 u_6\})$

   $G_2 = (\{v_1, v_2, v_3, v_4, v_5, v_6, v_7\}, \{v_1 v_2, v_1 v_3, v_1 v_4, v_1 v_5, v_1 v_6, v_1 v_7, v_2 v_3, v_2 v_5, v_3 v_4, v_3 v_5, v_3 v_7, v_4 v_5, v_5 v_6, v_6 v_7\})$

3. A graph G is called **self-complementary** provided G is isomorphic to its complement $\overline{G}$.

(a) Prove: If G is a self-complementary graph of order n, then $n \bmod 4 \in \{0, 1\}$.

⋆(b) Prove: If n is a positive integer and $n \bmod 4 \in \{0, 1\}$, then there exists a self-complementary graph of order n.

4. Let n, s, and t be integers with $0 \leq s \leq t < n$. We are interested in the question, Does there exist a graph G of order n such that $\delta(G) = s$ and $\Delta(G) = t$? The answer is obviously no if $s = 0$ and $t = n - 1$, or if $s = t$ and sn is odd, so we exclude these cases from consideration.

   (a) If $s = t$ and sn is even, then there exists an s-regular graph of order n. (See Section 8.2 of *Graphs and Digraphs*, 2nd ed., by Chartrand and Lesniak.) Give examples of (i) a 4-regular graph of order 9 and (ii) a 5-regular graph of order 10.

   (b) Using the result stated in part (a), answer the question for $s = t - 1$.

   (c) Answer the question for the remaining cases.

5. Let $d_1 \geq d_2 \geq \cdots \geq d_n$ be n positive integers. Show that the following two conditions are necessary for $(d_1, d_2, \ldots, d_n)$ to be a degree sequence of some graph G of order n (see Exercise 8 of Exercise Set 3.1):

   (a) $d_1 + d_2 + \cdots + d_n$ is even
   (b) For each p, $1 \leq p \leq n - 1$,

   $$\sum_{i=1}^{p} d_i \leq p(p-1) + \sum_{j=p+1}^{n} \min\{p, d_j\}$$

   (In fact, Erdös and Gallai (1960) have shown that conditions (a) and (b) are also sufficient for $(d_1, d_2, \ldots, d_n)$ to be a degree sequence of some graph G of order n.)

6. Let $G = (V, A)$, where $V = \{1, 2, 3, 4, 5, 6, 7, 8\}$ and $A = \{(1, 5), (2, 4), (2, 8), (3, 7),$ $(4, 6), (4, 8), (5, 2), (5, 4), (6, 2), (6, 3), (7, 5), (8, 1), (8, 6)\}$.

   (a) Draw the digraph G.
   (b) Find an eulerian circuit of G, preferably using Algorithm 3.3.1.

7. Let G be a graph of order n and size m.

   (a) Give a lower bound on m as a function of n if it is known that G is connected.
   (b) Give an upper bound on m as a function of n if it is known that G is disconnected.
   (c) Give examples to show that the bounds found in parts (a) and (b) are sharp (cannot be improved in general).

8. Let $G = (V, A)$, where $V = \{1, 2, 3, 4, 5, 6, 7, 8\}$ and $A = \{(1, 2), (2, 3), (3, 4), (4, 2),$ $(4, 5), (5, 6), (6, 7), (7, 5), (7, 8), (8, 6)\}$.

   (a) Draw the digraph G.
   (b) Give the adjacency matrix of G.
   (c) Apply Algorithm 3.4.1 to find the transitive closure $G^t = (V, A^t)$ of G.
   (d) Find the distance matrix of G, preferably using the algorithm developed in Exercise 20 of Exercise Set 3.4.

9. A graph $G = (V, E)$ is called a **k-partite graph** provided it is possible to partition V as $\{V_1, V_2, \ldots, V_k\}$ such that each edge $e \in E$ is incident with a vertex of V_i

and a vertex of V_j for some $i \neq j$. The sets $V_1, V_2, \ldots, V_k$ are called **partite sets** of G. For each the following graphs, find the minimum k for which it is k-partite.

(a) $G_1 = (\{v_1, v_2, v_3, v_4, v_5, v_6\}, \{v_1v_2, v_1v_5, v_1v_6, v_2v_3, v_2v_4, v_3v_4, v_4v_5, v_5v_6\})$

(b) $G_2 = (\{v_1, v_2, v_3, v_4, v_5, v_6, v_7\}, \{v_1v_2, v_1v_3, v_1v_6, v_1v_7, v_2v_5, v_3v_4, v_3v_5, v_4v_6, v_4v_7, v_5v_6\})$

(c) $G_3 = (\{v_1, v_2, v_3, v_4, v_5, v_6\}, \{v_1v_2, v_1v_5, v_1v_6, v_2v_3, v_2v_6, v_3v_4, v_3v_6, v_4v_5, v_4v_6, v_5v_6\})$

10. Let G be a k-partite graph with partite sets $V_1, V_2, \ldots, V_k$, where $|V_i| = n_i$, $1 \leq i \leq k$. If each vertex of V_i is adjacent with each vertex of V_j whenever $i \neq j$, then G is called a **complete k-partite graph**. Up to isomorphism, there is a unique complete k-partite graph with partite sets of cardinalities $n_1, n_2, \ldots, n_k$, and this graph is denoted by $K_{n_1, n_2, \ldots, n_k}$. Draw each of the following graphs.

(a) $K_{1,2,3}$ (b) $K_{2,2,2}$ (c) $K_{1,2,2,3}$ (d) $K_{2,2,2,2}$

11. Find the order and size of the complete k-partite graph $K_{n_1, n_2, \ldots, n_k}$.

12. Let G be a k-partite graph of order n and size m. In parts (a) through (d), give an upper bound on m.

(a) $k = 3$ and $n = 9$ (b) $k = 3$ and $n = 10$

(c) $k = 4$ and $n = 10$ (d) $k = 4$ and $n = 12$

$\star$(e) Generalize parts (a) through (d).

13. Let $G = (V, E)$, where $V = \{t, u, v, w, x, y, z\}$ and $E = \{tu, tv, tw, ux, vw, vy, vz, wx, wz, xy, yz\}$.

(a) Find a subgraph H_1 isomorphic to $K_{2,3}$.

(b) Find an induced subgraph H_2 isomorphic to $K_{1,3}$.

(c) Find nonisomorphic spanning subgraphs H_3 and H_4 such that each is regular of degree 2.

(d) Show that G does not contain a subgraph isomorphic to $K_{1,1,3}$. (See Problem 10.)

14. Let $H = (V, E)$, where $V = \{u, v, x, y, z\}$ and $E = \{uv, ux, uz, vx, vz, yz\}$. Find, up to isomorphism, all connected spanning subgraphs of H.

15. Characterize those graphs with the property that every induced subgraph is connected.

16. Characterize those graphs with the property that every induced subgraph on three or more vertices is connected.

17. Generalize Problems 15 and 16. (Hint: See Problem 10.)

18. Let G be a graph with $\Delta(G) = \Delta$.

(a) Prove: There exists a graph H such that G is an induced subgraph of H and H is Δ-regular.

(b) Disprove: There exists a graph H such that G is a spanning subgraph of H and H is Δ-regular.

19. Apply Algorithm 3.4.3 (Dijkstra's) to each of the following weighted graphs to find shortest 0-v paths. (Consider each edge to represent a pair of symmetric arcs.)

(a) $G_1 = (V, E, w)$, where $V = \{0, 1, \ldots, 9\}$, $E = \{01, 05, 07, 16, 27, 28, 35, 39, 45,$
46, 47, 58, 68, 69, 79\}$, and $w(01) = 2$, $w(05) = 3$, $w(07) = 1$, $w(16) = 5$,
$w(27) = 4$, $w(28) = 3$, $w(35) = 3$, $w(39) = 4$, $w(45) = 8$, $w(46) = 7$, $w(47) = 9$,
$w(58) = 6$, $w(68) = 5$, $w(69) = 3$, $w(79) = 2$.

(b) $G_2 = (V, E, w)$, where $V = \{0, 1, \ldots, 7\}$, $E = \{01, 03, 12, 13, 14, 15, 16, 24, 25,$
27, 46, 57\}$, and $w(01) = 1$, $w(03) = 9$, $w(12) = 3$, $w(13) = 7$, $w(14) = 5$,
$w(15) = 11$, $w(16) = 3$, $w(24) = 7$, $w(25) = 9$, $w(27) = 19$, $w(46) = 7$,
$w(57) = 13$.

20. Apply Algorithm 3.5.3 (Prim's) to each of the weighted graphs in Problem 19.

21. Adapt Algorithm 3.5.3 to solve the following problem: Given a connected weighted graph $G = (V, E, w)$ and $E_1 \subseteq E$, find a connected spanning subgraph $H = (V, E_0)$ of minimum weight such that $E_1 \subseteq E_0$.

22. Consider the acyclic digraph $G = (V, A)$, where $V = \{1, 2, \ldots, 8\}$ and $A = \{(1, 2), (1, 3), (1, 4), (2, 5), (2, 6), (3, 6), (3, 7), (4, 7), (5, 7), (5, 8), (6, 8)\}$. Each of the following parts lists the vertices of G; indicate, with justification, whether it is a valid result of applying Algorithm 3.5.1 (topological sort).

(a) $(1, 2, 3, 4, 5, 6, 7, 8)$ (b) $(1, 4, 3, 2, 6, 5, 8, 7)$

(c) $(1, 4, 2, 5, 7, 3, 6, 8)$

23. Consider the directed graph $G = (V, A)$, where $V = \{0, 1, \ldots, 7\}$ and $A = \{(0, 1), (0, 2), (1, 3), (1, 4), (1, 5), (2, 4), (2, 6), (3, 4), (4, 0), (5, 6), (5, 7), (6, 7), (7, 3)\}$. Apply Algorithm 3.5.2 to G to find a depth-first search tree with root 0.

24. Let $T = (V, A)$ be a rooted tree, with $V = \{v_1, v_2, \ldots, v_n\}$, and let k denote the number of leaves of T. Show that

$$k = \frac{1}{2}\left(1 + \sum_{i=1}^{n} |\text{od}(v_i) - 1|\right)$$

25. Prove: Every tree T of order $n \geq 4$ that is not a star is isomorphic to a spanning tree of $\overline{T}$.

26. Let n be a positive integer and let G be a graph with $\delta(G) \geq n - 1$. Prove: If T is any tree of order n, then G contains a subgraph isomorphic to T.

27. Let G be a connected graph and let $e = uv$ be an edge of G. Prove:

(a) The edge e is a bridge of G if and only if e is not on any cycle of G.

(b) If e is a bridge, then G has at least two vertices of odd degree.

(c) If e is a bridge, then G is not eulerian.

28. Prove: A graph G is a forest if and only if every induced subgraph of G contains a vertex of degree 0 or 1

29. A graph G is said to be **unicyclic** provided G is connected and contains a unique cycle. Prove: A graph G of order n and size m is unicyclic if and only if G is connected and $m = n$.

30. A tree $T = (V, E)$ of order n is termed **graceful** provided there is a function $f : V \to \{1, 2, \ldots, n\}$ such that

$$\{|f(u) - f(v)| \mid uv \in E\} = \{1, 2, \ldots, n - 1\}$$

It is conjectured that all trees are graceful.

 (a) Show that the star $K_{1,n-1}$ is graceful.
 (b) Show that the path P_n is graceful.

31. When is a graph isomorphic to the graph of an equivalence relation?

32. Let G be a connected graph that is not a tree. The minimum length among the cycles of G is called the **girth** of G. Consider the following problem: For $r \geq 2$ and $g \geq 3$, determine those graphs of smallest order among all r-regular graphs with girth g; such graphs are called (r, g)-**cages** and are guaranteed to exist by a result of Erdös and Sachs (1963).

 (a) Show that the cycle C_n is the unique $(2, n)$-cage.
 (b) Show that the complete graph K_{r+1} is the unique $(r, 3)$-cage.
 (c) Find the unique $(r, 4)$-cage.
 $\star$(d) Find a $(3, 5)$-cage.
 $\star$(e) Find a $(3, 6)$-cage.

33. For positive integers k and m, the **ramsey number** $r(k, m)$ (named for the mathematician Frank Ramsey) is the least positive integer n such that every graph of order n either contains an induced subgraph isomorphic to K_k (k mutually adjacent vertices) or contains an induced subgraph isomorphic to $\overline{K}_m$ (an independent set of m vertices). It can be remarked that $r(k, m) = r(m, k)$, $r(1, m) = 1$, and $r(2, m) = m$.

 (a) Show that $r(3, 3) = 6$.
 (b) Prove, for $k \geq 2$ and $m \geq 2$, that $r(k, m) \leq r(k - 1, m) + r(k, m - 1)$. (This shows that the ramsey numbers exist.)
 (c) It follows from part (b) that $r(3, 4) \leq r(2, 4) + r(3, 3) = 4 + 6 = 10$. Show, in fact, that $r(3, 4) \leq 9$.
 (d) Show that $r(3, 4) \geq 9$ by exhibiting a graph of order 8 that contains neither K_3 nor $\overline{K}_4$ as an induced subgraph.

Thus, $r(3, 4) = 9$. The only other known ramsey numbers are $r(3, 5) = 14$, $r(3, 6) = r(4, 4) = 18$, $r(3, 7) = 23$, and $r(3, 9) = 36$; moreover, $r(3, 8) = 28$ or 29.

34. Let $G = (V, A)$ be a digraph. For $k \geq 2$, the **kth power** of G is the digraph $G^k = (V, A_k)$, where

$$A_k = A \cup \{(u, v) \mid 2 \leq d(u, v) \leq k\}$$

 (a) Let $G = (V, A)$, where $V = \{s, t, u, v, w, x, y, z\}$ and $A = \{(s, t), (t, u), (u, v), (v, w), (w, t), (w, x), (x, y), (y, w), (y, z), (z, s), (z, x)\}$. Find G^2, G^3, and so on.
 (b) Define the analogous concepts for graphs

(c) A result of H. Fleischner [The square of every two-connected graph is hamiltonian, *Journal of Combinatorial Theory* 16B (1974): 29–34] says that if G is any 2-connected graph, then G^2 is hamiltonian. Give an example of a tree T such that T^2 is not hamiltonian.

(d) Prove: If T is any tree and u and v are vertices of T, then T^3 contains a spanning u-v path.

35. The 1990 NCAA basketball tournament began with 64 teams. In the first round, 32 games were played, eliminating 32 teams. In the second round, 16 games were played, eliminating 16 teams. The remaining 16 teams continued on to the next round of the tournament, and so on; in each round, half the remaining teams were eliminated, until the winner of the tournament was determined.

(a) Describe how such a single-elimination tournament can be represented as a binary tree.

(b) What is the height of the binary tree representing the 1990 NCAA basketball tournament?

(c) What was the total number of games played in the tournament?

36. An m-ary tree is called **ordered** provided each vertex x, other than the root, is assigned an integer $p(x)$ between 1 and m giving its "position" relative to its siblings, so that no two children of the same parent are assigned the same integer. We think of a child with a lower number as being "older" than a sibling with a higher number. For a binary tree, it is customary to assign the designations left and right, rather than the integers 1 and 2, respectively. Let $T = (V, A)$ be an ordered complete m-ary tree, where $m > 2$. We may then define an ordered binary tree $\widetilde{T} = (V, \widetilde{A})$ as follows:

1. The root of $\widetilde{T}$ is the root of T.
2. If q is a nonleaf vertex of T, then the eldest child of q becomes the left child of q in $\widetilde{T}$.
3. If q is not the root and is not the youngest child of its parent in T, then its next youngest sibling becomes its right child in $\widetilde{T}$.

Using this method, it is possible to convert an m-ary tree T, $m > 2$, into a binary tree $\widetilde{T}$ with the same vertex set. Let $T = (V, A)$, where

$$V = \{a, b, c, d, e, f, g, h, i, j, k, m, n\},$$
$$A = \{(a, b), (a, c), (a, d), (b, e), (b, f), (c, g), (d, h), (d, i), (d, j), (e, k), (g, m), (g, n)\}$$

(a is the root) $p(b) = 1$, $p(c) = 2$, $p(d) = 3$, $p(e) = 1$, $p(f) = 2$, $p(g) = 1$, $p(h) = 1$, $p(i) = 2$, $p(j) = 3$, $p(k) = 1$, $p(m) = 1$, and $p(n) = 2$. Draw the 3-ary tree T, and then find and draw $\widetilde{T}$.

37. With reference to the general ideas in Problem 36:

(a) When is a vertex of T a leaf of $\widetilde{T}$?

(b) Given that T has height h, give an upper bound on the height of $\widetilde{T}$ (in terms of m and h).

(c) Given $\widetilde{T}$, what information can be obtained about T?

38. A common operation on an ordered rooted tree is to **traverse** it—that is, to process or visit its vertices in some definite order. One method of traversing an ordered rooted tree is known as a **preorder traversal**. Let T be an ordered rooted tree with root r and let $T_1, T_2, \ldots, T_k$ be the subtrees of T whose roots are the children of r, in order from eldest to youngest. A preorder traversal of T is then defined recursively as follows: first visit r, and then perform preorder traversals of $T_1, T_2, \ldots, T_k$ in that order. For example, a preorder traversal of the tree T of Figure 3.5.1 visits the vertices in the order $a, b, e, f, h, c, d, g, i, j$.

   (a) Given the ordered rooted tree T of Problem 36, list the vertices of T in the order they are visited by a preorder traversal.

   (b) Informally describe an algorithm for performing a preorder traversal of a given ordered rooted tree T. Compare this algorithm with the depth-first search Algorithm 3.5.2.

39. Given integers k and n with $1 \le k \le n$, give an example of a graph G of order n such that $\chi(G) = k$ and $\chi(\overline{G}) = n + 1 - k$.

40. With reference to Problem 39, prove the following result of E. A. Nordhaus and J. W. Gaddum [On complementary graphs, *American Mathematical Monthly* 63 (1956): 175–177]: for any graph G of order n, $\chi(G) + \chi(\overline{G}) \le n + 1$.

41. A graph G is called k-minimal provided $\chi(G) = k$ and $\chi(G - e) = k - 1$ for every edge e of G.

   (a) Show that every connected k-minimal graph is k-critical.

   (b) Show that every 2-critical graph is 2-minimal.

   (c) Show that every 3-critical graph is 3-minimal.

   (d) Give an example of a 4-critical graph that is not 4-minimal.

42. Let $\omega(G)$ denote the largest order among the complete subgraphs of G. Then clearly $\chi(G) \ge \omega(G)$. Interestingly, there exists a graph G_k with $\omega(G_k) = 2$ and $\chi(G_k) = k$ for each $k \ge 2$. The following recursive method for constructing such graphs is due to J. Mycielski [Sur le coloriage des graphs, *Colloquium Mathematicum* 3 (1955): 161–162]. Suppose $G_k = (V_k, E_k)$ has been constructed, with $V_k = \{v_1, v_2, \ldots, v_n\}$. We construct G_{k+1} by adding $n + 1$ new vertices $u_0, u_1, \ldots, u_n$ to G_k, making u_0 adjacent with each u_i, and making u_i adjacent with each v_j that is adjacent to v_i in G_k.

   (a) Starting with $G_2 \simeq K_2$, show that $G_3 \simeq C_5$.

   (b) Show that G_4 is isomorphic to the Grötzsch graph of Figure 3.8.12(c).

   (c) Find the order of G_k.

   (d) Verify that $\omega(G_k) = 2$ and $\chi(G_k) = k$.

43. Prove that for any graph G, $\chi(G)$ is at most 1 more than the maximum length among the paths of G.

44. Let G be a graph of order n with vertex degrees $d_1 \ge d_2 \ge \cdots \ge d_n$. Prove that

$$\chi(G) \le \max\{\min\{i, d_i + 1\}\}$$

where the maximum is over all i, $1 \le i \le n$.

45. Given graphs $G_1 = (V_1, E_1)$ and $G_2 = (V_2, E_2)$ with $V_1 \cap V_2 = \{ \}$, their **cartesian product** is the graph $G_1 \times G_2$ with vertex set $V_1 \times V_2$ and edge set

$$\{(u_1, u_2)(v_1, v_2) \mid \text{ either } u_1 = v_1 \text{ and } u_2 v_2 \in E_2 \text{ or } u_2 = v_2 \text{ and } u_1 v_1 \in E_1\}$$

Each of the following parts gives two graphs G_1 and G_2. Find $G_1 \times G_2$ and draw the graphs G_1, G_2, and $G_1 \times G_2$.

(a) $G_1 = (\{u, v\}, \{uv\})$, $G_2 = (\{x, y, z\}, \{xy, xz, yz\})$
(b) $G_1 = (\{t, u, v\}, \{tu, uv\})$, $G_2 = (\{w, x, y, z\}, \{wx, xy, yz\})$
(c) $G_1 = (\{t, u, v\}, \{tu, uv\})$, $G_2 = (\{w, x, y, z\}, \{wx, wz, xy, yz\})$
(d) $G_1 = (\{s, t, u, v\}, \{st, su, sv\})$, $G_2 = (\{w, x, y, z\}, \{wx, wz, xy, yz\})$

46. The **n-cube** is the graph Q_n defined recursively as follows: $Q_1 = (\{0, 1\}, \{01\})$; for $n \geq 1$, $Q_{n+1} = Q_1 \times Q_n$ (see Problem 45).

(a) Find and draw the graphs Q_2, Q_3, and Q_4.
(b) Find the order and size of Q_n.
(c) Show that Q_n is bipartite for each n.
(d) Show that Q_n is hamiltonian for each $n \geq 2$.
(e) Find the maximum distance between two vertices of Q_n.
(f) For which values of n is Q_n planar?
(g) Show that Q_n has a perfect matching.

(The n-cubes are the underlying graphs for a certain class of computer networks known as "hypercube" networks.)

47. Show that the converse of Theorem 3.7.1 is false; that is, give an example of a nonhamiltonian graph $G = (V, E)$ having the property that, for each proper subset U of V, the subgraph of G induced by $V - U$ has at most $|U|$ components.

48. Give a necessary and sufficient condition for the complete tripartite graph K_{n_1, n_2, n_3} to be hamiltonian.

49. Determine all values of k and $n_1 \leq n_2 \leq \cdots \leq n_k$ for which the complete k-partite graph $K_{n_1, n_2, \ldots, n_k}$ is planar.

50. Let G be a connected graph of order n and size m.

(a) Prove: If $m \leq n + 2$, then G is planar.
(b) For each $n \geq 6$, give an example of a nonplanar connected graph of order n and size $n + 3$.

51. Let G be a planar graph with girth g (see Problem 32).

(a) Generalize the results of Exercises 22 and 23 of Exercise Set 3.8.
(b) Prove: If $g \geq 4$, then $\chi(G) \leq 4$.

52. Describe an algorithm for finding a maximal independent set W in a given graph $G = (V, E)$.

53. Consider the following algorithm for purportedly finding $n = \chi(G)$ for a given graph $G = (V, E)$:

```
begin
    n := 0;
    U := V;
    while U ≠ { } loop
        Find a maximal independent set W of ⟨U⟩
        (using the algorithm found in Problem 52);
        n := n + 1;
        U := U − W;
    end loop;
end;
```

Give an example to show that this algorithm does not find $\chi(G)$ in general. What does the algorithm produce?

54. Let $G = (V, E)$ be a bipartite graph with partite sets X and Y, where $|X| = 7$ and $|Y| = 4$. Suppose that G has the property that any four vertices of X can be matched to the vertices in Y. Find a lower bound for $|E|$.

55. A certain round-robin racquetball tournament involves n players.

   (a) If each player wins at least one match, show that there exist two players with the same number of wins.
   (b) Give an example to show the necessity of the hypothesis that each player wins at least one match in part (a).
   (c) Suppose that n is odd. Show that some player wins at least $(n+1)/2$ matches.

56. Let $A_1, A_2, \ldots, A_n$ be a family of finite sets with the property that, for each k, $1 \le k \le n$, and each choice of $i_1, i_2, \ldots, i_k$ with $1 \le i_1 < \cdots < i_k \le n$,

$$|A_{i_1} \cup A_{i_2} \cup \cdots \cup A_{i_k}| \ge 2k$$

Show that the family has two disjoint systems of distinct representatives.

57. A company has n distinct positions to fill and a pool of n applicants for these positions. Suppose that, for some integer k, $1 \le k \le n$, each applicant is qualified for exactly k positions and each position has exactly k applicants qualified to fill it. Show that each position can be filled by an applicant qualified for it.

58. Let G be the complete k-partite graph $K_{n_1, n_2, \ldots, n_k}$; find $\beta_1(G)$.

59. Determine whether the pruned chessboard represented below can be covered with dominoes. (Refer to Exercise 20 of Exercise Set 3.6.)

w_1	*	*	*	w_2	b_1	
b_2	w_3	b_3	w_4	b_4		*
	b_5	*	*	w_5		*
	w_6	b_6	w_7	b_7		*
w_8	b_8	*	b_9	w_9		*
b_{10}	*	*	w_{10}	b_{11}	w_{11}	

60. For each integer $n > 4$, give an example of a graph of order n that is:

    (a) Both eulerian and hamiltonian.

    (b) Eulerian, but not hamiltonian.

    (c) Hamiltonian, but not eulerian.

    (d) Neither eulerian nor hamiltonian.

Chapter *4*

Combinatorial Designs

4.1 *FINITE GEOMETRIES*

You are no doubt familiar with the geometry of the euclidean plane as treated in high school courses. Adding a coordinate system to this geometry gives us the cartesian plane and allows us to apply algebraic techniques to the study of geometry; this is the subject of analytic geometry and the calculus. The points of this geometry are ordered pairs of real numbers (x, y), and lines are sets of points that satisfy various axioms, or basic assumptions. For instance, given any two distinct points p_1 and p_2, it is an axiom of euclidean geometry that they determine a unique line; that is, there is a unique line α such that both p_1 and p_2 are on α. If $p_1 = (x_1, y_1)$ and $p_2 = (x_2, y_2)$, then the equation of the line determined by p_1 and p_2 is given by

$$(x_2 - x_1)(y - y_1) = (y_2 - y_1)(x - x_1)$$

or

$$(y_2 - y_1)x + (x_1 - x_2)y + (x_2 y_1 - x_1 y_2) = 0$$

Note that this last equation has the form

$$ax + by + c = 0$$

where a, b, and c are real numbers and a and b are not both 0. Conversely, any equation of this form is the equation of some line in the cartesian plane.

Another axiom of euclidean plane geometry, one that has generated a great deal of controversy and interest over several centuries, is the so-called parallel axiom or fifth postulate:

Given a point p and a line α such that p is not on α, there is a unique line β such that p is on β and β is parallel with α.

This axiom was controversial because it did not seem to be an obvious, self-evident truth, as it was thought axioms should be. Attempts were made to deduce it from the other axioms as a theorem, or to substitute for it an equivalent but more self-evident axiom. These attempts were unsuccessful. Finally, during the eighteenth and nineteenth centuries, mathematicians came to realize that there was nothing sacred about Euclid's parallel axiom—that other mathematically valid and interesting geometries are obtained by dropping the parallel axiom or substituting another axiom for it. For example, we can replace the parallel axiom with one saying that any two distinct lines intersect at a unique point. This gives a nice geometry in which points and lines have a symmetric, or "dual," relationship: two points determine a line and two lines determine a point.

At the end of the nineteenth century, the Italian mathematician Gino Fano found the first example of a finite geometry. This geometry, called the **Fano plane**, consists of a set of seven points $\{p_1, p_2, p_3, p_4, p_5, p_6, p_7\}$ and the seven lines $\alpha_1 = \{p_1, p_2, p_4\}$, $\alpha_2 = \{p_1, p_3, p_6\}$, $\alpha_3 = \{p_1, p_5, p_7\}$, $\alpha_4 = \{p_2, p_3, p_5\}$, $\alpha_5 = \{p_2, p_6, p_7\}$, $\alpha_6 = \{p_3, p_4, p_7\}$, and $\alpha_7 = \{p_4, p_5, p_6\}$. The Fano plane is often depicted as in Figure 4.1.1; here, for instance, the line segment through the points p_1, p_2, and p_4 represents the line α_1, whereas the circle through the points p_4, p_5, and p_6 represents the line α_7. By using this figure we can easily check that every two points determine a unique line and every two lines intersect at a unique point.

> **DEFINITION 4.1.1**   A **geometry** consists of a nonempty set P of elements called **points** together with a set L of nonempty subsets of P called **lines** satisfying the following two axioms:
>
> **A1.** Given two distinct points p_1 and p_2, there is a unique line α such that $p_1 \in \alpha$ and $p_2 \in \alpha$.
> **A2.** There is a subset of P containing four points, no three of which belong to the same line.   ∎

A geometry G with point set P and line set L is denoted by writing $G = (P, L)$. Points of an abstract geometry will usually be denoted by lowercase letters such as p, q, r, s and lines by lowercase Greek letters such as α, β, γ, δ. Axiom A2 is included in the definition of a geometry in order to avoid certain trivial situations.

We shall use standard geometric terminology—points are **on** lines, lines **contain** points, points on the same line are **colinear**, lines intersecting at a common point are **concurrent**, lines that do not intersect are called **parallel**, and so forth.

We wish to study two particular kinds of geometry, affine planes and projective planes; affine planes are defined by adding a parallel axiom to the definition of a geometry, whereas projective planes are defined by adding an axiom saying that two distinct lines intersect always at a unique point.

> **DEFINITION 4.1.2**   An **affine plane** is a geometry that obeys the following additional axiom:
>
> **A3.** Given a point p and a line α such that p is not on α, there is a unique line β such that p is on β and β is parallel with α.

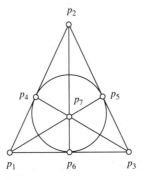

Figure 4.1.1 The Fano plane.

A **projective plane** is a geometry that satisfies the following additional axiom:

A3′. Given two distinct lines α_1 and α_2, there is a unique point p such that $p \in \alpha_1$ and $p \in \alpha_2$. ∎

- ■ *Example 4.1.1*

(a) The Fano plane is a projective plane with seven points and seven lines and can be shown to be the smallest projective plane; see Exercise 1.

(b) A geometry must have at least four points; let us try to construct an affine plane $G = (P, L)$ with four points. Let $P = \{0, 1, 2, 3\}$. Since every two points of P determine a line and no three points are colinear, it is clear that $|L| \geq C(4, 2) = 6$. In fact, a natural candidate for L is the set of all 2-element subsets of P:

$$L = \{\{0, 1\}, \{0, 2\}, \{0, 3\}, \{1, 2\}, \{1, 3\}, \{2, 3\}\}$$

It is a simple matter to check that G satisfies Axiom A3. For instance, take the case $p = 2$ and $\alpha = \{0, 1\}$. The only line parallel with α is $\beta = \{2, 3\}$, and we note that p is on β.

(c) The usual cartesian plane is an affine geometry. As already mentioned, the two distinct points $p_1 = (x_1, y_1)$ and $p_2 = (x_2, y_2)$ determine the line α with equation

$$(x_2 - x_1)(y - y_1) = (y_2 - y_1)(x - x_1)$$

To check Axiom A3, let α be the line with equation

$$ax + by + c = 0$$

and let the point $p = (x_1, y_1)$ not be on α; we wish to find the equation of the unique line β through p and parallel with α. Since β and α are parallel, they have the same slope; this means that the equation of β is

$$ax + by + d = 0$$

for some real number d. Using the fact that p is on β, we have

$$ax_1 + bx_2 + d = 0$$

implying that

$$d = -ax_1 - bx_2$$

(d) To give a concrete example of an infinite projective plane, consider a sphere S. We let p be a point on (the surface of) S and let $\bar{p}$ denote the point on S diametrically opposite p; that is, $\bar{p}$ is the unique point on S such that the line through p and $\bar{p}$ passes through the center of S. We define $G = (P, L)$ by letting P be the set of all pairs $\{p, \bar{p}\}$ of diametrically opposite points and letting L be the set of all "great circles" of S. (Recall that a great circle of a sphere is the intersection of the sphere with a plane containing its center.) Note that, if α is a great circle of S and p is on α, then $\bar{p}$ is also on α, so we can consider α to be a set of pairs of diametrically opposite points. Now let us check Axioms A1 and A3′. For A1, note that if p and q are distinct points on S and $q \neq \bar{p}$, then there is a unique great circle α of S that contains both $\{p, \bar{p}\}$ and $\{q, \bar{q}\}$. (It is interesting to note that the shorter portion of α between p and q gives the shortest path on S from p to q.) So A1 is satisfied. For A3′, we let α and β be distinct great circles of S. It is not difficult to verify that α and β intersect at a unique pair of diametrically opposite points, so that A3′ is satisfied. Thus, $G = (P, L)$ is a projective plane. ∎

The usual euclidean plane with cartesian coordinates is an affine plane constructed with respect to the field $\mathbb{R}$ of real numbers; a point is an ordered pair (x, y) with $x, y \in \mathbb{R}$, and a line is determined as the solution set of a linear equation

$$ax + by + c = 0$$

where $a, b, c \in \mathbb{R}$ and a and b are not both 0. We next describe how any field can be used to construct an affine plane.

To begin with a specific example, let us construct a finite affine plane over the field $\mathbb{Z}_3 = \{0, 1, 2\}$. The point set P is the set $\mathbb{Z}_3 \times \mathbb{Z}_3$ of all ordered pairs (x, y) with $x, y \in \mathbb{Z}_3$, so that

$$P = \{(0,0), (0,1), (0,2), (1,0), (1,1), (1,2), (2,0), (2,1), (2,2)\}$$

and has 9 elements. A line α is determined as the solution set of a linear equation

$$ax + by + c = 0$$

where $a, b, c \in \mathbb{Z}_3$ and a and b are not both 0. If $b \neq 0$, then the equation of α may be written in the form

$$y = mx + k$$

where $m = -ab^{-1}$ and $k = -cb^{-1}$. In this case the element $m \in \mathbb{Z}_3$ is called the **slope** of α. If $b = 0$, then the equation of α may be written as

$$x = k$$

where $k \in \mathbb{Z}_3$; in this case we say that the slope of α is "infinite." There are 3 choices for m when $b \neq 0$ and 3 choices of k for each such m. Hence, by the multiplication principle, there are 9 lines with slope m for some $m \in \mathbb{Z}_3$. In addition, there are three lines with infinite slope. Thus, by the addition principle, we obtain a total of 12 lines, grouped into 4 classes with 3 parallel lines in each class:

The parallel class with slope 0:

> The line $\alpha_0 = \{(0,0), (1,0), (2,0)\}$ has equation $y = 0$.
> The line $\alpha_1 = \{(0,1), (1,1), (2,1)\}$ has equation $y = 1$.
> The line $\alpha_2 = \{(0,2), (1,2), (2,2)\}$ has equation $y = 2$.

The parallel class with slope 1:

> The line $\beta_0 = \{(0,0), (1,1), (2,2)\}$ has equation $y = x$.
> The line $\beta_1 = \{(0,1), (1,2), (2,0)\}$ has equation $y = x + 1$.
> The line $\beta_2 = \{(0,2), (1,0), (2,1)\}$ has equation $y = x + 2$.

The parallel class with slope 2:

> The line $\gamma_0 = \{(0,0), (1,2), (2,1)\}$ has equation $y = 2x$.
> The line $\gamma_1 = \{(0,1), (1,0), (2,2)\}$ has equation $y = 2x + 1$.
> The line $\gamma_2 = \{(0,2), (1,1), (2,0)\}$ has equation $y = 2x + 2$.

The parallel class with infinite slope:

> The line $\delta_0 = \{(0,0), (0,1), (0,2)\}$ has equation $x = 0$.
> The line $\delta_1 = \{(1,0), (1,1), (1,2)\}$ has equation $x = 1$.
> The line $\delta_2 = \{(2,0), (2,1), (2,2)\}$ has equation $x = 2$.

In general, the affine plane constructed from a field F is called a **euclidean geometry** and is written $EG(F)$. If F is finite with order n, then $EG(F)$ is called a **finite euclidean geometry** and is written $EG(n)$; above we have constructed $EG(3)$. Since, given a prime power n, there is a unique field of order n, up to isomorphism, the geometry $EG(n)$ is uniquely determined. It has point set $F \times F$ and line set

$$\{\{(x,y) \mid ax + by + c = 0\} \mid a, b, c \in F \text{ and } (a,b) \neq (0,0)\}$$

where 0 denotes the additive identity of F.

Since $EG(n)$ has point set $F \times F$, the number of points is n^2. A line of $EG(n)$ is determined as the solution set of a linear equation $ax + by + c = 0$, where $a, b, c \in F$

and a and b are not both 0. Just as in the example for $F = \mathbb{Z}_3$, if $b \neq 0$, then such an equation may be written in the form

$$y = mx + k$$

where $m, k \in F$; the element m is called the **slope** of the line. If $b = 0$, then the equation may be written as

$$x = k$$

where $k \in F$; such a line is said to have **infinite slope**. Note that two lines are parallel if and only if they have the same slope (whether infinite or not). Hence, the lines of $EG(n)$ are partitioned into $n + 1$ sets of parallel lines; such a set is called a **parallel class**. For a given parallel class, there are n choices for the element k, so that each class contains n lines. Therefore, the number of lines of $EG(n)$ is $n(n + 1)$. We next formally show that $EG(n)$ is an affine plane.

THEOREM 4.1.1   Let F be a field. Then $EG(F)$ is an affine plane.

Proof   We must verify Axioms A1, A2, and A3.

To verify Axiom A1, let $p_1 = (x_1, y_1)$ and $p_2 = (x_2, y_2)$ be distinct points and let α be the line with equation

$$(x_2 - x_1)(y - y_1) = (y_2 - y_1)(x - x_1)$$

It is easy to check that $p_1 \in \alpha$ and $p_2 \in \alpha$. Now suppose that β is another line through p_1 and p_2, say, with equation $ax + by + c = 0$. Since $p_1 \in \beta$, we may write the equation of β in the form

$$b(y - y_1) = -a(x - x_1)$$

Also, since $p_2 \in \beta$, we have $b(y_2 - y_1) = -a(x_2 - x_1)$. If $x_1 = x_2$, then $b = 0$, and the equation of β is $x = x_1$. So $\beta = \alpha$. If $x_1 \neq x_2$, then we have

$$-a = b(y_2 - y_1)(x_2 - x_1)^{-1}$$

and the equation of β is then

$$b(y - y_1) = b(y_2 - y_1)(x_2 - x_1)^{-1}(x - x_1)$$

Multiplying both sides by $b^{-1}(x_2 - x_1)$ yields

$$(x_2 - x_1)(y - y_1) = (y_2 - y_1)(x - x_1)$$

so that $\beta = \alpha$.

For Axiom A2, let 1 denote the multiplicative identity of F. Then $1 \neq 0$, and the points $(0,0), (0,1), (1,0)$, and $(1,1)$ are four points of F of which no three are colinear.

To verify Axiom A3, let α be a line, say, with equation $ax + by + c = 0$, and let $p = (x_1, y_1)$ be a point not on α. Let β be a line with equation $ax + by + d = 0$, where $d \neq c$; it is easy then to show that α and β are parallel. We want a line β that contains p; by substituting the coordinates of p into the equation of β, we find that β contains p if and only if

$$ax_1 + by_1 + d = 0$$

or

$$d = -(ax_1 + by_1)$$

This shows that there is a unique line β through p and parallel with α. $\square$

Recall that there exists a finite field of order n if and only if $n = p^k$ for some prime p and positive integer k. This yields the following corollary:

COROLLARY 4.1.2 There exists a finite affine plane with n^2 points and $n^2 + n$ lines whenever $n = p^k$ for some prime p and positive integer k. $\square$

■ ***Example 4.1.2*** Construct the affine plane $EG(4)$.

Solution
Let $F = \{0, 1, t, 1 + t\}$, where addition and multiplication are done modulo the prime polynomial $1 + t + t^2$ (see Example 0.5.3). The point set of $EG(4)$ is $F \times F$ and has 16 elements. There are 20 lines, grouped into 5 classes with 4 parallel lines in each class:

The parallel class with slope 0:

$\alpha_0 = \{(0,0), (1,0), (t,0), (1+t,0)\}$ has equation $y = 0$.
$\alpha_1 = \{(0,1), (1,1), (t,1), (1+t,1)\}$ has equation $y = 1$.
$\alpha_2 = \{(0,t), (1,t), (t,t), (1+t,t)\}$ has equation $y = t$.
$\alpha_3 = \{(0,1+t), (1,1+t), (t,1+t), (1+t,1+t)\}$ has equation $y = 1+t$.

The parallel class with slope 1:

$\beta_0 = \{(0,0), (1,1), (t,t), (1+t,1+t)\}$ has equation $y = x$.
$\beta_1 = \{(0,1), (1,0), (t,1+t), (1+t,t)\}$ has equation $y = x+1$.
$\beta_2 = \{(0,t), (1,1+t), (t,0), (1+t,1)\}$ has equation $y = x+t$.
$\beta_3 = \{(0,1+t), (1,t), (t,1), (1+t,0)\}$ has equation $y = x+1+t$.

The parallel class with slope t:

$\gamma_0 = \{(0,0),(1,t),(t,1+t),(1+t,1)\}$ has equation $y = tx$.

$\gamma_1 = \{(0,1),(1,1+t),(t,t),(1+t,0)\}$ has equation $y = tx + 1$.

$\gamma_2 = \{(0,t),(1,0),(t,1),(1+t,1+t)\}$ has equation $y = tx + t$.

$\gamma_3 = \{(0,1+t),(1,1),(t,0),(1+t,t)\}$ has equation $y = tx + 1 + t$.

The parallel class with slope $1 + t$:

$\delta_0 = \{(0,0),(1,1+t),(t,1),(1+t,t)\}$ has equation $y = (1+t)x$.

$\delta_1 = \{(0,1),(1,t),(t,0),(1+t,1+t)\}$ has equation $y = (1+t)x + 1$.

$\delta_2 = \{(0,t),(1,1),(t,1+t),(1+t,0)\}$ has equation $y = (1+t)x + t$.

$\delta_3 = \{(0,1+t),(1,0),(t,t),(1+t,1)\}$ has equation $y = (1+t)x + 1 + t$.

The parallel class with infinite slope:

$\epsilon_0 = \{(0,0),(0,1),(0,t),(0,1+t)\}$ has equation $x = 0$.

$\epsilon_1 = \{(1,0),(1,1),(1,t),(1,1+t)\}$ has equation $x = 1$.

$\epsilon_2 = \{(t,0),(t,1),(t,t),(t,1+t)\}$ has equation $x = t$.

$\epsilon_3 = \{(1+t,0),(1+t,1),(1+t,t),(1+t,1+t)\}$ has equation $x = 1 + t$.   ∎

Since we have a method for constructing affine planes, we now wish to describe how to construct a projective plane from a given affine plane, and vice versa.

In euclidean plane geometry it is often convenient to postulate the existence of certain idealized "points at infinity" and an ideal "line at infinity" that contains the ideal points; parallel lines are said to intersect at one of the points at infinity. This interpretation allows certain theorems to be stated without treating parallel lines as a special case. Such a view was first put forth by the great mathematician and astronomer Johann Kepler (1571–1630), and the concept was extended in 1822 by the French geometer Jean Victor Poncelet (1788–1867), who is credited with the formal development of projective geometry.

Given an affine plane $G = (P, L)$, we define a relation $\sim$ on L by

$$\alpha \sim \beta \leftrightarrow (\alpha = \beta \text{ or } \alpha \text{ and } \beta \text{ are parallel})$$

Clearly, $\sim$ is an equivalence relation on L; its equivalence classes are the parallel classes of the affine plane G. We form a projective plane $G' = (P', L')$, where P' is obtained from P by adding a set of additional points, one corresponding to each equivalence class of the relation $\sim$. Corresponding to the equivalence class $[\alpha]$, we add the new point $p(\alpha)$; for each $\beta \in [\alpha]$, we define the line $\beta' \in L'$ by

$$\beta' = \beta \cup \{p(\alpha)\}$$

Thus, $p(\alpha) = p(\beta)$ if and only if $\alpha \sim \beta$; moreover, given that α and β are parallel in G, then in G' the lines α' and β' intersect at the point $p(\alpha)$. So

$$P' = P \cup P^*$$

and

$$L' = \{\beta' \mid \beta \in L\} \cup \{P^*\}$$

where $P^* = \{p(\alpha) \mid \alpha \in L\}$. For historical reasons such as those mentioned, we sometimes call the elements of P^* "points at infinity" and P^* the "line at infinity."

Let's check the axioms to make sure that G' is a projective plane. For A1, let p_1 and p_2 be two distinct points of G'. If $p_1, p_2 \in \alpha \in L$, then p_1 and p_2 both belong to $\alpha' \in L'$; furthermore, this is the only line of G' containing both p_1 and p_2. If $p_1, p_2 \in P^*$, then P^* is the only line of G' containing both p_1 and p_2. In the last case, suppose that $p_1 \in P$ and $p_2 \in P^*$, say, $p_2 = p(\alpha)$ for some $\alpha \in L$. If $p_1 \in \alpha$, then α' is the unique line of G' containing both p_1 and p_2. Otherwise, by A3, there is a unique line $\beta \in L$ such that $p_1 \in \beta$ and α is parallel with β; then β' is the unique line determined by p_1 and p_2. So G' satisfies A1.

It is obvious that G' satisfies A2. For A3', consider two distinct lines of G'. First suppose one of them is α' and the other is β', where $\alpha, \beta \in L$. If α and β are parallel, then $\alpha' \cap \beta' = p(\alpha)$. If α and β intersect at $p \in P$, then $\alpha' \cap \beta' = \{p\}$ also. Second, if one of the lines is α' and the other is P^*, then they intersect at $p(\alpha)$. So A3' is satisfied, and G' is a projective plane.

■ *Example 4.1.3* Let $G = (P, L)$ be the affine plane $EG(3)$. Construct the projective plane $G' = (P', L')$.

Solution
Recall that $P = \mathbb{Z}_3 \times \mathbb{Z}_3$ and that

$$L = \{\alpha_0, \alpha_1, \alpha_2, \beta_0, \beta_1, \beta_2, \gamma_0, \gamma_1, \gamma_2, \delta_0, \delta_1, \delta_2\}$$

where the 12 lines are grouped into the 4 parallel classes $\{\alpha_0, \alpha_1, \alpha_2\}$, $\{\beta_0, \beta_1, \beta_2\}$, $\{\gamma_0, \gamma_1, \gamma_2\}$, and $\{\delta_0, \delta_1, \delta_2\}$. We let $p = p(\alpha_0)$, $q = p(\beta_0)$, $r = p(\gamma_0)$, and $s = p(\delta_0)$; then

$$\alpha_i' = \alpha_i \cup \{p\} \qquad \beta_i' = \beta_i \cup \{q\} \qquad \gamma_i' = \gamma_i \cup \{r\} \qquad \delta_i' = \delta_i \cup \{s\}$$

for $i = 0, 1, 2$. So

$$P' = P \cup \{p, q, r, s\}$$

and

$$L' = \{\alpha_0', \alpha_1', \alpha_2', \beta_0', \beta_1', \beta_2', \gamma_0', \gamma_1', \gamma_2', \delta_0', \delta_1', \delta_2'\} \cup \{\{p, q, r, s\}\}$$

Note that G' has 13 points and 13 lines, each line contains 4 points, and each point is on 4 lines. ∎

Given an affine plane $G = (P, L)$, we have a method for constructing a projective plane $G' = (P', L')$. Conversely, suppose a projective plane $G' = (P', L')$ is given. Then we may let α be any line of G' and define

$$P = P' - \alpha$$

and

$$L = \{\beta - \alpha \mid \beta \in L' \text{ and } \beta \neq \alpha\}$$

Roughly speaking, $G = (P, L)$ is the geometry obtained by removing α and its points from G'. Then we can show that G is an affine plane; see Exercise 4. For example, let G' be the Fano projective plane consisting of the set of seven points $P' = \{p_1, p_2, p_3, p_4, p_5, p_6, p_7\}$ and the set of seven lines $L' = \{\alpha_1, \alpha_2, \alpha_3, \alpha_4, \alpha_5, \alpha_6, \alpha_7\}$, where $\alpha_1 = \{p_1, p_2, p_4\}$, $\alpha_2 = \{p_1, p_3, p_6\}$, $\alpha_3 = \{p_1, p_5, p_7\}$, $\alpha_4 = \{p_2, p_3, p_5\}$, $\alpha_5 = \{p_2, p_6, p_7\}$, $\alpha_6 = \{p_3, p_4, p_7\}$, and $\alpha_7 = \{p_4, p_5, p_6\}$. Let

$$P = P' - \alpha_7 = \{p_1, p_2, p_3, p_7\}$$

and

$$
\begin{aligned}
L &= \{\alpha_1 - \alpha_7, \alpha_2 - \alpha_7, \alpha_3 - \alpha_7, \alpha_4 - \alpha_7, \alpha_5 - \alpha_7, \alpha_6 - \alpha_7\} \\
&= \{\{p_1, p_2\}, \{p_1, p_3\}, \{p_1 p_7\}, \{p_2, p_3\}, \{p_2, p_7\}, \{p_3, p_7\}\}
\end{aligned}
$$

Note that this is essentially the affine plane of Example 4.1.1, part (b), except that the points have been renamed.

Therefore, there is a natural correspondence between affine planes and projective planes, so that we may choose to focus our study on either kind. For the remainder of this section, we concentrate on projective planes.

The projective plane constructed from the affine plane $EG(n)$ has $n^2 + n + 1$ points and $n^2 + n + 1$ lines for some integer $n \geq 2$; furthermore, each line contains $n + 1$ points and each point is on $n + 1$ lines. We wish to show that this result holds for any finite projective plane.

LEMMA 4.1.3   Let $G = (P, L)$ be a projective plane and let $\alpha_1, \alpha_2 \in L$. Then there is a bijection from α_1 to α_2; that is, any two lines have the same number of points.

Proof   Let q be a point not on either α_1 or α_2; let

$$L_q = \{\beta \in L \mid q \in \beta\}$$

Also, for distinct points p_1 and p_2, let $p_1 p_2$ denote the line through them. Define $h : \alpha_1 \to L_q$ and $g : L_q \to \alpha_2$ by

$$h(p) = pq \quad \text{and} \quad g(\beta) = \beta \cap \alpha_2$$

We claim that both h and g are bijections, in which case so is $f : \alpha_1 \to \alpha_2$ where $f = g \circ h$.

To show that h is one-to-one, let p_1 and p_2 be distinct points of α_1. Then $p_1 q$ and $p_2 q$ are distinct lines; hence, $h(p_1) \neq h(p_2)$. To show that h is onto, let $\beta \in L_q$. For $p = \alpha_1 \cap \beta$, note that $h(p) = pq = \beta$, so that h is onto.

To see that g is one-to-one, let β_1 and β_2 be distinct lines through q. Then $\beta_1 \cap \alpha_2 \neq \beta_2 \cap \alpha_2$, so that $g(\beta_1) \neq g(\beta_2)$. To see that g is onto, let $p \in \alpha_2$. Then $\beta = pq$ is such that $g(\beta) = p$. $\qquad\square$

It follows from Lemma 4.1.3 that if G is a finite projective plane, then each line contains $n + 1$ points for some positive integer n. The number n is called the **order** of G.

THEOREM 4.1.4 Let $G = (P, L)$ be a finite projective plane of order n. Then

$$|P| = |L| = n^2 + n + 1$$

and every point is on $n + 1$ lines.

Proof Let $q \in P$ and $\alpha \in L$ such that $q \notin \alpha$. As in the proof of the lemma, define $g : L_q \to \alpha$ by $g(\beta) = \beta \cap \alpha$. Since g is a bijection, we have $|L_q| = |\alpha| = n + 1$. Since q is an arbitrary point, this shows that every point is on $n + 1$ lines.

Suppose $\alpha = \{p_0, p_1, \ldots, p_n\}$ and let $\beta_i = qp_i$ (the line determined by q and p_i), $0 \leq i \leq n$. Note that each point of G other than q is on exactly one of the lines β_i, because $L_q = \{\beta_0, \beta_1, \ldots, \beta_n\}$, whereas q is on each β_i. Hence,

$$|P| = |\beta_0| + |\beta_1| + \cdots + |\beta_n| - n = (n + 1)^2 - n = n^2 + n + 1$$

Now suppose that $|P| = v$ and $|L| = b$. Count the number t of pairs (p, α) with $p \in P$, $\alpha \in L$, and $p \in \alpha$. Since each point is on $n + 1$ lines, we have $t = v(n + 1)$; since each line contains $n + 1$ points, we have $t = (n + 1)b$. Hence, $v = b$. $\qquad\square$

The projective plane constructed from $EG(F)$, where F is a field, is called a **field plane**, and it has order n if F is the finite field of order n. Thus, there exists a projective plane of order n whenever $n = p^k$ for some prime p and positive integer k. A long-standing conjecture is that these are the only possible orders.

The first value of n that is not a prime power is $n = 6$; it is known that there is no projective plane of order 6. In fact, we have the following result, due to R. H. Bruck and H. J. Ryser [The nonexistence of certain finite projective planes, *Canadian Journal of Mathematics* 1 (1949): 88–93]:

> **THEOREM 4.1.5**   Let n be an integer, $n \geq 2$, such that $n \bmod 4 \in \{1, 2\}$. If there is a projective plane of order n, then
>
> $$n = a^2 + b^2$$
>
> for some integers a and b. $\qquad\qquad\qquad\qquad\qquad\qquad\qquad\qquad\qquad\qquad\qquad$ ☐

Theorem 4.1.5 is a corollary to a more general theorem concerning the existence of symmetric balanced incomplete block designs; this theorem is given in Section 4.4.

Note that 6 mod 4 = 2 but that 6 cannot be expressed as the sum of two square integers; hence, there is no projective plane of order 6. Similarly, we can show that there is no projective plane of order 14, or 21, or 22, and so on.

The first value of n that is not a prime power and is not excluded by Theorem 4.1.5 is $n = 10$. (A projective plane of order 10 would have 111 points and 111 lines, with each line containing 11 points and each point on 11 lines.) In late 1988, C. Lam, L. Thiel, S. Swiercz, and J. McKay of Concordia University in Montreal, Canada, announced the conclusion of an extensive computer search showing that no projective plane of order 10 exists. Their work used a clever combination of theoretical and computer analysis, similar to the work of Appel and Haken on the four-color problem. Lam estimates that their proof involved about two orders of magnitude more computing power than did the proof of the four-color theorem—the final stage of the computation was performed on a Cray supercomputer, running on a low-priority, part-time basis from September 1986 through November 1988! Unfortunately, it does not appear that their techniques can be extended to settle the conjecture for $n = 12$, because a prohibitive amount of computation would be required. (For more information, see Barry Cipra's article in *Science* (December 1988): 1507–1508; and also Lam's article, The search for a finite projective plane of order 10, *American Mathematical Monthly* 98 (1991): 305–318.)

Lam and his colleagues have also identified the four different projective planes of order 9. In order to understand this statement, we need to say when two projective planes are considered "different." Just as for graphs and other mathematical structures, there is a notion of "isomorphism" for geometries. Let $G_1 = (P_1, L_1)$ and $G_2 = (P_2, L_2)$ be two geometries. We say that G_1 and G_2 are **isomorphic** provided there is a bijection $\phi : P_1 \rightarrow P_2$ such that the following two conditions are satisfied:

1. For each $\alpha \in L_1$, $\{\phi(p) \mid p \in \alpha\} \in L_2$.
2. For each $\beta \in L_2$, $\{\phi^{-1}(p) \mid p \in \beta\} \in L_1$.

The function ϕ is called an **isomorphism**; we describe properties 1 and 2 by saying that an isomorphism preserves incidence. Property 1 says that the image under ϕ of any line α of G_1 is a line of G_2; conversely, property 2 says that the inverse image under ϕ of any line β of G_2 is a line of G_1.

■ *Example 4.1.4* We are now ready to give an alternate construction for the field plane of order 3; that is, we construct a geometry $G = (P, L)$ and show that it is isomorphic to the field plane $G' = (P', L')$ of Example 4.1.3. We define $G = (P, L)$ by

$$P = \{(x, y, z) \in \mathbb{Z}_3 \times \mathbb{Z}_3 \times \mathbb{Z}_3 \mid ((x, y) \in \mathbb{Z}_3 \times \mathbb{Z}_3 \text{ and } z = 1)$$
$$\text{or } (x = 1 \text{ and } y \in \mathbb{Z}_3 \text{ and } z = 0)$$
$$\text{or } (x = 0 \text{ and } y = 1 \text{ and } z = 0)\}$$
$$= \{(x, y, 1) \mid x, y \in \mathbb{Z}_3\} \cup \{(1, y, 0) \mid y \in \mathbb{Z}_3\} \cup \{(0, 1, 0)\}$$

and

$$L = \{\{(x, y, z) \in P \mid ax + by + cz = 0\} \mid a, b, c \in \mathbb{Z}_3 \text{ and } (a, b, c) \neq (0, 0, 0)\}$$

Note that a line $\alpha \in L$ is determined by a homogeneous linear (nondegenerate) equation

$$ax + by + cz = 0$$

If $a = b = 0$, then $c \neq 0$ and we obtain the line whose equation is $z = 0$:

$$\{(0, 1, 0), (1, 0, 0), (1, 1, 0), (1, 2, 0)\}$$

This line corresponds to the "line at infinity" and its four points correspond to the "points at infinity" of the field plane. Now suppose that $(a, b) \neq (0, 0)$. If $z = 1$, then $(x, y, 1) \in \alpha$ if and only if (x, y) is a point on the line of $EG(3)$ with equation $ax + by + c = 0$. So the point $(x, y, 1)$ corresponds to the point (x, y) of $EG(3)$. If $z = 0$, then $(x, y, 0) \in \alpha$ if and only if (x, y) satisfies $ax + by = 0$. If $b = 0$, then $(0, 1, 0) \in \alpha$, so $(0, 1, 0)$ corresponds to the point at infinity for the parallel class of $EG(3)$ of lines with infinite slope. On the other hand, if $b \neq 0$, then $(1, -ab^{-1}, 0) \in \alpha$. So $(1, -ab^{-1}, 0)$ corresponds to the point at infinity for the parallel class of $EG(3)$ of lines with slope $-ab^{-1}$.

The preceding analysis shows us how to define an isomorphism $\phi : P' \to P$. For each $(x, y) \in \mathbb{Z}_3 \times \mathbb{Z}_3$, we let

$$\phi(x, y) = (x, y, 1)$$

In addition, we define

$$\phi(p) = (1, 0, 0) \quad \phi(q) = (1, 1, 0) \quad \phi(r) = (1, 2, 0) \quad \phi(s) = (0, 1, 0)$$

It is obvious that ϕ is a bijection. (You are asked to check that ϕ is an isomorphism. For example,

$$\phi(\alpha_0') = \phi(\{(0, 0), (1, 0), (2, 0), p\}) = \{(0, 0, 1), (1, 0, 1), (2, 0, 1), (1, 0, 0)\}$$

is the line of G with equation $y = 0$. Similarly,

$$\phi(\beta_1') = \phi(\{(0,1),(1,2),(2,0),q\}) = \{(0,1,1),(1,2,1),(2,0,1),(1,1,0)\}$$

is the line of G with equation $x + 2y + z = 0$.)

It can be checked directly that G is a projective plane; see Exercise 5. Alternately, since G' is a projective plane and G and G' are isomorphic, it follows that G is a projective plane; see Exercise 6. ∎

The idea of Example 4.1.4 can be generalized. Let F be a field. Then the **field plane over F** is the geometry $G = (P, L)$ defined by

$$P = \{(x,y,1) \mid x,y \in F\} \cup \{(1,y,0) \mid y \in F\} \cup \{(0,1,0)\}$$

and

$$L = \{\{(x,y,z) \in P \mid ax + by + cz = 0\} \mid a,b,c \in F \text{ and } (a,b,c) \neq (0,0,0)\}$$

Exercise 12 asks you to show that G is a projective plane. If F is finite with order $n = p^k$, where p is a prime and k is a positive integer, then G is called the **field plane of order n**.

For $n \in \{2,3,4,5,7,8\}$, there is a unique projective plane of order n, up to isomorphism, namely, the field plane of order n. As mentioned, Clement Lam and his colleagues found four nonisomorphic projective planes of order 9. Exercise 17 shows you how to construct a projective plane of order 9 that is not isomorphic to the field plane of this order.

As a consequence of the symmetry of Axioms A1 and A3′, projective planes obey the **principle of duality**. Suppose we have a statement about a projective plane. Roughly speaking, if we interchange the roles played by the terms *point* and *line*, then we obtain a **dual statement**. For example, the dual of Axiom A1 that two points are on a unique line is Axiom A3′ that two lines intersect at a unique point. The principle of duality says that a statement and its dual are either both true or both false. This implies that, if we prove a theorem about projective planes, then the dual statement of the theorem is also a theorem.

For instance, Lemma 4.1.3 states:

In a projective plane, any two lines contain the same number of points.

Thus, the dual of Lemma 4.1.3 is also a theorem:

In a projective plane, any two points are on the same number of lines.

To be more precise, let $G = (P, L)$ be a projective plane. The **dual** of G is the geometry $G^* = (P^*, L^*)$ defined by $P^* = L$ and

$$L^* = \{p^* \mid p \in P\}$$

where

$$p^* = \{\alpha \in L \mid p \in \alpha\}$$

In words, the points of G^* are the lines of G and a line of G^* is a set of all the lines concurrent at a given point of G. For example, let G be the Fano plane given by $P = \{p_1, p_2, \ldots, p_7\}$ and $L = \{\alpha_1, \alpha_2, \ldots, \alpha_7\}$, where $\alpha_1 = \{p_1, p_2, p_4\}$, $\alpha_2 = \{p_1, p_3, p_6\}$, $\alpha_3 = \{p_1, p_5, p_7\}$, $\alpha_4 = \{p_2, p_3, p_5\}$, $\alpha_5 = \{p_2, p_6, p_7\}$, $\alpha_6 = \{p_3, p_4, p_7\}$, and $\alpha_7 = \{p_4, p_5, p_6\}$. Then the lines of G^* are

$$p_1^* = \{\alpha_1, \alpha_2, \alpha_3\}, \quad p_2^* = \{\alpha_1, \alpha_4, \alpha_5\}, \quad p_3^* = \{\alpha_2, \alpha_4, \alpha_6\}, \quad p_4^* = \{\alpha_1, \alpha_6, \alpha_7\}$$
$$p_5^* = \{\alpha_3, \alpha_4, \alpha_7\}, \quad p_6^* = \{\alpha_2, \alpha_5, \alpha_7\}, \quad p_7^* = \{\alpha_3, \alpha_5, \alpha_6\}$$

In general, it can be shown that G^* is a projective plane and, if G has order n, then so does G^*; see Exercise 8.

For a nonempty set I, let $\{\alpha_i \mid i \in I\}$ be a collection of lines of G that are concurrent (intersect) at the point p. Then in G^* the points α_i, $i \in I$, are colinear, all lying on the line p^*. Similarly, if $\{p_i \mid i \in I\}$ is a collection of colinear points of G, say, $\{p_i \mid i \in I\} \subseteq \alpha \in L$, then the lines p_i^* are concurrent at α in G^*. Thus, the notions of "concurrent lines" and "colinear points" are dual.

To give a more complicated example of duality, let's define a **triangle** in a projective plane to be any set of three noncolinear points. Figure 4.1.2 shows two triangles $T_1 = \{q_1, r_1, s_1\}$ and $T_2 = \{q_2, r_2, s_2\}$. These two triangles are said to be **in perspective from a point**, namely, p, because the lines $q_1 q_2$, $r_1 r_2$, and $s_1 s_2$ are concurrent at p. Similarly, these two triangles are said to be **in perspective through a line**, namely, α, because the intersection points $q_1 r_1 \cap q_2 r_2$, $q_1 s_1 \cap q_2 s_2$, and $r_1 s_1 \cap r_2 s_2$ are colinear, all of them on α. Two triangles that are in perspective from a point and also in perspective through a line are said to form a **Desargues configuration**, named after the French mathematician Gérard Desargues (1593–1662).

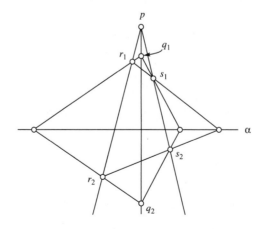

Figure 4.1.2 A Desargues configuration.

Desargues discovered the following result in the case of $EG(\mathbb{R})$.

THEOREM 4.1.6 Let T_1 and T_2 be two triangles of a field plane G. If T_1 and T_2 are in perspective from a point, then they are in perspective through a line (and hence form a Desargues configuration). □

The proof of Theorem 4.1.6 is not too difficult and is left for you to develop in Exercise 16. To aid in our discussion of this theorem and its dual, let us say that a projective plane G is **desarguesian** provided G satisfies the property that, whenever two triangles are in perspective from a point, then they are in perspective through a line. We may then restate Theorem 4.1.6 in the following form:

Every field plane is desarguesian.

Let $G = (P, L)$ be a projective plane and let $T_1 = \{q_1, r_1, s_1\}$ and $T_2 = \{q_2, r_2, s_2\}$ be two triangles. Suppose that T_1 and T_2 are in perspective from the point p. Let's see what this implies over in the dual projective plane G^*. Let $\beta_1 = q_1 r_1$, $\gamma_1 = q_1 s_1$, and $\delta_1 = r_1 s_1$ be the "sides" of T_1, and let $\beta_2 = q_2 r_2$, $\gamma_2 = q_2 s_2$, and $\delta_2 = r_2 s_2$ be the sides of T_2. Then $T_1^* = \{\beta_1, \gamma_1, \delta_1\}$ and $T_2^* = \{\beta_2, \gamma_2, \delta_2\}$ are triangles in G^*. Since $\beta_1 \cap \gamma_1 = q_1$ and $\beta_2 \cap \gamma_2 = q_2$ in G, we have in G^* that

$$\beta_1 \gamma_1 = q_1^* \qquad \text{and} \qquad \beta_2 \gamma_2 = q_2^* \qquad \text{so} \qquad \beta_1 \gamma_1 \cap \beta_2 \gamma_2 = q_1 q_2$$

In a similar fashion, note that

$$\beta_1 \delta_1 = r_1^* \qquad \text{and} \qquad \beta_2 \delta_2 = r_2^* \qquad \text{so} \qquad \beta_1 \delta_1 \cap \beta_2 \delta_2 = r_1 r_2$$

and

$$\gamma_1 \delta_1 = s_1^* \qquad \text{and} \qquad \gamma_2 \delta_2 = s_2^* \qquad \text{so} \qquad \gamma_1 \delta_1 \cap \gamma_2 \delta_2 = s_1 s_2$$

Now, since in G the lines $q_1 q_2$, $r_1 r_2$, and $s_1 s_2$ are concurrent at p, we have in G^* that the points $q_1 q_2$, $r_1 r_2$, and $s_1 s_2$ are colinear, all lying on the line p^*. Hence, in G^*, the triangles T_1^* and T_2^* are in perspective through the line p^*.

This means that the dual of the statement that two triangles are in perspective from a point is the statement that the two triangles are in perspective through a line. Therefore, it follows that the dual of Theorem 4.1.6 is:

Let T_1 and T_2 be two triangles of a field plane G. If T_1 and T_2 are in perspective through a line, then they are in perspective from a point.

Note that this is also the converse of Theorem 4.1.6.

EXERCISE SET 4.1

1. Show that a projective plane must have at least seven points.
2. Consider the affine plane $EG(2)$.
   (a) Find the points and lines of this plane.

(b) Construct the field plane of order 2.

(c) Show that the field plane of order 2 and the Fano plane of Figure 4.1.1 are isomorphic.

3. Construct the projective plane of order 4. (Hint: Use the affine plane $EG(4)$ as constructed in Example 4.1.2.)

4. Given a projective plane $G' = (P', L')$, let α be any line of G'; define

$$P = P' - \alpha$$

and

$$L = \{\beta - \alpha \mid \beta \in L' \text{ and } \beta \neq \alpha\}$$

Roughly speaking, $G = (P, L)$ is the geometry obtained by removing the points of α from G'. Show that G is an affine plane.

5. Verify directly that the geometry $G = (P, L)$ of Example 4.1.4 is a projective plane.

6. Let $G_1 = (P_1, L_1)$ and $G_2 = (P_2, L_2)$ be two geometries and let $\phi : P_1 \to P_2$ be an isomorphism. Prove:

(a) If G_1 is an affine plane, then so is G_2.

(b) If G_1 is a projective plane, then so is G_2.

7. The Learning Lab at a small college has seven mathematics tutors and must schedule three of them for each evening of any given week, in such a way that each tutor works the same number of evenings. Design such a schedule.

8. Let $G = (P, L)$ be a projective plane and let $G^* = (P^*, L^*)$ be the dual of G.

(a) Show that G^* is a projective plane.

(b) If G has order n, show that G^* also has order n.

9. Show that the Fano plane and its dual are isomorphic.

10. Show that there is a unique projective plane of order 2, up to isomorphism.

11. Show that there is a unique projective plane of order 3, up to isomorphism.

12. The purpose of this exercise is to generalize Example 4.1.4 and Exercise 5, giving an alternate construction for the field plane over the field F. Define $G = (P, L)$ as follows. First,

$$P = \{(x, y, 1) \mid x, y \in F\} \cup \{(1, y, 0) \mid y \in F\} \cup \{(0, 1, 0)\}$$

Alternately, we may take $P = (F \times F \times F) - \{(0, 0, 0)\}$, with the understanding that the points (x, y, z) and (dx, dy, dz) are equivalent for any $d \in F - \{0\}$. Second,

$$L = \{\{(x, y, z) \mid ax + by + cz = 0\} \mid a, b, c \in F \text{ and } (a, b, c) \neq (0, 0, 0)\}$$

(a) Verify that G satisfies Axiom A1 by showing that distinct points (x_1, y_1, z_1) and (x_2, y_2, z_2) determine the unique line with equation

$$(y_1 z_2 - y_2 z_1)x + (x_2 z_1 - x_1 z_2)y + (x_1 y_2 - x_2 y_1)z = 0$$

(b) Verify that G satisfies Axiom A2.

(c) Verify that G satisfies Axiom A3' by showing that distinct lines with equations

$$a_1 x + b_1 y + c_1 z = 0 \qquad \text{and} \qquad a_2 x + b_2 y + c_2 z = 0$$

determine the unique point equivalent to $(b_1 c_2 - b_2 c_1, a_2 c_1 - a_1 c_2, a_1 b_2 - a_2 b_1)$.

13. Let $G = (P, L)$ be the field plane over the field F (as constructed in Exercise 12) and let $p_1 = (x_1, y_1, z_1)$ and $p_2 = (x_2, y_2, z_2)$ be distinct points of G. Show that the line determined by p_1 and p_2 is

$$\{(k_1 x_1 + k_2 x_2, k_1 y_1 + k_2 y_2, k_1 z_1 + k_2 z_2) \mid k_1, k_2 \in F\}$$

14. Let $G = (P, L)$ be the field plane over the field F. Given an invertible 3 by 3 matrix M, define $\phi : P \to P$ by

$$\phi(x, y, z) = (x, y, z)M$$

Prove that ϕ is an isomorphism.

15. In a projective plane, define a **quadrangle** to be a set of four points, no three of which are colinear. Given the quadrangle $\{p, q, r, s\}$, the lines pq and rs are called **opposite sides**, as are pr and qs, and ps and qr; the three points $pq \cap rs$, $pr \cap qs$, and $ps \cap qr$ are called **diagonal points**.

   (a) Let G be the field plane of order 2^k, where k is a positive integer. Prove that the diagonal points of any quadrangle are colinear. (Hint: By using the result of Exercise 14, it suffices to prove that the diagonal points of $\{(1, 0, 0), (0, 1, 0), (0, 0, 1), (x_0, y_0, z_0)\}$ are colinear. Why?)

   (b) Show that the field plane of order 3 contains a quadrangle whose diagonal points are not colinear.

16. Prove Theorem 4.1.6 (or its dual). (Hint: Let $G = (P, L)$ be a field plane and let $T_1 = \{q_1, r_1, s_1\}$ and $T_2 = \{q_2, r_2, s_2\}$ be two triangles that are in perspective from the point p. By using the result of Exercise 14, we may assume that $q_1 = (1, 0, 0)$, $r_1 = (0, 1, 0)$, and $s_1 = (0, 0, 1)$. (Why?) Let $p = (x_0, y_0, z_0)$; if $p \in q_1 r_1$, there is nothing to prove, so we may assume that $z_0 \neq 0$. Similarly, we may assume that $x_0 \neq 0$ and $y_0 \neq 0$. Since $q_2 \in q_1 p$, there exists $h \in F - \{0\}$ such that $q_2 = (1 + h x_0, h y_0, h z_0)$, using the result of Exercise 13. Similarly, $r_2 = (j x_0, 1 + j y_0, j z_0)$ and $s_2 = (k x_0, k y_0, 1 + k z_0)$ for some $j, k \in F - \{0\}$. With this notation, show that $q_1 r_1 \cap q_2 r_2 = (j, -h, 0)$, $q_1 s_1 \cap q_2 s_2 = (-k, 0, h)$, and $r_1 s_1 \cap r_2 s_2 = (0, k, -j)$ and that these three points are each on the line $hx + jy + kz = 0$.)

17. Let $G' = (P', L')$, where

$$P' = \{p_0, p_1, \ldots, p_{12}\} \cup \{q_0, q_1, \ldots, q_{12}\} \cup \{r_0, r_1, \ldots, r_{12}\} \cup \{s_0, s_1, \ldots, s_{12}\}$$
$$\cup \{t_0, t_1, \ldots, t_{12}\} \cup \{u_0, u_1, \ldots, u_{12}\} \cup \{v_0, v_1, \ldots, v_{12}\}$$

and

$$L' = \{\alpha_0, \alpha_1, \ldots, \alpha_{12}\} \cup \{\beta_0, \beta_1, \ldots, \beta_{12}\} \cup \{\gamma_0, \gamma_1, \ldots, \gamma_{12}\}$$
$$\cup \{\delta_0, \delta_1, \ldots, \delta_{12}\} \cup \{\epsilon_0, \epsilon_1, \ldots, \epsilon_{12}\} \cup \{\zeta_0, \zeta_1, \ldots, \zeta_{12}\}$$
$$\cup \{\eta_0, \eta_1, \ldots, \eta_{12}\}$$

where

$$\alpha_i = \{p_{0+i}, p_{1+i}, p_{3+i}, p_{9+i}, q_{0+i}, r_{0+i}, s_{0+i}, t_{0+i}, u_{0+i}, v_{0+i}\}$$
$$\beta_i = \{p_{0+i}, q_{1+i}, q_{8+i}, s_{3+i}, s_{11+i}, t_{2+i}, t_{5+i}, t_{6+i}, v_{7+i}, v_{9+i}\}$$
$$\gamma_i = \{p_{0+i}, r_{1+i}, r_{8+i}, t_{7+i}, t_{9+i}, u_{3+i}, u_{11+i}, v_{2+i}, v_{5+i}, v_{6+i}\}$$
$$\delta_i = \{p_{0+i}, q_{7+i}, q_{9+i}, s_{1+i}, s_{8+i}, u_{2+i}, u_{5+i}, u_{6+i}, v_{3+i}, v_{11+i}\}$$
$$\epsilon_i = \{p_{0+i}, q_{2+i}, q_{5+i}, q_{6+i}, r_{3+i}, r_{11+i}, t_{1+i}, t_{8+i}, u_{7+i}, u_{9+i}\}$$
$$\zeta_i = \{p_{0+i}, r_{7+i}, r_{9+i}, s_{2+i}, s_{5+i}, s_{6+i}, t_{3+i}, t_{11+i}, u_{7+i}, u_{8+i}\}$$
$$\eta_i = \{p_{0+i}, q_{3+i}, q_{11+i}, r_{2+i}, r_{5+i}, r_{6+i}, s_{7+i}, s_{9+i}, v_{1+i}, v_{8+i}\}$$

(The subscripts on the points are always taken mod 13.) It can be shown that G is a projective plane of order 9. Now consider the triangles $T_1 = \{p_1, q_1, r_8\}$ and $T_2 = \{p_3, q_8, t_7\}$. Show that these triangles are in perspective from a point but are not in perspective through a line. Hence, G is not isomorphic to the field plane of order 9.

18. Find the dual statement A2* of Axiom A2. Given a system $G = (P, L)$ of points and lines satisfying Axioms A1 and A3', show that A2 and A2* are equivalent.

19. Consider the field plane of order 5 (as constructed by the method of Exercise 12).

    (a) Find the equation of the line containing the points $(3, 4, 1)$ and $(2, 2, 1)$.
    (b) Find the equation of the line containing the points $(3, 2, 1)$ and $(1, 4, 0)$.
    (c) Find the points on the line (with equation) $x + 2y + 3z = 0$.
    (d) Find the point of intersection of the lines $3x + 4y + z = 0$ and $2x + 2y + z = 0$.
    (e) Find the point of intersection of the lines $3x + 2y + z = 0$ and $x + 4y = 0$.

4.2 *LATIN SQUARES*

We began our study of combinatorics in Section 1.1 by considering the famous combinatorial problem known as the "problem of the 36 officers," studied in 1782 by Euler. His work on this problem led Euler to generalize the ideas involved by defining the term *Latin square*. Recall that for a positive integer n, a **Latin square of order** n is an arrangement of n distinct symbols, say, the integers $0, 1, \ldots, n - 1$, into a square array of n rows and n columns such that each number occurs exactly once in any row and in any column. Given a Latin square R of order n, we let r_{ij} denote the entry in row i and column j, where the rows and columns are indexed using some finite set of n elements; again let us use $\{0, 1, \ldots, n - 1\}$. Let S be another Latin square of order n (on the same set of symbols) with typical entry s_{ij}. We say that R and S are **orthogonal Latin squares** provided the n^2 ordered pairs

$$(r_{ij}, s_{ij})$$

are distinct for $0 \leq i, j \leq n - 1$, thus giving all n^2 elements in the product set $\{0, 1, \ldots, n - 1\} \times \{0, 1, \ldots, n - 1\}$. In this case, the n by n array (R, S) of ordered pairs (r_{ij}, s_{ij}) is called a **Graeco-Latin square of order n**. More generally, k Latin squares $A_1, A_2, \ldots, A_k$ of order n are said to be **pairwise-orthogonal** (or **mutually orthogonal**) provided A_i and A_j are orthogonal whenever $1 \leq i < j \leq k$. In this section we are interested in the following problem: Given $n \geq 2$, what is the maximum value of k such that there exists a set of k pairwise-orthogonal Latin squares of order n? (Recall that the problem of the 36 officers is equivalent to: Does there exist a pair of orthogonal Latin squares of order 6?)

Euler gave methods for constructing a pair of orthogonal Latin squares of order $n \geq 2$ whenever $n \bmod 4 \in \{0, 1, 3\}$ and conjectured that there does not exist a pair of orthogonal Latin squares of order n when $n \bmod 4 = 2$. It is easy to verify Euler's conjecture when $n = 2$ and, around 1900, G. Tarry showed that Euler's conjecture is true for $n = 6$. The cases $n = 10, 14, 18, 22, \ldots$ remained unsettled until around 1960, when R. C. Bose, E. T. Parker, and S. S. Shrikhande [Further results on the construction of mutually orthogonal Latin squares and the falsity of Euler's conjecture, *Canadian Journal of Mathematics* 12 (1960): 189–203] showed how to construct a pair of orthogonal Latin squares of order n whenever $n \geq 10$ and $n \bmod 4 = 2$. This amazing discovery showed that Euler's conjecture is false for $n > 6$, so that $n = 2$ and $n = 6$ are the only cases when a pair of orthogonal Latin squares fails to exist. However, although this result completely settles Euler's question, the more general problem mentioned in the preceding paragraph remains largely unsolved. For instance, it is not known whether there exists a set of three pairwise-orthogonal Latin squares of order 10. (A Graeco-Latin square of order 10 is shown in Figure 1.1.5.)

It is a simple matter to find a pair of orthogonal Latin squares of order 3 (see Exercise 3 of Exercise Set 1.1). Figure 4.2.1 shows three pairwise-orthogonal Latin squares of order 4. We claim that there does not exist a set of four pairwise-orthogonal Latin squares of order 4.

We proceed by contradiction: suppose that $\{A_1, A_2, A_3, A_4\}$ is such a set. Just as for the three Latin squares in the figure, we may assume that each A_i has first row $0, 1, 2, 3$. The reasoning for this is as follows: Suppose that A is a Latin square of order 4; in A replace each occurrence of the symbol a_{0j} by j for each j, $0 \leq j \leq 3$. Call the resulting square A'. It is an easy exercise to verify that A' is a Latin square of order 4 with first row $0, 1, 2, 3$, and, furthermore, that if B is a Latin square of order 4 orthogonal to A, then B and A' are orthogonal. Now let a_i be the $(1, 0)$-entry in A_i, $1 \leq i \leq 4$. Note that $a_i \neq 0$ since each A_i has 0 in the $(0, 0)$-entry. Thus, since a_1, a_2, a_3, and a_4 each belong to $\{1, 2, 3\}$, we have $a_i = a_j$ for some i, j, $1 \leq i < j \leq 4$. But then the $(0, a_i)$-entries of A_i and A_j yield the symbol pair (a_i, a_i), as do the $(1, 0)$-entries, contradicting the assumption that A_i and A_j are orthogonal.

$$
\begin{array}{cccc}
0 & 1 & 2 & 3 \\
1 & 0 & 3 & 2 \\
2 & 3 & 0 & 1 \\
3 & 2 & 1 & 0
\end{array}
\qquad
\begin{array}{cccc}
0 & 1 & 2 & 3 \\
2 & 3 & 0 & 1 \\
3 & 2 & 1 & 0 \\
1 & 0 & 3 & 2
\end{array}
$$

$$
\begin{array}{cccc}
0 & 1 & 2 & 3 \\
3 & 2 & 1 & 0 \\
1 & 0 & 3 & 2 \\
2 & 3 & 0 & 1
\end{array}
$$

Figure 4.2.1   Three pairwise-orthogonal Latin squares of order 4.

Generalizing the argument in the last paragraph yields the following result.

THEOREM 4.2.1   If there exists a set of k pairwise-orthogonal Latin squares of order n, then $k < n$.   □

This theorem implies that the best we can do is find $n-1$ pairwise-orthogonal Latin squares of order n. R. C. Bose [On the application of the properties of Galois fields to the problem of construction of hyper-Graeco-Latin squares, *Sankyha* 3 (1938): 323–338] and, independently, F. W. Levi, found an important case when this upper bound is achieved, namely, whenever there is a projective plane of order n. In particular, when n is a prime power, we can use the structure of the affine plane $EG(n)$ to construct $n - 1$ pairwise-orthogonal Latin squares of order n.

Let F be the field of order n. Recall that $EG(n)$ has point set $F \times F$ and its line set is partitioned into $n + 1$ parallel classes with n lines in each. In particular, there are n lines with slope 0 and n lines with infinite slope; we do not use them for the present construction. What we do use are the lines with finite, nonzero slope—for $m, k \in F$ and $m \neq 0$, we let α_{mk} denote the line with equation

$$y = mx + k$$

For each $m \in F - \{0\}$, we use the lines α_{mk} with slope m to define a Latin square A_m of order n; the n elements of F are the symbols of A_m and are also used to index its rows and columns. For a given point (a, b), there is a unique $k \in F$ for which $(a, b) \in \alpha_{mk}$, namely, $k = b - ma$. Thus, it is natural to let $b - ma$ be the (a, b)-entry of A_m.

To verify that A_m is a Latin square, note the following:

$$(a, b_1) \neq (a, b_2) \rightarrow b_1 \neq b_2 \rightarrow b_1 - ma \neq b_2 - ma$$
$$(a_1, b) \neq (a_2, b) \rightarrow a_1 \neq a_2 \rightarrow b - ma_1 \neq b - ma_2$$

Thus, A_m has distinct elements in the row indexed by a and in the column indexed by b. We next verify for $m_1, m_2 \in F - \{0\}$, $m_1 \neq m_2$, that A_{m_1} and A_{m_2} are orthogonal.

Let $a, b, c, d \in F$ and suppose that $(b - m_1 a, b - m_2 a) = (d - m_1 c, d - m_2 c)$. Then

$$b - m_1 a = d - m_1 c \qquad \text{and} \qquad b - m_2 a = d - m_2 c$$
$$\rightarrow b - d = m_1 (a - c) \qquad \text{and} \qquad b - d = m_2 (a - c)$$
$$\rightarrow m_1 (a - c) = m_2 (a - c)$$
$$\rightarrow (m_1 - m_2)(a - c) = 0$$
$$\rightarrow a - c = 0$$

since $m_1 - m_2 \neq 0$. Thus, $a = c$. Then, since $b - d = m_1 (a - c) = 0$, we obtain $b = d$. This shows that $(a, b) = (c, d)$, and so A_{m_1} and A_{m_2} are orthogonal. We have just proved:

THEOREM 4.2.2 If $n = p^k$ for some prime p and positive integer k, then there exists a set of $n - 1$ pairwise-orthogonal Latin squares of order n. $\qquad\qquad\square$

As mentioned, Theorem 4.2.2 can be generalized: There exists a projective plane of order n if and only if there is a set of $n - 1$ pairwise-orthogonal Latin squares of order n. See Exercises 6 and 8.

■ *Example 4.2.1* We use the affine plane $EG(4)$ to construct a set of three pairwise-orthogonal Latin squares of order 4. This affine plane is discussed in Example 4.1.2. Recall that $F = \{0, 1, t, 1 + t\}$, where addition and multiplication are done modulo the prime polynomial $1 + t + t^2$. There are four lines in each of five parallel classes; those with nonzero, finite slope are listed below, where α_{mk} (or $\alpha_{m,k}$) denotes the line with equation $y = mx + k$:

$$\alpha_{10} = \{(0, 0), (1, 1), (t, t), (1 + t, 1 + t)\}$$
$$\alpha_{11} = \{(0, 1), (1, 0), (t, 1 + t), (1 + t, t)\}$$
$$\alpha_{1t} = \{(0, t), (1, 1 + t), (t, 0), (1 + t, 1)\}$$
$$\alpha_{1,1+t} = \{(0, 1 + t), (1, t), (t, 1), (1 + t, 0)\}$$

$$\alpha_{t0} = \{(0, 0), (1, t), (t, 1 + t), (1 + t, 1)\}$$
$$\alpha_{t1} = \{(0, 1), (1, 1 + t), (t, t), (1 + t, 0)\}$$
$$\alpha_{tt} = \{(0, t), (1, 0), (t, 1), (1 + t, 1 + t)\}$$
$$\alpha_{t,1+t} = \{(0, 1 + t), (1, 1), (t, 0), (1 + t, t)\}$$

$$\alpha_{1+t,0} = \{(0, 0), (1, 1 + t), (t, 1), (1 + t, t)\}$$
$$\alpha_{1+t,1} = \{(0, 1), (1, t), (t, 0), (1 + t, 1 + t)\}$$
$$\alpha_{1+t,t} = \{(0, t), (1, 1), (t, 1 + t), (1 + t, 0)\}$$
$$\alpha_{1+t,1+t} = \{(0, 1 + t), (1, 0), (t, t), (1 + t, 1)\}$$

We use the lines in the class with slope 1 to form A_1—its rows and columns are indexed 0, 1, t, $1 + t$, and the (a, b)-entry is $b - a$ (which is the same as $b + a$):

$$A_1 = \begin{array}{cccc} 0 & 1 & t & 1+t \\ 1 & 0 & 1+t & t \\ t & 1+t & 0 & 1 \\ 1+t & t & 1 & 0 \end{array}$$

The lines in the class with slope t are used to form A_t—its (a, b)-entry is $b - at = b + at$:

$$A_2 = \begin{array}{cccc} 0 & 1 & t & 1+t \\ t & 1+t & 0 & 1 \\ 1+t & t & 1 & 0 \\ 1 & 0 & 1+t & t \end{array}$$

Similarly, the lines with slope $1 + t$ are used to construct A_{1+t}—its (a, b)-entry is $b - a(1 + t) = b + a(1 + t)$:

$$A_3 = \begin{array}{cccc} 0 & 1 & t & 1+t \\ 1+t & t & 1 & 0 \\ 1 & 0 & 1+t & t \\ t & 1+t & 0 & 1 \end{array}$$

You can check that A_1, A_2, and A_3 are pairwise-orthogonal. Note that, with t replaced by 2 and $1 + t$ replaced by 3, that A_1, A_2, and A_3 are the same as the three pairwise-orthogonal Latin squares of order 4 shown in Figure 4.2.1. ∎

What if n is not a prime power? Suppose that

$$n = n_1 n_2$$

where $1 < n_1 < n_2 < n$. Furthermore, suppose that $A_1, A_2, \ldots, A_{m_1}$ are m_1 pairwise-orthogonal Latin squares of order n_1 and $B_1, B_2, \ldots, B_{m_2}$ are m_2 pairwise-orthogonal Latin squares of order n_2. By using the idea of **composition** (introduced in Exercises 4 and 6 of Exercise Set 1.1), we can then find m pairwise-orthogonal Latin squares $C_1, C_2, \ldots, C_m$ of order n, where $m = \min\{m_1, m_2\}$.

Let A be a Latin square of order n_1 and B be a Latin square of order n_2. The **composition** of A with B is the Latin square $C = A \circ B$ of order $n = n_1 n_2$ defined as follows: For $0 \le i, j \le n - 1$, let $c_{ij} = a_{uv} + n_1 b_{xy}$, where $u = i$ div n_2, $v = j$ div n_2, $x = i$ mod n_2, and $y = j$ mod n_2. For example, Figure 4.2.2 shows Latin squares A, B, and C of orders 3, 4, and 12, respectively, with $C = A \circ B$.

(In forming C, notice that a typical entry a_{ij} of A is replaced by the n_2 by n_2 matrix

$$a_{ij}[1] + n_1 B$$

$$A = \begin{array}{cccc} 0 & 1 & 2 & 3 \\ 1 & 0 & 3 & 2 \\ 2 & 3 & 0 & 1 \\ 3 & 2 & 1 & 0 \end{array} \qquad B = \begin{array}{ccc} 0 & 1 & 2 \\ 1 & 2 & 0 \\ 2 & 0 & 1 \end{array}$$

$$C = \begin{array}{cccccccccccc} 0 & 4 & 8 & 1 & 5 & 9 & 2 & 6 & 10 & 3 & 7 & 11 \\ 4 & 8 & 0 & 5 & 9 & 1 & 6 & 10 & 2 & 7 & 11 & 3 \\ 8 & 0 & 4 & 9 & 1 & 5 & 10 & 2 & 6 & 11 & 3 & 7 \\ 1 & 5 & 9 & 0 & 4 & 8 & 3 & 7 & 11 & 2 & 6 & 10 \\ 5 & 9 & 1 & 4 & 8 & 0 & 7 & 11 & 3 & 6 & 10 & 2 \\ 9 & 1 & 5 & 8 & 0 & 4 & 11 & 3 & 7 & 10 & 2 & 6 \\ 2 & 6 & 10 & 3 & 7 & 11 & 0 & 4 & 8 & 1 & 5 & 9 \\ 6 & 10 & 2 & 7 & 11 & 3 & 4 & 8 & 0 & 5 & 9 & 1 \\ 10 & 2 & 6 & 11 & 3 & 7 & 8 & 0 & 4 & 9 & 1 & 5 \\ 3 & 7 & 11 & 2 & 6 & 10 & 1 & 5 & 9 & 0 & 4 & 8 \\ 7 & 11 & 3 & 6 & 10 & 2 & 5 & 9 & 1 & 4 & 8 & 0 \\ 11 & 3 & 7 & 10 & 2 & 6 & 9 & 1 & 5 & 8 & 0 & 4 \end{array}$$

Figure 4.2.2 Composition of Latin squares.

formed by scalar multiplying B by n_1 and then adding a_{ij} to each entry. Here $[1]$ denotes a matrix each of whose entries is 1; thus $a_{ij}[1]$ is a matrix each of whose entries is a_{ij}.)

We first verify that C is, indeed, a Latin square. For $0 \leq i \leq n - 1$ and $0 \leq j_1 \leq j_2 \leq n - 1$, we have

$$c_{ij_1} = c_{ij_2} \rightarrow a_{uv_1} + n_1 b_{xy_1} = a_{uv_2} + n_1 b_{xy_2}$$
$$\rightarrow a_{uv_1} = a_{uv_2} \text{ and } b_{xy_1} = b_{xy_2}$$
$$\rightarrow v_1 = v_2 \text{ and } y_1 = y_2$$
$$\rightarrow j_1 \text{ div } n_2 = j_2 \text{ div } n_2 \text{ and } j_1 \text{ mod } n_2 = j_2 \text{ mod } n_2$$
$$\rightarrow j_1 = j_2$$

In a similar fashion, for $0 \leq i_1 \leq i_2 \leq n - 1$ and $0 \leq j \leq n - 1$,

$$c_{i_1 j} = c_{i_2 j} \rightarrow i_1 = i_2$$

This shows that C is a Latin square. Next suppose that A and D are orthogonal Latin squares of order n_1 and that B and E are orthogonal Latin squares of order n_2. We show that $C = A \circ B$ and $F = D \circ E$ are orthogonal Latin squares of order n. For $0 \leq i_1 \leq i_2 \leq n - 1$ and $0 \leq j_1 \leq j_2 \leq n - 1$, let

$$u_1 = i_1 \text{ div } n_2, \qquad v_1 = j_1 \text{ div } n_2, \qquad u_2 = i_2 \text{ div } n_2, \qquad v_2 = j_2 \text{ div } n_2$$
$$x_1 = i_1 \text{ mod } n_2, \qquad y_1 = j_1 \text{ mod } n_2, \qquad x_2 = i_2 \text{ mod } n_2, \qquad y_2 = j_2 \text{ mod } n_2$$

Then we have

$$(c_{i_1 j_1}, f_{i_1 j_1}) = (c_{i_2 j_2}, f_{i_2 j_2})$$

$\rightarrow c_{i_1 j_1} = c_{i_2 j_2}$ and $f_{i_1 j_1} = f_{i_2 j_2}$

$\rightarrow (a_{u_1 v_1} = a_{u_2 v_2}$ and $b_{x_1 y_1} = b_{x_2 y_2})$ and $(d_{u_1 v_1} = d_{u_2 v_2}$ and $e_{x_1 y_1} = e_{x_2 y_2})$

$\rightarrow (a_{u_1 v_1} = a_{u_2 v_2}$ and $d_{u_1 v_1} = d_{u_2 v_2})$ and $(b_{x_1 y_1} = b_{x_2 y_2}$ and $e_{x_1 y_1} = e_{x_2 y_2})$

$\rightarrow (a_{u_1 v_1}, d_{u_1 v_1}) = (a_{u_2 v_2}, d_{u_2 v_2})$ and $(b_{x_1 y_1}, e_{x_1 y_1}) = (b_{x_2 y_2}, e_{x_2 y_2})$

$\rightarrow (u_1, v_1) = (u_2, v_2)$ and $(x_1, y_1) = (x_2, y_2)$

$\rightarrow u_1 = u_2, v_1 = v_2, x_1 = x_2,$ and $y_1 = y_2$

$\rightarrow i_1 = i_2$ and $j_1 = j_2$

$\rightarrow (i_1, j_1) = (i_2, j_2)$

as was to be shown. We have just proven the following result:

THEOREM 4.2.3   Suppose that

$$n = n_1 n_2$$

where $1 < n_1 < n_2 < n$. If there is a set of m_1 pairwise-orthogonal Latin squares of order n_1 and a set of m_2 pairwise-orthogonal Latin squares of order n_2, then there is a set of $m = \min\{m_1, m_2\}$ pairwise-orthogonal Latin squares of order n. $\square$

COROLLARY 4.2.4   Let n be an integer, $n > 2$. If $n \bmod 4 \in \{0, 1, 3\}$, then there is a pair of orthogonal Latin squares of order n. $\square$

Corollary 4.2.4 is Euler's famous result and its proof is left for you to develop in Exercise 10. As mentioned, Bose, Parker, and Shrikhande showed how to construct a pair of orthogonal Latin squares when $n > 6$ and $n \bmod 4 = 2$. Their method uses more complicated combinatorial designs and so is beyond the scope of this discussion.

As usual, no sooner is one mathematical problem resolved than another takes its place. What about the existence of a set of three pairwise-orthogonal Latin squares of order n? Certainly there does not exist such a set if $n = 2, 3,$ or 6. Theorems 4.2.2 and 4.2.3 can be applied to construct such a set unless either $n \bmod 4 = 2$ or $n \bmod 9 \in \{3, 6\}$; see Exercise 11. In all but one of these remaining cases it has been shown that there does exists a set of three pairwise-orthogonal Latin squares; the most recent case settled was $n = 14$ by V. Todorov [Three mutually orthogonal Latin squares of order fourteen, *Ars Combinatoria* 20 (1985): 45–47]. The one outstanding case is $n = 10$—no one has yet found a set of three pairwise-orthogonal Latin squares of order 10, nor has anyone shown that such a set is impossible!

For further study of this fascinating topic, see the book on combinatorial designs by W. D. Wallis.

EXERCISE SET 4.2

(Any Latin square of order n is assumed to use the symbols $0, 1, \ldots, n - 1$.)

1. Show that the number of Latin squares of order n is

$$n!(n - 1)! \cdots 2!1!$$

2. We say that two Latin squares of order n are **equivalent** provided it is possible to reorder the rows and/or columns of one so as to produce the other. A natural question is, Are any two Latin squares of the same order equivalent? Show that the answer is no by showing that the following Latin square (of order 4) is not equivalent to one of the Latin squares in Figure 4.2.1.

$$
\begin{array}{cccc}
0 & 1 & 2 & 3 \\
3 & 0 & 1 & 2 \\
2 & 3 & 0 & 1 \\
1 & 2 & 3 & 0
\end{array}
$$

3. Let B be a Latin square of order n. A set T of n entries of B is called a **transversal** provided there is one entry from each row and one entry from each column and $T = \{0, 1, \ldots, n - 1\}$. Let A be another Latin square of order n; show that A and B are orthogonal if and only if, for every $x \in \{0, 1, \ldots, n - 1\}$,

$$\{b_{ij} \mid a_{ij} = x\}$$

is a transversal of B.

4. Given a Latin square A of order n, $n > 2$ and $n \neq 6$, does there necessarily exist a Latin square B of order n such that A and B are orthogonal (that is, does A have an "orthogonal partner")? We can show that the answer to this question is no if n is even.

   (a) Define A by

   $$a_{ij} = (i + j) \bmod n$$

   Show that A has no transversal; hence, it has no orthogonal partner. (See Exercise 3.)

   (b) Show that the following Latin square (of order 5) has no orthogonal partner.

   $$
   \begin{array}{ccccc}
   0 & 1 & 2 & 3 & 4 \\
   1 & 0 & 4 & 2 & 3 \\
   2 & 3 & 0 & 4 & 1 \\
   3 & 4 & 1 & 0 & 2 \\
   4 & 2 & 3 & 1 & 0
   \end{array}
   $$

5. Let A be a Latin square of order n and let A^t be the transpose of A.

   (a) Prove: If A and A^t are orthogonal, then the diagonal elements of A form a transversal of A. (See Exercise 3.)

(b) Show that, if $n = 3$, then A and A^t cannot be orthogonal.

(c) Find a Latin square A of order 4 such that A and A^t are orthogonal.

(d) Find a Latin square A of order 5 such that A and A^t are orthogonal.

6. Let n be an integer, $n \geq 2$. Prove: If there exists a projective plane of order n, then there is a set of $n - 1$ pairwise-orthogonal Latin squares of order n.

7. Use the field plane $EG(5)$ to construct a set of four pairwise-orthogonal Latin squares of order 5.

8. Let n be an integer, $n \geq 2$. Prove: If there exists a set of $n-1$ pairwise-orthogonal Latin squares of order n, then there is a projective plane of order n.

9. Call the Latin squares of Figure 4.2.2 A_1, B_1, and C_1. Using appropriate Latin squares A_2 and B_2 of orders 4 and 3, respectively, construct $C_2 = A_2 \circ B_2$ such that C_1 and C_2 are orthogonal.

10. Prove Corollary 4.2.4.

11. Let n be an integer, $n \geq 2$. Prove: If $n \bmod 4 \neq 2$ and $n \bmod 9 \notin \{3,6\}$, then there is a set of three pairwise-orthogonal Latin squares of order n.

4.3 FACTORIZATIONS

Consider the problem of designing a schedule for a league with $2n$ teams; each week, for $2n - 1$ weeks, the teams are paired to play n games. It is desired that, over the course of the season, each team plays against every other team exactly once.

We associate the teams with the vertices $v_0, v_1, \ldots, v_{2n-1}$ of the complete graph K_{2n} of order $2n$, where the edge $v_i v_j$ represents the game between teams v_i and v_j. Then the set of n edges associated with the games to be played in a given week is a perfect matching of K_{2n}. We let E_i denote the perfect matching for week i, $1 \leq i \leq 2n - 1$, and require that $\{E_1, E_2, \ldots, E_{2n-1}\}$ be a partition of the edge set of K_{2n}.

A **factor** of a graph $G = (V, E)$ is a spanning subgraph of G. If $\{E_1, E_2, \ldots, E_n\}$ is a partition of E and $G_i = (V, E_i)$, then we say that $\{G_1, G_2, \ldots, G_n\}$ is a **factorization** of G and we write $G = G_1 \oplus G_2 \oplus \cdots \oplus G_n$. An r-regular factor of G is called an **r-factor**; if $G = G_1 \oplus G_2 \oplus \cdots \oplus G_n$ such that each G_i is an r-factor, then we say that G is **r-factorable** and call $\{G_1, G_2, \ldots, G_n\}$ an **r-factorization** of G. Note that a 1-factor is a perfect matching. Thus, our question about a schedule for a league with $2n$ teams reduces to the question of whether K_{2n} is 1-factorable.

THEOREM 4.3.1 For every positive integer n, the complete graph K_{2n} is 1-factorable.

Proof It is obvious that K_2 is 1-factorable. For $n \geq 2$, let the vertex set of K_{2n} be $\{v_0, v_1, \ldots, v_{2n-1}\}$. Draw the graph so that $v_1, v_2, \ldots, v_{2n-1}$ are the vertices of a regular $(2n - 1)$-gon, with v_0 at its center. For $1 \leq i \leq 2n - 1$, let the 1-factor G_i consist of the edge $v_0 v_i$ together with all the edges perpendicular with $v_0 v_i$. Then $G_1 \oplus G_2 \oplus \cdots \oplus G_{2n-1}$ is a 1-factorization of K_{2n}. $\square$

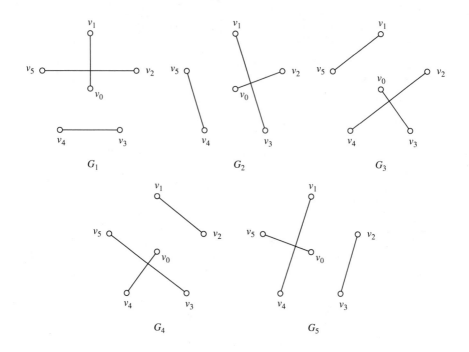

Figure 4.3.1   A 1-factorization of K_6.

Figure 4.3.1 shows the 1-factorization of K_6 constructed using the method given in the proof of Theorem 4.3.1. In general, note that the given 1-factorization of K_{2n} is constructed in a very straightforward way from the initial factor G_1—to obtain the factor G_i, we simply rotate G_1 clockwise about v_0 through the angle $2\pi(i-1)/(2n-1)$. For this reason, the factor G_1 is called the **starter** for the given 1-factorization.

The next result is due to D. König [Über graphen und ihre anwendung auf determinantentheorie und mengenlehre, *Mathematische Annalen* 77 (1916): 453–465].

THEOREM 4.3.2   Every r-regular bipartite graph, $r \geq 1$, is 1-factorable.

Proof   Let G be an r-regular bipartite graph with partite sets X and Y. We proceed by induction on r, the result being obvious for $r = 1$. For $r > 1$, note that it suffices to show that G has a 1-factor, for the removal of the edges of a 1-factor from G results in an $(r-1)$-regular bipartite graph, to which the induction hypothesis may be applied.

To show that G has a 1-factor, it suffices by Theorem 3.6.1 (Hall's Theorem) to show that

$$|S| \leq |N(S)|$$

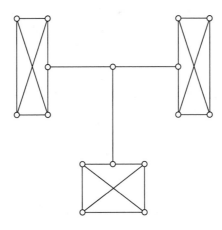

Figure 4.3.2   A cubic graph with no 1-factor.

for every subset S of X. (Note that, since G is regular, we have $|X| = |Y|$.) Let $S \subseteq X$. The number of edges of G incident with vertices of S is $r|S|$. Each such edge is also incident with a vertex of $N(S)$, and there are $r|N(S)|$ edges of G incident with vertices of $N(S)$. Hence, $r|N(S)| \geq r|S|$, so that $|N(S)| \geq |S|$. Therefore, by Theorem 3.6.1, X can be matched to Y, and so G has a 1-factor.
$\square$

The next corollary follows directly from Theorem 4.3.2; it can also be shown directly by noting an equivalence between 1-factorizations of $K_{n,n}$ and Latin squares of order n (see Exercise 2).

COROLLARY 4.3.3   The complete bipartite graph $K_{n,n}$ is 1-factorable. $\square$

Trivially, every 1-regular graph is 1-factorable. A 2-regular graph is 1-factorable if and only if each of its components is an even cycle. This brings us to the case of 3-regular graphs; such graphs are called **cubic** and necessarily must have even order. It is not the case that every connected cubic graph G contains a 1-factor, as is shown by the graph in Figure 4.3.2.

However, if we further require that G have no bridges, then we have the following famous result of J. Petersen [Die theorie der regulären graphs, *Acta Mathematica* 15 (1891): 193–220]; our proof uses Tutte's 1-factor theorem (Theorem 3.6.5).

THEOREM 4.3.4   If G is a bridgeless cubic graph, then G has a 1-factor and, hence, $G = G_1 \oplus G_2$ where G_1 is a 1-factor and G_2 is a 2-factor.

Proof   Let $G = (V, E)$ be a bridgeless cubic graph and let S be a nonempty proper subset of V with $|S| = s$. To show that G has a 1-factor, it suffices by Tutte's theorem (Theorem 3.6.5) to show that the number k of components of odd order of $G - S$ is such that $k \leq s$. This is certainly true if $k \leq 1$, so assume that $k \geq 2$ and let $C_1, C_2, \ldots, C_k$ be the odd components of $G - S$. For each i, $1 \leq i \leq k$, there is an edge between C_i and S, for otherwise C_i would be a cubic graph of odd order, a contradiction. In fact, since G is bridgeless, there must be at least two edges between each C_i and S. However, if there were exactly two edges joining some C_i and S, then again C_i would have an odd number of vertices of odd degree. Therefore, there exist at least three edges joining each C_i with S and, hence, at least $3k$ edges between S and $G - S$. Since G is cubic and $|S| = s$, the number of edges between S and $G - S$ is at most $3s$. This yields $3k \leq 3s$, or $k \leq s$. $\qquad\square$

In light of Petersen's theorem, it is natural to wonder whether every bridgeless cubic graph can, in fact, be 1-factored. Petersen showed that the answer is no, and his example is the famous Petersen graph that we have met before (see Figure 3.7.1). In Exercise 4 you are asked to show that this graph cannot be 1-factored.

Petersen also determined the 2-factorable graphs. Obviously, if G is 2-factorable, then G is $2r$-regular for some integer $r \geq 1$. It turns out that the converse holds as well.

THEOREM 4.3.5   A nonempty graph G is 2-factorable if and only if G is $2r$-regular for some positive integer r.

Proof   In view of the comment preceeding the theorem, it suffices to prove sufficiency. Let G be a $2r$-regular graph for $r \geq 1$. Since G is 2-factorable if and only if each of its components is 2-factorable, and since each component is $2r$-regular, we may assume that G is connected. In this case, we have that G is eulerian; let C be an eulerian circuit of G. Suppose that $G = (V, E)$, with $V = \{v_1, v_2, \ldots, v_n\}$; we define a bipartite graph B with partite sets $\{x_1, x_2, \ldots, x_n\}$ and $\{y_1, y_2, \ldots, y_n\}$ such that B has an edge joining x_i and y_j if and only if v_j immediately follows v_i on C. Since G is $2r$-regular, B is r-regular and so is 1-factorable by Theorem 4.3.2; so we let $\{F_1, F_2, \ldots, F_r\}$ be a 1-factorization of B.

Now consider F_k for k fixed. Starting with the vertex x_1, follow the edge $x_1 y_{j_1}$ of F_k incident with x_1. Note that $j_1 \neq 1$, so next go to x_{j_1} and follow the edge $x_{j_1} y_{j_2}$ of F_k incident with x_{j_1}. Note that $j_2 \neq 1$, for $j_2 = 1$ would imply that v_1 both immediately preceeds and follows v_{j_1} on C, contradicting the fact that C includes each edge of G exactly once. We may then go to x_{j_2} and follow the edge $x_{j_2} y_{j_3}$ of F_k, and so on. By continuing in this way we eventually come to a vertex x_i such that $x_i y_1$ is an edge of F_k. Then the edges $x_1 y_{j_1}, x_{j_1} y_{j_2}, \ldots, x_i y_1$ give rise to a cycle of G, namely,

$$v_1, v_{j_1}, v_{j_2}, \ldots, v_i, v_1$$

There is nothing special about the vertex x_1—we can start at any x_i and follow this process to arrive at a cycle of G. Each vertex of G is on exactly one such cycle, and thus we see that F_k corresponds to 2-factor G_k of G. Moreover, for $1 \le k_1 < k_2 \le r$, the 2-factors G_{k_1} and G_{k_2} have no edges in common, because the same is true of F_{k_1} and F_{k_2}. Therefore, $G = G_1 \oplus G_2 \oplus \cdots \oplus G_r$. $\qquad\square$

■ ***Example 4.3.1***   Apply the method given in the proof of Theorem 4.3.5 to construct a 2-factorization of the 4-regular graph $G = (\{v_1, v_2, v_3, v_4, v_5, v_6, v_7\}, \{v_1v_2, v_1v_3, v_1v_6, v_1v_7, v_2v_3, v_2v_4, v_2v_7, v_3v_4, v_3v_5, v_4v_5, v_4v_6, v_5v_6, v_5v_7, v_6v_7\})$, using the eulerian circuit C:

$$v_1, v_2, v_3, v_1, v_6, v_5, v_3, v_4, v_2, v_7, v_5, v_4, v_6, v_7, v_1$$

Solution

The bipartite graph B formed from G and C has vertex set $\{x_1, x_2, \ldots, x_7\} \cup \{y_1, y_2, \ldots, y_7\}$ and edge set

$$\{x_1y_2, x_1y_6, x_2y_3, x_2y_7, x_3y_1, x_3y_4, x_4y_2, x_4y_6, x_5y_3, x_5y_4, x_6y_5, x_6y_7, x_7y_1, x_7y_5\}$$

Note that B is 2-regular. Let $\{F_1, F_2\}$ be the 1-factorization of G with $F_1 = \langle\{x_1y_2, x_2y_3, x_3y_1, x_4y_6, x_5y_4, x_6y_7, x_7y_5\}\rangle$ and $F_2 = \langle\{x_1y_6, x_2y_7, x_3y_4, x_4y_2, x_5y_3, x_6y_5, x_7y_1\}\rangle$. The 2-factor G_1 of G corresponding to F_1 has edge set $\{v_1v_2, v_1v_3, v_2v_3, v_4v_5, v_4v_6, v_5v_7, v_6v_7\}$; note that G_1 is the union of a 3-cycle and a 4-cycle . The 2-factor G_2 of G corresponding to F_2 has edge set $\{v_1v_6, v_1v_7, v_2v_4, v_2v_7, v_3v_4, v_3v_5, v_5v_6\}$; note that G_2 is a hamiltonian cycle of G. ■

It follows from Theorem 4.3.5 that K_{2n+1} is 2-factorable. In fact, K_{2n+1} can be factored into n hamiltonian cycles.

THEOREM 4.3.6   For every positive integer n, the complete graph K_{2n+1} can be 2-factored so that each factor is a hamiltonian cycle.

Proof   It is obvious that K_3 is 2-factorable. For $n \ge 2$, let the vertex set of K_{2n+1} be $\{v_0, v_1, \ldots, v_{2n}\}$. Draw the graph so that $v_1, v_2, \ldots, v_{2n}$ are the vertices of a regular $2n$-gon, and place v_0 above the center of the $2n$-gon, at the apex of a regular $2n$-pyramid having the $2n$-gon as its base. (See Figure 4.3.3 for the case $n = 3$.) As in the proof of Theorem 4.3.1, we give a starter 2-factor G_1: it consists of the edges v_0v_1 and v_0v_{n+1}, all edges parallel with v_1v_2, and all edges parallel with v_2v_{2n}. Note that G_1 is the hamiltonian cycle

$$v_0, v_1, v_2, v_{2n}, v_3, v_{2n-1}, \ldots, v_n, v_{n+2}, v_{n+1}, v_0$$

The starter is used to obtain the factor G_i, $1 \le i \le n$—simply rotate G_1 clockwise about the axis of the pyramid through the angle $\pi(i - 1)/n$. Alternately, for

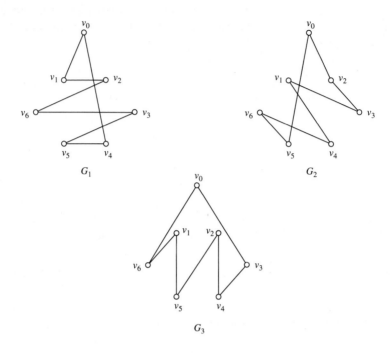

Figure 4.3.3   A factorization of K_7 into hamiltonian cycles.

$2 \leq i \leq n$, G_i consists of the edges $v_0 v_i$ and $v_0 v_{n+i}$, all edges parallel with $v_i v_{i+1}$, and all edges parallel with $v_{i-1} v_{i+1}$. Thus, $G = G_1 \oplus G_2 \oplus \cdots \oplus G_n$, where each G_i is a hamiltonian cycle.   □

Figure 4.3.3 shows the 2-factorization of K_7 constructed by using the method of the proof of Theorem 4.3.6.

In general, a spanning subgraph F of a graph G is called an **isofactor** of G provided G has a factorization, $G = F_1 \oplus F_2 \oplus \cdots \oplus F_n$, such that each F_i is isomorphic to F, $1 \leq i \leq n$. In this case we say that G is **F-factorable**. Thus, for example, Theorem 4.3.6 can be interpreted as saying that K_{2n+1} is C_{2n+1}-factorable. Further results on isofactors are left to the exercises and chapter problems.

EXERCISE SET 4.3 _____

1. Let n be a positive integer.

  (a) Show that K_{2n} can be factored into n hamiltonian paths.
  (b) Show that K_{2n} can be factored into $n - 1$ hamiltonian cycles and a 1-factor.

2. Let n be a positive integer; consider the complete bipartite graph G with partite sets $X = \{x_0, x_1, \ldots, x_{n-1}\}$ and $Y = \{y_0, y_1, \ldots, y_{n-1}\}$.

   (a) Let A be a Latin square of order n using the symbols $\{0, 1, \ldots, n-1\}$. Show how A may be used to define a 1-factorization of G.

   (b) Let $\{G_0, G_1, \ldots, G_{n-1}\}$ be a 1-factorization of G. Show how this 1-factorization may be used to define a Latin square of order n.

3. Recall, from Problem 46 of Chapter 3, the graph Q_n of the n-cube.

   (a) Prove that Q_n is 1-factorable.

   (b) Prove that Q_n is r-factorable if and only if r divides n.

4. Show that the Petersen graph cannot be 1-factored.

5. Use the method of the proof of Theorem 4.3.5 to give a 2-factorization of the graph $K_{2,2,2}$ of the octahedron.

6. Let $G_1 \oplus G_2 \oplus \cdots \oplus G_{2n-1}$ be the 1-factorization of K_{2n} given by the proof of Theorem 4.3.1. Define an array A of order $2n-1$ as follows: for $1 \le i, j \le 2n-1$,

$$a_{ij} = \begin{cases} i & \text{if } i = j \\ k & \text{if } i \neq j \text{ and } v_i v_j \in E(G_k) \end{cases}$$

Show that A is a symmetric Latin square of order $2n - 1$ with diagonal $(1, 2, \ldots, 2n - 1)$.

7. For $n \ge 1$, show that K_{2n+1} cannot be factored into hamiltonian paths.

8. Let A be a Latin square of order $2n - 1$ with symbol set $\{1, 2, \ldots, 2n - 1\}$. If necessary, permute the symbols of A so that the diagonal is $(1, 2, \ldots, 2n - 1)$. Show how A can be used to define a 1-factorization of K_{2n}.

9. Determine all isofactors of size 3 of the Petersen graph.

10. Using a technique similar to that used to prove Theorem 4.3.4, prove that any cubic graph having at most two bridges contains a 1-factor.

11. Prove or disprove: K_{10} is 3-factorable.

12. Prove or disprove: A nonempty graph G is 3-factorable if and only if G is $3r$-regular for some positive integer r.

13. Up to isomorphism, there is a unique graph of order 10 with degree sequence $(2, 1, \ldots, 1, 0)$; show that this graph is an isofactor of the Petersen graph.

14. Prove: If F is an isofactor of the Petersen graph, then F is a forest.

4.4 BLOCK DESIGNS

In 1850, the Reverend Thomas P. Kirkman posed the following problem in the *Lady's and Gentleman's Diary*: "Fifteen young ladies in a school walk out three abreast for seven days in succession; it is required to arrange them daily so that no two shall walk twice abreast." This problem is called **Kirkman's schoolgirls problem**.

Let us elucidate the problem. Suppose it is the girls' teacher who does the arranging. Each day for 7 days, the teacher leads the students on a walk, arranging the girls in 5 rows with 3 girls in each row; thus, each girl has 2 "walkmates" each day. Is it possible to plan the walk formations for the 7 days in such a way so that any pair of girls are walkmates on at most 1 day?

Let $X = \{x_1, x_2, \ldots, x_{15}\}$ be the set of girls. Notice that the walk formation for a given day is equivalent to a partition of X into five 3-element subsets, or **triples**. Doing this for all 7 days leads to a list of 35 triples, with the triples arranged into 7 partitions of X into five parts. Moreover, to yield a solution of the problem, it is required that any pair of girls belong together to at most one triple. Note that a given triple $\{x, y, z\}$ has three 2-element subsets: $\{x, y\}$, $\{x, z\}$, and $\{y, z\}$. Since there are 35 triples, a solution of the problem yields a total of $3 \cdot 35 = 105$ distinct pairs of girls. However, note that X has exactly $C(15, 2) = 105$ subsets of cardinality 2. Therefore, if Kirkman's problem has a solution, then any two girls are walkmates on exactly 1 day.

> **DEFINITION 4.4.1**   Let $X = \{x_1, x_2, \ldots, x_v\}$ be a set of $v \geq 3$ elements. A **Steiner triple system** on X is a set of triples $\{B_1, B_2, \ldots, B_b\}$ of X satisfying the following two conditions:
>
> 1.   There is a positive integer r such that each element of X belongs to exactly r triples.
> 2.   Any 2-element subset of X is contained in exactly one triple.   ∎

Steiner triple systems are named for J. Steiner who published a paper about them in 1853. He was interested in determining those values of v for which there exists a Steiner triple system on a set with v elements. Unknown to Steiner, Kirkman had solved this problem several years earlier [On a problem in combinations, *Cambridge and Dublin Mathematics Journal* 2 (1847): 191–204].

Suppose there exists a Steiner triple system $\{B_1, B_2, \ldots, B_b\}$ on a set X with v elements. Since X has $C(v, 2)$ 2-element subsets and each triple B_i contains 3 of them, we see that $3b = C(v, 2)$—that is,

$$6b = v(v - 1)$$

Let $x \in X$. Then the number of 2-element subsets to which x belongs is $v - 1$. Let B be a triple with $x \in B$, say, $B = \{x, y, z\}$. Then B accounts for exactly two of the 2-element subsets to which x belongs, namely, $\{x, y\}$ and $\{x, z\}$. Since x belongs to exactly r triples, we have

$$2r = v - 1$$

This implies that v must be odd, so that $v \bmod 6 \in \{1, 3, 5\}$. However, from the relation $6b = v(v - 1)$, we see that $v \bmod 6 = 5$ is impossible. Therefore, a necessary condition for the existence of a Steiner triple system on a set with $v \geq 3$ elements is $v \bmod 6 \in \{1, 3\}$. We wish to show that this condition is sufficient as well.

■ **Example 4.4.1** From the preceding discussion, we know that the first several values of v for which there may exist a Steiner triple system on $X = \{0, 1, \ldots, v-1\}$ are 3, 7, 9, 13, 15, 19, and so on. The Steiner triple system on $\{0, 1, 2\}$ is rather trivial; it has exactly 1 block—

$$\{0, 1, 2\}$$

A Steiner triple system on $\{0, 1, 2, 3, 4, 5, 6\}$ must have 7 blocks, each element must belong to exactly 3 blocks, and any pair of elements must determine a unique block. An example is

$$\{0, 1, 3\}, \{0, 2, 5\}, \{0, 4, 6\}, \{1, 2, 4\}, \{1, 5, 6\}, \{2, 3, 6\}, \{3, 4, 5\}$$

and is equivalent to the (Fano) projective plane of order 2. A Steiner triple system on $\{0, 1, \ldots, 8\}$ must have 12 blocks, each element must belong to 4 blocks, and any pair of elements must determine a unique block. An example is the following:

$$\{0, 1, 2\}, \{3, 4, 5\}, \{6, 7, 8\},$$
$$\{0, 3, 6\}, \{1, 4, 7\}, \{2, 5, 8\},$$
$$\{0, 4, 8\}, \{1, 5, 6\}, \{2, 3, 7\},$$
$$\{0, 5, 7\}, \{1, 3, 8\}, \{2, 4, 6\}$$ ■

We wish to prove that, if $v \geq 3$ and $v \bmod 6 \in \{1, 3\}$, then there is a Steiner triple system on v elements.

THEOREM 4.4.1 If $v \bmod 6 = 3$, then there is a Steiner triple system on v elements.

Proof The proof is only outlined, leaving verification of the details for you to provide in Exercise 4. We also illustrate the construction for the case $v = 15$.

Let $v = 3n$ with n odd; we construct a Steiner triple system on $\{0, 1, 2\} \times \{0, 1, \ldots, n-1\}$. For this we require a symmetric Latin square A of order n with diagonal $0, 1, \ldots, n-1$; such a Latin square is not difficult to construct; see Exercise 2. For the case $n = 5$, we have the following:

$$A = \begin{bmatrix} 0 & 3 & 1 & 4 & 2 \\ 3 & 1 & 4 & 2 & 0 \\ 1 & 4 & 2 & 0 & 3 \\ 4 & 2 & 0 & 3 & 1 \\ 2 & 0 & 3 & 1 & 4 \end{bmatrix}$$

The triples are of four types: type I consists of all triples of the form $\{(0,x),(0,y),(1,a_{xy})\}$, $0 \le x < y \le n-1$ (as usual, the rows and columns of A are indexed with $0,1,\ldots,n-1$); type II consists of all triples of the form $\{(1,x),(1,y),(2,a_{xy})\}$, $0 \le x < y \le n-1$; type III consists of all triples of the form $\{(2,x),(2,y),(0,a_{xy})\}$, $0 \le x < y \le n-1$; type IV consists of all triples of the form $\{(0,x),(1,x),(2,x)\}$, $0 \le x \le n-1$. Note that, to obtain a Steiner triple system on the set $V = \{0,1,2,\ldots,v-1\}$, we can simply replace the ordered pair (x,y) by the integer $nx + y$; with this replacement, the triples for the case $v = 15$ are listed as follows:

Type I: $\{0,1,8\}, \{0,2,6\}, \{0,3,9\}, \{0,4,7\}, \{1,2,9\},$
$\{1,3,7\}, \{1,4,5\}, \{2,3,5\}, \{2,4,8\}, \{3,4,6\}$
Type II: $\{5,6,13\}, \{5,7,11\}, \{5,8,14\}, \{5,9,12\}, \{6,7,14\},$
$\{6,8,12\}, \{6,9,10\}, \{7,8,10\}, \{7,9,13\}, \{8,9,11\}$
Type III: $\{10,11,3\}, \{10,12,1\}, \{10,13,4\}, \{10,14,2\}, \{11,12,4\},$
$\{11,13,2\}, \{11,14,0\}, \{12,13,0\}, \{12,14,3\}, \{13,14,1\}$
Type IV: $\{0,5,10\}, \{1,6,11\}, \{2,7,12\}, \{3,8,13\}, \{4,9,14\}$

In general, note that there are $C(n,2)$ triples of each of types I, II, and III, and n triples of type IV. Thus, $b = 3C(n,2) + n$, so that

$$6b = 9n(n-1) + 6n = 3n(3n-1) = v(v-1)$$

Now consider a typical element; by the symmetry of the triple types, it suffices to consider $(0,z)$. This element belongs to $n-1$ triples of type I, $(n-1)/2$ type III triples, and one type IV triple. Hence,

$$2r = 2(n-1) + (n-1) + 2 = 3n - 1 = v - 1$$

It remains to show that, for any two elements, there is a unique triple containing them both. We illustrate the general argument by considering the case of $(0,x)$ and $(1,z)$, $0 \le x, z \le n-1$, $x \ne z$. Note that a triple containing these two elements must be of type I. Indeed, there is a unique y, $0 \le y \le n-1$, $y \ne x$, such that $a_{xy} = z$; thus, $(0,x)$ and $(1,z)$ belong together only to the triple $\{(0,x),(0,y),(1,z)\}$. $\qquad \square$

THEOREM 4.4.2   If $v \bmod 6 = 1$, then there is a Steiner triple system on v elements.

Proof The proof is only outlined here, leaving verification of the details for you to develop in Exercise 8. We also illustrate the construction for the case $v = 13$.

Let $v = 3n + 1$ with $n = 2k$ even; we construct a Steiner triple system on $(\{0, 1, 2\} \times \{0, 1, \ldots, n - 1\}) \cup \{\infty\}$. (Here it is customary to use the special symbol ∞.) For this we require a symmetric Latin square A of order n with diagonal $0, 1, \ldots, k - 1, 0, 1, \ldots, k - 1$; such a Latin square is not difficult to construct; see Exercise 6. For $n = 4$, we have the following:

$$A = \begin{bmatrix} 0 & 2 & 1 & 3 \\ 2 & 1 & 3 & 0 \\ 1 & 3 & 0 & 2 \\ 3 & 0 & 2 & 1 \end{bmatrix}$$

The triples are of seven types: type I consists of the k triples of the form $\{(0, x), (1, x), (2, x)\}$, $0 \le x \le k - 1$; type II consists of the k triples of the form $\{\infty, (0, x), (1, x - k)\}$, $k \le x \le n - 1$; type III consists of the k triples of the form $\{\infty, (1, x), (2, x - k)\}$, $k \le x \le n - 1$; type IV consists of the k triples of the form $\{\infty, (2, x), (0, x - k)\}$, $k \le x \le n - 1$; type V consists of the $C(n, 2)$ triples of the form $\{(0, x), (0, y), (1, a_{xy})\}$, $0 \le x < y \le n - 1$; type VI consists of the $C(n, 2)$ triples of the form $\{(1, x), (1, y), (2, a_{xy})\}$, $0 \le x < y \le n - 1$; type VII consists of the $C(n, 2)$ triples of the form $\{(2, x), (2, y), (0, a_{xy})\}$, $0 \le x < y \le n - 1$. The triples for the case $v = 13$ are listed below; once again we replace the ordered pair (x, y) by the integer $nx + y = 4x + y$ to obtain a Steiner triple system on $V = \{0, 1, 2, \ldots, 11, \infty\}$.

Type I: $\{0, 4, 8\}, \{1, 5, 9\}$

Type II: $\{\infty, 2, 4\}, \{\infty, 3, 5\}$

Type III: $\{\infty, 6, 8\}, \{\infty, 7, 9\}$

Type IV: $\{\infty, 10, 0\}, \{\infty, 11, 1\}$

Type V: $\{0, 1, 6\}, \{0, 2, 5\}, \{0, 3, 7\}, \{1, 2, 7\}, \{1, 3, 4\}, \{2, 3, 6\}$

Type VI: $\{4, 5, 10\}, \{4, 6, 9\}, \{4, 7, 11\}, \{5, 6, 11\}, \{5, 7, 8\}, \{6, 7, 10\}$

Type VII: $\{8, 9, 2\}, \{8, 10, 1\}, \{8, 11, 3\}, \{9, 10, 3\}, \{9, 11, 0\}, \{10, 11, 2\}$

Note that $b = 4k + 3C(n, 2)$, so that

$$6b = 12n + 9n(n - 1) = 3n(3n + 1) = v(v - 1)$$

It is clear that the element ∞ belongs to $3k = (v - 1)/2$ triples. Consider a typical element different from ∞; by the symmetry of the triple types, it suffices to consider $(0, z)$. This element belongs to one type I triple, one type IV triple, $n - 1$ type V triples, and $(n - 2)/2$ type VII triples. Hence,

$$2r = 2 + 2 + 2(n - 1) + (n - 2) = 3n = v - 1$$

It remains to show that, for any two elements, there is a unique triple containing them both. We illustrate the general argument by considering the case of $(0, z)$ and $(1, z)$, $k \leq z \leq n - 1$. Note that a triple containing these two elements must be of type V. Indeed, since $z \geq k$, there is a unique y, $0 \leq y \leq n - 1$, $y \neq z$, such that $a_{zy} = z$; thus, $(0, z)$ and $(1, z)$ belong together only to the triple $\{(0, z), (0, y), (1, z)\}$. $\square$

Theorems 4.4.1 and 4.4.2 give us methods for constructing Steiner triple systems on v elements whenever $v \geq 3$ and $v \bmod 6 \in \{1, 3\}$. The next result gives a method for constructing a larger Steiner triple system from two smaller ones.

THEOREM 4.4.3   Let v_1 and v_2 be integers with $v_1 \geq 3$ and $v_2 \geq 3$. If there is a Steiner triple system on v_1 elements and a Steiner triple system on v_2 elements, then there is a Steiner triple system on $v_1 v_2$ elements.

Proof   Without loss of generality, we let $X = \{0, 1, \ldots, v_1 - 1\}$, $Y = \{0, 1, \ldots, v_2 - 1\}$, $\{B_1, B_2, \ldots, B_{b_1}\}$ be a Steiner triple system on X, and $\{C_1, C_2, \ldots, C_{b_2}\}$ be a Steiner triple system on Y. We construct a Steiner triple system on $X \times Y$. The triples are of three types: type I consists of the $b_1 v_2$ triples of the form $B_i \times \{y\}$, $1 \leq i \leq b_1$ and $0 \leq y \leq v_2 - 1$; type II consists of the $v_1 b_2$ triples of the form $\{x\} \times C_j$, $0 \leq x \leq v_1 - 1$ and $1 \leq j \leq b_2$; type III consists of the $6b_1 b_2$ triples of the form $\{(x_1, y_1), (x_2, y_2), (x_3, y_3)\}$, where (x_1, x_2, x_3) is a permutation of some B_i and (y_1, y_2, y_3) is a permutation of some C_j.

If b is the number of triples so defined, then $b = v_1 b_2 + v_2 b_1 + 6b_1 b_2$, so that

$$6b = 6v_1 b_2 + 6v_2 b_1 + 36b_1 b_2$$
$$= v_1 v_2 (v_2 - 1) + v_1 v_2 (v_1 - 1) + v_1 v_2 (v_1 - 1)(v_2 - 1)$$
$$= v_1 v_2 (v_1 v_2 - 1)$$

Let $(x, y) \in X \times Y$; it is easy to calculate that the number r of triples containing (x, y) is $(v_1 - 1)/2 + (v_2 - 1)/2 + (v_1 - 1)(v_2 - 1)/2$, so that

$$2r = (v_1 - 1) + (v_2 - 1) + (v_1 - 1)(v_2 - 1) = v_1 v_2 - 1$$

It remains for us to show that any two distinct elements of $X \times Y$ are contained in a unique triple. First we consider (x_1, y) and (x_2, y) with $0 \leq x_1 < x_2 \leq v_1$. Then there is a unique triple B_i of X such that $\{x_1, x_2\} \subseteq B_i$. Hence, the type I triple $B_i \times \{y\}$ is the unique triple containing both (x_1, y) and (x_2, y). A similar argument handles the case of (x, y_1) and (x, y_2), $0 \le y_1 < y_2 \le v_2 - 1$. Finally, we consider the two elements (x_1, y_1) and (x_2, y_2) with $0 \le x_1 < x_2 \le v_1 - 1$, and $0 \le y_1, y_2 \le v_2 - 1$, $y_1 \ne y_2$. So there is a unique triple B_i of X such that $\{x_1, x_2\} \subseteq B_i$ and a unique triple C_j of Y such that $\{y_1, y_2\} \subseteq C_j$. Suppose that $B_i = \{x_1, x_2, x_3\}$ and $C_j = \{y_1, y_2, y_3\}$. Then the type III triple $\{(x_1, y_1), (x_2, y_2), (x_3, y_3)\}$ is the unique triple containing both (x_1, y_1) and (x_2, y_2). □

Returning to Kirkman's schoolgirls problem, let $\{B_1, B_2, \dots, B_{35}\}$ be a Steiner triple system on $X = \{0, 1, 2, \dots, 14\}$. In order to solve Kirkman's problem, it must be possible to partition the 35 triples into seven groups of 5 triples each, such that the triples in each group form a partition of the set X. In general, let $\{B_1, B_2, \dots, B_b\}$ be a Steiner triple system on $X = \{0, 1, \dots, v - 1\}$. Note that X can be partitioned into triples if and only if $v \bmod 3 = 0$; let $v = 3s$, so that $b = s(v - 1)/2$. If the set of triples $\{B_1, B_2, \dots, B_b\}$ is partitioned into $(v - 1)/2$ groups of s triples each, such that the triples in each group form a partition of X, then we have what is called a **Kirkman triple system** on X. Thus, Kirkman's schoolgirls problem is equivalent to the problem of whether there exists a Kirkman triple system on $\{0, 1, 2, \dots, 14\}$.

We note that, if there exists a Kirkman triple system on $\{0, 1, 2, \dots, v - 1\}$, then $v \bmod 6 = 3$.

■ *Example 4.4.2* We use the result of Theorem 4.4.3 to help construct a Kirkman triple system on $V = \{0, 1, 2, \dots, 8\}$; here $v = 9 = v_1 v_2$ with $v_1 = v_2 = 3$. We begin with $X = Y = B_1 = C_1 = \{0, 1, 2\}$. We then apply the theorem to construct the following triples, making the usual replacement of (x, y) by $3x + y$.

Type I: $\{0, 3, 6\}, \{1, 4, 7\}, \{2, 5, 8\}$

Type II: $\{0, 1, 2\}, \{3, 4, 5\}, \{6, 7, 8\}$

Type III: $\{0, 4, 8\}, \{0, 5, 7\}, \{1, 3, 8\}, \{1, 5, 6\}, \{2, 3, 7\}, \{2, 4, 6\}$

Note that the triples of type I partition V, as do the triples of type II. In addition, the triples of type III may be divided into two partitions of V—

$$\{\{0, 4, 8\}, \{1, 5, 6\}, \{2, 3, 7\}\} \quad \text{and} \quad \{\{0, 5, 7\}, \{1, 3, 8\}, \{2, 4, 6\}\}$$

This yields the desired Kirkman triple system. ■

The general form of Kirkman's problem was solved relatively recently by D. K. Ray-Chaudhuri and R. M. Wilson [Solution of Kirkman's schoolgirl problem, *Proceedings of the Symposium on Pure Mathematics* 19 (American Mathematical Society, 1971): 187–203]. They showed that $v \bmod 6 = 3$ is both necessary and sufficient for the existence of a Kirkman triple system on $\{0, 1, 2, \dots, v - 1\}$. The proof of sufficiency is beyond the scope of this text; however, if you are interested, see the book on combinatorial designs by Wallis.

■   *Example 4.4.3*   The following Kirkman triple system on $\{0, 1, 2, \ldots, 14\}$ provides a solution to Kirkman's problem:

$$\{0, 1, 14\}, \{2, 5, 13\}, \{3, 7, 12\}, \{4, 6, 9\}, \{8, 10, 11\}$$
$$\{0, 2, 12\}, \{1, 5, 9\}, \{3, 4, 10\}, \{6, 7, 8\}, \{11, 13, 14\}$$
$$\{0, 3, 11\}, \{1, 7, 10\}, \{2, 4, 14\}, \{5, 6, 12\}, \{8, 9, 13\}$$
$$\{0, 4, 8\}, \{1, 6, 11\}, \{2, 3, 9\}, \{5, 7, 14\}, \{10, 12, 13\}$$
$$\{0, 5, 10\}, \{1, 2, 8\}, \{3, 6, 14\}, \{4, 7, 13\}, \{9, 11, 12\}$$
$$\{0, 6, 13\}, \{1, 4, 12\}, \{2, 7, 11\}, \{3, 5, 8\}, \{9, 10, 14\}$$
$$\{0, 7, 9\}, \{1, 3, 13\}, \{2, 6, 10\}, \{4, 5, 11\}, \{8, 12, 14\} \quad ■$$

A Steiner triple system is a special case of a more general type of combinatorial design known as a **balanced incomplete block design**.

DEFINITION 4.4.2   Let v, b, r, k, and λ be positive integers. A (v, b, r, k, λ)-**balanced incomplete block design**, or (v, b, r, k, λ)-**BIBD** for short, consists of a set X of cardinality v and a set (or multiset) $\{B_1, B_2, \ldots, B_b\}$ of k-element subsets of X satisfying the following two properties:

1.   For each $x \in X$, we have $x \in B_i$ for exactly r values of i, $1 \leq i \leq b$.
2.   For any $x, y \in X$, $x \neq y$, we have $\{x, y\} \subseteq B_i$ for exactly λ values of i, $1 \leq i \leq b$.   ■

Given a (v, b, r, k, λ)-BIBD, the elements of X are called **treatments** and the sets $B_1, B_2, \ldots, B_b$ are called **blocks**. Thus, each block contains k treatments, each treatment belongs to exactly r blocks, and any 2-element subset of X is contained in exactly λ blocks; λ is called the **index** of the design. We note that a Steiner triple system is a $(v, b, r, 3, 1)$-BIBD.

The word *balance* refers to condition 2 of the definition, whereas the word *incomplete* refers to the fact that, normally, $k < v$. If $k = v$, then each block is all of X, that is, each block is "complete." In this case, $b = r = \lambda$, and the design is not very interesting. So we henceforth assume that $k < v$.

In a Steiner triple system, or $(v, b, r, 3, 1)$-BIBD, we showed that

$$6b = v(v - 1) \qquad \text{and} \qquad 2r = v - 1$$

These results can be generalized for a (v, b, r, k, λ)-BIBD.

THEOREM 4.4.4   The identity

$$bk = vr$$

holds in any (v, b, r, k, λ)-BIBD.

Proof We count the number t of pairs (x, B) such that x is a treatment, B is a block, and $x \in B$. First, since each treatment belongs to r blocks, $t = vr$. Second, since each block contains k treatments, $t = bk$. This proves the result. □

THEOREM 4.4.5 The identity

$$r(k - 1) = (v - 1)\lambda$$

holds in any (v, b, r, k, λ)-BIBD.

Proof Let x_0 be a fixed treatment. We count the number t of pairs (y, B) such that y is a treatment, $y \neq x$, B is a block, and $\{x, y\} \subseteq B$. First, since x belongs to r blocks and each such block contains $k - 1$ elements besides x, we have $t = r(k - 1)$. Second, for any given treatment $y \neq x$, the pair $\{x, y\}$ is a subset of exactly λ blocks; since there are $v - 1$ choices for y, we have $t = (v - 1)\lambda$. This proves the result. □

DEFINITION 4.4.3 Given a (v, b, r, k, λ)-BIBD with treatment set $X = \{x_1, x_2, \ldots, x_v\}$ and blocks $B_1, B_2, \ldots, B_b$, its **incidence matrix** is the v by b $(0, 1)$-matrix A defined by

$$a_{ij} = \begin{cases} 0 & \text{if } x_i \notin B_j \\ 1 & \text{if } x_i \in B_j \end{cases}$$

■

■ ***Example 4.4.4*** A $(v, b, r, 3, 2)$-BIBD with treatment set $\{1, 2, \ldots, v\}$ has $3b = vr$ and $r = v - 1$. Hence,

$$3b = v(v - 1)$$

so that $v \bmod 3 \in \{0, 1\}$. Since $v > k = 3$, the smallest possible v is 4. In this case, $b = 4$ and $r = 3$; the blocks are:

$$B_1 = \{1, 2, 3\}, \quad B_2 = \{1, 2, 4\}, \quad B_3 = \{1, 3, 4\}, \quad B_4 = \{2, 3, 4\}$$

The incidence matrix of this $(4, 4, 3, 3, 2)$-BIBD is

$$A_1 = \begin{bmatrix} 1 & 1 & 1 & 0 \\ 1 & 1 & 0 & 1 \\ 1 & 0 & 1 & 1 \\ 0 & 1 & 1 & 1 \end{bmatrix}$$

The next smallest possible v is $v = 6$; in this case, $b = 10$ and $r = 5$. A possible collection of blocks for such a design is as follows:

$$B_1 = \{1,2,3\}, \quad B_2 = \{1,2,4\}, \quad B_3 = \{1,3,5\}, \quad B_4 = \{1,4,6\}, \quad B_5 = \{1,5,6\},$$
$$B_6 = \{2,3,6\}, \quad B_7 = \{2,4,5\}, \quad B_8 = \{2,5,6\}, \quad B_9 = \{3,4,5\}, \quad B_{10} = \{3,4,6\}$$

The incidence matrix for this $(6, 10, 5, 3, 2)$-BIBD is

$$A = \begin{bmatrix} 1 & 1 & 1 & 1 & 1 & 0 & 0 & 0 & 0 & 0 \\ 1 & 1 & 0 & 0 & 0 & 1 & 1 & 1 & 0 & 0 \\ 1 & 0 & 1 & 0 & 0 & 1 & 0 & 0 & 1 & 1 \\ 0 & 1 & 0 & 1 & 0 & 0 & 1 & 0 & 1 & 1 \\ 0 & 0 & 1 & 0 & 1 & 0 & 1 & 1 & 1 & 0 \\ 0 & 0 & 0 & 1 & 1 & 1 & 0 & 1 & 0 & 1 \end{bmatrix}$$

∎

THEOREM 4.4.6   Let A be the incidence matrix of a (v, b, r, k, λ)-BIBD, let A^t denote the transpose of A, let I denote the v by v identity matrix, and let J and J' denote the v by v and v by b matrices of all 1s, respectively. Then

$$AA^t = (r - \lambda)I + \lambda J \tag{4.4.1}$$

and

$$JA = kJ' \tag{4.4.2}$$

Proof   Let the treatment set be $X = \{1, 2, \ldots, v\}$ and let the blocks be $B_1, B_2, \ldots, B_b$. Since each block contains k treatments, each column of A has sum k. This means that each entry of JA is k, which proves identity (4.4.2).

For identity (4.4.1), note that the (i,j)-entry of AA^t is

$$a_{i1}a_{j1} + a_{i2}a_{j2} + \cdots + a_{ib}a_{jb}$$

The term $a_{in}a_{jn}$ is 1 if treatments i and j both belong to block B_n and is 0 otherwise. Since each treatment belongs to r blocks and any two distinct treatments belong together to λ blocks, we see that the (i,j)-entry of AA^t is r if $i = j$ and is λ if $i \neq j$. Thus,

$$AA^t = rI + \lambda(J - I) = (r - \lambda)I + \lambda J$$

This proves (4.4.1).   □

THEOREM 4.4.7   Given positive integers, v, b, r, k, and λ with $k < v$, if there exists a v by b $(0, 1)$-matrix A satisfying identities (4.4.1) and (4.4.2), then A is the incidence matrix of a (v, b, r, k, λ)-BIBD.

Proof Given A, let $X = \{x_1, x_2, \ldots, x_v\}$ be a set of v treatments and define blocks $B_1, B_2, \ldots, B_b$ by

$$x_i \in B_j \leftrightarrow a_{ij} = 1$$

By identity (4.4.2), we have that each column of A has sum k, so that each block has cardinality k. We let i and j be integers with $1 \leq i, j \leq v$ and $i \neq j$. Identity (4.4.1) implies that the (i, i)-entry of AA^t is r, whereas the (i, j)-entry is λ. As in the proof of Theorem 4.4.6, this means that x_i belongs to exactly r blocks, whereas x_i and x_j belong together to exactly λ blocks. Since i and j are arbitrary, this shows that we have a (v, b, r, k, λ)-BIBD. $\square$

We next wish to prove that in a balanced incomplete block design, the number of blocks is at least the number of treatments. This relation is known as **Fisher's inequality** and is surprisingly difficult to prove by direct means. We give a proof based on the determinant of AA^t. The proof of the following lemma is easy and is left for you to work out in Exercise 10.

LEMMA 4.4.8 With I and J as in Theorem 4.4.6, the determinant of the matrix $(r - \lambda)I + \lambda J$ is

$$(r - \lambda)^{v-1}[r + (v - 1)\lambda] \qquad \square$$

THEOREM 4.4.9 (Fisher's Inequality) In a (v, b, r, k, λ)-BIBD, the number of blocks is at least the number of treatments, namely,

$$b \geq v$$

Proof Let A be the incidence matrix of a (v, b, r, k, λ)-BIBD. Since $k < v$, we have $r > \lambda$ by Theorem 4.4.5. By identity (4.4.1) and Lemma 4.4.8, the determinant of AA^t equals

$$(r - \lambda)^{v-1}[r + (v - 1)\lambda]$$

and this is nonzero. Therefore, the matrix AA^t has rank v. Moreover, the rank of A is at least the rank of AA^t and is at most the minimum of v and b. Thus, $v \leq \min(v, b)$, so that $v \leq b$. $\square$

Now consider a (v, b, r, k, λ)-BIBD with treatment set $X = \{x_1, x_2, \ldots, x_v\}$ and blocks $B_1, B_2, \ldots, B_b$. If $b = v$, then we have a **symmetric balanced incomplete block design**. In this case $r = k$, and we may define a **dual design** as follows: The treatment set is $\{b_1, b_2, \ldots, b_v\}$ and the blocks $X_1, X_2, \ldots, X_b$ are defined by

$$b_j \in X_i \leftrightarrow x_i \in B_j$$

Note that, if A is the incidence matrix of the original BIBD, then A^t is the incidence matrix of the dual design. Furthermore, it follows from the proof of Theorem 4.4.9 that A has rank v; hence, A is invertible. We now show that the dual design is a (v, b, r, k, λ)-BIBD. First, in the original design, each treatment is an element of $r = k$ blocks. Thus, in the dual design, each block contains $k = r$ treatments. It follows that

$$JA^t = kJ = AJ$$

where, as before, J is the v by v matrix of all 1s. Second, in the original design, each block contains $k = r$ treatments. Hence, in the dual design, each treatment belongs to $r = k$ blocks. Now consider two distinct treatments b_{j_1} and b_{j_2} of the dual design; these correspond to blocks B_{j_1} and B_{j_2} of the original design. Note that for any i, $1 \le i \le b$,

$$\{b_{j_1}, b_{j_2}\} \subseteq X_i \leftrightarrow x_i \in B_{j_1} \cap B_{j_2}$$

so that the number of blocks containing both b_{j_1} and b_{j_2} is $|B_{j_1} \cap B_{j_2}|$, and this is the (j_1, j_2)-entry of $A^t A$. Then, by using identities (4.4.1) and (4.4.2),

$$
\begin{aligned}
A^t A &= A^{-1}(AA^t)A \\
&= (k - \lambda)I + \lambda A^{-1} JA \\
&= (k - \lambda)I + \lambda A^{-1}(kJ) \\
&= (k - \lambda)I + \lambda A^{-1}(AJ) \\
&= (k - \lambda)I + \lambda J
\end{aligned}
$$

It follows that every nondiagonal entry of $A^t A$ is λ.

■ ***Example 4.4.5***   A $(7, 7, 3, 3, 1)$-BIBD with treatment set $\{1, 2, \ldots, 7\}$ is a Steiner triple system; one example has the following blocks:

$$B_1 = \{1, 2, 4\}, \quad B_2 = \{1, 3, 6\}, \quad B_3 = \{1, 5, 7\}, \quad B_4 = \{2, 3, 5\}$$
$$B_5 = \{2, 6, 7\}, \quad B_6 = \{3, 4, 7\}, \quad B_7 = \{4, 5, 6\}$$

The incidence matrix of this BIBD is

$$
A = \begin{bmatrix}
1 & 1 & 1 & 0 & 0 & 0 & 0 \\
1 & 0 & 0 & 1 & 1 & 0 & 0 \\
0 & 1 & 0 & 1 & 0 & 1 & 0 \\
1 & 0 & 0 & 0 & 0 & 1 & 1 \\
0 & 0 & 1 & 1 & 0 & 0 & 1 \\
0 & 1 & 0 & 0 & 1 & 0 & 1 \\
0 & 0 & 1 & 0 & 1 & 1 & 0
\end{bmatrix}
$$

The transpose of A is

$$A^t = \begin{bmatrix} 1 & 1 & 0 & 1 & 0 & 0 & 0 \\ 1 & 0 & 1 & 0 & 0 & 1 & 0 \\ 1 & 0 & 0 & 0 & 1 & 0 & 1 \\ 0 & 1 & 1 & 0 & 1 & 0 & 0 \\ 0 & 1 & 0 & 0 & 0 & 1 & 1 \\ 0 & 0 & 1 & 1 & 0 & 0 & 1 \\ 0 & 0 & 0 & 1 & 1 & 1 & 0 \end{bmatrix}$$

and is the incidence matrix of the dual $(7, 7, 3, 3, 1)$-BIBD. If we again take the treatment set to be $\{1, 2, \ldots, 7\}$, then the blocks of this design are:

$$\{1, 2, 3\}, \{1, 4, 5\}, \{2, 4, 6\}, \{1, 6, 7\}, \{3, 4, 7\}, \{2, 5, 7\}, \{3, 5, 6\} \qquad \blacksquare$$

Another class of symmetric balanced incomplete block designs is the class of projective planes. Given a projective plane of order n, think of the points as treatments and the lines as blocks—then we have a symmetric BIBD with $v = b = n^2 + n + 1$, $r = k = n + 1$, and $\lambda = 1$ (see Exercise 12).

Theorem 4.1.5 gives, as a necessary condition for the existence of a projective plane of order n with $n \bmod 4 \in \{1, 2\}$, that n be the sum of two integer squares. We noted there that this result of R. H. Bruck and H. J. Ryser is a special case of a more general result concerning the existence of symmetric balanced incomplete block designs. Suppose we are given positive integers v, k, and λ with $k < v$ and $k(k - 1) = \lambda(v - 1)$. Does there exist a symmetric (v, v, k, k, λ)-BIBD? According to our next result, the answer is not necessarily. This is known as the Bruck–Ryser–Chowla theorem. The proof for v odd is difficult and was first given by S. Chowla and Ryser [Combinatorial problems, *Canadian Journal of Mathematics* 2(1950): 93–99]. The case for v even it not so hard and was given by M. P. Schutzenberger [A non-existence theorem for an infinite family of symmetrical block designs, *Annals of Eugenics* 14(1949): 286–287]; we shall give the proof in this case.

THEOREM 4.4.10 (Bruck–Chowla–Ryser Theorem) Let v, k, and λ be positive integers with $k < v$ and $k(k - 1) = \lambda(v - 1)$.

1. If v is even and there exists a symmetric (v, v, k, k, λ)-BIBD, then $k - \lambda$ is a perfect square.
2. If v is odd and there exists a symmetric (v, v, k, k, λ)-BIBD, then

$$x^2 = (k - \lambda)y^2 + (-1)^{(v-1)/2}\lambda z^2$$

for some integers x, y, and z, not all zero.

Proof As mentioned, we give the proof only for the case v even.

Let A be the incidence matrix of a symmetric (v, v, k, k, λ)-BIBD. Let $\det(A)$, $\det(A^t)$, and $\det(AA^t)$ denote the determinants of the matrices A, A^t, and AA^t, respectively. Since $\det(A) = \det(A^t)$ and $\det(AA^t) = \det(A) \cdot \det(A^t)$, we have, using Lemma 4.4.8, that

$$(\det(A))^2 = \det(AA^t) = k^2(k - \lambda)^{v-1}$$

Furthermore, A is a $(0, 1)$-matrix, so $\det(A)$ is an integer. Since $v - 1$ is odd, it must be that $k - \lambda$ is a perfect square.   □

■   *Example 4.4.6*   Let's use Theorem 4.4.10 for the case $\lambda = 2$.

For $v = b = 7$ and $r = k = 4$, a symmetric $(7, 7, 4, 4, 2)$-BIBD exists only if there exist integers x, y, z not all 0, such that $x^2 = 2y^2 - 2z^2$. This is satisfied, for example, when $x = 0$ and $y = z \neq 0$, so such a design is possible. In fact, such a design exists—see Exercise 9.

For $v = b = 16$, and $r = k = 6$, we have that a $(16, 16, 6, 6, 2)$-BIBD exists only if $k - \lambda = 4$ is a perfect square. Hence, such a design is possible.

For $v = b = 22$ and $r = k = 7$, we see that $k - \lambda = 5$ is not a perfect square. Thus, there does not exist a symmetric $(22, 22, 7, 7, 2)$-BIBD.   ■

■   *Example 4.4.7*   Does there exist a symmetric $(141, 141, 21, 21, 3)$-BIBD? If there does, then there exist integers x, y, z not all 0, such that

$$x^2 = 18y^2 + 3z^2 \tag{4.4.3}$$

We may assume, without loss of generality, that x, y, and z have no common positive factor greater than 1. Note that (4.4.3) implies that $x^2 \bmod 3 = 0$, and hence that $x \bmod 3 = 0$, say, $x = 3n$. Then we obtain from (4.4.3) that

$$3n^2 = 6y^2 + z^2 \tag{4.4.4}$$

Again, this implies that $z \bmod 3 = 0$, say, $z = 3m$. Now we obtain from (4.4.4) that

$$n^2 = 2y^2 + 3m^2$$

We have assumed that $y \bmod 3 \in \{1, 2\}$, so that $(2y^2 + 3m^2) \bmod 3 = 2$. However, $n^2 \bmod 3 \in \{0, 1\}$, so we have a contradiction. This proves that such a design does not exist.   ■

Given a balanced incomplete block design, a **parallel class** is a set of blocks that partition the set of treatments. The BIBD is called **resolvable** if the set of blocks can be partitioned into parallel classes; such a partition is called a **resolution**.

We have encountered several examples of resolvable designs. An affine plane with n^2 points has its $n^2 + n$ lines grouped into $n + 1$ parallel classes, with n lines in each class. Thus, an affine plane is a resolvable BIBD with parameter list

$$(n^2, n^2 + n, n + 1, n, 1)$$

As another example, a Kirkman triple system on $6n + 3$ treatments is a resolvable Steiner triple system; hence, it is a resolvable BIBD with parameter list

$$(6n + 3, (2n + 1)(3n + 1), 3n + 1, 3, 1)$$

Further examples of resolvable BIBD's are left to the exercises and chapter problems.

In this chapter we have attempted to provide an introduction to the subject of combinatorial designs, but have really just scratched the surface of a large and fascinating area of mathematics. It is hoped that your appetite for further study has been whetted; for this we recommend the fine recent book on combinatorial designs by Wallis.

EXERCISE SET 4.4

1. Construct a Steiner triple system on 21 elements:
   (a) Using Theorem 4.4.1.
   (b) Using Theorem 4.4.3.
2. For n odd, construct a symmetric Latin square of order n with diagonal $0, 1, \ldots, n - 1$.
3. Construct a Steiner triple system on 19 elements.
4. Provide the missing details for the proof of Theorem 4.4.1.
★5. Find a Kirkman triple system on 21 elements.
6. For n even, say, $n = 2k$, construct a symmetric Latin square A of order n with diagonal $0, 1, \ldots, k - 1, 0, 1, \ldots, k - 1$.
7. Verify that the incidence matrices A_1 and A of Example 4.4.4 satisfy identities (4.4.1), and (4.4.2).
8. Provide the missing details for the proof of Theorem 4.4.2.
9. Consider the matrix

$$A = \begin{bmatrix} 1 & 1 & 1 & 1 & 0 & 0 & 0 \\ 1 & 1 & 0 & 0 & 0 & 1 & 1 \\ 1 & 0 & 1 & 0 & 1 & 0 & 1 \\ 1 & 0 & 0 & 1 & 1 & 1 & 0 \\ 0 & 1 & 1 & 0 & 1 & 1 & 0 \\ 0 & 1 & 0 & 1 & 1 & 0 & 1 \\ 0 & 0 & 1 & 1 & 0 & 1 & 1 \end{bmatrix}$$

   (a) Show that A satisfies identities (4.4.1) and (4.4.2).

(b) Use A to construct a symmetric $(7, 7, 4, 4, 2)$-BIBD.

(c) Construct the dual design of the design in part (b).

10. Prove Lemma 4.4.8.

11. Define the idea of "isomorphism" for balanced incomplete block designs.

12. Verify that a projective plane of order n is a symmetric $(n^2 + n + 1, n^2 + n + 1, n + 1, n + 1, 1)$-BIBD.

$\star$**13.** Find two nonisomorphic $(6, 10, 5, 3, 2)$-BIBD's or prove that any two such designs are isomorphic.

14. Derive Theorem 4.1.5 as a corollary to Theorem 4.4.10. You may use the result that a positive integer can be represented as the sum of two integer squares if and only if, in its canonical representation, each prime factor p with $p \bmod 4 = 3$ occurs to an even power.

15. Show that the $(7, 7, 4, 4, 2)$-BIBD of Exercise 9 is unique, up to isomorphism.

16. Given a (v, b, r, k, λ)-BIBD, $b - 2r + \lambda > 0$, with treatment set X and blocks $B_1, B_2, \ldots, B_b$, the design on X with blocks $X - B_1, X - B_2, \ldots, X - B_b$ is called the **complementary design**.

(a) Show that the complementary design is a balanced incomplete block design, and find its parameters.

(b) Find the complementary design of the Kirkman triple system of Example 4.4.2.

(c) Describe how the incidence matrix of the original design and the incidence matrix of the complementary design are related.

CHAPTER 4 PROBLEMS

1. For each of the following statements, determine whether it is true, false, or whether its truth value is presently unknown.

(a) There exists an affine plane with 100 points and 110 lines.

(b) There exists an affine plane with 64 points and 72 lines.

(c) There exists a projective plane of order 21.

(d) There exists a projective plane of order 12.

(e) There exists a projective plane that is not isomorphic to a field plane.

(f) There exists a pair of orthogonal Latin squares of order 6.

(g) There exists a set of three pairwise-orthogonal Latin squares of order 10.

(h) There exists a set of seven pairwise-orthogonal Latin squares of order 8.

(i) There exists a set of four pairwise-orthogonal Latin squares of order 35.

(j) There exists a projective plane of order n if and only if there exists a set of $n - 1$ pairwise-orthogonal Latin squares of order n.

2. Five types of gasoline are to be tested to determine which gives the best mileage. Five different cars and five drivers are to be used. Each day for five consecutive days, the cars are paired with the gasoline types and given a fixed amount of fuel; then each car is assigned to a driver and driven a fixed route. Design a schedule for this experiment to make it as fair as possible.

3. Consider the affine plane $EG(7)$.

    (a) How many points are there?
    (b) How many lines are there?
    (c) Find the equation of the line containing the points $(2,4)$ and $(4,3)$.
    (d) Find all points on the line $y = 2x + 3$.
    (e) Do the lines $y = 2x + 3$ and $3x + 2y = 2$ intersect? If so, at what point?

⋆4. Show that the Petersen graph is not an isofactor of K_{10}.

5. Let F be the field of order 9:

$$F = \{0, 1, 2, t, 1 + t, 2 + t, 2t, 1 + 2t, 2 + 2t\}$$

constructed using the prime polynomial $x^2 + x + 2$ over $\mathbb{Z}_3$. Consider the projective plane of order 9 constructed from the affine plane $EG(9)$, that is, the field plane of order 9.

    (a) How many points are there?
    (b) How many lines are there?
    (c) Let α be the line containing the points $(0,1)$ and $(t, 2t + 2)$. Let β be the line containing the points $(2,2)$ and $(1+t, t)$. Find the point of intersection of α and β.

6. Show that no graph with degree sequence $(2,2,2,2,1,1,0,0,0,0)$ is an isofactor of the Petersen graph.

7. Determine which graphs with degree sequence $(2,2,2,1,1,1,1,0,0,0)$ are isofactors of the Petersen graph. (Hint: There are two of them.)

8. Up to isomorphism, there are two graphs with degree sequence $(2,2,1,1,1,1,1,1,0,0)$; show that both of them are isofactors of the Petersen graph.

9. Determine which trees are isofactors of the following complete graphs:
    (a) K_4 (b) K_5 (c) K_6 (⋆d) K_8

⋆10. Find, up to isomorphism, all isofactors of the Petersen graph with maximum degree 3.

11. Give an example of a connected graph G of composite size with the property that whenever F is a factor of G and the size of F divides the size of G, then F is an isofactor of G.

12. An isofactor F of a graph G is called **proper** provided its size is strictly between 1 and the size of G; a graph is called **prime** provided it has no proper isofactors.

    (a) Show that any graph of prime size is prime.
    (b) Give an example of a prime graph whose size is composite.

13. In each of the following parts, determine the missing parameters of a (v, b, r, k, λ)-BIBD, or else show that such a design is impossible.

    (a) $(v, 15, r, 4, 2)$
    (b) $(v, b, 7, 3, 2)$
    (c) $(v, b, 13, 6, 1)$
    (d) $(21, 70, r, 3, \lambda)$
    (e) $(17, b, 8, 5, \lambda)$
    (f) $(25, b, r, 5, 3)$

14. An **affine design** is a resolvable design with the property that, for some constant m, any two blocks belonging to different parallel classes have m treatments in common. Thus, an affine plane is an affine design with $m = 1$. Prove: An affine design has parameter list

    $$(s^2 m, s^2 m + \lambda m, sm + \lambda, sm, \lambda)$$

    for some positive integers s and m, where $\lambda = (sm - 1)/(s - 1)$. (Hint: To obtain the value of λ, let A be a fixed block and count, in two different ways, the number of triples (x, y, B) such that B is a block, $B \neq A$, $x \neq y$, and $\{x, y\} \subseteq A \cap B$.)

15. Consider the following question: For what values of v does there exist a resolvable $(v, b, r, 4, 1)$-BIBD?

    (a) Show that $v \bmod 12 = 4$ is necessary.
    ★(b) Sixteen golfers go to a resort for 5 days of golf. Each day, they are to be partitioned into 4 foursomes to play. It is required that each pair of golfers play together in the same foursome exactly once. Design a schedule for the golfers.

    (It has been shown that $v \bmod 12 = 4$ is sufficient for the existence of a resolvable $(v, b, r, 4, 1)$-BIBD; see H. Hanani, D. K. Ray-Chaudhuri, and R. M. Wilson, On resolvable designs, *Discrete Mathematics* 3 (1972): 343–357.)

16. Given positive integers v, b, r, k, and λ with $k < v$, let A be a v by b $(0, 1)$-matrix. Prove: A is the incidence matrix of a (v, b, r, k, λ)-BIBD if and only if

    $$JA = kJ' \qquad \text{and} \qquad AJ'' = rJ'$$

    where J, J', and J'' are the v by v, v by b, and b by b matrices of all 1s, respectively.

17. Given positive integers v and k with $k < v$, let X be a set of cardinality v. Show that the collection $\mathcal{P}_k(X)$ of all k-element subsets of X defines a BIBD on X and find its parameter list.

18. Given an integer $v > 2$, is there a unique $(v, b, r, 2, 1)$-BIBD, up to isomorphism?

19. Consider a (v, b, r, k, λ)-BIBD with treatment set X and block set $\mathcal{B} = \{B_1, B_2, \ldots, B_b\}$. Let $\mathcal{P}_k(X)$ be the set of all k-element subsets of X, and consider

$$\overline{\mathcal{B}} = \mathcal{P}_k(X) - \mathcal{B}$$

Show that, if $b < C(v, k)$, then $\overline{\mathcal{B}}$ defines a BIBD on X. Find its parameter list.

20. Consider the symmetric $(15, 15, 7, 7, 3)$-BIBD with the following blocks:

$$B_1 = \{1, 2, 3, 4, 5, 6, 7\}, \quad B_2 = \{1, 2, 3, 8, 9, 10, 11\}$$
$$B_3 = \{1, 2, 3, 12, 13, 14, 15\}, \quad B_4 = \{1, 4, 5, 8, 9, 12, 13\}$$
$$B_5 = \{1, 4, 5, 10, 11, 14, 15\}, \quad B_6 = \{1, 6, 7, 8, 9, 14, 15\}$$
$$B_7 = \{1, 6, 7, 10, 11, 12, 13\}, \quad B_8 = \{2, 4, 6, 8, 10, 12, 14\}$$
$$B_9 = \{2, 4, 7, 8, 11, 13, 15\}, \quad B_{10} = \{2, 5, 6, 9, 11, 12, 15\}$$
$$B_{11} = \{2, 5, 7, 9, 10, 13, 14\}, \quad B_{12} = \{3, 4, 6, 9, 11, 13, 14\}$$
$$B_{13} = \{3, 4, 7, 9, 10, 12, 15\}, \quad B_{14} = \{3, 5, 6, 8, 10, 13, 15\}$$
$$B_{15} = \{3, 5, 7, 8, 11, 12, 14\}$$

(a) Find the blocks of the dual design.

★(b) A **triplet** is a set of three blocks whose intersection has cardinality 3; for example, $\{B_1, B_2, B_3\}$ is a triplet. By considering triplets, show that this block design and its dual are not isomorphic.

21. Consider the design constructed as follows. The 16 treatments $1, 2, \ldots, 16$ are used to form a 4 by 4 matrix:

$$\begin{bmatrix} 1 & 2 & 3 & 4 \\ 5 & 6 & 7 & 8 \\ 9 & 10 & 11 & 12 \\ 13 & 14 & 15 & 16 \end{bmatrix}$$

To obtain a block, take the six elements, except for the common element, determined by a given row and column of the matrix. For example, the block determined by row 2 and column 3 is $\{3, 5, 6, 8, 11, 15\}$.

(a) Show that this construction defines a BIBD, and find its parameter list.

(b) Does this construction give a BIBD on treatment set $\{1, 2, \ldots, v\}$ for any other values of v, where v is a perfect square?

★**22.** This problem gives an alternate proof of Fisher's inequality (Theorem 4.4.9). Let $V_1, V_2, \ldots, V_v$ be the rows of the incidence matrix of a (v, b, r, k, λ)-BIBD (considered as $(0, 1)$-vectors). Define vectors $U_1, U_2, \ldots, U_v$ recursively by

$$U_1 = V_1, \quad U_2 = V_2 - \frac{\lambda}{r} V_1$$

$$U_{i+1} = V_{i+1} - \frac{\lambda}{r + (i - 1)\lambda}(V_1 + V_2 + \cdots + V_i) \quad 2 \le i \le v$$

Prove that $U_1, U_2, \ldots, U_v$ are orthogonal and hence independent. Thus, we have an independent set of v vectors in a vector space of dimension b, and it follows that $v \le b$.

23.   Consider a symmetric (v, v, k, k, λ)-BIBD with treatment set X and blocks $B_1, B_2, \ldots, B_v$.

   (a)   The design with treatment set $X - B_v$ and blocks $B_1 - B_v, B_2 - B_v, \ldots, B_{v-1} - B_v$ is called the **residual design** (with respect to block B_v) of the original. Show that this residual design is a $(v - k, v - 1, k, k - \lambda, \lambda)$-BIBD provided $\lambda \neq k - 1$.

   (b)   The design with treatment set B_v and blocks $B_1 \cap B_v, B_2 \cap B_v, \ldots, B_{v-1} \cap B_v$ is called the **derived design** (with respect to block B_v) of the original. Show that this derived design is a $(k, v - 1, k - 1, \lambda, \lambda - 1)$-BIBD provided $\lambda > 1$.

   (c)   Show that a finite affine plane is a residual design of a projective plane.

   (d)   Consider the symmetric $(11, 11, 6, 6, 3)$-BIBD with treatment set $\{1, 2, \ldots, 11\}$ and blocks

$$\{1, 2, 3, 4, 5, 6\}, \{1, 2, 3, 7, 9, 10\}, \{1, 2, 6, 8, 9, 11\}, \{1, 3, 5, 7, 8, 11\}$$

$$\{1, 4, 5, 9, 10, 11\}, \{1, 4, 6, 7, 8, 10\}, \{2, 3, 4, 8, 10, 11\}$$

$$\{2, 4, 5, 7, 8, 9\}, \{2, 5, 6, 7, 10, 11\}, \{3, 4, 6, 7, 9, 11\}, \{3, 5, 6, 8, 9, 10\}$$

   Find the residual and derived designs (with respect to the block $\{3, 5, 6, 8, 9, 10\}$).

24.   Consider a (v, b, r, k, λ)-BIBD.

   (a)   Prove that the number of blocks not disjoint from a given block B is at least

$$\frac{k(r - 1)^2}{(k - 1)(\lambda - 1) + r - 1}$$

   (b)   Show that equality holds in (a) if and only if any block not disjoint from B meets B in exactly

$$1 + \frac{(k - 1)(\lambda - 1)}{r - 1}$$

   treatments.

25.   Find residual and derived designs of the symmetric $(7, 7, 4, 4, 2)$-BIBD of Exercise 9 in Exercise Set 4.4.

26.   A winery has six different blends of chardonnay that it wishes to evaluate. It is desired that each of ten tasters taste three of the blends, that each blend be tasted by five of the tasters, and that each pair of the blends be tasted together by exactly two tasters. Is this possible?

27.   Show, in a (v, b, r, k, λ)-BIBD (with $k < v$), that $\lambda < r$.

28.   A subset B of cardinality k of $\mathbb{Z}_v = \{0, 1, \ldots, v - 1\}$ is called a **difference set** provided each element of $\mathbb{Z}_v - \{0\}$ occurs the same number λ of times among the $k(k - 1)$ differences $x - y$, $(x, y) \in B \times B$, $x \neq y$. For example, $\{0, 1, 3\}$ is a difference set for $\mathbb{Z}_7$, with $\lambda = 1$, since $0 - 1 = 6$, $0 - 3 = 4$, $1 - 0 = 1$, $1 - 3 = 5$, $3 - 0 = 3$, and $3 - 1 = 2$.

   (a)   Show that $\{0, 2, 3, 4, 8\}$ is a difference set for $\mathbb{Z}_{11}$.

   (b)   Show that $\{0, 1, 3, 9\}$ is a difference set for $\mathbb{Z}_{13}$.

Given a difference set B for $\mathbb{Z}_v$ of cardinality $k < v$, we may construct a symmetric BIBD on the treatment set $\mathbb{Z}_v$; the blocks of the design are $B_0, B_1, \ldots, B_{v-1}$, where B_i is obtained by adding $i \pmod{v}$ to each element of B. For example, by using the difference set $\{0, 1, 3\}$ for $\mathbb{Z}_7$, we obtain the following blocks: $B_0 = \{0, 1, 3\}$, $B_1 = \{1, 2, 4\}$, $B_2 = \{2, 3, 5\}$, $B_3 = \{3, 4, 6\}$, $B_4 = \{4, 5, 0\}$, $B_5 = \{5, 6, 1\}$, $B_6 = \{6, 0, 2\}$.

(c) Use the difference set of part (a) to construct a symmetric BIBD on $\mathbb{Z}_{11}$. What are the parameters of this design?

(d) Use the difference set of part (b) to construct a symmetric BIBD on $\mathbb{Z}_{13}$. What are the parameters of this design?

(e) Show that the above construction works in general. What are the parameters of the design so constructed?

29. Show that there is a Steiner triple system on 3^n elements for any positive integer n.

30. Show that if there is a Steiner triple system on v elements, then there is a Steiner triple system on v^n elements for any positive integer n.

References/Suggestions for Further Reading

Algorithms

The books by Baase and Wilf (1986) are general texts on the design and analysis of algorithms. The others primarily concern generating, ranking and unranking combinatorial objects.

Sara Baase. *Computer Algorithms: Introduction to Design and Analysis*, 2nd ed. Reading, MA: Addison-Wesley, 1988.

A. Nijenhuis and H. S. Wilf. *Combinatorial Algorithms*, 2nd ed. New York: Academic Press, 1978.

Dennis Stanton and Dennis White. *Constructive Combinatorics*. New York: Springer-Verlag, 1986.

Herbert S. Wilf. *Algorithms and Complexity*. Englewood Cliffs, NJ: Prentice-Hall, 1986.

Herbert S. Wilf. *Combinatorial Algorithms: An Update*. Philadelphia: Society for Industrial and Applied Mathematics, 1989.

Combinatorics

Kenneth P. Bogart. *Introductory Combinatorics*, 2nd ed. San Diego: Harcourt Brace Jovanovich, 1990.

Richard A. Brualdi. *Introductory Combinatorics*, 2nd ed. New York: North-Holland, 1992.

Alan Tucker. *Applied Combinatorics*, 2nd ed. New York: Wiley, 1984.

Designs

Anne Penfold Street and Deborah J. Street. *Combinatorics of Experimental Design*. Oxford: Clarendon Press, 1987.

W. D. Wallis. *Combinatorial Designs*. New York: Marcel Dekker, 1988.

Discrete Mathematics

Michael O. Albertson and Joan P. Hutchinson. *Discrete Mathematics with Algorithms*. New York: Wiley, 1988.

Albert D. Polimeni and H. Joseph Straight. *Foundations of Discrete Mathematics*, 2nd ed. Pacific Grove, CA: Brooks/Cole, 1990.

Kenneth A. Ross and Charles R. B. Wright. *Discrete Mathematics*, 3rd ed. Englewood Cliffs, NJ: Prentice-Hall, 1992.

Finite Fields

David M. Burton. *Abstract Algebra*. Dubuque, IA: W. C. Brown, 1988.

Joseph A. Gallian. *Contemporary Abstract Algebra*, 2nd ed. Lexington, MA: D. C. Heath, 1990.

Graphs

The book by Biggs, et al., presents graph theory from a historical perspective; Roberts's book accents applications of the subject to the social and life sciences. The rest are introductory texts, with the books by Gibbons and Gould having more emphasis on algorithms than the others.

N. L. Biggs, E. K. Lloyd, and R. J. Wilson. *Graph Theory 1736–1936*. Oxford: Clarendon Press, 1986.

Fred Buckley and Frank Harary. *Distance in Graphs*. Reading, MA: Addison-Wesley, 1990.

Gary Chartrand and Linda Lesniak. *Graphs and Digraphs*, 2nd ed. Pacific Grove, CA: Wadsworth and Brooks/Cole, 1986.

A. Gibbons. *Algorithmic Graph Theory*. Cambridge: Cambridge University Press, 1985.

Ronald J. Gould. *Graph Theory*. Redwood City, CA: Benjamin-Cummings, 1988.

Fred S. Roberts. *Discrete Mathematical Models, with Applications to Social, Biological, and Environmental Problems*. Englewood Cliffs, NJ: Prentice-Hall, 1976.

Robin J. Wilson and John J. Watkins. *Graphs: An Introductory Approach*. New York: Wiley, 1990.

Answers to Selected Exercises and Problems

Exercise Set 0.1

1. (a) false   (b) true   (c) true   (d) false   (e) true

    (f) false   (g) true   (h) true

3. (a) $\{\{\ \}\}$ (b) $\{\{\ \}, \{1\}\}$ (c) $\{\{\ \}, \{1\}, \{2\}, \{1,2\}\}$

    (d) $\{\{\ \}, \{1\}, \{2\}, \{3\}, \{1,2\}, \{1,3\}, \{2,3\}, \{1,2,3\}\}$

    (e) $\{\{\ \}, \{\{\ \}\}, \{\{\ 1\}\}, \{\{\ \}, \{1\}\}\}$

    (f) $\{\{\ \}, \{\{\ \}\}, \{\{\{\ \}\}\}, \{\{\ \}, \{\{\ \}\}\}\}$

5. 2^n

7. (a) $\{\ldots, -12, -6, 0, 6, 12, \ldots\}$ (b) $\{\ldots, -15, -9, -3, 3, 9, 15, \ldots\}$

    (c) $\{\ldots, -10, -6, -2, 2, 6, 10, \ldots\}$ (d) $\{\ldots, -10, -6, -2, 2, 6, 10, \ldots\}$

    (e) $\{\ \}$ (f) $\{m \mid m$ is a multiple of 3 or m is a multiple of 4$\}$

    (g) C (h) $\{m \mid m$ is both even or a multiple of 3 and not a multiple of 4$\}$

9. (a) A (b) B (c) $\{\ \}$ (d) $A \subset B \leftrightarrow (A - B = \{\ \}$ and $B - A \neq \{\ \})$

    (e) $A \subset B \leftrightarrow (A \cap B = A$ and $A \cap B \neq B)$ (f) $A = B$

11. (a) $\{(-1,0), (-1,1), (1,0), (1,1)\}$

    (b) $\{(0,-1), (0,0), (0,1), (1,-1), (1,0), (1,1)\}$

    (c) $\{(-1,0,-1), (-1,0,0), (-1,1,-1), (-1,1,0), (-1,1,1), (1,0,-1),$
       $(1,0,0), (1,0,1), (1,1,-1), (1,1,0), (1,1,1)\}$

    (e) $\{(-1,(0,-1)), (-1,(0,0)), (-1,(0,1)), (-1,(1,-1)), (-1,(1,0)),$
       $(-1,(1,1)), (1,(0,-1)), (1,(0,0)), (1,(0,1)), (1,(1,-1)), (1,(1,0)),$
       $(1,(1,1))\}$

    (f) $\{(0,0,0,0), (0,0,0,1), (0,0,1,0), (0,0,1,1), (0,1,0,0), (0,1,0,1),$
       $(0,1,1,0), (0,1,1,1), (1,0,0,0), (1,0,0,1), (1,0,1,0), (1,0,1,1),$
       $(1,1,0,0), (1,1,0,1), (1,1,1,0), (1,1,1,1)\}$

(g) $\{\{\ \}\}, \{(0,0)\}, \{(0,1)\}, \{(1,0)\}, \{(1,1)\}, \{(0,0),(0,1)\}, \{(0,0),(1,0)\},$
$\{(0,0),(1,1)\}, \{(0,1),(1,0)\}, \{(0,1),(1,1)\}, \{(1,0),(1,1)\},$
$\{(0,0),(0,1),(1,0)\}, \{(0,0),(0,1),(1,1)\}, \{(0,0),(1,0),(1,1)\},$
$\{(0,1),(1,0),(1,1)\}, \{(0,0),(0,1),(1,0),(1,1)\}$

(h) $\{(\{\ \},\{\ \}),(\{\ \},\{0\}),(\{\ \},\{1\}),(\{\ \},B),(\{0\},\{\ \}),(\{0\},\{0\}),(\{0\},\{1\}),$
$(\{0\},B),(\{1\},\{\ \}),(\{1\},\{0\}),(\{1\},\{1\}),(\{1\},B),(B,\{\ \}),(B,\{0\}),$
$(B,\{1\}),(B,B)\}$

25. The union is $(-1,100)$; the intersection is $(-1/50,2)$.

26. (a) The union is $\mathbb{Z}$; the intersection is $A_0 = \{0\}$.
    (b) The union is $(-1,2)$; the intersection is $\{0\}$.
    (c) The union is $\mathbb{N}$; the intersection is $\{\ \}$.
    (d) The union is $\mathbb{Z}$; the intersection is $\{0\}$.

27. The union is $(-1,1)$; the intersection is $\{0\}$.

28. The union is $\{(x,y) \mid x \neq 0\} \cup \{(0,0)\}$; the intersection is $\{(0,0)\}$.

30. The union is A_{100}; the intersection is A_1.

31. The union is $\mathbb{N}$; the intersection is $\{\ \}$.

32. The union is $\mathbb{Z}$; the intersection is A_0.

33. The union is $\mathbb{Q}$; the intersection is $\mathbb{Z}$.

34. The union is $(1/2,3)$; the intersection is $(1,2)$.

35. The union is $\{(0,0)\} \cup \{(x,y) \mid x \neq 0 \text{ and } y > 0\}$. The intersection is $\{(0,0)\}$.

Exercise Set 0.2

1. f_2 and f_4 are functions; f_1 and f_3 are not.

3. (a) no (b) yes (c) no (d) yes
   (e) one-to-one if and only if $|U| = 1$ (f) yes

5. (a) no (b) no (c) no (d) no (e) yes (f) yes

7. (a) no (b) yes (c) no (d) yes (e) yes (f) yes
   (g) no (h) yes

9. (a) yes (b) no (c) no (d) yes (e) yes (f) no
   (g) yes (h) yes

10. (a) any constant function (b) the doubling function $x \to 2x$
    (c) the function $x \to \lfloor x/2 \rfloor$ (d) the identity function

11. (a) yes (b) no (c) no (d) yes (e) yes
    (f) yes (g) no (h) no (i) yes (j) no

12. (a) f_1 is not a permutation; $f_1(A) = \{2,10\}$, $f_1^{-1}(B) = \{2,4,8,10\}$
    (b) f_2 is not a permutation; $f_2(A) = \{4,8\}$, $f_2^{-1}(B) = \{1,2,4,5,7,8,10,11\}$
    (c) f_3 is a permutation; $f_3(A) = \{3,5,7,11\}$, $f_3^{-1}(B) = B$

13. (a) yes (b) no (c) no (d) yes (e) yes (f) yes
    (g) no (h) no (i) no (j) no

15. (a) $f_1(A) = \{2,3\}, f_1^{-1}(B) = \{1,3,4\}$ (b) $f_2(A) = \{1,3\}, f_2^{-1}(B) = \{2\}$
    (c) $f_3(A) = \{2\}, f_3^{-1}(B) = \{\ \}$ (d) $f_4(A) = \{4\} = f_4^{-1}(B)$
    (e) $f_5(A) = A, f_5^{-1}(B) = \{\ldots, -3, -2, -1\}$
    (f) $f_6(A) = \{2,4,6,\ldots\}, f_6^{-1}(B) = \{\ldots -6, -4, -2, 0\} \cup \mathbb{Z}^+$
    (g) $f_7(A) = \mathbb{Z}^+, f_7^{-1}(B) = \{3,4,7,8,11,12,\ldots\}$ (h) $f_8(A) = \{1,3,5,\ldots\} = f_8^{-1}(B)$

17. (a) $f_1^{-1} : \mathbb{Q} \to \mathbb{Q}; f_1^{-1}(x) = (x-2)/4$ (b) $f_2^{-1} = f_2$ (c) $f_3^{-1} = f_3$
    (d) $f_4^{-1} : \mathbb{Z}_{39} \to \mathbb{Z}_{39}; f_4^{-1}(x) = 8(x-2) \bmod 39$
    (e) $f_5^{-1} : \mathbb{Z} \to \mathbb{Z}; f_5^{-1}(m) = m - 1$
    (f) $f_6^{-1} : \mathbb{N} \to \mathbb{Z}; f_6^{-1}(m) = \begin{cases} (m+1)/2 & \text{if } m \text{ is odd} \\ -m/2 & \text{if } m \text{ is even} \end{cases}$
    (g) $f_7^{-1} : \{1,2,3,4\} \to \{1,2,3,4\}; f_7^{-1}(1) = 2, f_7^{-1}(2) = 3, f_7^{-1}(3) = 4, f_7^{-1}(4) = 1$
    (h) $f_8^{-1} : \{1,2,3,4\} \to \{1,2,3,4\}; f_8^{-1}(1) = 3, f_8^{-1}(2) = 4, f_8^{-1}(3) = 1, f_8^{-1}(4) = 2$

18. Define $f : \{0,1\} \to \{0,1,2\}$ by $f(0) = 0$, $f(1) = 1$, and $g : \{0,1,2\} \to \{0,1\}$ by $g(0) = 0, g(1) = 1 = g(2)$.

19. (a) $g \circ f : \mathbb{Z} \to (0,\infty); (g \circ f)(m) = 1/(|m| + 1)$
    (b) $g \circ f : \mathbb{R} \to (0,1); (g \circ f)(x) = x^2/(x^2 + 1)$
    (c) $g \circ f : \mathbb{Q} - \{2\} \to \mathbb{Q} - \{0\}; (g \circ f)(x) = x - 2$
    (d) $g \circ f : \mathbb{R} \to [0,\infty); (g \circ f)(x) = |x|$
    (e) $g \circ f : \mathbb{Q} - \{10/3\} \to \mathbb{Q} - \{2\}; (g \circ f)(r) = (6r - 14)/(3r - 10)$
    (f) $g \circ f : \mathbb{Z} \to \mathbb{Z}_5; (g \circ f)(m) = (m + 1) \bmod 5$
    (g) $g \circ f : \mathbb{Z}_8 \to \mathbb{Z}_6; (g \circ f)(m) = 0$
    (h) $g \circ f : \{1,2,3,4\} \to \{1,2,3,4\}; (g \circ f)(1) = 2, (g \circ f)(2) = 3, (g \circ f)(3) = 4, (g \circ f)(4) = 1$

20. See the answer to Exercise 18.

21. (a) All are functions on $\mathbb{Z}$: $f^{-1}(m) = m - 1, g^{-1}(m) = 2 - m, (f \circ g)(m) = 3 - m$,
    $(f \circ g)^{-1}(m) = 3 - m = (g^{-1} \circ f^{-1})(m)$.
    (b) All are functions on $\mathbb{Z}_7$: $f^{-1}(m) = (m + 4) \bmod 7, g^{-1}(m) = 4m \bmod 7$,
    $(f \circ g)(m) = (2m + 3) \bmod 7, (f \circ g)^{-1}(m) = (4m + 2) \bmod 7 = (g^{-1} \circ f^{-1})(m)$.
    (c) All are functions on $\{1,2,3,4\}$: $f^{-1}(1) = 2, f^{-1}(2) = 3, f^{-1}(3) = 4, f^{-1}(4) = 1$;
    $g^{-1}(1) = 3, g^{-1}(2) = 4, g^{-1}(3) = 1, g^{-1}(4) = 2; (f \circ g)(1) = 2, (f \circ g)(2) = 3$,
    $(f \circ g)(3) = 4, (f \circ g)(4) = 1; (f \circ g)^{-1}(1) = 4 = (g^{-1} \circ f^{-1})(1), (f \circ g)^{-1}(2) =$
    $1 = (g^{-1} \circ f^{-1})(2), (f \circ g)^{-1}(3) = 2 = (g^{-1} \circ f^{-1})(3), (f \circ g)^{-1}(4) = 3 = (g^{-1} \circ f^{-1})(4)$.
    (d) All are functions on $\{1,2,3,4\}$: $f^{-1}(1) = 4, f^{-1}(2) = 1, f^{-1}(3) = 3, f^{-1}(4) = 2$;
    $g^{-1}(1) = 1, g^{-1}(2) = 4, g^{-1}(3) = 2, g^{-1}(4) = 3; (f \circ g)(1) = 2, (f \circ g)(2) = 3$,
    $(f \circ g)(3) = 1, (f \circ g)(4) = 4; (f \circ g)^{-1}(1) = 3, (f \circ g)^{-1}(2) = 1, (f \circ g)^{-1}(3) = 2$,
    $(f \circ g)^{-1}(4) = 4$. Again, $(f \circ g)^{-1} = g^{-1} \circ f^{-1}$.
    (e) All are functions on $\mathbb{Q}$: $f^{-1}(x) = x/4, g^{-1}(x) = 2x + 3, (f \circ g)(x) = 2(x - 3)$,
    $(f \circ g)^{-1}(x) = (x + 6)/2 = (g^{-1} \circ f^{-1})(x)$.
    (f) All are functions on $\mathbb{Q} - \{1\}$: $f^{-1}(x) = (x + 1)/2, g^{-1}(x) = x/(x - 1)$,
    $(f \circ g)(x) = (x + 1)/(x - 1), (f \circ g)^{-1}(x) = (x + 1)/(x - 1) = (g^{-1} \circ f^{-1})(x)$.

23. (a) $(g \circ f)(m) = 1 - m = (f^{-1} \circ g^{-1})(m)$
    (b) $(g \circ f)(m) = (2m + 6) \bmod 7, (f^{-1} \circ g^{-1})(m) = (4m - 3) \bmod 7$
    (c) $(g \circ f)(1) = 2, (g \circ f)(2) = 3, (g \circ f)(3) = 4, (g \circ f)(4) = 1; (f^{-1} \circ g^{-1})(1) = 4$,
    $(f^{-1} \circ g^{-1})(2) = 1, (f^{-1} \circ g^{-1})(3) = 2, (f^{-1} \circ g^{-1})(4) = 3$
    (d) $(g \circ f)(1) = 3, (g \circ f)(2) = 2, (g \circ f)(3) = 4, (g \circ f)(4) = 1; (f^{-1} \circ g^{-1})(1) = 4$,
    $(f^{-1} \circ g^{-1})(2) = 2, (f^{-1} \circ g^{-1})(3) = 1, (f^{-1} \circ g^{-1})(4) = 3$
    (e) $(g \circ f)(x) = (4x - 3)/2, (f^{-1} \circ g^{-1})(x) = (2x + 3)/4$
    (f) $(g \circ f)(x) = (2x - 1)/(2x - 2) = (f^{-1} \circ g^{-1})(x)$

25. (a) $(f \circ f)(x)$ = the paternal grandfather of x (b) $(f \circ g)(x) = f(x)$
(c) $(g \circ f)(x)$ = the eldest sibling of x (d) $(g \circ g)(x) = g(x)$

27. (a) $f(x) = 1 - x$ (b) $f(x) = x/2$
(c) $f(x) = 4(x - 0.5)^2$ (d) $f(x) = 1/2$

33. (a) $(f \circ g)(m) = 2m + 1$ (b) $(g \circ f)(m) = 2(m + 1)$

(c) $(f \circ h)(m) = \begin{cases} 1 & \text{if } m \text{ is even} \\ 2 & \text{if } m \text{ is odd} \end{cases}$ (d) $(h \circ f)(m) = \begin{cases} 1 & \text{if } m \text{ is even} \\ 0 & \text{if } m \text{ is odd} \end{cases}$

(e) $(g \circ h)(m) = \begin{cases} 2 & \text{if } m \text{ is even} \\ 0 & \text{if } m \text{ is odd} \end{cases}$ (f) $(h \circ g)(m) = 1$ (g) $(g \circ g)(m) = 4m$

(h) $(h \circ f \circ g)(m) = 0$

37. $f^{-1}(y) = 8y \bmod 119$

38. $f(x, y) = nx + y$, for example

39. $g \circ f : \mathbb{Q} - \{1/4\} \to \mathbb{Q} - \{3/2\}, (g \circ f)(x) = (1 - 6x)/(1 - 4x);$
$(g \circ f)^{-1} : \mathbb{Q} - \{3/2\} \to \mathbb{Q} - \{1/4\}, (g \circ f)^{-1}(x) = (1 - x)/(6 - 4x);$
$f^{-1} : \mathbb{Q} - \{0\} \to \mathbb{Q} - \{1/4\}, f^{-1}(x) = -(x + 1)/4; g^{-1} : \mathbb{Q} - \{3/2\} \to \mathbb{Q} - \{0\},$
$g^{-1}(x) = 1/(3 - 2x); f^{-1} \circ g^{-1} = (g \circ f)^{-1}$

40. (a) $f(x) = x/(1 - x^2)$ or $f(x) = \tan(\pi x/2)$, for example

(b) $h(x) = \begin{cases} 2x - 1 & \text{if } x > 0 \\ -2x & \text{if } x \leq 0 \end{cases}$

41. $g(x) = (b - a)x + a$, for instance

43. See the answer to Exercise 40(a).

45. $g(x) = (d - c)(x - a)/(b - a) + c$, for example

46. $h(x) = \begin{cases} f(x) & \text{if } x \in A_1 \\ g(x) & \text{if } x \in B_1 \end{cases}$

47. $h(x, y) = (f(x), g(y))$

49. (a) O   (b) Θ   (c) O   (d) O   (e) O   (f) Θ
(g) O   (h) Θ   (i) O   (j) Θ

55. (a) $\gcd(119, 154) = 7 = (119)(-9) + (154)(7)$
(b) $\gcd(357, 629) = 17 = (357)(-7) + (629)(4)$
(c) $\gcd(405, 1380) = 15 = (405)(-17) + (1380)(5)$
(d) $\gcd(812, 2800) = 28 = (812)(-31) + (2800)(9)$

Exercise Set 0.3

3. Only the matrices are given.

(a) $\begin{bmatrix} 1 & 1 & 1 & 1 \\ 0 & 1 & 0 & 1 \\ 0 & 0 & 1 & 1 \\ 0 & 0 & 0 & 1 \end{bmatrix}$ (b) $\begin{bmatrix} 0 & 1 & 1 & 1 \\ 0 & 0 & 1 & 1 \\ 0 & 0 & 0 & 1 \\ 0 & 0 & 0 & 0 \end{bmatrix}$

(c) $$\begin{bmatrix} 1 & 0 & 0 & 0 \\ 0 & 1 & 0 & 0 \\ 0 & 0 & 1 & 0 \\ 0 & 0 & 0 & 1 \end{bmatrix}$$
(d) $$\begin{bmatrix} 0 & 1 & 1 & 1 \\ 1 & 0 & 1 & 1 \\ 1 & 1 & 0 & 1 \\ 1 & 1 & 1 & 0 \end{bmatrix}$$

(e) $$\begin{bmatrix} 1 & 1 & 1 & 1 \\ 0 & 0 & 0 & 1 \\ 0 & 0 & 0 & 0 \\ 0 & 0 & 0 & 0 \end{bmatrix}$$
(f) $$\begin{bmatrix} 0 & 1 & 0 & 1 \\ 1 & 0 & 1 & 0 \\ 0 & 1 & 0 & 1 \\ 1 & 0 & 1 & 0 \end{bmatrix}$$

6. The adjacency matrix is

$$\begin{bmatrix} 0 & 1 & 1 & 1 & 1 & 1 \\ 0 & 0 & 1 & 0 & 1 & 1 \\ 0 & 0 & 0 & 0 & 0 & 1 \\ 0 & 0 & 0 & 0 & 1 & 1 \\ 0 & 0 & 0 & 0 & 0 & 1 \\ 0 & 0 & 0 & 0 & 0 & 0 \end{bmatrix}$$

5. (a) reflexive, symmetric   (b) irreflexive, antisymmetric, transitive
   (c) irreflexive, symmetric   (d) reflexive, symmetric, transitive
   (e) transitive

7. (a) reflexive, symmetric, transitive   (b) irreflexive, antisymmetric
   (c) reflexive   (d) reflexive, symmetric, transitive

9. (a) irreflexive, symmetric   (b) reflexive, symmetric
   (c) reflexive, antisymmetric, transitive   (d) reflexive, symmetric, transitive
   (e) irreflexive, symmetric   (f) symmetric, transitive

11. (a) reflexive, antisymmetric, transitive   (b) reflexive, symmetric, transitive
    (c) reflexive, transitive   (d) irreflexive, symmetric
    (e) reflexive, symmetric, transitive   (f) reflexive, symmetric, transitive

13. (a) reflexive, transitive   (b) irreflexive, symmetric
    (c) reflexive, symmetric, transitive   (d) irreflexive, antisymmetric, transitive
    (e) reflexive, antisymmetric, transitive   (f) none

14. Only (k), (n), and (o) are false.

15. The relation A_4 is an equivalence relation; the distinct equivalence classes are $[1] = \{1, 3, 5\}$ and $[2] = \{2, 4\}$.

17. The relation A_1, being the complete relation on $\{-2, -1, 0, 1, 2\}$, is an equivalence relation; the only equivalence class is $[0] = \{-2, -1, 0, 1, 2\}$. The relation A_4 is also an equivalence relation; the distinct equivalence classes are $[0] = \{0\}$, $[1] = \{-1, 1\}$, and $[2] = \{-2, 2\}$.

19. The relation A_4 is an equivalence relation; the distinct equivalence classes are $[2] = \{2, 4, 6, 8, \ldots\}$ and $[3] = \{3, 5, 7, 9, \ldots\}$.

20. The set of equivalence classes is $\{\{2\}, \{1, 3\}, \{0, 4\}, \{-1, 5\}, \{-2, 6\}, \ldots\}$.

21. The relations A_2, A_5, and A_6 are equivalence relations. For A_2, the distinct equivalence classes are $[1] = \{\ldots, -3, -1, 1, 3, \ldots\}$ and $[2] = \{\ldots, -4, -2, 2, 4, \ldots\}$. For A_5, the distinct equivalence classes are $[1] = \{\ldots, -5, -2, 1, 4, 7, \ldots\}$, $[2] = \{\ldots, -4, -1, 2, 5, 8, \ldots\}$, and $[3] = \{\ldots, -6, -3, 3, 6, \ldots\}$. For A_6, the distinct equivalence classes are $[1] = \{1, 2, 3, \ldots\}$ and $[-1] = \{\ldots, -3, -2, -1\}$.

23. The relation of part (c) is the only equivalence relation; $[1] = \{1\}$ and $[2] = \{2, 3, 4\}$.

24. (a) true   (b) false   (c) false

25. (b) $[X] = \{\{2,5\}, \{2,3,5\}, \{2,4,5\}, \{2,3,4,5\}\}$

26. (a) $A_1 = \{(1,1), (2,2), (3,3), (4,4), (5,5), (6,6)\}$
   (b) $A_2 = \{(1,1), (2,2), (2,3), 3,2), (3,3), (4,4), (4,5), (4,6), (5,4), (5,5), (5,6),$
   $(6,4), (6,5), (6,6)\}$
   (c) $A_3 = \{(1,1), (1,3), (1,5), (2,2), (2,4), (2,6), (3,1), (3,3), (3,5), (4,2),$
   $(4,4), (4,6), (5,1), (5,3), (5,5), (6,2), (6,4), (6,6)\}$
   (d) $A_4 = V \times V$

27. (a) 1   (b) 2   (c) 5   (d) 15

29. (a) Yes; the arc set of the Hasse diagram is $\{(f,e), (f,d), (e,c), (d,b), (c,a), (b,a)\}$.
   (b) Yes; the arc set of the Hasse diagram is $\{(f,d), (e,d), (d,c), (c,b), (b,a)\}$.
   (c) Yes; the arc set of the Hasse diagram is $\{(f,e), (f,d), (e,c), (e,b), (d,c), (d,b),$
   $(c,a), (b,a)\}$.
   (d) Yes; the arc set of the Hasse diagram is $\{(f,e), (e,c), (e,d), (d,b), (c,b), (b,a)\}$.
   (e) no, not reflexive
   (f) yes; no arcs in the Hasse diagram.
   (g) Yes; the arc set of the Hasse diagram is $\{(f,d), (e,c), (d,b), (c,a)\}$.
   (h) no, not transitive
   (i) Yes; the arc set of the Hasse diagram is $\{(f,d), (e,c), (d,c), (d,b), (c,a)\}$.
   (j) No, not antisymmetric.

31. The relation of part (e) is a partial-order relation; the arc set of the Hasse diagram is $\{(1,2), (1,3), (2,4)\}$.

39. (a) A is not reflexive provided $(v,v) \notin A$ for some $v \in V$.
   (b) A is not irreflexive provided $(v,v) \in A$ for some $v \in V$.
   (c) A is not symmetric provided, for some $u,v \in V$, $(u,v) \in A$ and $(v,u) \notin A$.
   (d) A is not antisymmetric provided, for some $u,v \in V$, $u \neq v$, $(u,v) \in A$ and $(v,u) \in A$.
   (e) A is not transitive provided, for some $u,v,w \in V$, $(u,v) \in A$ and $(v,w) \in A$ and $(u,w) \notin A$.

41. Only (g), (h), (i), (m), and (n) are false.

43. (a) Yes; $[(a,b)]$ is the line $y = x + (b - a)$ with slope 1 and y-intercept $b - a$.
   (b) Yes; $[(1,0)] = \{(1,0)\}$; for $(a,b) \neq (1,0)$, $[(a,b)]$ is the circle centered at $(1,0)$ with radius $(a - 1)^2 + b^2$.
   (c) no
   (d) Yes; $[(0,0)] = \{(0,0)\}$; for $(a,b) \neq (0,0)$, $[(a,b)]$ is the square with vertices at $(r,0)$, $(0,r)$, $(-r,0)$, and $(0,-r)$, where $r = |a| + |b|$.
   (e) Yes; $[(0,0)] = \{(x,y) \mid x = 0 \text{ or } y = 0\}$; for $ab \neq 0$, $[(a,b)]$ is the hyperbola $y = ab/x$.

44. (a) For $m \in \mathbb{Z}$, $[m] = [m, m+1)$ (b) For $m \in \mathbb{Z}$, $[m] = [(2m - 1)/2, (2m+1)/2)$
   (c) $[0] = \{0\}$; for $x > 0$, $[x] = \{-x, x\}$
   (d) $[0] = \mathbb{Z}$; for $x \in \mathbb{R} - \mathbb{Z}$, $[x] = \{\ldots, x - 2, x - 1, x, x + 1, x + 2, \ldots\}$
   (e) $[0] = \mathbb{Q}$; for $x \in \mathbb{R} - \mathbb{Q}$, $[x] = \{x + r \mid r \in \mathbb{Q}\}$

51. (a) reflexive, transitive   (b) irreflexive, symmetric
   (c) irreflexive, symmetric   (d) reflexive, symmetric
   (e) symmetric

52. (a) $\neq$ (b) $<$ (c) Define A by $(m,n) \in A \leftrightarrow |m - n| \leq 1$.   (d) $\leq$

53. (b) $A^r = A \leftrightarrow A$ is reflexive   (d) $A_1 \cup \{(3,3),(4,4),(5,5)\}$ (e) $\leq$
(f) $\sim \cup \{\ldots,(-2,-2),(0,0),(2,2),\ldots\}$

54. (b) $A^s = A \leftrightarrow A$ is symmetric   (d) $A_1 \cup \{(2,1),(3,1),(4,1)\}$ (e) $\neq$

55. (a) $A_1 \cup \{(1,1),(1,3),(1,4),(2,2),(2,4)\}$ (b) $<$

58. All are true.

Exercise Set 0.4

1. (a) yes   (b) no   (c) no   (d) yes

3. (a) 3   (b) 2   (c) 2   (d) 1

Exercise Set 0.5

1.

$+$	1	r	$1+r$	r^2	$1+r^2$	$r+r^2$	$1+r+r^2$
1	0	$1+r$	r	$1+r^2$	r^2	$1+r+r^2$	$r+r^2$
r	$1+r$	0	1	$r+r^2$	$1+r+r^2$	r^2	$1+r^2$
$1+r$	r	1	0	$1+r+r^2$	$r+r^2$	$1+r^2$	r^2
r^2	$1+r^2$	$r+r^2$	$1+r+r^2$	0	1	r	$1+r$
$1+r^2$	r^2	$1+r+r^2$	$r+r^2$	1	0	$1+r$	r
$r+r^2$	$1+r+r^2$	r^2	$1+r^2$	r	$1+r$	0	1
$1+r+r^2$	$r+r^2$	$1+r^2$	r^2	$1+r$	r	1	0

$\cdot$	r	$1+r$	r^2	$1+r^2$	$r+r^2$	$1+r+r^2$
r	r^2	$r+r^2$	$1+r$	1	$1+r+r^2$	$1+r^2$
$1+r$	$r+r^2$	$1+r^2$	$1+r+r^2$	r^2	1	r
r^2	$1+r$	$1+r+r^2$	$r+r^2$	r	$1+r^2$	1
$1+r^2$	1	r^2	r	$1+r+r^2$	$1+r$	$r+r^2$
$r+r^2$	$1+r+r^2$	1	$1+r^2$	$1+r$	r	r^2
$1+r+r^2$	$1+r^2$	r	1	$r+r^2$	r^2	$1+r$

3. The addition table stays the same; here is the revised multiplication table:

$\cdot$	r	$1+r$	r^2	$1+r^2$	$r+r^2$	$1+r+r^2$
r	r^2	$r+r^2$	$1+r^2$	$1+r+r^2$	1	$1+r$
$1+r$	$r+r^2$	$1+r^2$	1	r	$1+r+r^2$	r^2
r^2	$1+r^2$	1	$1+r+r^2$	$1+r$	r	$r+r^2$
$1+r^2$	$1+r+r^2$	r	$1+r$	$r+r^2$	r^2	1
$r+r^2$	1	$1+r+r^2$	r	r^2	$1+r$	$1+r^2$
$1+r+r^2$	$1+r$	r^2	$r+r^2$	1	$1+r^2$	r

5. $F_9 = \{0,1,2,r,1+r,2+r,2r,1+2r,2+2r\}$; multiplication is based on the prime polynomial $1+x^2$, so that $r^2 = 2$. The addition and multiplication tables are given.

+	1	2	r	1+r	2+r	2r	1+2r	2+2r
1	2	0	1+r	2+r	r	1+2r	2+2r	2r
2	0	1	2+r	r	1+r	2+2r	2r	1+2r
r	1+r	2+r	2r	1+2r	2+2r	0	1	2
1+r	2+r	r	1+2r	2+2r	2r	1	2	0
2+r	r	1+r	2+2r	2r	1+2r	2	0	1
2r	1+2r	2+2r	0	1	2	r	1+r	2+r
1+2r	2+2r	2r	1	2	0	1+r	2+r	r
2+2r	2r	1+2r	2	0	1	2+r	r	1+r

·	2	r	1+r	2+r	2r	1+2r	2+2r
2	1	2r	2+2r	1+2r	r	2+r	1+r
r	2r	2	2+r	2+2r	1	1+r	1+2r
1+r	2+2r	2+r	2r	1	1+2r	2	r
2+r	1+2r	2+2r	1	r	1+r	2r	2
2r	r	1	1+2r	1+r	2	2+2r	2+r
1+2r	2+r	1+r	2	2r	2+2r	r	1
2+2r	1+r	1+2r	r	2	2+r	1	2r

Exercise Set 1.1

1. The only two Latin squares of order 2 are

$$\begin{bmatrix} 0 & 1 \\ 1 & 0 \end{bmatrix} \quad \text{and} \quad \begin{bmatrix} 1 & 0 \\ 0 & 1 \end{bmatrix}$$

and they are not orthogonal.

3. (a)

$$R = \begin{bmatrix} 0 & 1 & 2 & 3 & 4 \\ 1 & 2 & 3 & 4 & 0 \\ 2 & 3 & 4 & 0 & 1 \\ 3 & 4 & 0 & 1 & 2 \\ 4 & 0 & 1 & 2 & 3 \end{bmatrix}$$

and

$$S = \begin{bmatrix} 0 & 1 & 2 & 3 & 4 \\ 2 & 3 & 4 & 0 & 1 \\ 4 & 0 & 1 & 2 & 3 \\ 1 & 2 & 3 & 4 & 0 \\ 3 & 4 & 0 & 1 & 2 \end{bmatrix}$$

5. Denote the ranks and suits using the integers 0, 1, 2, 3, where jack = 0, queen = 1, king = 2, ace = 3, and clubs = 0, diamonds = 1, hearts = 2, spades = 3.

(a) You can use the Graeco-Latin square of order 4 in Figure 1.1.4(b).

(b)

$$\begin{bmatrix} (0,0) & (1,1) & (2,2) & (3,3) \\ (2,3) & (3,2) & (0,1) & (1,0) \\ (3,1) & (2,0) & (1,3) & (0,2) \\ (1,2) & (0,3) & (3,0) & (2,1) \end{bmatrix}$$

Exercise Set 1.2

1. 70
2. 107
3. (a) 10^9 (b) 9^9 (c) $9^2 10^7$ (d) 10!
4. 112
5. (a) $34,632$ (b) $1,280,448$ (c) $47,309,184$ (d) $33,696$
   (e) $47,342,880$
6. 20
7. (a) $7,311,616$ (b) $2,313,441$ (c) $28,561$ (d) $114,244$
   (e) $519,168$ (f) $4,998,175$
8. 1024   976
9. (a) 27   (b) 12   (c) 7
10. (a) 1   (b) 4   (c) 10   (d) 20
11. (a) 3125   (b) 120   (c) 1200   (d) 240
13. (a) 9000   (b) 4500   (c) 2673   (d) 3168   (e) 90
15. (a) 1296   (b) 360
16. (a) 1728   (b) 60   (c) 100   (d) 525   (e) 27   (f) 216
    (g) 999
17. (a) $125,000$ (b) $120,050$
18. $\prod_{i=1}^{k}(a_i + 1)$
19. (a) $16,807$ (b) 2520   (c) 0   (d) $16,564$ (e) $10,633$
    (f) 1296

Exercise Set 1.3

1. 210
2. 4410
3. (a) 6   (b) 210   (c) 720   (d) 360
4. $n \geq 7$
5. $n \geq 10$
6. 1296   360
7. (a) $40,320$ (b) $1,373,568$
9. (a) 1035   (b) 576
10. $121,080,960$
11. (a) 720   (b) 120   (c) 20   (d) 64
12. (a) 24   (b) 6   (c) 60

13. 8008

14. 209

15. 685,464

16. (a) 48   (b) 36

17. 720 (assuming the orientation of the keys is not important)

18. (a) 5040   (b) 144   (c) 96   (d) 576

19. (a) 1,098,240   (b) 54,912   (c) 3744   (d) 36   (e) 9180
    (f) 5112

20. $C(52,6)C(48,4)$; $C(52,6)C(48,4)C(6,1)C(4,1)$

21. (a) 128   (b) 35   (c) 16

22. 2^{33}

23. (a) 2^{mn} (b) 2^{m^2}

24. (a) $C(n,1)C(n+1,2)(n-1)!$
    (b) $C(n,1)C(n+2,3)(n-1)! + C(n,2)C(n+2,2)C(n,2)(n-2)!$
    (c) $C(n,1)C(n+3,4)(n-1)! + C(n,1)C(n+3,3)C(n-1,1)C(n,2)(n-2)! +$
    $C(n,3)C(n+3,2)C(n+1,2)C(n-1,2)(n-3)!$

25. (a) $(n-1)!$ (b) $(n-1)!/2$

26. (a) at least 7   (b) at least 10

27. (a) 621,075   (b) 225,225   (c) 525,525

28. (b) $(1,2,3,4),(1,2,4,3),(1,3,2,4),(1,3,4,2),(1,4,2,3),(1,4,3,2),(2,1,3,4),$
    $(2,1,4,3),(2,3,1,4),(2,3,4,1),(2,4,1,3),(2,4,3,1),(3,1,2,4),(3,1,4,2),$
    $(3,2,1,4),(3,2,4,1),(3,4,1,2),(3,4,2,1),(4,1,2,3),(4,1,3,2),$
    $(4,2,1,3),(4,2,3,1),(4,3,1,2),(4,3,2,1)$
    (c) $(1,2,3,4,7,5,6)$ (d) $(1,6,3,2,4,5,7)$

29. 2520

31. 24,676,704

32. (b) $\{1,2,3\}$, $\{1,2,4\}$, $\{1,2,5\}$, $\{1,2,6\}$, $\{1,3,4\}$, $\{1,3,5\}$,$\{1,3,6\}$, $\{1,4,5\}$, $\{1,4,6\}$,
    $\{1,5,6\}$, $\{2,3,4\}$, $\{2,3,5\}$, $\{2,3,6\}$, $\{2,4,5\}$, $\{2,4,6\}$, $\{2,5,6\}$, $\{3,4,5\}$, $\{3,4,6\}$,
    $\{3,5,6\}$, $\{4,5,6\}$
    (c) $\{1,3,4,6\}$ (d) $\{2,5,6,7\}$

35. (a) n^m (b) $P(n,m)$

36. (b) $\{\ \}$, $\{1\}$, $\{1,2\}$, $\{1,2,3\}$, $\{1,2,3,4\}$, $\{1,2,4\}$, $\{1,3\}$, $\{1,3,4\}$, $\{1,4\}$, $\{2\}$, $\{2,3\}$,
    $\{2,3,4\}$, $\{2,4\}$, $\{3\}$, $\{3,4\}$, $\{4\}$
    (c) $\{1,3,4,5,6\}$ (d) $\{1,2,3,4,5,7\}$ (e) $\{2\}$

37. (a) $C(m+n-1,n-1)$ (b) $C(n,m)$

Exercise Set 1.5

1. 6188

2. (a) $2n$ (b) $2n(n-1)$ (c) $4n(n-1)(n-2)/3$ (d) $C(n,x)2^x$

3. 495

4. (a) 91, 125　　　　　　(b) 60

5. (a) 279, 936　　　　(b) 15, 120　　　　(c) 210　　　　(d) 1890

6. (a) 35　　　　(b) 1680　　　　(c) 165

7. (a) 16, 777, 216　　　　(b) 455　　　　(c) 10, 670, 040　　　　(d) 165

    (e) 369, 600　　　　(f) 1

8. (a) 12, 600　　　　(b) 20

9. (a) $C(52; 5, 5, 5, 5, 32)$　　　　(b) $C(4; 1, 1, 1, 1)C(48; 4, 4, 4, 4, 32)$　　　　(c) $C(13, 5)^4$

10. (a) 756　　　　(b) 90　　　　(c) 6

11. (a) 14, 080　　　　(b) 1344　　　　(c) 64

12. 84

13. 252

15. $(2n!)/(2^n n!)$

17. (a) $C(35, 9)$　　　　(b) $C(26, 3)C(9, 2)$　　　　(c) $C(26, 10)$

19. (a) 46, 376　　　　(b) 23, 751　　　　(c) 3876

Exercise Set 1.6

1. (a) $C(4, 2) = C(3, 2) + C(3, 1) = C(2, 2) + C(2, 1) + C(2, 1) + C(2, 0) = 1 + C(1, 1) + C(1, 0) +$
    $C(1, 1) + C(1, 0) + 1 = 6$

3. (a) $(x + y)^5 = x^5 + 5x^4y + 10x^3y^2 + 10x^2y^3 + 5xy^4 + y^5$
    (b) $(x + y)^6 = x^6 + 6x^5y + 15x^4y^2 + 20x^3y^3 + 15x^2y^4 + 6xy^5 + y^6$
    (c) $(x + 3y)^7 = x^7 + 21x^6y + 189x^5y^2 + 945x^4y^3 + 2835x^3y^4 + 5103x^2y^5 + 5103xy^6 + 2187y^7$

7. (a) 1365　　　　(b) 13, 440　　　　(c) $-48, 384$　　　　(d) 41

9. (a) $C(13, 5)$　　　　(b) $C(12, 6)$　　　　(c) 2^{10}　　　　(d) $9 \cdot 2^8$

13. (b) $C(6, 3) = 6C(5, 3)/3 = 2C(5, 3) = 2 \cdot 5 \cdot C(4, 3)/2 = 5C(4, 3) = 5 \cdot 4 \cdot C(3, 3) = 20$

14. (b) $n^2(n + 1)^2/4$

17. (b) $(1 + r)^n$

19. $(n + 1)2^n$

25. (a) 12, 600　　　　(b) 504　　　　(c) $-967, 680$　　　　(d) 49

Chapter 1 Problems

1. (a) 120　　　　(b) 84

2. (a) $C(52, 13)$　　　　(b) $C(4, 2)C(48, 11)$　　　　(c) $C(4, 3)C(4, 2)C(4, 1)C(4, 0)C(36, 7)$

    (d) $C(52, 13) - C(39, 13)$　　　　(e) $C(13, 4)C(13, 3)^3$　　　　(f) $C(4, 1)C(13, 4)C(13, 3)^3$

3. 11, 550

4. 50, 000

5. (a) $C(8, 3)25^5$　　　　(b) $C(8, 3)5^3 21^5 + C(8, 4)5^4 21^4$　　　　(c) $P(26, 8)$

(d) $25^8 + C(8,2)25^6 + C(8,4)25^4 + C(8,6)25^2 + 1$

6. 961

7. (a) 8568   (b) 4806   (c) 7098   (d) 2184

8. $m(m+1)^n$

9. (a) 288   (b) 360   (c) 3039   (d) 2925

10. (a) $2^{n(n-1)}$ (b) $2^{n(n-1)}$ (c) $2^{n(n+1)/2}$ (d) $2^n 3^{n(n-1)/2}$

11. (a) $P(26,26)$ (b) $P(21,21)P(22,5)$

    (c) $P(25,25)$ (d) $P(20,20)P(21,5)$

12. (a) 462   (b) 434

13. (a) 495   (b) 105

14. (a) 12   (b) 4320   (c) $n! - (n - m + 1)!m!$

15. (a) 55   (b) 52

18. (a) 10^{20} (b) $C(20; 2, 2, \ldots, 2)$

19. (a) $362,880$ (b) $22,680$ (c) 1680   (d) 36

20. (a) $490,314$ (b) $12,870$ (c) 2520   (d) 35

    (e) If order matters, the answer is 4096; if order does not matter, 816.

21. (a) 84   (b) 35

22. (a) 30   (b) 720   (c) 180   (d) 474   (e) 66

    (f) $130,636,800$

24. (a) $3,628,800$ (b) 8640

27. $43,750$

28. $208 + 120\sqrt{3}$

29. -99

31. (a) 1 7 21 35 35 21 7 1   (b) 1 8 28 56 70 56 28 8 1

32. $2C(n+1,3)$

36. $1/(n+1)$

38. $C(2n, n-1)$

45. $C(n+m-1, 2n-1)$

46. $2C(17; 10, 6, 1) + 2^3 C(17; 13, 1, 3)$

49. $132 + 72\sqrt{2} + 48\sqrt{5} + 40\sqrt{10}$

54. $43,589,145,600$

55. (a) 165   (b) 35   (c) $65,536$ (d) $6,652,800$

    (e) $1,411,200$ (f) $40,320$

56. $m!C(m+n-1, n-1)$

57. $26,345,088,000$

58. $m!C(m-1, n-1)$

59. $479,001,600$

60. $m!$

Exercise Set 2.1

1. 200

2. (a) 18   (b) 27

3. 6,792,448

4. (a) 184   (b) 41

5. 15

6. 1236

7. (a) 42   (b) 78

8. 1918

9. 2304

10. 49,974,120

11. 62,064

12. 65,280

13. 366

14. 126

15. $\sum_{k=1}^{t}(-1)^{k-1}C(t,k)C(nt-nk+k; 1,\ldots,1,n,\ldots,n)$

16. 746,856

17. (a) $(n-1)^{(n+1)/2}n^{(n-1)/2}$ (b) $\sum_{k=0}^{(n+1)/2}(-1)^kC((n+1)/2,k)(n-k)!$

18. (a) 2   (b) 32   (c) 1,488   (d) $\sum_{k=0}^{n}(-2)^kC(n,k)(2n-k-1)!$

19. 999,998,990,100

21. 197,131,200

22. 315

23. $C(n,k)d(n-k)$

25. $C(n,k)\sum_{j=0}^{n-k}(-1)^j2^{k+j}C(n-k,j)(2n-k-j-1)!$

26. $\sum_{k=0}^{m}(-1)^kC(m,k)P(n-k,m-k)$

27. (a) 101,644   (b) 44

28. $(n-1)!d(n)$

29. 53

31. 9

Exercise Set 2.2

1. $k(m) = (n - m + 1)k(m - 1)/m$

2. (a) $h_1(0) = 1$, $h_1(1) = 3$, $h_1(n) = 2(h_1(n - 1) + h_1(n - 2))$, $n \geq 2$
   (b) $h_2(0) = 1$, $h_2(1) = 3$, $h_2(n) = 2h_2(n - 1) + h_2(n - 2)$, $n \geq 2$
   (c) $h_3(0) = 1$, $h_3(1) = 3$, $h_3(2) = 9$, $h_3(n) = 2(h_3(n - 1) + h_3(n - 2) + h_3(n - 3))$, $n \geq 3$

   (d) $h_4(0) = 1$, $h_4(1) = 3$, $h_4(n) = 1 + 2h_4(n - 1) + \sum_{k=0}^{n-2} h_4(k)$, $n \geq 2$

3. $t(1) = 1$, $t(n) = (2n - 1)t(n - 1)$, $n \geq 2$

4. $v(n) = (1.1)v(n - 1) + 100$

5. $w(n) = w(n - 1) + w(n - 2)$

6. $r(n) = r(n - 1) + n$

7. For convenience, define $t(2) = 1$; then $t(n) = \sum_{k=3}^{n} t(k - 1)t(n - k + 2)$, $n \geq 3$

8. $b(n) = b(n - 1) + b(n - 2)$

9. $p(n) = (1.1)p(n - 1) - 6000$

10. (a) $r(n) = r(n - 1) + 1$
    (b) $m(n) = m(n - 1) + 2^{n-1}$
    (c) $p(n) = (100)2^{n-1} + 2p(n - 1)$

11. $b(n) = 2b(n - 1) + 1$

12. (a) $g_1(n) = 2(n - 1)g_1(n - 1)$, $n \geq 3$

15. (a) $s(m) = s(m - 1) + 1$
    (b) $s(m) = 2s(m/2) + 1$

17. (a) $t(m) = t(m - 1) + (m - 1)$

19. (a) $h_1(n) = (4^n - (-1)^n)/5$ (b) $h_2(n) = 2$
    (c) $h_3(n) = (1 + (-1)^n)/2$
    (d) $h_4(n) = [(-1)^n + (2 - \sqrt{3})^n + (2 + \sqrt{3})^n]/3$
    (e) $h_5(n) = (1 - (-1)^n)2^{n-2}$ (f) $h_6(n) = (n + 1)!$
    (g) $h_7(n) = [(4)3^n - 3 - (-3)^n]/12$ (h) $h_8(n) = (n - 2)4^{n-1}$
    (i) $h_9(n) = [(-2)^n + 8 - 6n]/9$ (j) $h_0(n) = (-1)^n + (n^2 - n - 1)2^n$

21. (b) $g(n) = 3^n - 2^{n+1}$

22. (b) $h(n) = [i^n - (-i)^n]/(2i)$

23. (a) $a_1(n) = (-1)^n + (3n - 1)2^n$
    (b) $a_2(n) = 2(-1)^n + (4n - 1)3^n$
    (c) $a_3(n) = 3^n + 3(i^n + (-i)^n)$

25. $s(m, 1) = 1 = s(m, m)$, $s(m, n) = s(m - 1, n - 1) + s(m - 1, n)$, $1 < n < m$

26. (a) $f_1(n) = an + b$ (b) $f_2(n) = an^2 + c$ (c) $f_3(n) = an^2 + bn + c$
    (d) $f_4 = a^n$ (e) $f_5(n) = a^n + b$ (f) $f_6(n) = a$

27. (b) $f(2n + 2) - 1$ (c) $f(2n + 1)$ (d) $(-1)^n f(n - 1) + 1$, $n \geq 1$

Exercise Set 2.3

1. $(1 + x + x^2 + x^3)^2(1 + x + x^2 + x^3 + x^4)^2$

3. $(x^2 + x^4 + x^6 + x^8)^3(x + x^2 + x^3 + x^4)^2$

4. (a) $x^k g(x)$ (b) $[g(x) - (c_0 + c_1 x + \cdots + c_{k-1} x^{k-1})]/x^k$

5. (a) $(1 - x)^m$ (b) $x^4/(1 + x)$ (c) $2x/(1 - x)^2$
    (d) $(1 + x)/(1 - x)^2$ (e) $x(1 + x)/(1 - x)^3$ (f) $(4 - 3x)/(1 - x)^2$
    (g) e^x (h) $x(1 + 4x + x^2)/(1 - x)^4$ (i) $x^3/(1 - x)^4$
    (j) $(\ln(1 + x))/x$

7. (a) The generating function is $g_1(x) = (1 - 7x)/[(1 - 2x)(1 - 3x)]$; $f_1(n) = 5 \cdot 2^n - 4 \cdot 3^n$
    (b) $g_2(x) = x/(1 - 4x^2)$; $f_2(n) = 0$ if n is even, $f_2(n) = 2^{n-1}$ if n is odd
    (c) $g_3(x) = x(1 + x)/[(1 + 3x)(1 - 3x)(1 - x)]$; $f_3(n) = (4 \cdot 3^{n-1} + (-3)^{n-1} - 1)/4$
    (d) $g_4(x) = (8x - 1)/(1 - 4x)^2$; $f_4(n) = (n - 1)4^n$
    (e) $g_5(x) = (1 - 3x^2)/[(1 + 2x)(1 - x)^2]$; $f_5(n) = ((-2)^n - 6n + 8)/9$

8. $x/[(1 - x)(1 - x^2)^2]$

11. $1/[(1 - x)(1 - x^5)(1 - x^{10})(1 - x^{25})]$

12. $(1 - x^{30})/[(1 - x)(1 - x^{25})]$

13. $x^5(1 - x^{10})^{10}/(1 - x^2)^{10}$

15. $1/[(1 - x)(1 - x^2)(1 - x^3)]$

17. $1/(1 - x)^m$

19. $(1 + x)^8/(1 - x)^2$

20. $x^m(1 - x^6)^m/(1 - x)^m$

21. $(1 - x^5)^5/[x^{10}(1 - x)^5]$

22. $x^7/(1 - x)^5$

24. (a) 4^n (b) $\sqrt{\frac{m}{m-4}}$

Exercise Set 2.4

1.

$m = 8$:	1	127	966	1701	1050	266	28	1	
$m = 9$:	1	255	3025	7770	6951	2646	462	36	1

3. $B(7) = 877$, $B(8) = 4140$, $B(9) = 21147$

5.

$m = 1$:	1								
$m = 2$:	1	1							
$m = 3$:	1	1	1						
$m = 4$:	1	2	1	1					
$m = 5$:	1	2	2	1	1				
$m = 6$:	1	3	3	2	1	1			
$m = 7$:	1	3	4	3	2	1	1		
$m = 8$:	1	4	5	5	3	2	1	1	
$m = 9$:	1	4	7	6	5	3	2	1	1

7. $n^4 = 1(n)_1 + 7(n)_2 + 6(n)_3 + 1(n)_4$

9. $\{\{1\}, \{2\}, \{3, 4, 5\}\}$; $\{\{1\}, \{2, 3\}, \{4, 5\}\}$; $\{\{1\}, \{2, 3, 4\}, \{5\}\}$; $\{\{1\}, \{2, 3, 5\}, \{4\}\}$;
$\{\{1\}, \{2, 4\}, \{3, 5\}\}$; $\{\{1\}, \{2, 4, 5\}, \{3\}\}$; $\{\{1\}, \{2, 5\}, \{3, 4\}\}$; $\{\{1, 2\}, \{3\}, \{4, 5\}\}$;
$\{\{1, 2\}, \{3, 4\}, \{5\}\}$; $\{\{1, 2\}, \{3, 5\}, \{4\}\}$; $\{\{1, 2, 3\}, \{4\}, \{5\}\}$; $\{\{1, 2, 4\}, \{3\}, \{5\}\}$;
$\{\{1, 2, 5\}, \{3\}, \{4\}\}$; $\{\{1, 3\}, \{2\}, \{4, 5\}\}$; $\{\{1, 3\}, \{2, 4\}, \{5\}\}$; $\{\{1, 3\}, \{2, 5\}, \{4\}\}$;
$\{\{1, 3, 4\}, \{2\}, \{5\}\}$; $\{\{1, 3, 5\}, \{2\}, \{4\}\}$; $\{\{1, 4\}, \{2\}, \{3, 5\}\}$; $\{\{1, 4\}, \{2, 3\}, \{5\}\}$;
$\{\{1, 4\}, \{2, 5\}, \{3\}\}$; $\{\{1, 4, 5\}, \{2\}, \{3\}\}$; $\{\{1, 5\}, \{2\}, \{3, 4\}\}$; $\{\{1, 5\}, \{2, 3\}, \{4\}\}$;
$\{\{1, 5\}, \{2, 4\}, \{3\}\}$;

10. (a) $2^{m-1} - 1$ (b) $C(m, 2)$

11. (a) $(0, 3, 1, 0, 0, 1)$ (c) $5 + 5 + 2 + 1 + 1 + 1$ (d) $(3, 1, 0, 0, 2)$

13. (a) $700, 075$ (b) 34   (c) $611, 501$ (d) 15

(e) $15, 400$ (f) 1

15. 350

17. 6; $5 + 1$; $4 + 2$; $3 + 3$; $4 + 1 + 1$; $3 + 2 + 1$; $2 + 2 + 2$; $3 + 1 + 1 + 1$; $2 + 2 + 1 + 1$;
$2 + 1 + 1 + 1 + 1$; $1 + 1 + 1 + 1 + 1 + 1$

19. 2

23. (a) $\lfloor m/2 \rfloor$

25. (a) 1 (b) 2 (c) 3 (d) 5

26. $\prod_{k=1}^{m} 1/(1 - x^k)$

Exercise Set 2.5

1. (a) $(1, 2, 4, 5, 8)$ (b) $(1, 5, 3, 7, 2)$ (c) $(2, 5, 1, 3, 4)$ (d) $(1, 2, 3, 4, 5)$

3. (a) 0   (b) 391   (c) 1199   (d) 6719

5. (a) $(1, 2, 3, 4, 5)$ (b) $(1, 2, 4, 6, 7)$ (c) $(2, 4, 8, 7, 6)$ (d) $(8, 7, 6, 5, 4)$

Exercise Set 2.6

1. (a) $\{1, 3, 5, 6\}$ (b) $\{1, 2, 5\}$ (c) $\{2, 4, 6\}$ (d) $\{\ \}$

3. (a) 0   (b) 10   (c) 23   (d) 31

5. (a) $\{\ \}$ (b) $\{1, 3, 4\}$ (c) $\{2, 5\}$ (d) $\{5\}$

7. (a) $\{1, 2, 3, 5\}$ (b) $\{1, 3, 4, 8\}$ (c) $\{2, 5, 6, 7\}$ (d) $\{1, 2, 3, 4\}$

9. (a) 0   (b) 17   (c) 50   (d) 69

11. (a) $\{1, 2, 3, 4\}$ (b) $\{1, 4, 5, 8\}$ (c) $\{3, 5, 7, 8\}$ (d) $\{5, 6, 7, 8\}$

Chapter 2 Problems

1. 110

2. 19

3. (a) $q(1) = q(2) = 1$, $q(3) = 3$ (b) $q(4) = 11$

   (c) $q(n) = \sum_{k=0}^{n-1} (-1)^k C(n-1,k)(n-k)!$

4. (a) $14,833$ (b) 216   (c) $30,960$

5. (a) $q^*(1) = 1$, $q^*(2) = 0$, $q^*(3) = q^*(4) = 1$ (b) $q^*(5) = 8$

   (c) For $n > 1$, $q^*(n) = (-1)^n + \sum_{k=0}^{n-1} (-1)^k C(n,k)(n-k-1)!$

10. (a) 360   (b) 216   (c) 120

11. 60

12. $1,235,520$

13. 20

15. 800

18. $F(m, 1) = 1$, $F(m, n) = 0$, $m < 2n$

    $F(m, n) = F(m-2, n-1) + F(m-3, n-1) + F(m-4, n-1)$, $4 \le 2n \le m$

19. (a) $w(1) = w(2) = 1$, $w(3) = 2$
    (b) $w(n) = w(n-1) + w(n-3)$, $n \ge 4$

20. $t(0) = 1$, $t(n) = t(n-1) + 3^{n-1}$, $n \ge 1$

21. $g_3(1) = 1$, $g_3(n) = C(3n-1,2)g_3(n-1)$, $n \ge 2$

22. $g_k(1) = 1$, $g_k(n) = C(kn-1,k-1)g_k(n-1)$, $n \ge 2$

23. (a) $h_1(n) = 5 \cdot 2^n - 4 \cdot 3^n$ (b) $h_2(n) = 2^{n-2}(1 + (-1)^{n+1})$

    (c) $h_3(n) = [4 \cdot 3^{n-1} + (-3)^{n-1} - 1]/4$ (d) $h_4(n) = (n-1)4^n$

    (e) $h_5(n) = [(-2)^n + 8 - 6n]/9$ (f) $h_6(n) = 2(-3)^n + 4n - 1$

    (g) $h_7(n) = 2^n \cdot n!$ (h) $h_8(n) = 8 + (20n - 21)3^{n-1}$

24. (a) $f_7(n) = 3^n + 4$ (b) $f_8(n) = n + 2 - 2^n$

    (c) $f_9(n) = (-2)^n + 9n^2 + 21n + 8$ (d) $f_0(n) = [10(-3)^n + 4^{n+1}]/7$

27. (a) $1/(1-2x)$ (b) $(1 + 2x - x^2)/(1-x)^3$ (c) $1/(1-x)^5$

    (d) $(1+x)/(1-x)^2$

28. (b) $b(n) = 3 \cdot 2^n - n - 2$

29. (b) $c(n) = (-2)^n + 9n^2 + 21n + 8$

30. (b) $a(n) = 4 \cdot 3^n - 2n^2 - 4n - 3$

35. (a) 877   (b) $21,147$

48. $n + 1$

49. $x^k/(1-x^2)^k$

54. $(3^{m-1} - 2^m + 1)/2$

57. $c(0) = 1$, $c(n) = \sum_{k=1}^{n} c(k-1)c(n-k)$, $n \ge 1$

Exercise Set 3.1

1. $G_1 = (\{u, v, w, x, y, z\}, \{uv, vw, vz, wx, xy, xz, yz\})$; 7

2. $\Delta(G) \leq 1$

5. (a) K_n

   (b) Let $G_n = (\{v_1, v_2, \ldots, v_{2n}\}, \{v_1 v_2, v_3 v_4, \ldots, v_{2n-1} v_{2n}\})$

   (c) C_n

   (d) Let $H_n = (\{v_1, v_2, \ldots, v_{2n}\}, \{v_1 v_2, v_2 v_3, \ldots, v_{2n-1} v_{2n}, v_{2n} v_1\} \cup \{v_1 v_{n+1}, v_2 v_{n+2}, \ldots, v_n v_{2n}\})$

7. (a) Define $\phi \colon V(G_1) \rightarrow V(G_2)$ by $\phi(u_1) = v_4$, $\phi(u_2) = v_2$, $\phi(u_3) = v_5$, $\phi(u_4) = v_1$, $\phi(u_5) = v_6$, $\phi(u_6) = v_3$; then ϕ is an isomorphism. The graph G_3 is not isomorphic to G_1 because G_3 has size 6 whereas G_1 has size 7.

   (b) Define $\phi \colon V(G_4) \rightarrow V(G_6)$ by $\phi(u_1) = x_1$, $\phi(u_2) = x_2$, $\phi(u_3) = x_4$, $\phi(u_4) = x_3$, $\phi(u_5) = x_5$; then ϕ is an isomorphism. The graph G_5 is not isomorphic to G_4 because G_5 has a vertex of degree 2 and G_4 does not.

   (c) Define $\phi \colon V(G_7) \rightarrow V(G_9)$ by $\phi(u_1) = x_2$, $\phi(u_2) = x_3$, $\phi(u_3) = x_4$, $\phi(u_4) = x_6$, $\phi(u_5) = x_1$, $\phi(u_6) = x_5$; then ϕ is an isomorphism. The graph G_8 contains a cycle of length 4 whereas G_7 does not, so G_8 is not isomorphic to G_7.

8. (a) Let $G_1 = (\{u, v, w, x, y\}, \{uv, vw, vy, wx, xy\})$.

   (b) Such a graph would have three vertices of odd degree, contradicting Corollary 3.1.2.

   (c) Let G be a graph of order 5 and let $x \in V(G)$ with $\deg(x) = 4$. Then $vx \in E(G)$ for all $v \in V(G) - \{x\}$, so that $\delta(G) \geq 1$.

   (d) Let $G_2 = (\{u, v, w, x, y, z\}, \{uv, uz, vw, wx, wy, wz, xy, xz, yz\})$.

   (e) Let $G_3 = (\{u, v, w, x, y, z\}, \{uz, vy, vz, wx, wy, wz, xy, xz, yz\})$.

   (f) Let $G_4 = (\{t, u, v, w, x, y, z\}, \{ty, tz, ux, uy, uz, vw, vx, vy, vz, wx, wy, wz, xy, xz\})$.

9. (a) Let $G_1 = (\{u, v\}, \{uv\})$, $G_2 = (\{u, v\}, \{\ \})$.

   (b) Let $G_1 = (\{u, v, x, y\}, \{uv, xy\})$, $G_2 = (\{u, v, x, y\}, \{uv, ux\})$.

   (c) Let $G_1 = (\{u_1, u_2, u_3, u_4, u_5\}, \{u_1 u_2, u_2 u_3, u_2 u_5, u_3 u_4, u_4 u_5\})$, $G_2 = (\{u_1, u_2, u_3, u_4, u_5\}, \{u_1 u_2, u_2 u_3, u_3 u_4, u_3 u_5, u_4 u_5\})$.

16. (d) Let $G_r = (\{u_1, u_2, \ldots, u_{r+1}\} \cup \{v_1, v_2, \ldots, v_{r+1}\}, \{u_i v_j \mid i \neq j\})$.

17. (a) Either G is isomorphic to $K_{3,3}$ or G is isomorphic to $(\{u_1, u_2, u_3, v_1, v_2, v_3\}, \{u_1 u_2, u_2 u_3, u_3 u_1, v_1 v_2, v_2 v_3, v_3 v_1, u_1 v_1, u_2 v_2, u_3 v_3\})$.

    (b) G is isomorphic to K_5.

19. Hint: Let n_m be the number of graphs of order 5 and size m, up to isomorphism. Then $n_m = n_{10-m}$, $0 \leq m \leq 4$ (Why?), and $n_0 = n_1 = 1$, $n_2 = 2$, $n_3 = 4$, $n_4 = n_5 = 6$.

21. $E(\overline{G_1}) = \{u_1 u_3, u_1 u_4, u_2 u_4, u_2 u_6, u_3 u_5, u_3 u_6, u_4 u_6, u_5 u_6\}$; $E(\overline{G_4}) = \{u_2 u_3, u_4 u_5\}$;

    $E(\overline{G_7}) = \{u_1 u_3, u_1 u_6, u_2 u_4, u_2 u_5, u_2 u_6, u_3 u_5, u_4 u_5, u_4 u_6\}$

22. (a) $C(n, 2) - m$ (b) $n - 1 - d$ (c) $n - 1 - \Delta$ (d) $n - 1 - \delta$

23. (a) $u, x, w, z, y, x, w, t, v$ (b) u, t, w, z, y, x, w, v (c) u, t, v (d) $u, t, w, x, y, z,$

    (e) $u, t, v, z, y, v, w, x, u$ (f) u, t, w, x, u (g) u, t, v, w, z, y, x, u

Exercise Set 3.2

1. The graphs G_3 and G_6 are eulerian.

   (c) $u_1, u_2, u_3, u_1, u_4, u_3, u_6, u_5, u_1$ is an eulerian circuit of G_3.

(f) $v_1, v_2, v_3, v_4, v_5, v_6, v_3, v_1, v_4, v_2, v_6, v_1$ is an eulerian circuit of G_6.

3. The graphs G_2 and G_4 are edge-traceable.

   (b) $v_3, v_2, v_1, v_5, v_2, v_4, v_3, v_5, v_4$ is an eulerian trail of G_2.
   (d) $v_2, v_1, v_6, v_2, v_3, v_4, v_5, v_3$ is an eulerian trail of G_4.

4. (a) Note that vertices a, b, c, and d all have odd degree.
   (b) Cross from region a to region b using one bridge and then cross back to region a using the other bridge; next, cross from a to c and then back to a using the two bridges joining regions a and c; then cross to region d; then cross from d to b and back using the two bridges joining these regions; finally, cross from d to c and then from c to region b.

Exercise Set 3.3

1. (a) u, x, w, v, u (b) u, z, w, y, v, x (c) $z, u, z, w, x, w, v, u, x$
   (d) One possibility is the walk $u, z, u, z, y, x, w, x, w, v, u, z, w, v, y, x, w, v, u, x$.
   (a) The distance matrix (see Exercise 3.1.20) is as follows, where the rows and columns correspond to u, v, x, y, and z, respectively.

$$\begin{bmatrix} 0 & 1 & 4 & 3 & 2 \\ 2 & 0 & 3 & 2 & 1 \\ 3 & 1 & 0 & 3 & 2 \\ 4 & 2 & 1 & 0 & 3 \\ 1 & 2 & 2 & 1 & 0 \end{bmatrix}$$

2. The distance matrix is as follows, where the rows and columns correspond to u, v, w, x, y, and z, respectively.

$$\begin{bmatrix} 0 & 3 & 2 & 1 & 2 & 1 \\ 1 & 0 & 3 & 2 & 1 & 2 \\ 2 & 1 & 0 & 1 & 2 & 3 \\ 3 & 2 & 1 & 0 & 3 & 4 \\ 4 & 3 & 2 & 1 & 0 & 5 \\ 1 & 2 & 1 & 2 & 1 & 0 \end{bmatrix}$$

4. $G = (\{u, v, x, y\}, \{(u, x), (v, u), (x, v), (x, y), (y, u)\})$

5. $\text{id}(u) = 1$, $\text{id}(v) = 2$, $\text{id}(w) = 2$, $\text{id}(x) = 3$, $\text{id}(y) = 2$, $\text{id}(z) = 1$,
   $\text{od}(u) = 3$, $\text{od}(v) = 1$, $\text{od}(w) = 2$, $\text{od}(x) = 1$, $\text{od}(y) = 1$, $\text{od}(z) = 3$

6. This is false; see the digraph G in the answer to Exercise 4.

7. Let $A = \{(v_i, v_j) \mid j < i\}$.

11. (a) The arc set of the complement is $\{(u, u), (u, v), (u, x), (u, y), (u, z), (v, v), (v, y),$
    $(v, z), (x, u), (x, v), (x, x), (x, z), (y, v), (y, y), (y, z), (z, v), (z, x), (z, y)\}$.
    (b) The arc set of the complement is $\{(u, u), (u, w), (u, y), (v, u), (v, v), (v, x), (v, z), (w, u),$
    $(w, w), (w, y), (w, z), (x, u), (x, v), (x, x), (x, y), (x, z), (y, u), (y, v), (y, w), (y, y), (y, z), (z, v),$
    $(z, x), (z, z)\}$.

13. (a) Define $\phi \colon V(G_1) \to V(G_2)$ by $\phi(a) = y$, $\phi(b) = v$, $\phi(c) = x$, $\phi(d) = z$, $\phi(e) = u$; verify that ϕ is an isomorphism.

(b) Suppose that $\phi: V(G_3) \rightarrow V(G)$ is an isomorphism. Since G_3 has exactly one loop at a and G has a loop at z, it must be that $\phi(a) = z$. Then, since $(a, e) \in A(G_3)$ and $(z, u) \in A(G)$, we have that $\phi(e) = u$. But then $(e, b) \in A(G_3)$ implies $(u, \phi(b)) \in A(G)$, a contradiction since u has outdegree 0 in G.

15. (a) $G_1 = (\{u\}, \{\ \})$, $G_2 = (\{u\}, \{(u, u)\})$
   (b) $G_1 = (\{u, v\}, \{(u, v)\})$, $G_2 = (\{u, v\}, \{(u, u)\})$
   (c) $G_1 = (\{u, v\}, \{(u, v), (v, u)\})$, $G_2 = (\{u, v\}, \{(u, u), (v, v)\})$; or, if loops are not allowed, let $G_1 = (\{u, v, x, y\}, \{(u, v), (v, x), (x, y), (y, u)\})$ and $G_2 = (\{u, v, x, y\}, \{(u, v), (v, u), (x, y), (y, x)\})$.

17. Hint: There are 16 of them.

20. (a) $1, 2, 1, 4, 3, 2, 5, 4, 7, 3, 7, 6, 5, 6, 3, 1$
   (b) $1, 2, 4, 3, 2, 5, 1, 6, 3, 4, 5, 6, 1$
   (c) $1, 2, 3, 4, 5, 2, 6, 7, 2, 8, 3, 9, 7, 8, 9, 6, 5, 1$

21. $t, z, y, x, y, v, z, v, u, x, w, v, t, z, u, t$

23. (a) The arc set of the converse is $\{(u, v), (u, y), (u, z), (x, v), (x, y), (y, x), (z, z)\}$.
   (b) The arc set of the converse is $\{(u, z), (v, u), (v, w), (w, x), (w, z), (x, u), (x, w), (x, y), (y, v), (y, z), (z, u)\}$.

Exercise Set 3.4

1. $A^t = (V \times V) - \{(v, u), (v, z), (w, u), (w, z), (x, u), (x, z), (y, u), (y, z)\}$

3. Note that each digraph in Figure 3.4.7 is strong, so the transitive closure is the complete digraph in each case.

5. (a) $H_1 = (\{u, v, w, x, y\}, \{(u, v), (u, x), (v, y), (w, v), (w, x), (x, w), (y, x)\})$
   (b) $H_2 = (\{u, v, x, y\}, \{(u, v), (u, x), (v, y), (y, x)\})$
   (c) $H_3 = (\{u, v, x, y\}, \{(u, v), (u, x), (v, y), (y, x)\})$
   (d) $H_4 = (\{u, v, w, x, y, z\}, \{(u, v), (u, x), (u, z), (v, y), (w, v), (x, w), (y, x), (z, u), (z, w), (z, y)\})$
   (e) $H_5 = (\{u, v, x, y, z\}, \{(u, v), (u, z), (v, y), (y, x)\})$

7. We give the condensation.

   (a) $(\{\{1, 2, 5, 6\}, \{3, 4, 7\}, \{8\}\}, \{(\{3, 4, 7\}, \{1, 2, 5, 6\}), (\{8\}, \{1, 2, 5, 6\}), (\{8\}, \{3, 4, 7\})\})$
   (b) $(\{\{1, 2, 3\}, \{4, 5, 6\}, \{7\}, \{8, 9\}\}, \{(\{1, 2, 3\}, \{4, 5, 6\}), (\{7\}, \{8, 9\}), (\{8, 9\}, \{1, 2, 3\}), (\{8, 9\}, \{4, 5, 6\})\})$

8. (a) True; for $U \subset V$, the subgraph $\langle U \rangle$ can be obtained by deleting the vertices of $V - U$.
   (b) False; note that a subgraph obtained by deleting arcs must be a spanning subgraph, whereas an arc-induced subgraph need not be spanning.

13. The shortest c_0-c_i paths, $2 \le i \le 6$, are as follows:

$$c_0, c_2; \quad c_0, c_3; \quad c_0, c_3, c_4; \quad c_0, c_3, c_4, c_5; \quad c_0, c_3, c_6$$

14. The ith row sum gives $od(v_i)$; the jth column sum gives $id(v_j)$.

15. Shortest 0-v paths, $2 \le v \le 5$, are:

$$0, 3, 1, 2; \quad 0, 3; \quad 0, 4; \quad 0, 4, 5$$

17. We give the final computed values in each part.

(a) $d(2) = 2$, $d(3) = 3$, $d(4) = 5$, $d(5) = 4$, $p(2) = 1$, $p(3) = 1$, $p(4) = 1$, $p(5) = 3$

(b) $d(2) = 14$, $d(3) = 5$, $d(4) = 10$, $d(5) = 8$, $d(6) = 1$, $p(2) = 3$, $p(3) = 6$, $p(4) = 5$, $p(5) = 6$, $p(6) = 1$

(c) $d(2) = 5$, $d(3) = 2$, $d(4) = 7$, $d(5) = 7$, $d(6) = 17$, $p(2) = 1$, $p(3) = 1$, $p(4) = 3$, $p(5) = 1$, $p(6) = 4$

(d) $d(2) = 16$, $d(3) = 17$, $d(4) = 11$, $d(5) = 16$, $d(6) = 18$, $d(7) = 9$, $p(2) = 4$, $p(3) = 2$, $p(4) = 1$, $p(5) = 7$, $p(6) = 7$, $p(7) = 1$

Exercise Set 3.5

1. (a) $(\{1, 3, 4, 5, 6, 7, 8, 9\}, \{(1, 3), (3, 6), (3, 7), (3, 8), (4, 5), (5, 9), (7, 4)\})$
   (b) $(\{4, 5, 9\}, \{(4, 5), (5, 9)\})$ (c) $(\{4, 5, 7, 9\}, \{(4, 5), (5, 9), (7, 4)\})$
   (d) $(\{1, 3, 4, 5, 6, 7, 8, 9\}, \{(1, 3), (3, 6), (3, 7), (4, 5), (5, 9), (7, 4), (8, 1)\})$

3. (a) $(\{0, 1, 2, 3, 4, 5, 6, 7\}, \{(0, 1), (0, 2), (0, 3), (1, 4), (2, 5), (3, 6), (4, 7)\})$
   (b) $(\{0, 1, 2, 3, 4, 5, 6, 7, 8\}, \{(0, 1), (1, 3), (1, 4), (2, 5), (2, 6), (4, 7), (7, 8), (8, 2)\})$

7. (a) k (b) $(m^{h-k+1} - 1)/(m - 1)$, $m > 1$

8. (a) Let $T = (\{1, 2, \ldots, 15\}, \{(m, n) | n = 2m \text{ or } n = 2m + 1\})$.
   (b) Let $T = (\{1, 2, \ldots, 10\}, \{(m, n) | m \bmod 3 = 1 \text{ and } (n = m+1 \text{ or } n = m+2 \text{ or } n = m+3)\})$.
   (c) $mh + 1$

9. $(m^{h+1} - 1)/(m - 1)$, $m > 1$

11. The list $(1, 2, 3, 4, 5, 6, 7, 8, 9)$ is returned.

13. There are two, isomorphic to either $(\{0, 1, 2, 3, 4, 5, 6\}, \{01, 02, 03, 04, 15, 16\})$ or $(\{0, 1, 2, 3, 4, 5, 6\}, \{01, 02, 03, 04, 15, 26\})$.

14. Hint: There are 10 of them.

15. There are five of them, isomorphic to the spanning trees T_1, T_2, T_3, T_4, and T_5, where $E(T_1) = \{uv, ux, uy, uz, wx\}$, $E(T_2) = \{uv, ux, uz, wz, yz\}$, $E(T_3) = \{uv, ux, uy, wx, yz\}$, $E(T_4) = \{uv, ux, uy, wz, yz\}$, and $E(T_5) = \{uv, vy, wx, wz, xy\}$.

16. We give the edge set of a minimum spanning tree.

    (a) $\{st, tx, xz, yz\}$ (b) $\{rz, sy, tx, tz, xy\}$ (c) $\{rt, st, tx, ty, xz\}$
    (d) $\{qz, rs, st, sx, tz, xy\}$

17. There are two minimum spanning trees with weight 25; one has edge set $\{rx, ry, sx, tx, yz\}$ and the other has edge set $\{ry, sx, sy, tx, yz\}$.

19. One possibility is the tree with edge set $\{ad, au, az, bv, cx, uv, vx, vy\}$.

21. It is isomorphic to $(\{u_1, u_2, \ldots, u_n\}, \{u_1u_2, u_1u_3, \ldots, u_1u_{n-1}, u_{n-1}u_n\})$.

22. They are $T_1 = (V, E_1)$, $T_2 = (V, E_2)$, and $T_3 = (V, E_3)$, where $V = \{u_1, u_2, \ldots, u_n\}$, $E_1 = \{u_1u_2, u_1u_3, \ldots, u_1u_{n-2}, u_{n-2}u_{n-1}, u_{n-2}u_n\}$, $E_2 = \{u_1u_2, u_1u_3, \ldots, u_1u_{n-2}, u_{n-3}u_{n-1}, u_{n-2}u_n\}$, and $E_3 = \{u_1u_2, u_1u_3, \ldots, u_1u_{n-2}, u_{n-2}u_{n-1}, u_{n-1}u_n\}$.

31. $m = n - k$

33. We give the edge set of a minimum spanning tree.

    (a) $\{st, tx, xz, yz\}$ (b) $\{rz, sy, tx, tz, xy\}$ (c) $\{rt, st, tx, ty, xz\}$
    (d) $\{qz, rs, st, sx, tz, xy\}$

34. There are two minimum spanning trees with weight 25; one has edge set $\{rx, ry, sx, tx, yz\}$ and the other has edge set $\{ry, sx, sy, tx, yz\}$.

Exercise Set 3.6

1. (a) $(2, 1, 3, 5, 4)$ (b) No; $X_1 \cup X_2 \cup X_3 \cup X_4$ has only 3 elements.

   (c) No; $X_1 \cup X_2 \cup X_3 \cup X_4 \cup X_5$ has only 4 elements.   (d) $(1, 2, 3, 4, 5)$

6. $P(m, n)$

8. $\{x_1y_1, x_2y_2, x_3y_5, x_4y_4, x_5y_3\}$

9. C_5

12. $n = 8$; an SDR for $(X_0, X_1, \ldots, X_7)$ is $(0, 4, 1, 8, 7, 3, 9, 5)$.

13. An SDR is $(2, 1, 3, 5, 4)$, i.e., assign applicant 2 to position 1, applicant 1 to position 2, etc.

17. The number of maximum matchings of K_{2n} is $(2n)!/(2^n n!)$; for K_{2n+1}, the number is $(2n + 1)!/(2^n n!)$.

19. $P(n, m)$

20. $\{b_1w_1, b_2w_3, b_3w_2, b_4w_4, b_5w_5, b_6w_6\}$

21. $\{x_1, x_2, x_3, y_4, x_5\}$

23. $\{x_1, x_5, y_2, y_3\}$

Exercise Set 3.7

4. (a) $E(Cl(G)) = E(G) \cup \{sw, sy, tz, uy\}$ (b) $Cl(G) = G$ (c) $Cl(G) \simeq K_8$

5. The closure is isomorphic to a complete graph in each part.

6. (a) $Cl(G_1) = (V, E_1 \cup \{ty\})$; G_1 is not hamiltonian, since $G_1 - y$ has two components.
   (b) $Cl(G_2) = G_2$; the cycle t, w, z, v, y, u, x, t is a hamiltonian cycle of G_2.
   (c) $Cl(G_3) \simeq K_7$; thus, G_3 is hamiltonian.

9. (a) Let G_3 have order 6 and degree sequence $(2, 2, 3, 4, 4, 5)$.
   (b) Let G_4 have order 5 and degree sequence $(2, 2, 3, 3, 4)$.
   (c) Let G_5 have order 6 and degree sequence $(2, 3, 3, 3, 3, 4)$.
   (d) Let G_6 have order 5 and degree sequence $(2, 3, 3, 4, 4)$.

10. This digraph is not strong and hence not hamiltonian. It is also not traceable.

11. (a) hamiltonian   (b) traceable, not hamiltonian   (c) hamiltonian
    (b) traceable, not hamiltonian

Exercise Set 3.8

5. (a) 5   (b) 4   (c) 4

7. 3

9. 2

10. (b) $G = (\{u, v, w, x, y, z\}, \{uv, uw, ux, uy, uz, vw, vx, vy, vz, wx, xy, xz\})$

12. (b) Use the graph G of Figure 3.8.12(c).
    (c) Use the result of Exercise 14; thus, for the graph G of Figure 3.8.12(c), $K_1 + G$ is 5-critical.

13. (a) 3
    (b) Each color class is a pairwise-disjoint collection of subsets of U.
    (c) We require a maximal independent set S of H such that the union of the subsets in S is U. Since each of the subsets $A, B, \dots, G$ has at most 3 elements, S is required to have cardinality at least 4. However, no such S exists.

26. An example is the graph of the icosahedron shown in Figure 3.7.5.

28. (a) planar   (b) nonplanar (see Chapter 4, Problem 51)
    (c) nonplanar   (d) planar

29. The maximal independent sets are $\{t, u\}$, $\{t, v\}$, $\{u, w, y\}$, $\{u, z\}$, and $\{x, z\}$.

Chapter 3 Problems

2. (a) These are not isomorphic, because $G_1 \simeq K_{3,3}$ whereas G_2 is not bipartite.
   (b) An isomorphism $\phi: V(G_1) \to V(G_2)$ is defined by $\phi(u_1) = v_1$, $\phi(u_2) = v_6$, $\phi(u_3) = v_2$, $\phi(u_4) = v_3$, $\phi(u_5) = v_5$, $\phi(u_6) = v_4$, $\phi(u_7) = v_7$.

6. (b) $1, 5, 2, 4, 6, 2, 8, 6, 3, 7, 5, 4, 8, 1$

7. (a) $m \geq n - 1$ (b) $m \leq C(n - 1, 2)$
   (c) Any tree of order n shows that the bound in (a) is sharp. For (b), let $G = (\{v_1, v_2, \dots, v_n\}, E)$, such that $\deg(v_1) = \cdots = \deg(v_{n-1}) = n - 2$ and $\deg(v_n) = 0$.

9. (a) 3-partite   (b) 2-partite   (c) 4-partite

11. The order is $n = n_1 + n_2 + \cdots + n_k$ and the size is $[n_1(n - n_1) + n_2(n - n_2) + \cdots + n_k(n - n_k)]/2 = (n^2 - n_1^2 - n_2^2 - \cdots - n_k^2)/2$.

12. (e) Let $n = kq + r$, $0 \leq r < n$. Let $n_i = q$, $1 \leq i \leq k - r$ and $n_i = q + 1$, $k - r < i \leq k$. Show that m is at most the size of $K_{n_1, n_2, \dots, n_k}$.

13. (a) $H_1 = (\{v, w, x, y, z\}, \{vw, vy, wx, wz, xy, yz\})$
    (b) $H_2 = \langle\{u, w, x, y\}\rangle$
    (c) $H_3 = (V, \{tu, tw, ux, vz, vw, xy, yz\})$; $H_4 = (V, \{tu, tw, ux, vy, vz, wx, yz\})$
    (d) $K_{1,1,3}$ has two nonadjacent vertices of degree 4, whereas G does not.

14. They are H, $\langle\{uv, ux, uz, vx, yz\}\rangle$, $\langle\{uv, ux, uz, vz, yz\}\rangle$, $\langle\{ux, uz, vx, vz, yz\}\rangle$, $\langle\{uv, ux, uz, yz\}\rangle$, and $\langle\{uv, uz, vx, yz\}\rangle$.

15. A graph G of order n has the property that every induced subgraph is connected if and only if $G \simeq K_n$.

17. Let $t \geq 2$. A graph G of order n has the property that every induced subgraph on t or more vertices is connected if and only if $G \simeq K_{n_1, n_2, \dots, n_k}$ for some $k \geq 2$, where $n_1 \leq n_2 \leq \cdots \leq n_k < t$ and $n_1 + n_2 + \cdots + n_k = n$.

19. (a) $d(1) = 2$, $d(2) = 5$, $d(3) = 6$, $d(4) = 10$, $d(5) = 3$, $d(6) = 6$, $d(7) = 1$, $d(8) = 8$, $d(9) = 3$; $p(1) = p(5) = p(7) = 0$, $p(2) = p(4) = p(9) = 7$, $p(3) = 5$, $p(6) = 9$, $p(8) = 2$
    (b) $d(1) = 1$, $d(2) = 4$, $d(3) = 8$, $d(4) = 6$, $d(5) = 12$, $d(6) = 4$, $d(7) = 23$; $p(1) = 0$, $p(2) = p(3) = p(4) = p(5) = p(6) = 1$, $p(7) = 2$

20. (a) A possible edge set is $\{01, 05, 07, 27, 28, 35, 46, 69, 79\}$.

(b) A possible edge set is $\{01, 12, 13, 14, 16, 25, 57\}$.

22. (a) yes   (b) yes   (c) no; 3 must occur before 7

23. A possibility is the spanning rooted tree $(V, \{01, 02, 13, 15, 34, 56, 67\})$.

31. A graph G is isomorphic to the graph of an equivalence relation if and only if each component of G is complete.

32. (a) If G is a $(2, n)$-cage, then G contains an n-cycle, and, since C_n is a 2-regular graph with girth n, it must be that $G \simeq C_n$.

    (b) Let G be an $(r, 3)$-cage. Then G has order at least $r + 1$. Furthermore, K_{r+1} is an r-regular graph with girth 3. Thus, $G \simeq K_{r+1}$.

    (c) The unique $(r, 4)$-cage is $K_{r,r}$.

    (d) The Petersen graph is the unique $(3, 5)$-cage.

    (e) Let $H = (\{a, b, c, d, e, f, s, t, u, v, w, x, y, z\}, \{ab, af, az, bc, bw, cd, ct, de, dy, ef, ev, fs, st, sx, uv, uz, vw, wx, xy, yz\})$. The graph H is called the **Heawood graph** and, up to isomorphism, is the unique $(3, 6)$-cage.

34. (a) $G^2 = (V, A_2)$, where $A_2 = A \cup \{(s, u), (t, v), (u, w), (v, t), (v, x), (w, u), (w, y), (x, w),$
$(x, z), (y, s), (y, t), (y, x), (z, t), (z, y)\}$; $G^3 = (V, A_3)$, where $A_3 = A_2 \cup \{(s, v), (t, w),$
$(u, t), (u, x), (v, u), (v, y), (w, v), (w, z), (x, s), (x, t), (y, u), (z, u), (z, w)\}$;
$G^4 = (V, A_4)$, where $A_4 = A_3 \cup \{(s, w), (t, x), (u, y), (v, z), (w, s), (x, u), (y, v),$
$(z, v)\}$; $G^5 = (V, A_5)$, where $A_5 = A_4 \cup \{(s, x), (t, y), (u, z), (v, s), (x, v)\}$; $G^6 = (V, A_6)$,
where $A_6 = A_5 \cup \{(s, y), (t, z), (u, s)\}$; $G^7 = (V, A_7)$, where $A_7 = A_6 \cup \{(s, z), (t, s)\}$.

    (c) Let $T = (\{s, u, v, w, x, y, z\}, \{su, sv, sw, ux, vy, wz\})$.

36. $\tilde{A} = \{(a, b), (b, c), (b, e), (c, d), (c, g), (d, h), (e, f), (e, k), (g, m), (h, i), (i, j), (m, n)\}$; $p(b) = $ left, $p(c) = $ right, $p(d) = $ right, $p(e) = $ left, $p(f) = $ right, $p(g) = $ left, $p(h) = $ left, $p(i) = $ right, $p(j) = $ right, $p(k) = $ left, $p(m) = $ left, $p(n) = $ right.

37. (a) A vertex v of T is a leaf of $\tilde{T}$ if and only if v is a leaf of T and v is the youngest child of its parent in T.

    (b) *mh*   (c) We can recover T completely.

38. (a) $a, b, e, h, f, c, g, m, n, d, h, i, j$

39. Let $G = (\{v_1, v_2, \ldots, v_n\}, E)$, where $\langle \{v_1, v_2, \ldots, v_k\}\rangle \simeq K_k$ and $\deg(v_i) = 0$, $k < i \le n$.

45. (a) $G_1 \times G_2 = \{(u, x), (u, y), (u, z), (v, x), (v, y), (v, z)\}, \{(u, x)(u, y), (u, x)(u, z), (u, x)(v, x),$
$(u, y)(u, z), (u, y)(v, y), (u, z)(v, z)\})$

    (b) $G_1 \times G_2 = (\{(t, w), (t, x), (t, y), (u, w), (u, x), (u, y), (u, z), (v, w), (v, x), (v, y), (v, z)\},$
$\{(t, w)(t, x), (t, w)(t, x), (t, w)(u, w), (t, x)(t, y), (t, x)(u, x), (t, y)(t, z), (t, y)(u, y), (t, z)(u, z),$
$(u, w)(u, x), (u, w)(v, w), (u, x)(u, y), (u, x)(v, x), (u, y)(u, z), (u, y)(v, y), (u, z)(v, z),$
$(v, w)(v, x), (v, x)(v, y), (v, y)(v, z)\}$

46. (b) Q_n has order 2^n and size $n2^{n-1}$.   (e) n

    (f) Q_n is planar if and only if $n \le 3$.

47. Use the graph T^2 of Problem 34(c).

48. Assume $n_1 \le n_2 \le n_3$. Then K_{n_1, n_2, n_3} is hamiltonian if and only if $n_1 + n_2 \ge n_3$.

49. Assume, as usual, that $n_1 \le n_2 \le \cdots \le n_k$. This graph is planar if and only if either $n = 2$ and $n_1 \le 2$, or $n = 3$ and $n_2 = 1$ or $n_2 = n_3 = 2$, or $n = 4$, $n_3 = 1$, and $n_4 \le 2$.

53. Let $G = (\{u, v, x, y\}, \{uv, vx, xy\}) \simeq P_4$. If the first W selected is $W = \{u, y\}$, then the "algorithm" will compute $\chi(G) = 3$, whereas the correct value is $\chi(G) = 2$. However, this method does produce an upper bound, namely n, on $\chi(G)$.

54. 25

59. $\{b_1w_2, b_2w_1, b_3w_4, b_4w_5, b_5w_3, b_6w_7, b_7w_9, b_8w_6, b_9w_{10}, b_{10}w_8, b_{11}w_{11}\}$

60. (a) C_n
    (b) For $n = 5$, use $(\{u, v, x, y, z\}, \{ux, uz, vy, vz, xz, yz\})$; for $n = 7$, use $(\{s, t, u, v, x, y, z\}, \{su, sx, tv, ty, uz, vz, xz, yz\})$; for n even, use $K_{2,n-2}$; for n odd, $n \geq 9$, use $K_{1,1,1,n-4}$.
    (c) For n even, use K_n; for n odd, use W_n.
    (d) P_n

Exercise Set 4.1

2. (a) $P = \{(0,0), (0,1), (1,0), (1,1)\}$,
   $L = \{\{(0,0), (1,0)\}, \{(0,1), (1,1)\}, \{(0,0), (1,1)\}, \{(0,1), (1,0)\}, \{(0,0), (0,1)\}, \{(1,0), (1,1)\}\}$
   (b) $P' = \{(0,0), (0,1), (1,0), (1,1), p, q, r\}$,
   $L' = \{\{(0,0), (1,0), p\}, \{(0,1), (1,1), p\}, \{(0,0), (1,1), q\}, \{(0,1), (1,0), q\}, \{(0,0), (0,1), r\}, \{(1,0), (1,1), r\}, \{p, q, r\}\}$

7. Hint: Use the Fano plane.

19. (a) $2x + 4y + 3z = 0$ (b) $x + y = 0$
    (c) $(0,1,1), (1,3,1), (2,0,1), (3,2,1), (4,4,1), (1,2,0)$
    (d) $(4,3,1)$ (e) $(1,1,0)$

Exercise Set 4.2

5.

(c)
0	2	3	1
3	1	0	2
1	3	2	0
2	0	1	3

(d)
0	4	3	2	1
3	1	4	0	2
1	0	2	4	3
4	2	1	3	0
2	3	0	1	4

7.

0	1	2	3	4
4	0	1	2	3
3	4	0	1	2
2	3	4	0	1
1	2	3	4	0

0	1	2	3	4
3	4	0	1	2
1	2	3	4	0
4	0	1	2	3
2	3	4	0	1

0	1	2	3	4
2	3	4	0	1
4	0	1	2	3
1	2	3	4	0
3	4	0	1	2

0	1	2	3	4
1	2	3	4	0
2	3	4	0	1
3	4	0	1	2
4	0	1	2	3

9.

$$A_2 = \begin{matrix} 0 & 1 & 2 & 3 \\ 2 & 3 & 0 & 1 \\ 3 & 2 & 1 & 0 \\ 1 & 0 & 3 & 2 \end{matrix} \qquad B_2 = \begin{matrix} 0 & 1 & 2 \\ 2 & 0 & 1 \\ 1 & 2 & 0 \end{matrix}$$

$$C_2 = \begin{matrix}
0 & 4 & 8 & 1 & 5 & 9 & 2 & 6 & 10 & 3 & 7 & 11 \\
8 & 0 & 4 & 9 & 1 & 5 & 10 & 2 & 6 & 11 & 3 & 7 \\
4 & 8 & 0 & 5 & 9 & 1 & 6 & 10 & 2 & 7 & 11 & 3 \\
2 & 6 & 10 & 3 & 7 & 11 & 0 & 4 & 8 & 1 & 5 & 9 \\
10 & 2 & 6 & 11 & 3 & 7 & 8 & 0 & 4 & 9 & 1 & 5 \\
6 & 10 & 2 & 7 & 11 & 3 & 4 & 8 & 0 & 5 & 9 & 1 \\
3 & 7 & 11 & 2 & 6 & 10 & 1 & 5 & 9 & 0 & 4 & 8 \\
11 & 3 & 7 & 10 & 2 & 6 & 9 & 1 & 5 & 8 & 0 & 4 \\
7 & 11 & 3 & 6 & 10 & 2 & 5 & 9 & 1 & 4 & 8 & 0 \\
1 & 5 & 9 & 0 & 4 & 8 & 3 & 7 & 11 & 2 & 6 & 10 \\
9 & 1 & 5 & 8 & 0 & 4 & 11 & 3 & 7 & 10 & 2 & 6 \\
5 & 9 & 1 & 4 & 8 & 0 & 7 & 11 & 3 & 6 & 10 & 2
\end{matrix}$$

Exercise Set 4.3

1. (a) Hint: Use the result of Theorem 4.3.6.

   (b) Hint: Use the result of Theorem 4.3.1.

5. Let the graph be $(\{v_1, v_2, v_3, v_4, v_5, v_6\}, \{v_1v_3, v_1v_4, v_1v_5, v_1v_6, v_2v_3, v_2v_4, v_2v_5, v_2v_6, v_3v_5, v_3v_6,$ $v_4v_5, v_4v_6\})$ and use the following eulerian circuit: $v_1, v_3, v_5, v_1, v_4, v_5, v_2, v_3, v_6, v_2, v_4, v_6, v_1$. This leads to the 2-factorization $G_1 \oplus G_2$, where G_1 has edge set $\{v_1v_3, v_1v_5, v_2v_4, v_2v_6, v_3v_5,$ $v_4v_6\}$ and G_2 has edge set $\{v_1v_4, v_1v_6, v_2v_3, v_2v_5, v_3v_6, v_4v_5\}$.

9. The graphs of order 10 and size 3 with the following degree sequences are isofactors of the Petersen graph: $(2, 2, 1, 1, 0, 0, 0, 0, 0, 0)$, $(2, 1, 1, 1, 1, 0, 0, 0, 0, 0)$, $(1, 1, 1, 1, 1, 1, 0, 0, 0, 0)$.

11. True; use Theorem 4.3.1.

12. False; consider K_7.

13. Let the Petersen graph be labeled as in Figure 3.8.7 and consider the factorization $G_1 \oplus G_2 \oplus G_3$, where G_1 has edge set $\{au, ay, cd, xw, vz\}$, G_2 has edge set $\{bc, cz, du, vw, xy\}$, and G_3 has edge set $\{ab, bw, dx, uv, yz\}$.

Exercise Set 4.4

1. (a) Use the Latin square

$$A = \begin{matrix}
0 & 4 & 1 & 5 & 2 & 6 & 3 \\
4 & 1 & 5 & 2 & 6 & 3 & 0 \\
1 & 5 & 2 & 6 & 3 & 0 & 4 \\
5 & 2 & 6 & 3 & 0 & 4 & 1 \\
2 & 6 & 3 & 0 & 4 & 1 & 5 \\
6 & 3 & 0 & 4 & 1 & 5 & 2 \\
3 & 0 & 4 & 1 & 5 & 2 & 6
\end{matrix}$$

(As usual, we replace the element (x, y) by $7x + y$ to obtain a Steiner triple system on the treatment set $\{0, 1, 2, \ldots, 20\}$.) There are 21 type I triples: $\{x, y, 7 + a_{xy}\}$, $0 \leq x <$ $y \leq 6$; 21 type II triples: $\{7 + x, 7 + y, 14 + a_{xy}\}$, $0 \leq x < y \leq 6$; 21 type III triples: $\{14 + x, 14 + y, a_{xy}\}$, $0 \leq x < y \leq 6$; and 7 type IV triples: $\{0, 7, 14\}$, $\{1, 8, 15\}$, $\{2, 9, 16\}$, $\{3, 10, 17\}$, $\{4, 11, 18\}$, $\{5, 12, 19\}$, $\{6, 13, 20\}$.

2. See Exercise 6 of Exercise Set 4.3.

3. Use the Latin square

$$A = \begin{matrix} 0 & 3 & 1 & 4 & 2 & 5 \\ 3 & 1 & 4 & 2 & 5 & 0 \\ 1 & 4 & 2 & 5 & 0 & 3 \\ 4 & 2 & 5 & 0 & 3 & 1 \\ 2 & 5 & 0 & 3 & 1 & 4 \\ 5 & 0 & 3 & 1 & 4 & 2 \end{matrix}$$

(As usual, we replace the element (x, y) by $6x + y$ to obtain a Steiner triple system on the treatment set $\{0, 1, 2, \ldots, 17, \infty\}$.) There are 3 type I triples: $\{0, 6, 12\}$, $\{1, 7, 13\}$, $\{2, 8, 14\}$; 3 type II triples: $\{\infty, 3, 6\}$, $\{\infty, 4, 7\}$, $\{\infty, 5, 8\}$; 3 type III triples: $\{\infty, 9, 12\}$, $\{\infty, 11, 13\}$, $\{\infty, 11, 14\}$; 3 type IV triples: $\{\infty, 15, 0\}$, $\{\infty, 16, 1\}$, $\{\infty, 17, 2\}$; 15 type V triples: $\{x, y, 6 + a_{xy}\}$, $0 \le x < y \le 5$; 15 type VI triples: $\{6 + x, 6 + y, 12 + a_{xy}\}$, $0 \le x < y \le 5$; and 15 type VII triples: $\{12 + x, 12 + y, a_{xy}\}$, $0 \le x < y \le 5$.

9. Let the treatment set be $V = \{1, 2, 3, 4, 5, 6, 7\}$.

(b) The blocks are: $\{1, 2, 3, 4\}, \{1, 2, 5, 6\}, \{1, 3, 5, 7\}, \{1, 4, 6, 7\}, \{3, 4, 5, 6\}, \{2, 4, 5, 7\}, \{2, 3, 6, 7\}$.

(c) The blocks are: $\{1, 2, 3, 4\}, \{1, 2, 6, 7\}, \{1, 3, 5, 7\}, \{1, 4, 5, 6\}, \{2, 3, 5, 6\}, \{2, 4, 5, 7\}, \{3, 4, 6, 7\}$.

Chapter 4 Problems

1. (a) F(alse)   (b) T(rue)   (c) F   (d) U(nknown)   (e) T
   (f) F   (g) U   (h) T   (i) T   (j) T

2. Let A and B be two orthogonal Latin squares of order 5; let a_{ij} be the type of gasoline to be used in car i on day j and let b_{ij} be the driver for car i on day j.

3. (a) 49   (b) 56   (c) $y = 3x + 5$

   (d) $(0, 3), (1, 5), (2, 0), (3, 2), (4, 4), (5, 6), (6, 1)$ (e) no

5. (a) 91   (b) 91

   (c) The lines α and β intersect at the point at infinity corresponding to the parallel class with slope t.

7. Let the Petersen graph be labeled as in Figure 3.8.7. One isofactor is isomorphic to the subgraph induced by $\{ab, bc, cd, vw, vz\}$; the other is isomorphic to the subgraph induced by $\{ab, au, bw, cd, uv\}$.

9. (a) P_4 (b) none
   (c) P_6 and the tree isomorphic to $(\{1, 2, 3, 4, 5, 6\}, \{12, 23, 34, 36, 45\})$.

11. P_7

13. (a) $(10, 15, 6, 4, 2)$ (b) not possible   (c) $(66, 143, 13, 6, 1)$
    (d) $(21, 70, 10, 3, 1)$ (e) not possible   (f) $(25, 90, 18, 5, 3)$

23. (d) The residual design has treatment set $\{1, 2, 4, 7, 11\}$ and the following blocks: $\{1, 2, 4\}$, $\{1, 2, 7\}, \{1, 2, 11\}, \{1, 7, 11\}, \{1, 4, 11\}, \{1, 4, 7\}, \{2, 4, 11\}$, $\{2, 4, 7\}, \{2, 7, 11\}, \{4, 7, 11\}$. The derived design has treatment set $\{3, 5, 6, 8, 9, 10\}$ and the following blocks: $\{3, 5, 6\}, \{3, 9, 10\}, \{6, 8, 9\}$, $\{3, 5, 8\}, \{5, 9, 10\}, \{6, 8, 10\}, \{3, 8, 10\}, \{5, 8, 9\}, \{5, 6, 10\}, \{3, 6, 9\}$.

25. The residual design (with respect to the block $\{2,3,6,7\}$) has treatment set $\{1,4,5\}$ and the following blocks: $\{1,4\},\{1,5\},\{1,5\},\{1,4\},\{4,5\},\{4,5\}$. The derived design has treatment set $\{2,3,6,7\}$ and the following blocks: $\{2,3\},\{2,6\},\{3,7\},\{6,7\},\{3,6\},\{2,7\}$.

26. Use a $(6,10,5,3,2)$-BIBD; see Example 4.4.4, for instance.

28. (c) The construction results in the $(11,11,5,5,2)$-BIBD with the following blocks:
    $\{0,2,3,4,8\},\{1,3,4,5,9\},\{2,4,5,6,10\},\{0,3,5,6,7\},\{1,4,6,7,8\},$
    $\{2,5,7,8,9\},\{3,6,8,9,10\},\{0,4,7,9,10\},\{0,1,5,8,10\},\{0,1,2,6,9\},$
    $\{1,2,3,7,10\}$.

    (d) The construction results in the $(13,13,4,4,1)$-BIBD with the following blocks:
    $\{0,1,3,9\},\{1,2,4,10\},\{2,3,5,11\},\{3,4,6,12\},\{0,4,5,7\},\{1,5,6,8\},$
    $\{2,6,7,9\},\{3,7,8,10\},\{4,8,9,11\},\{5,9,10,12\},\{0,6,10,11\},$
    $\{1,7,11,12\},\{0,2,8,12\}$.

Index